Chemistry
FIFTH EDITION

Selected material from
General, Organic, and Biochemistry
FIFTH EDITION

Katherine J. Denniston
Towson University

Joseph J. Topping
Towson University

Robert L. Caret
Towson University

Boston   Burr Ridge, IL   Dubuque, IA   New York   San Francisco   St. Louis
Bangkok   Bogotá   Caracas   Lisbon   London   Madrid
Mexico City   Milan   New Delhi   Seoul   Singapore   Sydney   Taipei   Toronto

Chemistry
FIFTH EDITION

Copyright © 2007 by The McGraw-Hill Companies, Inc. All rights reserved. Printed in the United States of America. Except as permitted under the United States Copyright Act of 1976, no part of this publication may be reproduced or distributed in any form or by any means, or stored in a data base retrieval system, without prior written permission of the publisher.

This book is a McGraw-Hill Custom Publishing textbook and contains select material from *General, Organic, and Biochemistry*, Fifth Edition by Katherine J. Denniston, Joseph J. Topping, and Robert L. Caret. Copyright © 2007 by The McGraw-Hill Companies, Inc. Reprinted with permission of the publisher. Many custom published texts are modified versions or adaptations of our best-selling textbooks. Some adaptations are printed in black and white to keep prices at a minimum, while others are in color.

3 4 5 6 7 8 9 0 CDC CDC 0 9 8 7 6

ISBN 13: 978-0-07-329819-1
ISBN 10: 0-07-329819-0

Editor: Shirley Grall
Production Editor: Jessica Portz
Cover Photo: Molecule: PhotoLink/Getty Images
Cover Design: Gregory L. Crippen
Printer/Binder: C-DOC Services, Inc.

Brief Contents

Chemistry Connections and Perspectives xvii   Preface xix

GENERAL CHEMISTRY

1. Chemistry: Methods and Measurement 1
2. The Structure of the Atom and the Periodic Table 37
3. Structure and Properties of Ionic and Covalent Compounds 77
4. Calculations and the Chemical Equation 117
5. States of Matter: Gases, Liquids, and Solids 151
6. Solutions 177
7. Energy, Rate, and Equilibrium 207
8. Acids and Bases and Oxidation-Reduction 239
9. The Nucleus, Radioactivity, and Nuclear Medicine 275

ORGANIC CHEMISTRY

10. An Introduction to Organic Chemistry: The Saturated Hydrocarbons 303
11. The Unsaturated Hydrocarbons: Alkenes, Alkynes, and Aromatics 339
12. Alcohols, Phenols, Thiols, and Ethers 381
13. Aldehydes and Ketones 415
14. Carboxylic Acids and Carboxylic Acid Derivatives 447
15. Amines and Amides 487

BIOCHEMISTRY

16. Carbohydrates 523
17. Lipids and Their Functions in Biochemical Systems 555
18. Protein Structure and Function 595
19. Enzymes 629
20. Introduction to Molecular Genetics 667
21. Carbohydrate Metabolism 711
22. Aerobic Respiration and Energy Production 745
23. Fatty Acid Metabolism 777

Glossary G-1   Answers to Selected Problems AP-1   Credits C-1   Index I-1

Contents

Chemistry Connections and Perspectives xvii

Preface xix

GENERAL CHEMISTRY

1 Chemistry: Methods and Measurements 1

Chemistry Connection: *Chance Favors the Prepared Mind* 2

1.1 The Discovery Process 3
  Chemistry 3
  Major Areas of Chemistry 3
  The Scientific Method 3

A Human Perspective: *The Scientific Method* 4
  Models in Chemistry 5

A Medical Perspective: *Curiosity, Science, and Medicine* 5

1.2 Matter and Properties 7
  Matter and Physical Properties 7
  Matter and Chemical Properties 8
  Intensive and Extensive Properties 8
  Classification of Matter 9

1.3 Measurement in Chemistry 11
  Data, Results, and Units 11
  English and Metric Units 11
  Unit Conversion: English and Metric Systems 13
  Conversion of Units Within the Same System 13
  Conversion of Units from One System to Another 15

1.4 Significant Figures and Scientific Notation 16
  Significant Figures 16
  Recognition of Significant Figures 17
  Scientific Notation 18
  Error, Accuracy, Precision, and Uncertainty 19
  Significant Figures in Calculation of Results 20
  Exact (Counted) and Inexact Numbers 22
  Rounding Off Numbers 22

1.5 Experimental Quantities 23
  Mass 23
  Length 24
  Volume 24
  Time 25
  Temperature 25
  Energy 27
  Concentration 27
  Density and Specific Gravity 28

A Human Perspective: *Food Calories* 28

A Medical Perspective: *Diagnosis Based on Waste* 31

Summary 32
Key Terms 33
Questions and Problems 33
Critical Thinking Problems 35

2 The Structure of the Atom and the Periodic Table 37

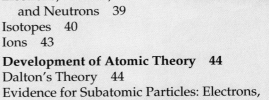

Chemistry Connection: *Managing Mountains of Information* 38

2.1 Composition of the Atom 39
  Electrons, Protons, and Neutrons 39
  Isotopes 40
  Ions 43

2.2 Development of Atomic Theory 44
  Dalton's Theory 44
  Evidence for Subatomic Particles: Electrons, Protons, and Neutrons 45
  Evidence for the Nucleus 45

2.3 Light, Atomic Structure, and the Bohr Atom 46
  Light and Atomic Structure 46

An Environmental Perspective: *Electromagnetic Radiation and Its Effects on Our Everyday Lives* 48
  The Bohr Atom 49
  Modern Atomic Theory 51

2.4 The Periodic Law and the Periodic Table 52

A Human Perspective: *Atomic Spectra and the Fourth of July* 52
  Numbering Groups in the Periodic Table 54
  Periods and Groups 55

Metals and Nonmetals 55
Atomic Number and Atomic Mass 55

2.5 Electron Arrangement and the Periodic Table 56

A Medical Perspective: *Copper Deficiency and Wilson's Disease* 57

Valence Electrons 58
The Quantum Mechanical Atom 60
Energy Levels and Sublevels 61
Electron Configuration and the Aufbau Principle 62
Shorthand Electron Configurations 64

2.6 The Octet Rule 65

Ion Formation and the Octet Rule 65

A Medical Perspective: *Dietary Calcium* 67

2.7 Trends in the Periodic Table 68

Atomic Size 68
Ion Size 68
Ionization Energy 69
Electron Affinity 70

Summary 71
Key Terms 72
Questions and Problems 72
Critical Thinking Problems 75

3 Structure and Properties of Ionic and Covalent Compounds 77

3.1 Chemical Bonding 78

Chemistry Connection: *Magnets and Migration* 78

Lewis Symbols 79
Principal Types of Chemical Bonds: Ionic and Covalent 79
Polar Covalent Bonding and Electronegativity 82

3.2 Naming Compounds and Writing Formulas of Compounds 84

Ionic Compounds 84
Covalent Compounds 89

3.3 Properties of Ionic and Covalent Compounds 92

Physical State 92

A Human Perspective: *Origin of the Elements* 92

Melting and Boiling Points 93
Structure of Compounds in the Solid State 93
Solutions of Ionic and Covalent Compounds 93

3.4 Drawing Lewis Structures of Molecules and Polyatomic Ions 93

Lewis Structures of Molecules 93

A Medical Perspective: *Blood Pressure and the Sodium Ion/Potassium Ion Ratio* 94

Lewis Structures of Polyatomic Ions 97
Lewis Structure, Stability, Multiple Bonds, and Bond Energies 100

Lewis Structures and Resonance 101
Lewis Structures and Exceptions to the Octet Rule 103
Lewis Structures and Molecular Geometry; VSEPR Theory 104
Lewis Structures and Polarity 109

3.5 Properties Based on Electronic Structure and Molecular Geometry 111

Solubility 111
Boiling Points of Liquids and Melting Points of Solids 112

Summary 113
Key Terms 114
Questions and Problems 115
Critical Thinking Problems 116

4 Calculations and the Chemical Equation 117

Chemistry Connection: *The Chemistry of Automobile Air Bags* 118

4.1 The Mole Concept and Atoms 119

The Mole and Avogadro's Number 119
Calculating Atoms, Moles, and Mass 121

4.2 The Chemical Formula, Formula Weight, and Molar Mass 123

The Chemical Formula 123
Formula Weight and Molar Mass 124

4.3 The Chemical Equation and the Information It Conveys 126

A Recipe for Chemical Change 126
Features of a Chemical Equation 127
The Experimental Basis of a Chemical Equation 127
Writing Chemical Reactions 128
Types of Chemical Reactions 129

4.4 Balancing Chemical Equations 132

4.5 Calculations Using the Chemical Equation 136

General Principles 136

A Medical Perspective: *Carbon Monoxide Poisoning: A Case of Combining Ratios* 136

Use of Conversion Factors 137
Theoretical and Percent Yield 143

A Medical Perspective: *Pharmaceutical Chemistry: The Practical Significance of Percent Yield* 144

Summary 145
Key Terms 146
Questions and Problems 146
Critical Thinking Problems 149

5 States of Matter: Gases, Liquids, and Solids 151

Chemistry Connection: *The Demise of the Hindenburg* 152

5.1 The Gaseous State 153
Ideal Gas Concept 153
Measurement of Gases 153
Kinetic Molecular Theory of Gases 154
Boyle's Law 155
Charles's Law 157
Combined Gas Law 160
Avogadro's Law 161
Molar Volume of a Gas 163
Gas Densities 163
The Ideal Gas Law 163
Dalton's Law of Partial Pressures 166

An Environmental Perspective: *The Greenhouse Effect and Global Warming* 166

Ideal Gases Versus Real Gases 167

5.2 The Liquid State 167
Compressibility 167
Viscosity 167
Surface Tension 168

A Medical Perspective: *Blood Gases and Respiration* 168

Vapor Pressure of a Liquid 169
Van der Waals Forces 170
Hydrogen Bonding 170

5.3 The Solid State 171
Properties of Solids 171
Types of Crystalline Solids 172

Summary 173
Key Terms 174
Questions and Problems 174
Critical Thinking Problems 176

6 Solutions 177

6.1 Properties of Solutions 178
Chemistry Connection: *Seeing a Thought* 178

General Properties of Liquid Solutions 179
Solutions and Colloids 179
Degree of Solubility 180
Solubility and Equilibrium 181
Solubility of Gases: Henry's Law 181

6.2 Concentration Based on Mass 182
Weight/Volume Percent 182

A Human Perspective: *Scuba Diving: Nitrogen and the Bends* 183

Weight/Weight Percent 185
Parts Per Thousand (ppt) and Parts Per Million (ppm) 185

6.3 Concentration of Solutions: Moles and Equivalents 187
Molarity 187
Dilution 188
Representation of Concentration of Ions in Solution 190

6.4 Concentration-Dependent Solution Properties 191
Vapor Pressure Lowering 192
Freezing Point Depression and Boiling Point Elevation 192
Osmotic Pressure 193

A Medical Perspective: *Oral Rehydration Therapy* 197

6.5 Water as a Solvent 198
A Human Perspective: *An Extraordinary Molecule* 199

6.6 Electrolytes in Body Fluids 200
A Medical Perspective: *Hemodialysis* 201

Summary 202
Key Terms 203
Questions and Problems 204
Critical Thinking Problems 205

7 Energy, Rate, and Equilibrium 207

7.1 Thermodynamics 208
Chemistry Connection: *The Cost of Energy? More Than You Imagine* 208

The Chemical Reaction and Energy 209
Exothermic and Endothermic Reactions 209
Enthalpy 211
Spontaneous and Nonspontaneous Reactions 211
Entropy 211

A Human Perspective: *Triboluminescence: Sparks in the Dark with Candy* 213

Free Energy 214

7.2 Experimental Determination of Energy Change in Reactions 215

7.3 Kinetics 218
The Chemical Reaction 219
Activation Energy and the Activated Complex 220
Factors That Affect Reaction Rate 221

A Medical Perspective: *Hot and Cold Packs* 221

Mathematical Representation of Reaction Rate 224

7.4 Equilibrium 226
Rate and Reversibility of Reactions 226
Physical Equilibrium 226
Chemical Equilibrium 227

The Generalized Equilibrium-Constant Expression for a Chemical Reaction 228
LeChatelier's Principle 232

Summary 235
Key Terms 236
Questions and Problems 236
Critical Thinking Problems 237

8 Acids and Bases and Oxidation-Reduction 239

Chemistry Connection: *Drug Delivery* 240

8.1 Acids and Bases 241
Arrhenius Theory of Acids and Bases 241
Brønsted-Lowry Theory of Acids and Bases 241
Acid-Base Properties of Water 242
Acid and Base Strength 242
Conjugate Acids and Bases 243
The Dissociation of Water 246

8.2 pH: A Measurement Scale for Acids and Bases 246
A Definition of pH 246
Measuring pH 247
Calculating pH 247
The Importance of pH and pH Control 252

8.3 Reactions Between Acids and Bases 252
Neutralization 252
Polyprotic Substances 255

8.4 Acid-Base Buffers 256
An Environmental Perspective: *Acid Rain* 256
The Buffer Process 257
Addition of Base (OH$^-$) to a Buffer Solution 258
Addition of Acid (H$_3$O$^+$) to a Buffer Solution 258
Preparation of a Buffer Solution 259
The Henderson-Hasselbalch Equation 261
A Medical Perspective: *Control of Blood pH* 262

8.5 Oxidation-Reduction Processes 263
Oxidation and Reduction 263
A Medical Perspective: *Oxidizing Agents for Chemical Control of Microbes* 264
Applications of Oxidation and Reduction 265
A Medical Perspective: *Electrochemical Reactions in the Statue of Liberty and in Dental Fillings* 266
Biological Processes 267
Voltaic Cells 267
Electrolysis 269
A Medical Perspective: *Turning the Human Body into a Battery* 270

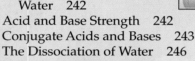

Summary 271
Key Terms 272
Questions and Problems 272
Critical Thinking Problems 274

9 The Nucleus, Radioactivity, and Nuclear Medicine 275

Chemistry Connection: *An Extraordinary Woman in Science* 276

9.1 Natural Radioactivity 277
Alpha Particles 278
Beta Particles 278
Gamma Rays 278
Properties of Alpha, Beta, and Gamma Radiation 279

9.2 Writing a Balanced Nuclear Equation 279
Alpha Decay 280
Beta Decay 280
Position Emission 280
Gamma Production 280
Predicting Products of Nuclear Decay 281

9.3 Properties of Radioisotopes 282
Nuclear Structure and Stability 282
Half-Life 282

9.4 Nuclear Power 284
Energy Production 284
An Environmental Perspective: *Nuclear Waste Disposal* 285
Nuclear Fission 286
Nuclear Fusion 287
Breeder Reactors 287

9.5 Radiocarbon Dating 288

9.6 Medical Applications of Radioactivity 288
Cancer Therapy Using Radiation 289
Nuclear Medicine 289
Making Isotopes for Medical Applications 291
A Medical Perspective: *Magnetic Resonance Imaging* 292

9.7 Biological Effects of Radiation 293
Radiation Exposure and Safety 293

9.8 Measurement of Radiation 295
An Environmental Perspective: *Radon and Indoor Air Pollution* 295
Nuclear Imaging 296
Computer Imaging 296
The Geiger Counter 296
Film Badges 296
Units of Radiation Measurement 297

Summary 298
Key Terms 299
Questions and Problems 299
Critical Thinking Problems 301

ORGANIC CHEMISTRY

10 An Introduction to Organic Chemistry: The Saturated Hydrocarbons 303

Chemistry Connection: *The Origin of Organic Compounds* 304

10.1 The Chemistry of Carbon 305
  Important Differences Between Organic and Inorganic Compounds 306

An Environmental Perspective: *Frozen Methane: Treasure or Threat?* 307
  Families of Organic Compounds 308

10.2 Alkanes 310
  Structure and Physical Properties 310
  Alkyl Groups 313
  Nomenclature 315

An Environmental Perspective: *Oil-Eating Microbes* 317
  Constitutional or Structural Isomers 319

10.3 Cycloalkanes 321
  cis-trans Isomerism in Cycloalkanes 322

10.4 Conformations of Alkanes and Cycloalkanes 325
  Alkanes 325
  Cycloalkanes 325

An Environmental Perspective: *The Petroleum Industry and Gasoline Production* 326

10.5 Reactions of Alkanes and Cycloalkanes 327
  Combustion 327

A Medical Perspective: *Polyhalogenated Hydrocarbons Used as Anesthetics* 328
  Halogenation 329

A Medical Perspective: *Chloroform in Your Swimming Pool?* 331

Summary of Reactions 332
Summary 332
Key Terms 332
Questions and Problems 333
Critical Thinking Problems 337

11 The Unsaturated Hydrocarbons: Alkenes, Alkynes, and Aromatics 339

Chemistry Connection: *A Cautionary Tale: DDT and Biological Magnification* 340

11.1 Alkenes and Alkynes: Structure and Physical Properties 342

11.2 Alkenes and Alkynes: Nomenclature 343

11.3 Geometric Isomers: A Consequence of Unsaturation 346

A Medical Perspective: *Killer Alkynes in Nature* 346

11.4 Alkenes in Nature 352

11.5 Reactions Involving Alkenes and Alkynes 354
  Hydrogenation: Addition of H_2 354
  Halogenation: Addition of X_2 356
  Hydration: Addition of H_2O 358
  Hydrohalogenation: Addition of HX 361

A Human Perspective: *Folklore, Science, and Technology* 362
  Addition Polymers of Alkenes 364

A Human Perspective: *Life Without Polymers?* 364

11.6 Aromatic Hydrocarbons 366

An Environmental Perspective: *Plastic Recycling* 366
  Structure and Properties 368
  Nomenclature 368
  Polynuclear Aromatic Hydrocarbons 371
  Reactions Involving Benzene 372

11.7 Heterocyclic Aromatic Compounds 373

Summary of Reactions 374
Summary 375
Key Terms 375
Questions and Problems 375
Critical Thinking Problems 379

12 Alcohols, Phenols, Thiols, and Ethers 381

Chemistry Connection: *Polyols for the Sweet Tooth* 382

12.1 Alcohols: Structure and Physical Properties 384

12.2 Alcohols: Nomenclature 385
  I.U.P.A.C. Names 385
  Common Names 386

12.3 Medically Important Alcohols 387

A Medical Perspective: *Fetal Alcohol Syndrome* 388

12.4 Classification of Alcohols 389

12.5 Reactions Involving Alcohols 391
  Preparation of Alcohols 391
  Dehydration of Alcohols 394
  Oxidation Reactions 397

12.6 Oxidation and Reduction in Living Systems 400

A Human Perspective: *Alcohol Consumption and the Breathalyzer Test* 401

12.7 Phenols 402

12.8 Ethers 403

12.9 Thiols 405

Summary of Reactions 409
Summary 409
Key Terms 410
Questions and Problems 410
Critical Thinking Problems 413

13 Aldehydes and Ketones 415

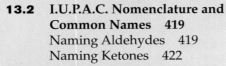

Chemistry Connection: *Genetic Complexity from Simple Molecules 416*

13.1 Structure and Physical Properties 417

13.2 I.U.P.A.C. Nomenclature and Common Names 419
Naming Aldehydes 419
Naming Ketones 422

13.3 Important Aldehydes and Ketones 424

13.4 Reactions Involving Aldehydes and Ketones 424
Preparation of Aldehydes and Ketones 424

A Medical Perspective: *Formaldehyde and Methanol Poisoning 426*
Oxidation Reactions 427

A Human Perspective: *Alcohol Abuse and Antabuse 430*
Reduction Reactions 431

A Medical Perspective: *That Golden Tan Without the Fear of Skin Cancer 432*
Addition Reactions 434
Keto-Enol Tautomers 436
Aldol Condensation 438

A Human Perspective: *The Chemistry of Vision 440*

Summary of Reactions 440
Summary 442
Key Terms 443
Questions and Problems 443
Critical Thinking Problems 445

14 Carboxylic Acids and Carboxylic Acid Derivatives 447

Chemistry Connection: *Wake Up, Sleeping Gene 448*

14.1 Carboxylic Acids 449
Structure and Physical Properties 449
Nomenclature 451
Some Important Carboxylic Acids 455

An Environmental Perspective: *Garbage Bags from Potato Peels 456*
Reactions Involving Carboxylic Acids 457

14.2 Esters 461
Structure and Physical Properties 461
Nomenclature 461
Reactions Involving Esters 462

A Human Perspective: *The Chemistry of Flavor and Fragrance 464*

14.3 Acid Chlorides and Acid Anhydrides 470
Acid Chlorides 470
Acid Anhydrides 473

14.4 Nature's High-Energy Compounds: Phosphoesters and Thioesters 476

A Human Perspective: *Carboxylic Acid Derivatives of Special Interest 478*

Summary of Reactions 480
Summary 481
Key Terms 481
Questions and Problems 481
Critical Thinking Problems 485

15 Amines and Amides 487

Chemistry Connection: *The Nicotine Patch 488*

15.1 Amines 489
Structure and Physical Properties 489
Nomenclature 493
Medically Important Amines 495
Reactions Involving Amines 497

A Human Perspective: *Methamphetamine 500*
Quaternary Ammonium Salts 501

15.2 Heterocyclic Amines 502

15.3 Amides 504
Structure and Physical Properties 504
Nomenclature 505
Medically Important Amides 505
Reactions Involving Amides 506

A Medical Perspective: *Semisynthetic Penicillins 507*

15.4 A Preview of Amino Acids, Proteins, and Protein Synthesis 510

15.5 Neurotransmitters 511
Catecholamines 511
Serotonin 511

A Medical Perspective: *Opiate Biosynthesis and the Mutant Poppy 512*
Histamine 514
γ-Aminobutyric Acid and Glycine 515
Acetylcholine 515
Nitric Oxide and Glutamate 516

Summary of Reactions 517
Summary 518
Key Terms 518
Questions and Problems 518
Critical Thinking Problems 521

BIOCHEMISTRY

16 Carbohydrates 523

Chemistry Connection: *Chemistry Through the Looking Glass* 524

16.1 Types of Carbohydrates 526
16.2 Monosaccharides 527

A Human Perspective: *Tooth Decay and Simple Sugars* 528

16.3 Stereoisomers and Stereochemistry 529
    Stereoisomers 529
    Rotation of Plane-Polarized Light 530
    The Relationship Between Molecular Structure and Optical Activity 532
    Fischer Projection Formulas 532
    The D- and L- System of Nomenclature 534

16.4 Biologically Important Monosaccharides 535
    Glucose 535
    Fructose 539
    Galactose 540
    Ribose and Deoxyribose, Five-Carbon Sugars 541
    Reducing Sugars 541

16.5 Biologically Important Disaccharides 543
    Maltose 543
    Lactose 544
    Sucrose 545

A Human Perspective: *Blood Transfusions and the Blood Group Antigens* 546

16.6 Polysaccharides 548
    Starch 548
    Glycogen 548
    Cellulose 549

A Medical Perspective: *Monosaccharide Derivatives and Heteropolysaccharides of Medical Interest* 550

Summary 552
Key Terms 552
Questions and Problems 553
Critical Thinking Problems 554

17 Lipids and Their Functions in Biochemical Systems 555

Chemistry Connection: *Lifesaving Lipids* 556

17.1 Biological Functions of Lipids 557
17.2 Fatty Acids 558
    Structure and Properties 558
    Chemical Reactions of Fatty Acids 561

A Human Perspective: *Mummies Made of Soap* 563
    Eicosanoids: Prostaglandins, Leukotrienes, and Thromboxanes 564

17.3 Glycerides 567
    Neutral Glycerides 567
    Phosphoglycerides 568

17.4 Nonglyceride Lipids 570
    Sphingolipids 570
    Steroids 572

A Medical Perspective: *Disorders of Sphingolipid Metabolism* 573

A Medical Perspective: *Steroids and the Treatment of Heart Disease* 574
    Waxes 577

17.5 Complex Lipids 577
17.6 The Structure of Biological Membranes 581
    Fluid Mosaic Structure of Biological Membranes 582
    Membrane Transport 583

A Medical Perspective: *Liposome Delivery Systems* 586

A Medical Perspective: *Antibiotics That Destroy Membrane Integrity* 588
    Energy Requirements for Transport 590

Summary 591
Key Terms 591
Questions and Problems 592
Critical Thinking Problems 593

18 Protein Structure and Function 595

Chemistry Connection: *Angiogenesis Inhibitors: Proteins That Inhibit Tumor Growth* 596

18.1 Cellular Functions of Proteins 597
18.2 The α-Amino Acids 597
    Structure of Amino Acids 597
    Stereoisomers of Amino Acids 598

A Medical Perspective: *Proteins in the Blood* 599
    Classes of Amino Acids 600

18.3 The Peptide Bond 602

A Human Perspective: *The Opium Poppy and Peptides in the Brain* 604

18.4 The Primary Structure of Proteins 606
18.5 The Secondary Structure of Proteins 606
    α-Helix 607
    β-Pleated Sheet 609

18.6 The Tertiary Structure of Proteins 609

18.7 The Quaternary Structure of Proteins 611
A Human Perspective: *Collagen: A Protein That Holds Us Together* 612

18.8 An Overview of Protein Structure and Function 613

18.9 Myoglobin and Hemoglobin 615
Myoglobin and Oxygen Storage 615
Hemoglobin and Oxygen Transport 616
Oxygen Transport from Mother to Fetus 616
Sickle Cell Anemia 617

18.10 Denaturation of Proteins 617
A Medical Perspective: *Immunoglobulins: Proteins That Defend the Body* 618
Temperature 618
pH 620
Organic Solvents 621
Detergents 621
Heavy Metals 622
Mechanical Stress 622

18.11 Dietary Protein and Protein Digestion 622

Summary 624
Key Terms 625
Questions and Problems 625
Critical Thinking Problems 627

19 Enzymes 629

Chemistry Connection: *Super Hot Enzymes and the Origin of Life* 630

19.1 Nomenclature and Classification 631
Classification of Enzymes 631
Nomenclature of Enzymes 635

19.2 The Effect of Enzymes on the Activation Energy of a Reaction 636

19.3 The Effect of Substrate Concentration on Enzyme-Catalyzed Reactions 637

19.4 The Enzyme-Substrate Complex 637

19.5 Specificity of the Enzyme-Substrate Complex 639

19.6 The Transition State and Product Formation 639

A Medical Perspective: *HIV Protease Inhibitors and Pharmaceutical Drug Design* 641

19.7 Cofactors and Coenzymes 643

19.8 Environmental Effects 646
Effect of pH 646
Effect of Temperature 647

A Medical Perspective: *α_1-Antitrypsin and Familial Emphysema* 648

19.9 Regulation of Enzyme Activity 649
Allosteric Enzymes 649
Feedback Inhibition 650
Proenzymes 651
Protein Modification 651

19.10 Inhibition of Enzyme Activity 652
Irreversible Inhibitors 652
Reversible, Competitive Inhibitors 652
Reversible, Noncompetitive Inhibitors 653

A Medical Perspective: *Enzymes, Nerve Transmission, and Nerve Agents* 654

19.11 Proteolytic Enzymes 656
A Medical Perspective: *Enzymes, Isoenzymes, and Myocardial Infarction* 658

19.12 Uses of Enzymes in Medicine 660

Summary 661
Key Terms 662
Questions and Problems 663
Critical Thinking Problems 665

20 Introduction to Molecular Genetics 667

Chemistry Connection: *Molecular Genetics and Detection of Human Genetic Disease* 668

20.1 The Structure of the Nucleotide 669
Nucleotide Structure 669

20.2 The Structure of DNA and RNA 671
DNA Structure: The Double Helix 671
Chromosomes 673
RNA Structure 674

20.3 DNA Replication 674
A Medical Perspective: *Fooling the AIDS Virus with "Look-Alike" Nucleotides* 676
Bacterial DNA Replication 677
Eukaryotic DNA Replication 681

20.4 Information Flow in Biological Systems 681
Classes of RNA Molecules 681
Transcription 682
Post-transcriptional Processing of RNA 684

20.5 The Genetic Code 685

20.6 Protein Synthesis 687
The Role of Transfer RNA 688
The Process of Translation 690

20.7 Mutation, Ultraviolet Light, and DNA Repair 692
The Nature of Mutations 692
The Results of Mutations 692
Mutagens and Carcinogens 693
Ultraviolet Light Damage and DNA Repair 694

A **Medical Perspective:** *The Ames Test for Carcinogens* 694
Consequences of Defects in DNA Repair 695
20.8 Recombinant DNA 696
Tools Used in the Study of DNA 696
Genetic Engineering 699
20.9 Polymerase Chain Reaction 701
A **Human Perspective:** *DNA Fingerprinting* 702
20.10 The Human Genome Project 703
Genetic Strategies for Genome Analysis 704
DNA Sequencing 704
A **Medical Perspective:** *A Genetic Approach to Familial Emphysema* 705

Summary 706
Key Terms 707
Questions and Problems 707
Critical Thinking Problems 709

21 Carbohydrate Metabolism 711

Chemistry Connection: *The Man Who Got Tipsy from Eating Pasta* 712
21.1 ATP: The Cellular Energy Currency 713
21.2 Overview of Catabolic Processes 715
Stage I: Hydrolysis of Dietary Macromolecules into Small Subunits 715
Stage II: Conversion of Monomers into a Form That Can Be Completely Oxidized 717
Stage III: The Complete Oxidation of Nutrients and the Production of ATP 718
21.3 Glycolysis 719
An Overview 719
Reactions of Glycolysis 721
A **Medical Perspective:** *Genetic Disorders of Glycolysis* 722
Regulation of Glycolysis 726
21.4 Fermentations 726
Lactate Fermentation 727
Alcohol Fermentation 727
A **Human Perspective:** *Fermentations: The Good, the Bad, and the Ugly* 728
21.5 The Pentose Phosphate Pathway 730
21.6 Gluconeogenesis: The Synthesis of Glucose 730
21.7 Glycogen Synthesis and Degradation 733
The Structure of Glycogen 733
Glycogenolysis: Glycogen Degradation 733
Glycogenesis: Glycogen Synthesis 735
A **Medical Perspective:** *Diagnosing Diabetes* 738
Compatibility of Glycogenesis and Glycogenolysis 739
A **Human Perspective:** *Glycogen Storage Diseases* 740

Summary 741
Key Terms 742
Questions and Problems 742
Critical Thinking Problems 744

22 Aerobic Respiration and Energy Production 745

Chemistry Connection: *Mitochondria from Mom* 746
22.1 The Mitochondria 747
Structure and Function 747
Origin of the Mitochondria 747
A **Human Perspective:** *Exercise and Energy Metabolism* 748
22.2 Conversion of Pyruvate to Acetyl CoA 750
22.3 An Overview of Aerobic Respiration 752
22.4 The Citric Acid Cycle (The Krebs Cycle) 753
Reactions of the Citric Acid Cycle 753
22.5 Control of the Citric Acid Cycle 756
22.6 Oxidative Phosphorylation 757
A **Human Perspective:** *Brown Fat: The Fat That Makes You Thin?* 758
Electron Transport Systems and the Hydrogen Ion Gradient 760
ATP Synthase and the Production of ATP 760
Summary of the Energy Yield 761
22.7 The Degradation of Amino Acids 762
Removal of α-Amino Groups: Transamination 762
Removal of α-Amino Groups: Oxidative Deamination 764
The Fate of Amino Acid Carbon Skeletons 766
22.8 The Urea Cycle 766
Reactions of the Urea Cycle 766
A **Medical Perspective:** *Pyruvate Carboxylase Deficiency* 769
22.9 Overview of Anabolism: The Citric Acid Cycle as a Source of Biosynthetic Intermediates 770

Summary 772
Key Terms 773
Questions and Problems 773
Critical Thinking Problems 775

23 Fatty Acid Metabolism 777

Chemistry Connection: *Obesity: A Genetic Disorder?* 778
23.1 Lipid Metabolism in Animals 779
Digestion and Absorption of Dietary Triglycerides 779
Lipid Storage 781

23.2	**Fatty Acid Degradation** 782	
	An Overview of Fatty Acid Degradation 782	

A Human Perspective: *Losing Those Unwanted Pounds of Adipose Tissue* 784

      The Reactions of β-Oxidation 786

23.3 Ketone Bodies 789
    Ketosis 789
    Ketogenesis 789

23.4 Fatty Acid Synthesis 791
    A Comparison of Fatty Acid Synthesis and Degradation 791

23.5 The Regulation of Lipid and Carbohydrate Metabolism 793
    The Liver 793

A Medical Perspective: *Diabetes Mellitus and Ketone Bodies* 794

    Adipose Tissue 796
    Muscle Tissue 796
    The Brain 796

23.6 The Effects of Insulin and Glucagon on Cellular Metabolism 797

Summary 799
Key Terms 799
Questions and Problems 799
Critical Thinking Problems 801

Glossary   *G-1*
Answers to Odd-Numbered Problems   *AP-1*
Credits   *C-1*
Index   *I-1*

Chemistry Connections and Perspectives

Chemistry Connection

Chance Favors the Prepared Mind	2
Managing Mountains of Information	38
Magnets and Migration	78
The Chemistry of Automobile Air Bags	118
The Demise of the Hindenburg	152
Seeing a Thought	178
The Cost of Energy? More Than You Imagine	208
Drug Delivery	240
An Extraordinary Woman in Science	276
The Origin of Organic Compounds	304
A Cautionary Tale: DDT and Biological Magnification	340
Polyols for the Sweet Tooth	382
Genetic Complexity from Simple Molecules	416
Wake Up, Sleeping Gene	448
The Nicotine Patch	488
Chemistry Through the Looking Glass	524
Lifesaving Lipids	556
Angiogenesis Inhibitors: Proteins That Inhibit Tumor Growth	596
Super Hot Enzymes and the Origin of Life	630
Molecular Genetics and Detection of Human Genetic Disease	668
The Man Who Got Tipsy from Eating Pasta	712
Mitochondria from Mom	746
Obesity: A Genetic Disorder?	778

A Human Perspective

The Scientific Method	4
Food Calories	28
Atomic Spectra and the Fourth of July	52
Origin of the Elements	92
Scuba Diving: Nitrogen and the Bends	183
An Extraordinary Molecule	199
Triboluminescence: Sparks in the Dark with Candy	213
Folklore, Science, and Technology	362
Life Without Polymers?	364
Alcohol Consumption and the Breathalyzer Test	401
Alcohol Abuse and Antabuse	430
The Chemistry of Vision	440
The Chemistry of Flavor and Fragrance	464
Carboxylic Acid Derivatives of Special Interest	478
Methamphetamine	500
Tooth Decay and Simple Sugars	528
Blood Transfusions and the Blood Group Antigens	546
Mummies Made of Soap	563
The Opium Poppy and Peptides in the Brain	604
Collagen: A Protein That Holds Us Together	612
DNA Fingerprinting	702
Fermentations: The Good, the Bad, and the Ugly	728
Glycogen Storage Diseases	740
Exercise and Energy Metabolism	748
Brown Fat: The Fat That Makes You Thin?	758
Losing Those Unwanted Pounds of Adipose Tissue	784

A Medical Perspective

Curiosity, Science, and Medicine	5
Diagnosis Based on Waste	31
Copper Deficiency and Wilson's Disease	57
Dietary Calcium	67
Blood Pressure and the Sodium Ion/Potassium Ion Ratio	94
Carbon Monoxide Poisoning: A Case of Combining Ratios	136
Pharmaceutical Chemistry: The Practical Significance of Percent Yield	144
Blood Gases and Respiration	168
Oral Rehydration Therapy	197
Hemodialysis	201
Hot and Cold Packs	221
Control of Blood pH	262
Oxidizing Agents for Chemical Control of Microbes	264
Electrochemical Reactions in the Statue of Liberty and in Dental Fillings	266
Turning the Human Body into a Battery	270
Magnetic Resonance Imaging	292
Polyhalogenated Hydrocarbons Used as Anesthetics	328
Chloroform in Your Swimming Pool?	331
Killer Alkynes in Nature	346
Fetal Alcohol Syndrome	388
Formaldehyde and Methanol Poisoning	426
That Golden Tan Without the Fear of Skin Cancer	432
Semisynthetic Penicillins	507
Opiate Biosynthesis and the Mutant Poppy	512
Monosaccharide Derivatives and Heteropolysaccharides of Medical Interest	550
Disorders of Sphingolipid Metabolism	573
Steroids and the Treatment of Heart Disease	574
Liposome Delivery Systems	586
Antibiotics That Destroy Membrane Integrity	588
Proteins in the Blood	599
Immunoglobulins: Proteins That Defend the Body	618
HIV Protease Inhibitors and Pharmaceutical Drug Design	641
α_1-Antitrypsin and Familial Emphysema	648
Enzymes, Nerve Transmission, and Nerve Agents	654
Enzymes, Isoenzymes, and Myocardial Infarction	658
Fooling the AIDS Virus with "Look-Alike" Nucleotides	676
The Ames Test for Carcinogens	694
A Genetic Approach to Familial Emphysema	705
Genetic Disorders of Glycolysis	722
Diagnosing Diabetes	738
Pyruvate Carboxylase Deficiency	769
Diabetes Mellitus and Ketone Bodies	794

An Environmental Perspective

Electromagnetic Radiation and Its Effects on Our Everyday Lives	48
The Greenhouse Effect and Global Warming	166
Acid Rain	256
Nuclear Waste Disposal	285
Radon and Indoor Air Pollution	295
Frozen Methane: Treasure or Threat?	307
Oil-Eating Microbes	317
The Petroleum Industry and Gasoline Production	326
Plastic Recycling	366
Garbage Bags from Potato Peels	456

Preface

The fifth edition of *General, Organic, and Biochemistry*, like our earlier editions, has been designed to help undergraduate majors in health-related fields understand key concepts and appreciate the significant connections between chemistry, health, and the treatment of disease. We have tried to strike a balance between theoretical and practical chemistry, while emphasizing material that is unique to health-related studies. We have written at a level intended for students whose professional goals do not include a mastery of chemistry, but for whom an understanding of the principles and practice of chemistry is a necessity.

While we have stressed the importance of chemistry to the health-related professions, this book was written for all students who need a one- or two-semester introduction to chemistry. Our focus on the relationship between chemistry, the environment, medicine, and the function of the human body is an approach that can engage students in a variety of majors. We have integrated the individual disciplines of inorganic, organic, and biochemistry to emphasize their interrelatedness rather than their differences. This approach provides a sound foundation in chemistry and teaches students that life is not a magical property, but rather the result of a set of chemical reactions that obey the scientific laws.

Key Features of the Fifth Edition

In preparing the fifth edition, we have been guided by the collective wisdom of over fifty reviewers who are experts in one of the three sub-disciplines covered in the book and who represent a diversity of experience, in community colleges and in four-year colleges and universities. We have retained the core approach of our successful earlier editions, updated material where necessary, and expanded or removed material consistent with retention of the original focus and mission of the book. Throughout the project, we have been careful to ensure that the final product is as student-oriented and readable as its predecessors.

New Features
- Chapters 2 and 3 of the fourth edition have been combined in this edition to provide more integrated and concise coverage of atomic structure and periodicity.
- Each boxed topic has been enhanced by questions intended to motivate the student to go beyond what is written and/or solidify the relationship between the boxed topic and the chapter material.
- Twenty to thirty new in-chapter and end-of-chapter questions have been added to each chapter to allow instructors greater flexibility in assigning problems and to give students more opportunity to test themselves. Most chapters now include at least 100 questions and problems.
- End-of-chapter problems are now organized according to the level of understanding required of the student. *Foundations* questions provide students the opportunity to review factual information, emphasizing basic concepts, definitions, and drill. *Applications* questions are more complex or relate more directly to real-world situations. They require students to understand the information and to apply that knowledge to higher-order problems.
- The art program has undergone significant revision. New figures have been added and many others revised to create a common style and pedagogical strategy. The efforts of our Art Consultant, Dr. Ann Eakes of Northwest Vista College, have been invaluable in providing a new perspective on the art program.
- The ARIS website and other media supplements, as described later in this Preface, have been enhanced. Appendices, formerly at the end of the textbook, can now be readily accessed on the website. The instructors' Digital Content Manager CD-ROM contains electronic files of text figures and tables as well as PowerPoint lecture slides.

We designed the fifth edition to promote student learning and facilitate teaching. It is important to engage students, to appeal to visual learners, and to provide a variety of pedagogical tools to help them organize and summarize information. We have utilized a variety of strategies to accomplish our goals.

Engaging Students
Students learn better when they can see a clear relationship between the subject material they are studying and real life. We wrote the text to help students make connections between the principles of chemistry and their previous life experiences or their future professional experiences. Our strategy to accomplish this integration includes the following:

- **Boxed Readings—"Chemistry Connection":** Introductory vignettes allow students to see the significance of chemistry in their daily lives and in their future professions.
- **Boxed Perspectives:** These short stories present real-world situations that involve one or more topics that students will encounter in the chapter. The "Medical Perspectives" relate chemistry to a health concern or a diagnostic application. The "Environmental Perspectives" deal with issues, including the impact of chemistry on the ecosystem and how these environmental changes affect human health. "Human Perspectives" delve into chemistry and society and include such topics as gender issues in science and historical viewpoints. In the fifth edition, we have added a number of new boxed topics and, where needed, updated all of those perspectives retained from the fourth edition. We have included topics, such as self-tanning lotions and sugar substitutes, which are of interest to students today, as well as the most recent strategies for the treatment of HIV/AIDS. New perspectives on opioid drugs, methamphetamines, alcohol abuse, and drugs to treat chemical addiction have been added to this edition.

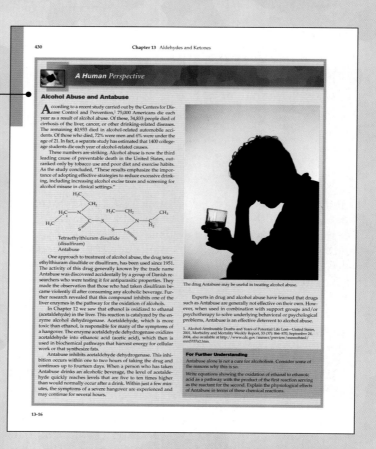

Learning Tools

In designing the original learning system we asked ourselves the question: "If we were students, what would help us organize and understand the material covered in this chapter?" With valuable suggestions from our reviewers, we have made some modifications to improve the learning system. However, with the blessings of those reviewers, we have retained all the elements of the system that have been shown to support student learning:

- **Learning Goals:** A set of chapter objectives at the beginning of each chapter previews concepts that will be covered in the chapter. Icons locate text material that supports the learning goals.
- **Detailed Chapter Outline:** A listing of topic headings is provided for each chapter. Topics are arranged in outline form to help students organize the material in their own minds.
- **Chapter Cross-References:** To help students locate the pertinent background material, references to previous chapters, sections, and perspectives are noted in the margins of the text. These marginal cross-references also alert students to upcoming topics related to the information currently being studied.

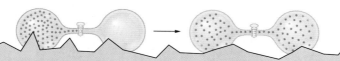

- **Summary of Reactions:** In the organic chemistry chapters, each major reaction type is highlighted on a green background. Major equations are summarized at the end of the chapter, facilitating review.
- **Chapter Summary:** Each major topic of the chapter is briefly reviewed in paragraph form in the end-of-chapter summary. These summaries serve as a mini-study guide, covering the major concepts in the chapter.
- **Key Terms:** Key terms are printed in boldface in the text, defined immediately, and listed at the end of the chapter. Each end-of-chapter key term is accompanied by a section number for rapid reference.
- **Glossary of Key Terms:** In addition to being listed at the end of the chapter, each key term from the text is also defined in the alphabetical glossary at the end of the book.

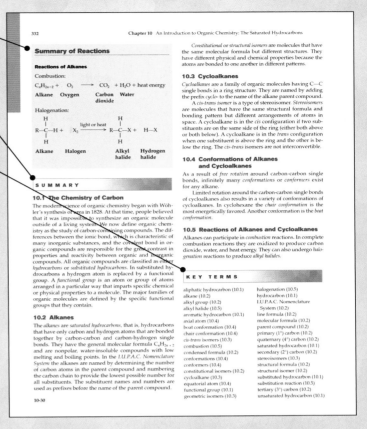

Detailed List of Changes

Changes and updates are evident in every chapter of this fifth edition. Major changes to individual chapters include:

- Chapters 2 and 3 have been combined to give seamless coverage of atomic structure and periodicity, all in one unit.
- The balancing of chemical equations, previously touched on in two sections of the book, has been combined and placed in one section of the book.
- Chapter 10, "Alkanes," now includes an updated perspective on oil-eating microbes and an additional example problem.
- Chapter 11, "Alkenes and Alkynes," now has reactions of alkynes added, including example problems of each.
- Chapter 12, "Alcohols," now includes new perspectives on methanol poisoning, alcohol abuse, and the use of Antabuse.
- Chapter 14, "Carboxylic Acids," offers students an updated perspective on biodegradable plastics.
- Chapter 15, "Amines and Amides," provides new perspectives on Methamphetamine, as well as one on opiate biosynthesis and mutant poppies.
- Chapter 16, "Carbohydrates," gives an updated perspective on the sucrose/tooth decay connection. We have also removed the old food pyramid from the chapter.
- Chapter 18, "Proteins," gives students improved coverage of amino acid structure, properties, and stereoisomers, as well as properties of the peptide bond.
- Chapter 19, "Enzymes," includes a completely reworked section on enzyme nomenclature.
- Chapter 20, "Molecular Genetics," has been highly reorganized. This had been the last chapter in the fourth edition, but is now placed with the other chapters devoted to macromolecules. This chapter also includes a new, more detailed section on bacterial DNA replication.

The Art Program

Today's students are much more visually oriented than any previous generation. Television and the computer represent alternate modes of learning. We have built upon this observation through expanded use of color, figures, and three-dimensional computer-generated models. This art program enhances the readability of the text and provides alternative pathways to learning.

Dynamic Illustrations

Each chapter is amply illustrated using figures, tables, and chemical formulas. All of these illustrations are carefully annotated for clarity. Approximately 220 full-color illustrations have been revised for this edition, in addition to over 30 new illustrations and 50 new photos, to help students better understand difficult concepts. In many cases, illustrations have been redrawn to be more realistic, and have been color-enhanced.

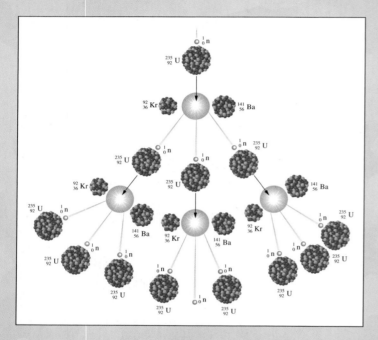

Color-Coding Scheme

We have color-coded reactions so that chemical groups being added or removed in a reaction can be quickly recognized.

- Red print is used in chemical equations or formulas to draw the reader's eye to key elements or properties in a reaction or structure.
- Blue print is used when additional features must be highlighted.

> Aldehydes and ketones can be distinguished on the basis of differences in their reactivity. The most common laboratory test for aldehydes is the **Tollens' test.** When exposed to the Tollens' reagent, a basic solution of $Ag(NH_3)_2^+$, an aldehyde undergoes oxidation. The silver ion (Ag^+) is reduced to silver metal (Ag^0) as the aldehyde is oxidized to a carboxylic acid anion.
>
> $$\underset{\text{Aldehyde}}{R-\overset{O}{\underset{\|}{C}}-H} + \underset{\substack{\text{Silver}\\\text{ammonia complex—}\\\text{Tollens' reagent}}}{Ag(NH_3)_2^+} \longrightarrow \underset{\substack{\text{Carboxylate}\\\text{anion}}}{R-\overset{O}{\underset{\|}{C}}-O^-} + \underset{\substack{\text{Silver}\\\text{metal}\\\text{mirror}}}{Ag^0}$$
>
> Silver metal precipitates from solution and coats the flask, producing a smooth silver mirror, as seen in Figure 13.4. The test is therefore often called the Tollens' silver mirror test. The commercial manufacture of silver mirrors uses a similar process. Ketones cannot be oxidized to carboxylic acids and do not react with the Tollens' reagent.

- Green background screens denote generalized chemical and mathematical equations. In the organic chemistry chapters, the Summary of Reactions at the end of these chapters is also highlighted with a green background screen for ease of recognition.

> **Beta Particles**
>
> The **beta particle** (β), in contrast, is a fast-moving electron traveling at approximately 90% of the speed of light as it leaves the nucleus. It is formed in the nucleus by the conversion of a neutron into a proton. The beta particle is represented as
>
> $$_{-1}^{0}e \quad \text{or} \quad _{-1}^{0}\beta \quad \text{or} \quad \beta$$

- Yellow background in the general and biochemistry sections of the text illustrates energy, either as energy stored in electrons or in groups of atoms. In the organic chemistry section of the text, yellow background screens also reveal the parent chain of an organic compound.

> **Ionization Energy**
>
> The energy required to remove an electron from an isolated atom is the **ionization energy.** The process for sodium is represented as follows:
>
> $$\text{ionization energy} + Na \longrightarrow Na^+ + e^-$$

- There are certain situations in which it is necessary to adopt a unique color convention tailored to the material in a particular chapter. For example, in Chapter 18, the structures of amino acids require four colors to draw attention to key features of these molecules. For consistency, red is used to denote the acid portion of an amino acid, and blue is used to denote the basic portion of an amino acid. Green print is used to denote the R groups, and a yellow background screen directs the eye to the α-carbon.

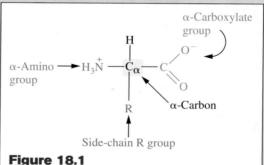

Figure 18.1
General structure of an α-amino acid. All amino acids isolated from proteins, with the exception of proline, have this general structure.

Computer-Generated Models

The students' ability to understand the geometry and three-dimensional structure of molecules is essential to the understanding of organic and biochemical reactions. Computer generated models are used throughout the text because they are both accurate and easily visualized.

Problem Solving and Critical Thinking

Perhaps the best preparation for a successful and productive career is the development of problem-solving and critical thinking skills. To this end, we created a variety of problems that require recall, fundamental calculations, and complex reasoning. In this edition, we have used suggestions from our reviewers, as well as from our own experience, to enhance the problem sets to include more practice problems for difficult concepts and further integration of the subject areas.

In-Chapter Examples, Solutions, and Questions

Each chapter includes a number of examples that show the student, step-by-step, how to properly reach the correct solution to model problems. Whenever possible, the examples are followed by in-text questions that allow students to test their mastery of information and to build self-confidence.

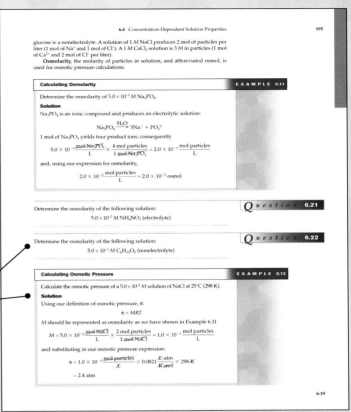

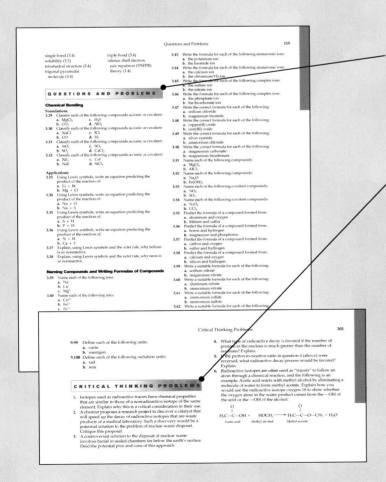

In-Chapter and End-of-Chapter Problems

We have created a wide variety of paired concept problems. The answers to the odd-numbered questions are found at the back of the book as reinforcement for students as they develop problem-solving skills. However, students must then be able to apply the same principles to the related even-numbered problems.

Critical Thinking Problems

Each chapter includes a set of critical thinking problems. These problems are intended to challenge students to integrate concepts to solve more complex problems. They make a perfect complement to the classroom lecture because they provide an opportunity for in-class discussion of complex problems dealing with daily life and the health care sciences.

Over the course of the last four editions, hundreds of reviewers have shared their knowledge and wisdom with us, as well as the reaction of their students to elements of this book. Their contributions, as well as our own continuing experience in the area of teaching and learning science, have resulted in a text that we are confident will provide a strong foundation in chemistry, while enhancing the learning experience of students.

Supplementary Materials

This text is supported by a complete package for instructors and students. Several print and media supplements have been prepared to accompany the text and make learning as meaningful and up-to-date as possible.

For the Instructor

- **Digital Content Manager CD/DVD:** This primary instructor supplement offers over 800 visual images, including illustrations, photos, examples, boxed readings, and tables from the text. These images are in full color and can be readily incorporated into lecture presentations, exams, or classroom materials. Also on the Digital Content Manager are PowerPoint Lecture Outline slides, prepared by Dr. Ann Eakes of Northwest Vista College, that cover all 23 chapters and can be modified according to instructor preference.

- **Instructor's Manual:** Written by the authors and also Dr. Timothy Dwyer of Towson University, this ancillary contains suggestions for organizing lectures, instructional objectives, perspectives on boxed readings from the text, a list of each chapter's key problems and concepts, and more. The Instructor's Manual is a part of the Instructor's Testing and Resource CD, and is also available through the ARIS website for this text.
- **Transparencies:** A set of 100 transparencies is available to help the instructor coordinate the lecture with key illustrations from the text.
- **Computerized Classroom Management System:** This Instructor's Testing and Resource CD includes a database of test questions, reproducible student self-quizzes, and a grade-recording program. Also found on this CD is the Instructor's Manual to accompany this text.

- *A Laboratory Manual for General, Organic, and Biochemistry,* Fifth Edition, by Charles H. Henrickson, Larry C. Byrd, and Norman W. Hunter of Western Kentucky University, offers clear and concise laboratory experiments that reinforce students' understanding of concepts. Prelaboratory exercises, questions, and report sheets are coordinated with each experiment to ensure active student involvement and comprehension. A new student tutorial on graphing with Excel has been added to this edition.
- **Laboratory Resource Guide:** Written by Charles H. Henrickson, Larry C. Byrd, and Norman W. Hunter of Western Kentucky University, this helpful prep guide contains the hints that the authors have learned over the years to ensure students' success in the laboratory. This Resource Guide is available through the ARIS course website for this text.
- **ARIS** (McGraw-Hill's Assessment, Review and Instruction System) for *General, Organic, and Biochemistry*—a complete electronic homework and course management system—is designed for greater ease of use than any other system available. ARIS enables instructors to create and share course materials and assignments with colleagues with a few clicks of the mouse. Instructors can edit questions, import their own content, and create announcements and due dates for assignments. ARIS has automatic grading and reporting of easy-to-assign homework, quizzing, and testing. Once a student is registered in the course, all student activity within McGraw-Hill's ARIS is automatically recorded and available to the instructor through a fully integrated grade book that can be downloaded to Excel. This book-specific website is found at www.mhhe.com/denniston5e.

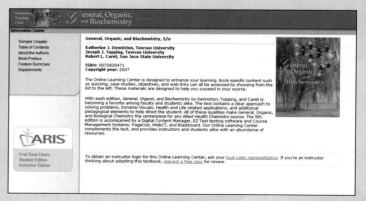

- **Course Management Systems—PageOut, WebCT, and Blackboard:** The course cartridge that accompanies *General, Organic, and Biochemistry,* Fifth Edition, includes all ARIS website content, and the entire test bank that accompanies this new edition.

For the Student

- **Student Study Guide/Solutions Manual:** A separate Student Study Guide/Solutions Manual, prepared by Dr. Timothy Dwyer and the authors of this text, is available. It contains the answers and complete solutions for the odd-numbered problems. It also offers students a variety of exercises and keys for testing their comprehension of basic, as well as difficult, concepts.
- **Schaum's Outline of General, Organic, and Biological Chemistry:** Written by George Odian and Ira Blei, this supplement provides students with over 1400 solved problems with complete solutions. It also teaches effective problem-solving techniques.
- **ARIS:** McGraw-Hill's Assessment, Review, and Instruction System for *General, Organic, and Biochemistry* is available to students and instructors using this text. The website offers quizzes, key definitions, a review of mathematics applied to problem solving, important tables, definitions, and more. This book-specific website can be found at www.mhhe.com/denniston5e.

Acknowledgments

We are thankful to our families, whose patience and support made it possible for us to undertake this project. We are also grateful to our many colleagues at McGraw-Hill for their support, guidance, and assistance.

A revision cannot move forward without the feedback of professors teaching the course. The reviewers have our gratitude and assurance that their comments received serious consideration. The following professors provided reviews, participated in a focus group, or gave valuable advice for the preparation of the fifth edition:

Ayoni F. Akinyele   *Howard University, Washington, D.C.*
James Armstrong   *City College of San Francisco*
Maher Atteya   *Georgia Perimeter College*
Teresa L. Brown   *Rochester Community and Technical College*
Alan J. Bruha   *Lewis and Clark Community College*
Kathleen Brunke   *Christopher Newport University*
Derald Chriss   *Southern University*
Sharon W. Chriss   *Southern University*
Ana Ciereszko   *Miami–Dade College*
Rajeev B. Dabke   *Columbus State University*
Philip Denton   *Fresno City College*
Brahmadeo Dewprashad   *Borough of Manhattan Community College*
Ronald P. Drucker   *City College of San Francisco*
Fredesvinda B. Cand. Dura   *LaGuardia Community College/CUNY*
Ann Eakes   *Northwest Vista College*
Barbara L. Edgar   *University of Minnesota, Twin Cities*
Amber Flynn-Charlebois   *William Paterson University*
Wes Fritz   *College of DuPage*
Elizabeth Gardner   *University of Texas at El Paso*
Galen G. George   *Santa Rosa Junior College*
Walter E. Godwin   *University of Arkansas at Monticello*
Cliff Gottlieb   *Shasta College*
James K. Hardy   *The University of Akron*
Jonathan Heath   *Horry-Georgetown Technical College*
T.G. Jackson   *University of South Alabama*
Richard H. Jarman   *College of DuPage*
James T. Johnson   *Sinclair Community College*
David A. Katz   *Pima Community College*
Colleen Kelley   *Pima Community College*
Laura Kibler-Herzog   *Georgia State University*
Edith Preciosa Klingberg   *University of Toledo*
Terry L. Lampe   *Georgia Perimeter College*
Richard H. Langley   *Stephen F. Austin State University*
David Lippmann   *Texas State University, San Marcos*
Jeanette C. Madea   *Broward Community College*
Tammy J. Melton   *Middle Tennessee State University*
Melvin Merken   *Worcester State College*
Li-June Ming   *University of South Florida*
John T. Moore   *Stephen F. Austin State University*
Joseph C. Muscarella   *Henry Ford Community College*
Elva Mae Nicholson   *Eastern Michigan University*
Thomas J. Nycz   *Broward Community College, North Campus*
Beng Guat Ooi   *Middle Tennessee State University*
Alicia Paterno Parsi   *Duquesne University*
Jerry Poteat   *Georgia Perimeter College*
Parris F. Powers   *Volunteer State Community College*
Betsy Ratcliff   *West Virginia University*
Douglas E. Raynie   *South Dakota State University*
Susan S. Reid   *North Hennepin Community College*
Lynette M. Rushton   *South Puget Sound Community College*
Howard Theodore Silverstein   *Georgia Perimeter College*
Melinda M. Sorensson   *University of Louisana*
Luise Strange de Soria   *Georgia Perimeter College*
Melissa R. Synder   *Rochester Community and Technical College*
Sheryl K. Wallace   *South Plains College*
Marcy Whitney   *University of Alabama*
Catherine Woytowicz   *The George Washington University*
Clara Wu   *LaGuardia Community College, City University of New York*
Vaneica Young   *University of Florida*

GENERAL CHEMISTRY

Chemistry

Methods and Measurement

Learning Goals

1. Describe the interrelationship of chemistry with other fields of science and medicine.

2. Discuss the approach to science, the scientific method, and distinguish among the terms *hypothesis*, *theory*, and *scientific law*.

3. Describe the properties of the solid, liquid, and gaseous states.

4. Provide specific examples of physical and chemical properties and physical and chemical change.

5. Distinguish between intensive and extensive properties.

6. Classify matter as element, compound, or mixture.

7. Distinguish between data and results.

8. Learn the major units of measure in the English and metric systems, and be able to convert from one system to another.

9. Report data and results using scientific notation and the proper number of significant figures.

10. Use appropriate experimental quantities in problem solving.

11. Calculate the density of an object from mass and volume data and calculate the specific gravity of an object from its density.

Name the forms of measurement that apply to this activity.

Outline

Chemistry Connection:
Chance Favors the Prepared Mind

Introduction
1.1 The Discovery Process

A Human Perspective:
The Scientific Method

A Medical Perspective:
Curiosity, Science, and Medicine

1.2 Matter and Properties

1.3 Measurement in Chemistry

1.4 Significant Figures and Scientific Notation

1.5 Experimental Quantities

A Human Perspective:
Food Calories

A Medical Perspective:
Diagnosis Based on Waste

Chemistry Connection

Chance Favors the Prepared Mind

Most of you have chosen a career in medicine because you want to help others. In medicine, helping others means easing pain and suffering by treating or curing diseases. One important part of the practice of medicine involves observation. The physician must carefully observe the patient and listen to his or her description of symptoms to arrive at a preliminary diagnosis. Then appropriate tests must be done to determine whether the diagnosis is correct. During recovery the patient must be carefully observed for changes in behavior or symptoms. These changes are clues that the treatment or medication needs to be modified.

These practices are also important in science. The scientist makes an observation and develops a preliminary hypothesis or explanation for the observed phenomenon. Experiments are then carried out to determine whether the hypothesis is reasonable. When performing the experiment and analyzing the data, the scientist must look for any unexpected results that indicate that the original hypothesis must be modified.

Several important discoveries in medicine and the sciences have arisen from accidental observations. A health care worker or scientist may see something quite unexpected. Whether this results in an important discovery or is ignored depends on the training and preparedness of the observer.

It was Louis Pasteur, a chemist and microbiologist, who said, "Chance favors the prepared mind." In the history of science and medicine there are many examples of individuals who have made important discoveries because they recognized the value of an unexpected observation.

One such example is the use of ultraviolet (UV) light to treat infant jaundice. Infant jaundice is a condition in which the skin and the whites of the eyes appear yellow because of high levels of the bile pigment bilirubin in the blood. Bilirubin is a breakdown product of the oxygen-carrying blood protein hemoglobin. If bilirubin accumulates in the body, it can cause brain damage and death. The immature liver of the baby cannot remove the bilirubin.

An observant nurse in England noticed that when jaundiced babies were exposed to sunlight, the jaundice faded. Research based on her observation showed that the UV light changes the bilirubin into another substance that can be excreted. To this day, jaundiced newborns are treated with UV light.

The Pap smear test for the early detection of cervical and uterine cancer was also developed because of an accidental observation. Dr. George Papanicolaou, affectionately called Dr. Pap, was studying changes in the cells of the vagina during the stages of the menstrual cycle. In one sample he recognized cells that looked like cancer cells. Within five years, Dr. Pap had perfected a technique for staining cells from vaginal fluid and observing them microscopically for the presence of any abnormal cells. The lives of countless women have been saved because a routine Pap smear showed early stages of cancer.

In this first chapter of your study of chemistry you will learn more about the importance of observation and accurate, precise measurement in medical practice and scientific study. You will also study the scientific method, the process of developing hypotheses to explain observations, and the design of experiments to test those hypotheses.

Introduction

When you awoke this morning, a flood of chemicals called neurotransmitters was sent from cell to cell in your nervous system. As these chemical signals accumulated, you gradually became aware of your surroundings. Chemical signals from your nerves to your muscles propelled you out of your warm bed to prepare for your day.

For breakfast you had a glass of milk, two eggs, and buttered toast, thus providing your body with needed molecules in the form of carbohydrates, proteins, lipids, vitamins, and minerals. As you ran out the door, enzymes of your digestive tract were dismantling the macromolecules of your breakfast. Other enzymes in your cells were busy converting the chemical energy of food molecules into adenosine triphosphate (ATP), the universal energy currency of all cells.

As you continue through your day, thousands of biochemical reactions will keep your cells functioning optimally. Hormones and other chemical signals will regulate the conditions within your body. They will let you know if you are hungry or thirsty. If you injure yourself or come into contact with a disease-causing microorganism, chemicals in your body will signal cells to begin the necessary repair or defense processes.

Life is an organized array of large, carbon-based molecules maintained by biochemical reactions. To understand and appreciate the nature of a living being, we must understand the principles of science and chemistry as they apply to biological molecules.

1.1 The Discovery Process

Chemistry

Chemistry is the study of matter, its chemical and physical properties, the chemical and physical changes it undergoes, and the energy changes that accompany those processes. **Matter** is anything that has mass and occupies space. The changes that matter undergoes always involve either gain or loss of energy. **Energy** is the ability to do work to accomplish some change. The study of chemistry involves matter, energy, and their interrelationship. Matter and energy are at the heart of chemistry.

Major Areas of Chemistry

Chemistry is a broad area of study covering everything from the basic parts of an atom to interactions between huge biological molecules. Because of this, chemistry encompasses the following specialties.

Biochemistry is the study of life at the molecular level and the processes associated with life, such as reproduction, growth, and respiration. *Organic chemistry* is the study of matter that is composed principally of carbon and hydrogen. Organic chemists study methods of preparing such diverse substances as plastics, drugs, solvents, and a host of industrial chemicals. *Inorganic chemistry* is the study of matter that consists of all of the elements other than carbon and hydrogen and their combinations. Inorganic chemists have been responsible for the development of unique substances such as semiconductors and high-temperature ceramics for industrial use. *Analytical chemistry* involves the analysis of matter to determine its composition and the quantity of each kind of matter that is present. Analytical chemists detect traces of toxic chemicals in water and air. They also develop methods to analyze human body fluids for drugs, poisons, and levels of medication. *Physical chemistry* is a discipline that attempts to explain the way in which matter behaves. Physical chemists develop theoretical concepts and try to prove them experimentally. This helps us understand how chemical systems behave.

Over the last thirty years, the boundaries between the traditional sciences of chemistry, physics, and biology, as well as mathematics and computer science have gradually faded. Medical practitioners, physicians, nurses, and medical technologists use therapies that contain elements of all these disciplines. The rapid expansion of the pharmaceutical industry is based on a recognition of the relationship between the function of an organism and its basic chemical makeup. Function is a consequence of changes that chemical substances undergo.

For these reasons, an understanding of basic chemical principles is essential for anyone considering a medically related career; indeed, a worker in any science-related field will benefit from an understanding of the principles and applications of chemistry.

The Scientific Method

The **scientific method** is a systematic approach to the discovery of new information. How do we learn about the properties of matter, the way it behaves in nature,

A Human Perspective

The Scientific Method

The discovery of penicillin by Alexander Fleming is an example of the scientific method at work. Fleming was studying the growth of bacteria. One day, his experiment was ruined because colonies of mold were growing on his plates. From this failed experiment, Fleming made an observation that would change the practice of medicine: Bacterial colonies could not grow in the area around the mold colonies. Fleming hypothesized that the mold was making a chemical compound that inhibited the growth of the bacteria. He performed a series of experiments designed to test this hypothesis.

The key to the scientific method is the design of carefully controlled experiments that will either support or disprove the hypothesis. This is exactly what Fleming did.

In one experiment he used two sets of tubes containing sterile nutrient broth. To one set he added mold cells. The second set (the control tubes) remained sterile. The mold was allowed to grow for several days. Then the broth from each of the tubes (experimental and control) was passed through a filter to remove any mold cells. Next, bacteria were placed in each tube. If Fleming's hypothesis was correct, the tubes in which the mold had grown would contain the chemical that inhibits growth, and the bacteria would not grow. On the other hand, the control tubes (which were never used to grow mold) would allow bacterial growth. This is exactly what Fleming observed.

Within a few years this *antibiotic*, penicillin, was being used to treat bacterial infections in patients.

For Further Understanding

What is the purpose of the control tubes used in this experiment?

What common characteristics do you find in this story and the Medical Perspective on page 5?

and how it can be modified to make useful products? Chemists do this by using the scientific method to study the way in which matter changes under carefully controlled conditions.

The scientific method is not a "cookbook recipe" that, if followed faithfully, will yield new discoveries; rather, it is an organized approach to solving scientific problems. Every scientist brings his or her own curiosity, creativity, and imagination to scientific study. But scientific inquiry still involves some of the "cookbook approach."

Characteristics of the scientific process include the following:

1. *Observation.* The description of, for example, the color, taste, or odor of a substance is a result of observation. The measurement of the temperature of a liquid or the size or mass of a solid results from observation.
2. *Formulation of a question.* Humankind's fundamental curiosity motivates questions of why and how things work.
3. *Pattern recognition.* If a scientist finds a cause-and-effect relationship, it may be the basis of a generalized explanation of substances and their behavior.
4. *Developing theories.* When scientists observe a phenomenon, they want to explain it. The process of explaining observed behavior begins with a hypothesis. A **hypothesis** is simply an attempt to explain an observation, or series of observations, in a commonsense way. If many experiments support a hypothesis, it may attain the status of a theory. A **theory** is a hypothesis supported by extensive testing (experimentation) that explains scientific facts and can predict new facts.
5. *Experimentation.* Demonstrating the correctness of hypotheses and theories is at the heart of the scientific method. This is done by carrying out carefully designed experiments that will either support or disprove the theory or hypothesis.

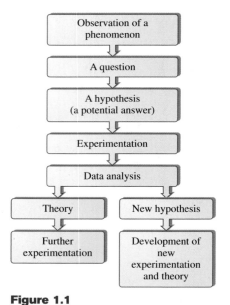

Figure 1.1
The scientific method, an organized way of doing science. A degree of trial and error is apparent here. If experimentation does not support the hypothesis, one must begin the cycle again.

A Medical Perspective

Curiosity, Science, and Medicine

Curiosity is one of the most important human traits. Small children constantly ask "why?". As we get older, our questions become more complex, but the curiosity remains.

Curiosity is also the basis of the scientific method. A scientist observes an event, wonders why it happens, and sets out to answer the question. Dr. Michael Zasloff's curiosity may lead to the development of an entirely new class of antibiotics. When he was a geneticist at the National Institutes of Health, his experiments involved the surgical removal of the ovaries of African clawed frogs. After surgery he sutured (sewed up) the incision and put the frogs back in their tanks. These water-filled tanks were teeming with bacteria, but the frogs healed quickly, and the incisions did not become infected!

Of all the scientists to observe this remarkable healing, only Zasloff was curious enough to ask whether there were chemicals in the frogs' skin that defended the frogs against bacterial infections—a new type of antibiotic. All currently used antibiotics are produced by fungi or are synthesized in the laboratory. One big problem in medicine today is more and more pathogenic (disease-causing) bacteria are becoming resistant to these antibiotics. Zasloff hoped to find an antibiotic that worked in an entirely new way so the current problems with antibiotic resistance might be overcome.

Dr. Zasloff found two molecules in frog skin that can kill bacteria. Both are small proteins. Zasloff named them *magainins*, from the Hebrew word for shield. Most of the antibiotics that we now use enter bacteria and kill them by stopping some biochemical process inside the cell. Magainins are more direct; they simply punch holes in the bacterial membrane, and the bacteria explode.

One of the magainins, now chemically synthesized in the laboratory so that no frogs are harmed, may be available to the public in the near future. This magainin can kill a wide variety of bacteria (broad-spectrum antibiotic), and it has passed the Phase I human trials. If this compound passes all the remaining tests, it will be used in treating deep infected wounds and ulcers, providing an alternative to traditional therapy.

The curiosity that enabled Zasloff to advance the field of medicine also catalyzed the development of chemistry. We will see the product of this fundamental human characteristic as we study the work of many extraordinary chemists throughout this chapter.

For Further Understanding
Why is it important for researchers to continually design and develop new antibacterial substances?

What common characteristics do you find in this work and the discovery discussed in the Chemistry Connection on page 2?

6. *Summarizing information.* A scientific **law** is nothing more than the summary of a large quantity of information. For example, the law of conservation of matter states that matter cannot be created or destroyed, only converted from one form to another. This statement represents a massive body of chemical information gathered from experiments.

The scientific method involves the interactive use of hypotheses, development of theories, and thorough testing of theories using well-designed experiments and is summarized in Figure 1.1.

Models in Chemistry

Hypotheses, theories, and laws are frequently expressed using mathematical equations. These equations may confuse all but the best of mathematicians. For this reason a *model* of a chemical unit or system is often used to make ideas more clear. A good model based on everyday experience, although imperfect, gives a great deal of information in a simple fashion. Consider the fundamental unit of methane, the major component of natural gas, which is composed of one carbon atom (symbolized by C) and four hydrogen atoms (symbolized by H).

A geometrically correct model of methane can be constructed from balls and sticks. The balls represent the individual units (atoms) of hydrogen and carbon, and the sticks correspond to the attractive forces that hold the hydrogen and carbon together. The model consists of four balls representing hydrogen symmetrically arranged around a center ball representing carbon. The "carbon" ball is attached to each "hydrogen" ball by sticks, as shown:

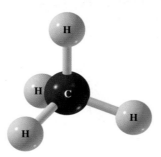

Color-coding the balls distinguishes one type of matter from another; the geometrical form of the model, all of the angles and dimensions of a tetrahedron, are the same for each methane unit found in nature. Methane is certainly not a collection of balls and sticks; but such models are valuable because they help us understand the chemical behavior of methane and other, more complex substances.

Chemists and physicists have used the observed properties of matter to develop models of the individual units of matter. These models collectively make up what we now know as the atomic theory of matter.

These models have developed from experimental observations over the past two hundred years. Thus theory and experiment reinforce each other. We must gain some insight into atomic structure to appreciate the behavior of the atoms themselves as well as larger aggregates of atoms: compounds.

The structure-properties concept has advanced so far that compounds are designed and synthesized in the laboratory with the hope that they will perform very specific functions, such as curing diseases that have been resistant to other forms of treatment. Figure 1.2 shows some of the variety of modern technology that has its roots in the understanding of the atom.

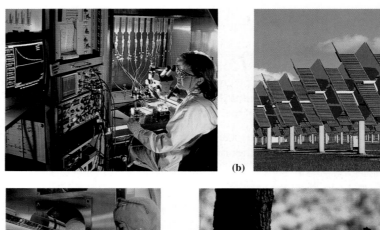
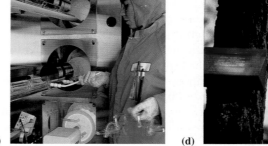

Figure 1.2
Examples of technology originating from scientific inquiry: (a) synthesis of a new drug, (b) solar energy cells, (c) preparation of solid-state electronics, (d) use of a gypsy moth sex attractant for insect control.

1.2 Matter and Properties

Properties are characteristics of matter and are classified as either physical or chemical. In this section we will learn the meaning of physical and chemical properties and how they are used to characterize matter.

Matter and Physical Properties

There are three *states of matter*: the **gaseous state,** the **liquid state,** and the **solid state.** A gas is made up of particles that are widely separated. In fact, a gas will expand to fill any container; it has no definite shape or volume. In contrast, particles of a liquid are closer together; a liquid has a definite volume but no definite shape; it takes on the shape of its container. A solid consists of particles that are close together and that often have a regular and predictable pattern of particle arrangement (crystalline). A solid has both fixed volume and fixed shape. Attractive forces, which exist between all particles, are very pronounced in solids and much less so in gases.

Water is the most common example of a substance that can exist in all three states over a reasonable temperature range (Figure 1.3). Conversion of water from one state to another constitutes a *physical change*. A **physical change** produces a recognizable difference in the appearance of a substance without causing any change in its composition or identity. For example, we can warm an ice cube and it will melt, forming liquid water. Clearly its appearance has changed; it has been transformed from the solid to the liquid state. It is, however, still water; its composition and identity remain unchanged. A physical change has occurred. We could in fact demonstrate the constancy of composition and identity by refreezing the liquid water, re-forming the ice cube. This melting and freezing cycle could be repeated over and over. This very process is a hallmark of our global weather changes. The continual interconversion of the three states of water in the environment (snow, rain, and humidity) clearly demonstrates the retention of the identity of water particles or *molecules*.

A **physical property** can be observed or measured without changing the composition or identity of a substance. As we have seen, melting ice is a physical change. We can measure the temperature when melting occurs; this is the *melting point* of water. We can also measure the *boiling point* of water, when liquid water becomes a gas. Both the melting and boiling points of water, and of any other substance, are physical properties.

A practical application of separation of materials based upon their differences in physical properties is shown in Figure 1.4.

(a)

(b)

(c)

Figure 1.3
The three states of matter exhibited by water: (a) solid, as ice; (b) liquid, as ocean water; (c) gas, as humidity in the air.

Figure 1.4
An example of separation based on differences in physical properties. Magnetic iron is separated from other nonmagnetic substances. A large-scale version of this process is important in the recycling industry.

Chapter 1 Chemistry: Methods and Measurement

Matter and Chemical Properties

We have noted that physical properties can be exhibited, measured, or observed without any change in identity or composition. In contrast, **chemical properties** do result in a change in composition and can be observed only through chemical reactions. A **chemical reaction** is a process of rearranging, removing, replacing, or adding atoms to produce new substances. For example, the process of photosynthesis can be shown as

$$\text{carbon dioxide} + \text{water} \xrightarrow[\text{Chlorophyll}]{\text{Light}} \text{sugar} + \text{oxygen}$$

Light is the energy needed to make the reaction happen. Chlorophyll is the energy absorber, converting light energy to chemical energy.

Chapter 7 discusses the role of energy in chemical reactions.

This chemical reaction involves the conversion of carbon dioxide and water (the *reactants*) to a sugar and oxygen (the *products*). The products and reactants are clearly different. We know that carbon dioxide and oxygen are gases at room temperature and water is a liquid at this temperature; the sugar is a solid white powder. A chemical property of carbon dioxide is its ability to form sugar under certain conditions. The process of formation of this sugar is the *chemical change*.

EXAMPLE 1.1 Identifying Properties

Can the process that takes place when an egg is fried be described as a physical or chemical change?

Solution

Examine the characteristics of the egg before and after frying. Clearly, some significant change has occurred. Furthermore, the change appears irreversible. More than a simple physical change has taken place. A chemical reaction (actually, several) must be responsible; hence chemical change.

Question 1.1

Classify each of the following as either a chemical property or a physical property:

a. color
b. flammability
c. hardness
d. odor
e. taste

Question 1.2

Classify each of the following as either a chemical change or a physical change:

a. water boiling to become steam
b. butter becoming rancid
c. combustion of wood
d. melting of ice in spring
e. decay of leaves in winter

Intensive and Extensive Properties

See page 28 for a discussion of density and specific gravity.

It is important to recognize that properties can also be classified according to whether they depend on the size of the sample. Consequently, there is a fundamental difference between properties such as density and specific gravity and properties such as mass and volume.

An **intensive property** is a property of matter that is *independent* of the *quantity* of the substance. Density, boiling and melting points, and specific gravity are intensive properties. For example, the boiling point of one single drop of water is exactly the same as the boiling point of a liter of water.

An **extensive property** *depends* on the *quantity* of a substance. Mass and volume are extensive properties. There is an obvious difference between 1 g of silver and 1 kg of silver; the quantities and, incidentally, the value, differ substantially.

EXAMPLE 1.2

Differentiating Between Intensive and Extensive Properties

Is temperature an extensive or intensive property?

Solution

Imagine two glasses each containing 100 g of water, and each at 25°C. Now pour the contents of the two glasses into a larger glass. You would predict that the mass of the water in the larger glass would be 200 g (100 g + 100 g) because mass is an extensive property, dependent on quantity. However, we would expect the temperature of the water to remain the same (not 25°C + 25°C); hence temperature is *an intensive property* . . . independent of quantity.

Classification of Matter

Chemists look for similarities in properties among various types of materials. Recognizing these likenesses simplifies learning the subject and allows us to predict the behavior of new substances on the basis of their relationship to substances already known and characterized.

Many classification systems exist. The most useful system, based on composition, is described in the following paragraphs (see also Figure 1.5).

All matter is either a *pure substance* or a *mixture*. A **pure substance** has only one component. Pure water is a pure substance. It is made up only of particles containing two hydrogen atoms and one oxygen atom, that is, water molecules (H_2O).

There are different types of pure substances. Elements and compounds are both pure substances. An **element** is a pure substance that cannot be changed into a simpler form of matter by any chemical reaction. Hydrogen and oxygen, for example, are elements. Alternatively, a **compound** is a substance resulting from the combination of two or more elements in a definite, reproducible way. The elements hydrogen and oxygen, as noted earlier, may combine to form the compound water, H_2O.

A **mixture** is a combination of two or more pure substances in which each substance retains its own identity. Alcohol and water can be combined in a mixture. They coexist as pure substances because they do not undergo a chemical reaction; they exist as thoroughly mixed discrete molecules. This collection of dissimilar particles is the mixture. A mixture has variable composition; there are an infinite

At present, more than one hundred elements have been characterized. A complete listing of the elements and their symbols is found on the inside front cover of this textbook.

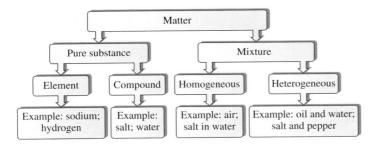

Figure 1.5
Classification of matter. All matter is either a pure substance or a mixture of pure substances. Pure substances are either elements or compounds, and mixtures may be either homogeneous (uniform composition) or heterogeneous (nonuniform composition).

Figure 1.6
Schematic representation of some classes of matter. (a) A pure substance, water, consists of a single component. (b) A homogeneous mixture, ethanol and water, has a uniform distribution of components. (c) A heterogeneous mixture, marble, has a nonuniform distribution of components. The lack of homogeneity is readily apparent.

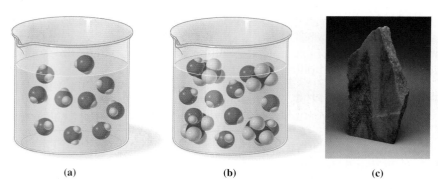

(a) (b) (c)

A detailed discussion of solutions (homogeneous mixtures) and their properties is presented in Chapter 6.

number of combinations of quantities of alcohol and water that can be mixed. For example, the mixture may contain a small amount of alcohol and a large amount of water or vice versa. Each is, however, an alcohol–water mixture.

A mixture may be either *homogeneous* or *heterogeneous* (Figure 1.6). A **homogeneous mixture** has uniform composition. Its particles are well mixed, or thoroughly intermingled. A homogeneous mixture, such as alcohol and water, is described as a *solution*. Air, a mixture of gases, is an example of a gaseous solution. A **heterogeneous mixture** has a nonuniform composition. A mixture of salt and pepper is a good example of a heterogeneous mixture. Concrete is also composed of a heterogeneous mixture of materials (various types and sizes of stone and sand present with cement in a nonuniform mixture).

EXAMPLE 1.3 Categorizing Matter

Is seawater a pure substance, a homogeneous mixture, or a heterogeneous mixture?

Solution

Imagine yourself at the beach, filling a container with a sample of water from the ocean. Examine it. You would see a variety of solid particles suspended in the water: sand, green vegetation, perhaps even a small fish! Clearly, it is a mixture, and one in which the particles are not uniformly distributed throughout the water; hence a heterogeneous mixture.

Question 1.3

Is each of the following materials a pure substance, a homogeneous mixture, or a heterogeneous mixture?

a. ethyl alcohol
b. blood
c. Alka-Seltzer dissolved in water
d. oxygen in a hospital oxygen tank

Question 1.4

Is each of the following materials a pure substance, a homogeneous mixture, or a heterogeneous mixture?

a. air
b. paint
c. perfume
d. carbon monoxide

1.3 Measurement in Chemistry

Data, Results, and Units

A scientific experiment produces **data.** Each piece of data is the individual result of a single measurement or observation. Examples include the *mass* of a sample and the *time* required for a chemical reaction to occur. Mass, length, volume, time, temperature, and energy are common types of data obtained from chemical experiments.

Results are the outcome of an experiment. Data and results may be identical, but more often several related pieces of data are combined, and logic is used to produce a result.

> **Distinguishing Between Data and Results**   EXAMPLE 1.4
>
> In many cases, a drug is less stable if moisture is present, and excess moisture can hasten the breakdown of the active ingredient, leading to loss of potency. Therefore we may wish to know how much water a certain quantity of a drug gains when exposed to air. To do this experiment, we must first weigh the drug sample, then expose it to the air for a period and reweigh it. The change in weight,
>
> $$[\text{weight}_{\text{final}} - \text{weight}_{\text{initial}}] = \text{weight difference}$$
>
> indicates the weight of water taken up by the drug formulation. The initial and final weights are individual bits of *data*; by themselves they do not answer the question, but they do provide the information necessary to calculate the answer: the results. The difference in weight and the conclusions based on the observed change in weight are the *results* of the experiment.

The experiment described in Example 1.4 was really not a very good experiment because many other environmental conditions were not measured. Measurement of the temperature and humidity of the atmosphere and the length of time that the drug was exposed to the air (the creation of a more complete set of data) would make the results less ambiguous.

Any measurement made in the experiment must also specify the units of that measurement. An initial weight of three *ounces* is clearly quite different than three *pounds*. A **unit** defines the basic quantity of mass, volume, time, or whatever quantity is being measured. A number that is not followed by the correct unit usually conveys no useful information.

Proper use of units is central to all aspects of science. The following sections are designed to develop a fundamental understanding of this vital topic.

English and Metric Units

The *English system* is a collection of functionally unrelated units. In the *English system of measurement*, the standard *pound* (lb) is the basic unit of weight. The fundamental unit of *length* is the standard *yard* (yd), and the basic unit of *volume* is the standard *gallon* (gal). The English system is used in the United States in business and industry. However, it is not used in scientific work, primarily because it is difficult to convert from one unit to another. For example,

$$1 \text{ foot} = 12 \text{ inches} = 0.33 \text{ yard} = \frac{1}{5280} \text{ mile} = \frac{1}{6} \text{ fathom}$$

Clearly, operations such as the conversion of 1.62 yards to units of miles are not straightforward. In fact, the English "system" is not really a system at all. It is simply a collection of measures accumulated throughout English history. Because they have no common origin, it is not surprising that conversion from one unit to another is not straightforward.

The United States, the last major industrial country to retain the English system, has begun efforts to convert to the metric system. The *metric system* is truly "systematic." It is composed of a set of units that are related to each other decimally, in other words, as powers of ten. Because the *metric system* is a decimal-based system, it is inherently simpler to use and less ambiguous. For example, the length of an object may be represented as

$$1 \text{ meter} = 10 \text{ decimeters} = 100 \text{ centimeters} = 1000 \text{ millimeters}$$

The metric system was originally developed in France just before the French Revolution in 1789. The more extensive version of this system is the *Système International*, or *S.I. system*. Although the S.I. system has been in existence for over forty years, it has yet to gain widespread acceptance. To make the S.I. system truly systematic, it utilizes certain units, especially those for pressure, that many find unwieldy.

In this text we will use the metric system, not the S.I. system, and we will use the English system only to the extent of converting *from* it to the more scientifically useful metric system.

In the metric system, there are three basic units. Mass is represented as the *gram*, length as the *meter*, and volume as the *liter*. Any subunit or multiple unit contains one of these units preceded by a prefix indicating the power of ten by which the base unit is to be multiplied to form the subunit or multiple unit. The most common metric prefixes are shown in Table 1.1.

Other metric units, for time, temperature, and energy, will be treated in Section 1.5.

The same prefix may be used for volume, mass, length, time, and so forth. Consider the following examples:

$$1 \text{ milliliter (mL)} = \frac{1}{1000} \text{ liter} = 0.001 \text{ liter} = 10^{-3} \text{ liter}$$

You can find further information online at www.mhhe.com/denniston5e in "A Review of Mathematics."

A volume unit is indicated by the base unit, liter, and the prefix *milli-*, which indicates that the unit is one thousandth of the base unit. In the same way,

$$1 \text{ milligram (mg)} = \frac{1}{1000} \text{ gram} = 0.001 \text{ gram} = 10^{-3} \text{ gram}$$

and

The representation of numbers as powers of ten may be unfamiliar to you. This useful notation is discussed in Section 1.4.

$$1 \text{ millimeter (mm)} = \frac{1}{1000} \text{ meter} = 0.001 \text{ meter} = 10^{-3} \text{ meter}$$

TABLE 1.1	Some Common Prefixes Used in the Metric System	
Prefix	**Multiple**	**Decimal Equivalent**
mega (M)	10^6	1,000,000.
kilo (k)	10^3	1,000.
deka (da)	10^1	10.
deci (d)	10^{-1}	0.1
centi (c)	10^{-2}	0.01
milli (m)	10^{-3}	0.001
micro (µ)	10^{-6}	0.000001
nano (n)	10^{-9}	0.000000001

Unit Conversion: English and Metric Systems

To convert from one unit to another, we must have a *conversion factor* or series of conversion factors that relate two units. The proper use of these conversion factors is called the *factor-label method*. This method is also termed *dimensional analysis*.

This method is used for two kinds of conversions: to convert from one unit to another within the *same system* or to convert units from *one system to another*.

Conversion of Units Within the Same System

We know, for example, that in the English system,

$$1 \text{ gallon} = 4 \text{ quarts}$$

Because dividing both sides of the equation by the same term does not change its identity,

$$\frac{1 \text{ gallon}}{1 \text{ gallon}} = \frac{4 \text{ quarts}}{1 \text{ gallon}}$$

The expression on the left is equal to unity (1); therefore

$$1 = \frac{4 \text{ quarts}}{1 \text{ gallon}} \quad \text{or} \quad 1 = \frac{1 \text{ gallon}}{4 \text{ quarts}}$$

Now, multiplying any other expression by the ratio 4 quarts/1 gallon or 1 gallon/ 4 quarts will not change the value of the term, because multiplication of any number by 1 produces the original value. However, there is one important difference: The units will have changed.

EXAMPLE 1.5 Using Conversion Factors

Convert 12 gallons to units of quarts.

Solution

$$12 \text{ gal} \times \frac{4 \text{ qt}}{1 \text{ gal}} = 48 \text{ qt}$$

The conversion factor, 4 qt/1 gal, serves as a bridge, or linkage, between the unit that was given (gallons) and the unit that was sought (quarts).

The conversion factor in Example 1.5 may be written as 4 qt/1 gal or 1 gal/4 qt, because both are equal to 1. However, only the first factor, 4 qt/1 gal, will give us the units we need to solve the problem. If we had set up the problem incorrectly, we would obtain

$$12 \text{ gal} \times \frac{1 \text{ gal}}{4 \text{ qt}} = 3 \frac{\text{gal}^2}{\text{qt}}$$

Incorrect units

Clearly, units of gal^2/qt are not those asked for in the problem, nor are they reasonable units. The factor-label method is therefore a self-indicating system; the correct units (those required by the problem) will result only if the factor is set up properly.

Table 1.2 lists a variety of commonly used English system relationships that may serve as the basis for useful conversion factors.

Conversion of units within the metric system may be accomplished by using the factor-label method as well. Unit prefixes that dictate the conversion factor facilitate unit conversion (refer to Table 1.1).

TABLE 1.2 Some Common Relationships Used in the English System

A. Weight	1 pound = 16 ounces
	1 ton = 2000 pounds
B. Length	1 foot = 12 inches
	1 yard = 3 feet
	1 mile = 5280 feet
C. Volume	1 gallon = 4 quarts
	1 quart = 2 pints
	1 quart = 32 fluid ounces

EXAMPLE 1.6 Using Conversion Factors

Convert 10.0 centimeters to meters.

Solution

First, recognize that the prefix *centi-* means $1/100$ of the base unit, the meter (m), just as one cent is $1/100$ of a dollar. There are 100 cents in a dollar and there are 100 cm in one meter. Thus, our conversion factor is either

$$\frac{1 \text{ m}}{100 \text{ cm}} \quad \text{or} \quad \frac{100 \text{ cm}}{1 \text{ m}}$$

each being equal to 1. Only one, however, will result in proper cancellation of units, producing the correct answer to the problem. If we proceed as follows:

$$10.0 \text{ cm} \times \frac{1 \text{ m}}{100 \text{ cm}} = 0.100 \text{ m}$$

Data given   Conversion factor   Desired result

we obtain the desired units, meters. If we had used the conversion factor 100 cm/1 m, the resulting units would be meaningless and the answer would have been incorrect:

$$10.0 \text{ cm} \times \frac{100 \text{ cm}}{1 \text{ m}} = 1000 \frac{\text{cm}^2}{\text{m}}$$

Incorrect units

Question 1.5

Convert 1.0 liter to each of the following units, using the factor-label method:
a. milliliters
b. microliters
c. kiloliters
d. centiliters
e. dekaliters

Question 1.6

Convert 1.0 gram to each of the following units:
a. micrograms
b. milligrams
c. kilograms
d. centigrams
e. decigrams

TABLE 1.3 Commonly Used "Bridging" Units for Intersystem Conversions

Quantity	English		Metric
Mass	1 pound	=	454 grams
	2.2 pounds	=	1 kilogram
Length	1 inch	=	2.54 centimeters
	1 yard	=	0.91 meter
Volume	1 quart	=	0.946 liter
	1 gallon	=	3.78 liters

Conversion of Units from One System to Another

The conversion of a quantity expressed in units of one system to an equivalent quantity in the other system (English to metric or metric to English) requires a *bridging* conversion unit. Examples are shown in Table 1.3.

The conversion may be represented as a three-step process:

1. Conversion from the units given in the problem to a bridging unit.
2. Conversion to the other system using the bridge.
3. Conversion within the desired system to units required by the problem.

English and metric conversions are shown in Tables 1.1 and 1.2.

EXAMPLE 1.7 Using Conversion Factors Between Systems

Convert 4.00 ounces to kilograms.

Solution

Step 1. A convenient bridging unit for mass is 1 pound = 454 grams. To use this conversion factor, we relate ounces (given in the problem) to pounds:

$$4.00 \text{ ounces} \times \frac{1 \text{ pound}}{16 \text{ ounces}} = 0.250 \text{ pound}$$

Step 2. Using the bridging unit conversion, we get

$$0.250 \text{ pound} \times \frac{454 \text{ grams}}{1 \text{ pound}} = 114 \text{ grams}$$

Step 3. Grams may then be directly converted to kilograms, the desired unit:

$$114 \text{ grams} \times \frac{1 \text{ kilogram}}{1000 \text{ grams}} = 0.114 \text{ kilogram}$$

The calculation may also be done in a single step by arranging the factors in a chain:

$$4.00 \text{ oz} \times \frac{1 \text{ lb}}{16 \text{ oz}} \times \frac{454 \text{ g}}{1 \text{ lb}} \times \frac{1 \text{ kg}}{1000 \text{ g}} = 0.114 \text{ kg}$$

Helpful Hint: Refer to the discussion of rounding off numbers on page 22.

EXAMPLE 1.8 Using Conversion Factors

Convert 1.5 meters² to centimeters².

Solution

The problem is similar to the conversion performed in Example 1.6. However, we must remember to include the exponent in the units. Thus

$$1.5 \text{ m}^2 \times \left(\frac{10^2 \text{ cm}}{1 \text{ m}}\right)^2 = 1.5 \text{ m}^2 \times \frac{10^4 \text{ cm}^2}{1 \text{ m}^2} = 1.5 \times 10^4 \text{ cm}^2$$

Note: The exponent affects both the number *and* unit within the parentheses.

Question 1.7

a. Convert 0.50 inch to meters.
b. Convert 0.75 quart to liters.
c. Convert 56.8 grams to ounces.
d. Convert 1.5 cm² to m².

Question 1.8

a. Convert 0.50 inch to centimeters.
b. Convert 0.75 quart to milliliters.
c. Convert 56.8 milligrams to ounces.
d. Convert 3.6 m² to cm².

1.4 Significant Figures and Scientific Notation

Information-bearing figures in a number are termed *significant figures*. Data and results arising from a scientific experiment convey information about the way in which the experiment was conducted. The degree of uncertainty or doubt associated with a measurement or series of measurements is indicated by the number of figures used to represent the information.

Significant Figures

LEARNING GOAL

Consider the following situation: A student was asked to obtain the length of a section of wire. In the chemistry laboratory, several different types of measuring devices are usually available. Not knowing which was most appropriate, the student decided to measure the object using each device that was available in the laboratory. The following data were obtained:

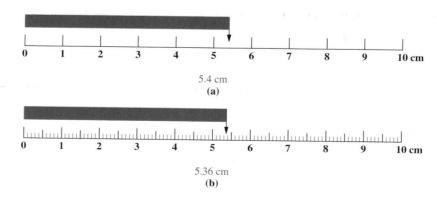

5.4 cm
(a)

5.36 cm
(b)

1.4 Significant Figures and Scientific Notation

Two questions should immediately come to mind:

Are the two answers equivalent?

If not, which answer is correct?

In fact, the two answers are *not* equivalent, but *both* are correct. How do we explain this apparent contradiction?

The data are not equivalent because each is known to a different degree of certainty. The answer 5.36 cm, containing three significant figures, specifies the length of the object more exactly than 5.4 cm, which contains only two significant figures. The term **significant figures** is defined to be all digits in a number representing data or results that are known with certainty *plus one uncertain digit*.

In case (a), we are certain that the object is at least 5 cm long and equally certain that it is *not* 6 cm long because the end of the object falls between the calibration lines 5 and 6. We can only estimate between 5 and 6, because there are no calibration indicators between 5 and 6. The end of the wire appears to be approximately four-tenths of the way between 5 and 6, hence 5.4 cm. The 5 is known with certainty, and 4 is estimated; there are two significant figures.

In case (b), the ruler is calibrated in tenths of centimeters. The end of the wire is at least 5.3 cm and not 5.4 cm. Estimation of the second decimal place between the two closest calibration marks leads to 5.36 cm. In this case, 5.3 is certain, and the 6 is estimated (or uncertain), leading to three significant digits.

Both answers are correct because each is consistent with the measuring device used to generate the data. An answer of 5.36 cm obtained from a measurement using ruler (a) would be *incorrect* because the measuring device is not capable of that exact specification. On the other hand, a value of 5.4 cm obtained from ruler (b) would be erroneous as well; in that case the measuring device is capable of generating a higher level of certainty (more significant digits) than is actually reported.

In summary, the number of significant figures associated with a measurement is determined by the measuring device. Conversely, the number of significant figures reported is an indication of the sophistication of the measurement itself.

> The uncertain digit represents the degree of doubt in a single measurement.

> The uncertain digit results from an estimation.

Recognition of Significant Figures

Only *significant* digits should be reported as data or results. However, are all digits, as written, significant digits? Let's look at a few examples illustrating the rules that are used to represent data and results with the proper number of significant digits.

EXAMPLE 1.9

RULE: All nonzero digits are significant. ■

7.314 has *four* significant digits.

EXAMPLE 1.10

RULE: The number of significant digits is independent of the position of the decimal point. ■

73.14 has *four* significant digits, as does 7.314.

EXAMPLE 1.11

RULE: Zeros located between nonzero digits are significant. ■

60.052 has *five* significant figures.

EXAMPLE 1.12

RULE: Zeros at the end of a number (often referred to as trailing zeros) are significant if the number contains a decimal point. ■

4.70 has *three* significant figures.

Helpful Hint: Trailing zeros are ambiguous; the next section offers a solution for this ambiguity.

EXAMPLE 1.13

RULE: Trailing zeros are insignificant if the number does not contain a decimal point and are significant if a decimal point is indicated. ■

100 has *one* significant figure; 100. has three significant figures.

EXAMPLE 1.14

RULE: Zeros to the left of the first nonzero integer are not significant; they serve only to locate the position of the decimal point. ■

0.0032 has *two* significant figures.

Question 1.9

How many significant figures are contained in each of the following numbers?

a. 7.26   b. 726   c. 700.2   d. 7.0   e. 0.0720

Question 1.10

How many significant figures are contained in each of the following numbers?

a. 0.042   b. 4.20   c. 24.0   d. 240   e. 204

Scientific Notation

LEARNING GOAL 9

It is often difficult to express very large numbers to the proper number of significant figures using conventional notation. The solution to this problem lies in the use of **scientific notation,** also referred to as *exponential notation,* which involves the representation of a number as a power of ten.

The speed of light is 299,792,458 m/s. For many calculations, two or three significant figures are sufficient. Using scientific notation, two significant figures, and rounding (p. 22), the speed of light is 3.0×10^8 m/s. The conversion is illustrated using simpler numbers:

$$6200 = 6.2 \times 1000 = 6.2 \times 10^3$$

or

$$5340 = 5.34 \times 1000 = 5.34 \times 10^3$$

RULE: To convert a number greater than 1 to scientific notation, the original decimal point is moved x places to the left, and the resulting number is multiplied by 10^x. The exponent (x) is a *positive* number equal to the number of places the original decimal point was moved. ■

You can find further information online at www.mhhe.com/denniston5e in "A Review of Mathematics."

1.4 Significant Figures and Scientific Notation

Scientific notation is also useful in representing numbers less than 1. For example, the mass of a single helium atom is

$$0.0000000000000000000000006692 \text{ gram}$$

a rather cumbersome number as written. Scientific notation would represent the mass of a single helium atom as 6.692×10^{-24} gram. The conversion is illustrated by using simpler numbers:

$$0.0062 = 6.2 \times \frac{1}{1000} = 6.2 \times \frac{1}{10^3} = 6.2 \times 10^{-3}$$

or

$$0.0534 = 5.34 \times \frac{1}{100} = 5.34 \times \frac{1}{10^2} = 5.34 \times 10^{-2}$$

RULE: To convert a number less than 1 to scientific notation, the original decimal point is moved x places to the right, and the resulting number is multiplied by 10^{-x}. The exponent $(-x)$ is a *negative* number equal to the number of places the original decimal point was moved. ■

Question 1.11

Represent each of the following numbers in scientific notation, showing only significant digits:

a. 0.0024   b. 0.0180   c. 224

Question 1.12

Represent each of the following numbers in scientific notation, showing only significant digits:

a. 48.20   b. 480.0   c. 0.126

Error, Accuracy, Precision, and Uncertainty

Error is the difference between the true value and our estimation, or measurement, of the value. Some degree of error is associated with any measurement. Two types of error exist: random error and systematic error. *Random error* causes data from multiple measurements of the same quantity to be scattered in a more or less uniform way around some average value. *Systematic error* causes data to be either smaller or larger than the accepted value. Random error is inherent in the experimental approach to the study of matter and its behavior; systematic error can be found and, in many cases, removed or corrected.

Examples of systematic error include such situations as:

- Dust on the balance pan, which causes all objects weighed to appear heavier than they really are.
- Impurities in chemicals used for the analysis of materials, which may interfere with (or block) the desired process.

Accuracy is the degree of agreement between the true value and the measured value. **Uncertainty** is the degree of doubt in a single measurement.

When measuring quantities that show continuous variation, for example, the weight of this page or the volume of one of your quarters, some doubt or uncertainty is present because the answer cannot be expressed with an infinite number of meaningful digits. The number of meaningful digits is determined by the measuring device. The presence of some error is a natural consequence of any measurement.

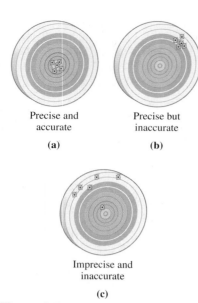

Figure 1.7
An illustration of precision and accuracy in replicate experiments.

Remember the distinction between the words *zero* and *nothing*. Zero is one of the ten digits and conveys as much information as 1, 2, and so forth. *Nothing* implies no information; the digits in the positions indicated by *x*'s could be 0, 1, 2, or any other.

The simple process of converting the fraction ²/₃ to its decimal equivalent can produce a variety of answers that depend on the device used to perform the calculation: pencil and paper, calculator, computer. The answer might be

$$0.67$$
$$0.667$$
$$0.6667$$

and so forth. All are correct, but each value has a different level of uncertainty. The first number listed, 0.67, has the greatest uncertainty.

It is always best to measure a quantity several times. Modern scientific instruments are designed to perform measurements rapidly; this allows many more measurements to be completed in a reasonable period. Replicate measurements of the same quantity minimize the uncertainty of the result. **Precision** is a measure of the agreement of replicate measurements.

It is important to recognize that accuracy and precision are not the same thing. It is possible to have one without the other. However, when scientific measurements are carefully made, the two most often go hand in hand; high-quality data are characterized by high levels of precision and accuracy.

In Figure 1.7, bull's-eye (a) shows the goal of all experimentation: accuracy *and* precision. Bull's-eye (b) shows the results to be repeatable (good precision); however, some error in the experimental procedure has caused the results to center on an incorrect value. This error is systematic, occurring in each replicate measurement. Occasionally, an experiment may show "accidental" accuracy. The precision is poor, but the average of these replicate measurements leads to a correct value. We don't want to rely on accidental success; the experiment should be repeated until the precision inspires faith in the accuracy of the method. Modern measuring devices in chemistry, equipped with powerful computers with immense storage capacity, are capable of making literally thousands of individual replicate measurements to enhance the quality of the result. Bull's-eye (c) describes the most common situation. A low level of precision is all too often associated with poor accuracy.

Significant Figures in Calculation of Results
Addition and Subtraction

If we combine the following numbers:

37.68	liters
108.428	liters
6.71862	liters

our calculator will show a final result of

152.82662	liters

Clearly, the answer, with eight digits, defines the volume of total material much more accurately than *any* of the individual quantities being combined. This cannot be correct; *the answer cannot have greater significance than any of the quantities that produced the answer.* We rewrite the problem:

37.68xxx	liters
108.428xx	liters
+ 6.71862	liters
152.82662	(should be 152.83) liters

where x = no information; x may be any integer from 0 to 9. Adding 2 to two unknown numbers (in the right column) produces no information. Similar logic prevails for the next two columns. Thus, five digits remain, all of which are significant. Conventional rules for rounding off would dictate a final answer of 152.83.

See rules for rounding off discussed on page 22.

Question 1.13

Report the result of each of the following to the proper number of significant figures:

a. $4.26 + 3.831 =$
b. $8.321 - 2.4 =$
c. $16.262 + 4.33 - 0.40 =$

Question 1.14

Report the result of each of the following to the proper number of significant figures:

a. $7.939 + 6.26 =$
b. $2.4 - 8.321 =$
c. $2.333 + 1.56 - 0.29 =$

Multiplication and Division

In the preceding discussion of addition and subtraction, the position of the decimal point in the quantities being combined has a bearing on the number of significant figures in the answer. In multiplication and division this is not the case. The decimal point position is irrelevant when determining the number of significant figures in the answer. It is the number of significant figures in the data that is important. Consider

$$\frac{4.237 \times 1.21 \times 10^{-3} \times 0.00273}{11.125} = 1.26 \times 10^{-6}$$

The answer is limited to three significant figures; the answer can have *only* three significant figures because two numbers in the calculation, 1.21×10^{-3} and 0.00273, have three significant figures and "limit" the answer. Remember, *the answer can be no more precise than the least precise number from which the answer is derived*. The *least precise number* is the number with the fewest significant figures.

Question 1.15

Report the results of each of the following operations using the proper number of significant figures:

a. $63.8 \times 0.80 =$
b. $\dfrac{63.8}{0.80} =$
c. $\dfrac{53.8 \times 0.90}{0.3025} =$

> **Question 1.16** Report the results of each of the following operations using the proper number of significant figures:
>
> a. $\dfrac{27.2 \times 15.63}{1.84} =$
>
> b. $\dfrac{13.6}{18.02 \times 1.6} =$
>
> c. $\dfrac{12.24 \times 6.2}{18.02 \times 1.6} =$

You can find further information online at www.mhhe.com/denniston5e in "A Review of Mathematics."

Exponents

Now consider the determination of the proper number of significant digits in the results when a value is multiplied by any power of ten. In each case the number of significant figures in the answer is identical to the number contained in the original term. Therefore

$$(8.314 \times 10^2)^3 = 574.7 \times 10^6 = 5.747 \times 10^8$$

and

$$(8.314 \times 10^2)^{1/2} = 2.883 \times 10^1$$

Each answer contains four significant figures.

Exact (Counted) and Inexact Numbers

Inexact numbers, by definition, have uncertainty (the degree of doubt in the final significant digit). *Exact numbers*, on the other hand, have no uncertainty. Exact numbers may arise from a definition; there are *exactly* 60 minutes in 1 hour or there are exactly 1000 mL in 1 liter.

Exact numbers are a consequence of counting. Counting the number of dimes in your pocket or the number of letters in the alphabet are common examples. The fact that exact numbers have no uncertainty means that they do not limit the number of significant figures in the result of a calculation.

For example,

$$4.00 \text{ oz} \times 1 \text{ lb}/16 \text{ oz} = 0.250 \text{ lb (3 significant figures)}$$

or

$$2568 \text{ oz} \times 1 \text{ lb}/16 \text{ oz} = 160.5 \text{ lb (4 significant figures)}$$

In both examples, the number of significant figures in the result is governed by the data (the number of ounces) not the conversion factor, which is *exact*, because it is defined.

A good rule of thumb to follow is: In the metric system the quantity being converted, not the conversion factor, generally determines the number of significant figures.

Rounding Off Numbers

The use of an electronic calculator generally produces more digits for a result than are justified by the rules of significant figures on the basis of the data input. For example, on your calculator,

$$3.84 \times 6.72 = 25.8048$$

The most correct answer would be 25.8, dropping 048.

A number of acceptable conventions for rounding exist. Throughout this book we will use the following:

RULE: When the number to be dropped is less than 5, the preceding number is not changed. When the number to be dropped is 5 or larger, the preceding number is increased by one unit. ■

> **EXAMPLE 1.15**
>
> **Rounding Numbers**
>
> Round off each of the following to three significant figures.
>
> **Solution**
>
> a. 63.6<u>6</u>9 becomes 63.7. *Rationale:* 6 > 5.
> b. 8.77<u>1</u>5 becomes 8.77. *Rationale:* 1 < 5.
> c. 2.22<u>4</u>5 becomes 2.22. *Rationale:* 4 < 5.
> d. 0.000410<u>9</u> becomes 0.000411. *Rationale:* 9 > 5.
>
> *Helpful Hint:* Symbol $x > y$ implies "x greater than y." Symbol $x < y$ implies "x less than y."

Question 1.17

Round off each of the following numbers to three significant figures.

a. 61.40   b. 6.171   c. 0.066494

Question 1.18

Round off each of the following numbers to three significant figures.

a. 6.2262   b. 3895   c. 6.885

1.5 Experimental Quantities

Thus far we have discussed the scientific method and its role in acquiring data and converting the data to obtain the results of the experiment. We have seen that such data must be reported in the proper units with the appropriate number of significant figures. The quantities that are most often determined include mass, length, volume, time, temperature, and energy. Now let's look at each of these quantities in more detail.

10 LEARNING GOAL

Mass

Mass describes the quantity of matter in an object. The terms *weight* and *mass*, in common usage, are often considered synonymous. They are not, in fact. **Weight** is the force of gravity on an object:

Weight = mass × acceleration due to gravity

When gravity is constant, mass and weight are directly proportional. But gravity is not constant; it varies as a function of the distance from the center of the earth. Therefore weight cannot be used for scientific measurement because the weight of an object may vary from one place on the earth to the next.

Mass, on the other hand, is independent of gravity; it is a result of a comparison of an unknown mass with a known mass called a *standard mass*. Balances are instruments used to measure the mass of materials.

Examples of common balances used for the determination of mass are shown in Figure 1.8.

Figure 1.8
Three common balances that are useful for the measurement of mass. (a) A two-pan comparison balance for approximate mass measurement suitable for routine work requiring accuracy to 0.1 g (or perhaps 0.01 g). (b) A top-loading single-pan electronic balance that is similar in accuracy to (a) but has the advantages of speed and ease of operation. The revolution in electronics over the past twenty years has resulted in electronic balances largely supplanting the two-pan comparison balance in routine laboratory usage. (c) An analytical balance that is capable of precise mass measurement (three to five significant figures beyond the decimal point). A balance of this type is used when the highest level of precision and accuracy is required.

(a)

(b)

(c)

The common conversion units for mass are as follows:

$$1 \text{ gram (g)} = 10^{-3} \text{ kilogram (kg)} = \frac{1}{454} \text{ pound (lb)}$$

In chemistry, when we talk about incredibly small bits of matter such as individual atoms or molecules, units such as grams and even micrograms are much too large. We don't say that a 100-pound individual weighs 0.0500 ton; the unit does not fit the quantity being described. Similarly, an atom of a substance such as hydrogen is very tiny. Its mass is only 1.661×10^{-24} gram.

One *atomic mass unit* (amu) is a more convenient way to represent the mass of one hydrogen atom, rather than 1.661×10^{-24} gram:

$$1 \text{ amu} = 1.661 \times 10^{-24} \text{ g}$$

Units should be chosen to suit the quantity being described. This can easily be done by choosing a unit that gives an exponential term closest to 10^0.

Length

The standard metric unit of *length*, the distance between two points, is the meter. Large distances are measured in kilometers; smaller distances are measured in millimeters or centimeters. Very small distances such as the distances between atoms on a surface are measured in *nanometers* (nm):

$$1 \text{ nm} = 10^{-7} \text{ cm} = 10^{-9} \text{ m}$$

Common conversions for length are as follows:

$$1 \text{ meter (m)} = 10^2 \text{ centimeters (cm)} = 3.94 \times 10^1 \text{ inch (in.)}$$

Volume

The standard metric unit of *volume*, the space occupied by an object, is the liter. A liter is the volume occupied by 1000 grams of water at 4 degrees Celsius (°C). The volume, 1 liter, also corresponds to:

$$1 \text{ liter (L)} = 10^3 \text{ milliliters (mL)} = 1.06 \text{ quarts (qt)}$$

The relationship between the liter and the milliliter is shown in Figure 1.9.

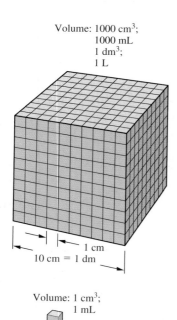

Figure 1.9
The relationships among various volume units.

1.5 Experimental Quantities

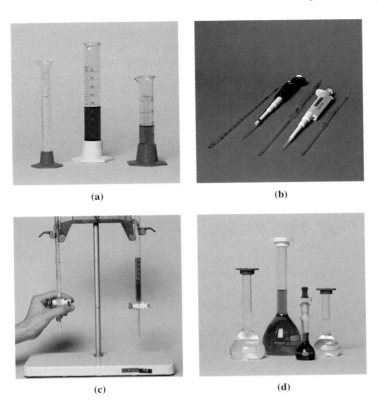

Figure 1.10
Common laboratory equipment used for the measurement of volume. Graduated (a) cylinders, (b) pipets, and (c) burets are used for the delivery of liquids. (d) Volumetric flasks are used to contain a specific volume. A graduated cylinder is usually used for measurement of approximate volume; it is less accurate and precise than either pipets or burets.

Typical laboratory glassware used for volume measurement is shown in Figure 1.10. The volumetric flask is designed to *contain* a specified volume, and the graduated cylinder, pipet, and buret *dispense* a desired volume of liquid.

Time

The standard metric unit of time is the second. The need for accurate measurement of time by chemists may not be as apparent as that associated with mass, length, and volume. It is necessary, however, in many applications. In fact, matter may be characterized by measuring the time required for a certain process to occur. The rate of a chemical reaction is a measure of change as a function of time.

Temperature

Temperature is the degree of "hotness" of an object. This may not sound like a very "scientific" definition, and, in a sense, it is not. We know intuitively the difference between a "hot" and a "cold" object, but developing a precise definition to explain this is not easy. We may think of the temperature of an object as a measure of the amount of heat in the object. However, this is not strictly true. An object increases in temperature because its heat content has increased and vice versa; however, the relationship between heat content and temperature depends on the quantity and composition of the material.

Many substances, such as mercury, expand as their temperature increases, and this expansion provides us with a way to measure temperature and temperature changes. If the mercury is contained within a sealed tube, as it is in a thermometer, the height of the mercury is proportional to the temperature. A mercury thermometer may be calibrated, or scaled, in different units, just as a ruler can be. Three common temperature scales are *Fahrenheit (°F), Celsius (°C),* and *Kelvin (K)*. Two convenient reference temperatures that are used to calibrate a thermometer are the freezing and boiling temperatures of water. Figure 1.11 shows the relationship between the scales and these reference temperatures.

The Kelvin scale is of particular importance because it is directly related to molecular motion. As molecular speed increases, the Kelvin temperature proportionately increases.

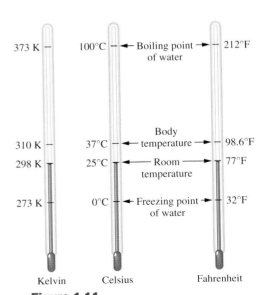

Figure 1.11
The freezing point and boiling point of water, body temperature, and room temperature expressed in the three common units of temperature.

You can find further information online at www.mhhe.com/denniston5e in "A Review of Mathematics."

The Kelvin symbol does not have a degree sign. The degree sign implies a value that is *relative* to some standard. Kelvin is an *absolute* scale.

Although Fahrenheit temperature is most familiar to us, Celsius and Kelvin temperatures are used exclusively in scientific measurements. It is often necessary to convert a temperature reading from one scale to another. To convert from Fahrenheit to Celsius, we use the following formula:

$$°C = \frac{°F - 32}{1.8}$$

To convert from Celsius to Fahrenheit, we solve the formula (above) for °F, resulting in

$$°F = 1.8°C + 32$$

To convert from Celsius to Kelvin, we use the formula

$$K = °C + 273.15$$

EXAMPLE 1.16 Converting from Fahrenheit to Celsius and Kelvin

Normal body temperature is 98.6°F. Calculate the corresponding temperature in degrees Celsius:

Solution

Using the expression relating °C and °F,

$$°C = \frac{°F - 32}{1.8}$$

Substituting the information provided,

$$= \frac{98.6 - 32}{1.8} = \frac{66.6}{1.8}$$

results in:

$$= 37.0°C$$

Calculate the corresponding temperature in Kelvin units:

Solution

Using the expression relating K and °C,

$$K = °C + 273.15$$

substituting the value obtained in the first part,

$$= 37.0 + 273.15$$

results in:

$$= 310.2 \text{ K}$$

Question 1.19

The freezing temperature of water is 32°F. Calculate the freezing temperature of water in:

a. Celsius units
b. Kelvin units

Question 1.20

When a patient is ill, his or her temperature may increase to 104°F. Calculate the temperature of this patient in:

a. Celsius units
b. Kelvin units

Energy

Energy, the ability to do work, may be categorized as either **kinetic energy,** the energy of motion, or **potential energy,** the energy of position. Kinetic energy may be considered as energy in process; potential energy is stored energy. All energy is either kinetic or potential.

Another useful way of classifying energy is by form. The principal forms of energy include light, heat, electrical, mechanical, and chemical energy. All of these forms of energy share the following set of characteristics:

- In chemical reactions, energy cannot be created or destroyed.
- Energy may be converted from one form to another.
- Conversion of energy from one form to another always occurs with less than 100% efficiency. Energy is not lost (remember, energy cannot be destroyed) but rather, is not useful. We buy gasoline to move our car from place to place; however, much of the energy stored in the gasoline is released as heat.
- All chemical reactions involve either a "gain" or a "loss" of energy.

Energy absorbed or liberated in chemical reactions is usually in the form of heat energy. Heat energy may be represented in units of *calories* or *joules,* their relationship being

$$1 \text{ calorie (cal)} = 4.18 \text{ joules (J)}$$

One calorie is defined as the amount of heat energy required to increase the temperature of 1 gram of water 1°C.

Heat energy measurement is a quantitative measure of heat content. It is an extensive property, dependent upon the quantity of material. Temperature, as we have mentioned, is an intensive property, independent of quantity.

Not all substances have the same capacity for holding heat; 1 gram of iron and 1 gram of water, even if they are at the same temperature, do *not* contain the same amount of heat energy. One gram of iron will absorb and store 0.108 calorie of heat energy when the temperature is raised 1°C. In contrast, 1 gram of water will absorb almost ten times as much energy, 1.00 calorie, when the temperature is increased an equivalent amount.

Units for other forms of energy will be introduced in later chapters.

> The *kilocalorie* (kcal) is the familiar nutritional calorie. It is also known as the large Calorie; note that in this term the C is uppercase to distinguish it from the normal calorie. The large calorie is 1000 small calories. Refer to Section 7.2 and A Human Perspective: Food Calories for more information.

> Water in the environment (lakes, oceans, and streams) has a powerful effect on the climate because of its ability to store large quantities of energy. In summer, water stores heat energy, moderating temperatures of the surrounding area. In winter, some of this stored energy is released to the air as the water temperature falls; this prevents the surroundings from experiencing extreme changes in temperature.

Concentration

Concentration is a measure of the number of particles of a substance, or the mass of those particles, that are contained in a specified volume. Concentration is a widely used way of representing mixtures of different substances. Examples include:

- The concentration of oxygen in the air
- Pollen counts, given during the hay fever seasons, which are simply the number of grains of pollen contained in a measured volume of air
- The amount of an illegal drug in a certain volume of blood, indicating the extent of drug abuse
- The proper dose of an antibiotic, based on a patient's weight.

A Human Perspective

Food Calories

The body gets its energy through the processes known collectively as metabolism, which will be discussed in detail in subsequent chapters on biochemistry and nutrition. The primary energy sources for the body are carbohydrates, fats, and proteins, which we obtain from the foods we eat. The amount of energy available from a given foodstuff is related to the Calories (C) available in the food. Calories are a measure of the energy and heat content that can be derived from the food. One (food) Calorie (symbolized by C) equals 1000 (metric) calories (symbolized by c):

1 Calorie = 1000 calories = 1 kilocalorie

The energy available in food can be measured by totally burning the food; in other words, we are using the food as a fuel. The energy given off in the form of heat is directly related to the amount of chemical energy, energy stored in chemical bonds, that is available in the food and that the food could provide to the body through the various metabolic pathways.

The classes of food molecules are not equally energy rich. For instance, when oxidized via metabolic pathways, carbohydrates and proteins provide the cell with 4 Calories per gram, whereas fats generate approximately 9 Calories per gram.

In addition, as with all processes, not all the available energy can be efficiently extracted from the food; a certain percentage is always lost. The average person requires between 2000 and 3000 Calories per day to maintain normal body functions such as the regulation of body temperature, muscle movement, and so on. If a person takes in more Calories than the body uses, the Calorie-containing substances will be stored as fat, and the person will gain weight. Conversely, if a person uses more Calories than are ingested, the individual will lose weight.

Excess Calories are stored in the form of fat, the form that provides the greatest amount of energy per gram. Too many Calories lead to too much fat. Similarly, a lack of Calories (in the form of food) forces the body to raid its storehouse, the fat. Weight is lost in this process as the fat is consumed. Unfortunately, it always seems easier to add fat to the storehouse than to remove it.

The "rule of thumb" is that 3500 Calories are equivalent to approximately 1 pound of body weight. You have to take in 3500 Calories more than you use to gain a pound, and you have to expend 3500 Calories more than you normally use to lose a pound. If you eat as little as 100 Calories a day above your body's needs, you could gain about 10–11 pounds per year:

$$\frac{100 \cancel{C}}{\cancel{day}} \times \frac{365 \cancel{day}}{1 \text{ year}} \times \frac{1 \text{ lb}}{3500 \cancel{C}} = \frac{10.4 \text{ lb}}{\text{year}}$$

A frequently recommended procedure for increasing the rate of weight loss involves a combination of dieting (taking in fewer Calories) and exercise. The numbers of Calories used in several activities are:

Activity	Energy Output (Calories/minute-pound of body weight)
Running	0.110
Swimming	0.045
Jogging	0.070
Bicycling	0.050
Tennis	0.045
Walking	0.028
Golfing	0.036
Driving a car	0.015
Standing or sitting	0.011
Sleeping	0.008

For Further Understanding

Sarah runs 1 hour each day, and Nancy swims 2 hours each day. Assuming that Sarah and Nancy are the same weight, which girl burns more calories in 1 week?

Would you expect a runner to burn more calories in summer or winter? Why?

We will describe many situations in which concentration is used to predict useful information about chemical reactions (Sections 6.4 and 8.2, for example). In Chapter 6 we calculate a numerical value for concentration from experimental data.

Density and Specific Gravity

Both mass and volume are a function of the *amount* of material present (extensive property). **Density,** the ratio of mass to volume,

$$d = \frac{\text{mass}}{\text{volume}} = \frac{m}{V}$$

1.5 Experimental Quantities

TABLE 1.4   Densities of Some Common Materials

Substance	Density (g/mL)	Substance	Density (g/mL)
Air	0.00129 (at 0°C)	Methyl alcohol	0.792
Ammonia	0.000771 (at 0°C)	Milk	1.028–1.035
Benzene	0.879	Oxygen	0.00143 (at 0°C)
Bone	1.7–2.0	Rubber	0.9–1.1
Carbon dioxide	0.001963 (at 0°C)	Turpentine	0.87
Ethyl alcohol	0.789	Urine	1.010–1.030
Gasoline	0.66–0.69	Water	1.000 (at 4°C)
Gold	19.3	Water	0.998 (at 20°C)
Hydrogen	0.000090 (at 0°C)	Wood	0.3–0.98
Kerosene	0.82	(balsa, least dense; ebony	
Lead	11.3	and teak, most dense)	
Mercury	13.6		

Figure 1.12
Density (mass/volume) is a unique property of a material. A mixture of wood, water, brass, and mercury is shown, with the cork—the least dense—floating on water. Additionally, brass, with a density greater than water but less than liquid mercury, floats on the interface between these two liquids.

is *independent* of the amount of material (intensive property). Density is a useful way to characterize or identify a substance because each substance has a unique density (Figure 1.12).

One milliliter of air and 1 milliliter of iron do not weigh the same amount. There is much more mass in 1 milliliter of iron; its density is greater.

Density measurements were used to discriminate between real gold and "fool's gold" during the gold rush era. Today the measurement of the density of a substance is still a valuable analytical technique. The densities of a number of common substances are shown in Table 1.4.

In density calculations, the mass is usually represented in grams, and volume is given in either milliliters (mL) or cubic centimeters (cm³ or cc):

$$1 \text{ mL} = 1 \text{ cm}^3 = 1 \text{ cc}$$

The unit of density would therefore be g/mL, g/cm³, or g/cc.

 **LEARNING GOAL**

Intensive and extensive properties are described on page 9.

EXAMPLE 1.17

Calculating the Density of a Solid

2.00 cm³ of aluminum are found to weigh 5.40 g. Calculate the density of aluminum in units of g/cm³.

Solution

The density expression is:

$$d = \frac{m}{V} = \frac{\text{g}}{\text{cm}^3}$$

Substituting the information given in the problem,

$$= \frac{5.40 \text{ g}}{2.00 \text{ cm}^3}$$

results in:

$$= 2.70 \text{ g/cm}^3$$

EXAMPLE 1.18 Calculating the Mass of a Gas from Its Density

Air has a density of 0.0013 g/mL. What is the mass of a 6.0-L sample of air?

Solution

$$0.0013 \text{ g/mL} = 1.3 \times 10^{-3} \text{ g/mL}$$

(The decimal point is moved three positions to the right.) This problem can be solved by using conversion factors:

$$6.0 \text{ L air} \times \frac{10^3 \text{ mL air}}{1 \text{ L air}} \times \frac{1.3 \times 10^{-3} \text{ g air}}{\text{mL air}} = 7.8 \text{ g air}$$

EXAMPLE 1.19 Using the Density to Calculate the Mass of a Liquid

Calculate the mass, in grams, of 10.0 mL of mercury (symbolized Hg) if the density of mercury is 13.6 g/mL.

Solution

Using the density as a conversion factor from volume to mass, we have

$$m = (10.0 \text{ mL Hg})\left(13.6 \frac{\text{g Hg}}{\text{mL Hg}}\right)$$

Cancellation of units results in:

$$= 136 \text{ g Hg}$$

EXAMPLE 1.20 Using the Density to Calculate the Volume of a Liquid

Calculate the volume, in milliliters, of a liquid that has a density of 1.20 g/mL and a mass of 5.00 grams.

Solution

Using the density as a conversion factor from mass to volume, we have

$$V = (5.00 \text{ g liquid})\left(\frac{1 \text{ mL liquid}}{1.20 \text{ g liquid}}\right)$$

Cancellation of units results in:

$$= 4.17 \text{ mL liquid}$$

Question 1.21

The density of ethyl alcohol (200 proof, or pure alcohol) is 0.789 g/mL at 20°C. Calculate the mass of a 30.0-mL sample.

Question 1.22

Calculate the volume, in milliliters, of 10.0 g of a saline solution that has a density of 1.05 g/mL.

A Medical Perspective

Diagnosis Based on Waste

Any archaeologist would say that you can learn a great deal about the activities and attitudes of a society by finding the remains of their dump sites and studying their waste.

Similarly, urine, a waste product consisting of a wide variety of metabolites, may be analyzed to indicate abnormalities in various metabolic processes or even unacceptable behavior (recall the steroid tests in Olympic competition).

Many of these tests must be performed by using sophisticated and sensitive instrumentation. However, a very simple test, the measurement of the specific gravity of urine, can be an indicator of diabetes mellitus or Bright's disease. The normal range for human urine specific gravity is 1.010–1.030.

A hydrometer, a weighted glass bulb inserted in a liquid, may be used to determine specific gravity. The higher it floats in the liquid, the more dense the liquid. A hydrometer that is calibrated to indicate the specific gravity of urine is called a urinometer.

Although hydrometers have been replaced by more modern measuring devices that use smaller samples, these newer instruments operate on the same principles as the hydrometer.

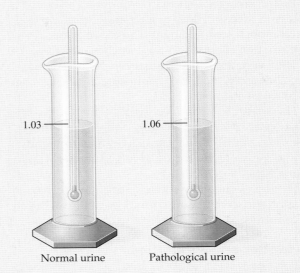

A hydrometer, used in the measurement of the specific gravity of urine.

For Further Understanding

Give reasons that may account for such a broad range of "normal" values.

Could results depend on food or medicine consumed prior to the test?

For convenience, values of density are often related to a standard, well-known reference, the density of pure water at 4°C. This "referenced" density is called the **specific gravity**, the ratio of the density of the object in question to the density of pure water at 4°C.

$$\text{specific gravity} = \frac{\text{density of object (g/mL)}}{\text{density of water (g/mL)}}$$

Specific gravity is a *unitless* term. Because the density of water at 4.0°C is 1.00 g/mL, the numerical values for the density and specific gravity of a substance are equal. That is, an object with a density of 2.00 g/mL has a specific gravity of 2.00 at 4°C.

Routine hospital tests involving the measurement of the specific gravity of urine and blood samples are frequently used as diagnostic tools. For example, diseases such as kidney disorders and diabetes change the composition of urine. This compositional change results in a corresponding change in the specific gravity. This change is easily measured and provides the basis for a quick preliminary diagnosis. This topic is discussed in greater detail in A Medical Perspective: Diagnosis Based on Waste.

> Specific gravity is frequently referenced to water at 4°C, its temperature of maximum density (1.000 g/mL). Other reference temperatures may be used. However, the temperature must be specified.

SUMMARY

1.1 The Discovery Process

Chemistry is the study of matter and the changes that matter undergoes. *Matter* is anything that has mass and occupies space. The changes that matter undergoes always involve either gain or loss of energy. *Energy* is the ability to do work (to accomplish some change). Thus a study of chemistry involves matter, energy, and their interrelationship.

The major areas of chemistry include biochemistry, organic chemistry, inorganic chemistry, analytical chemistry, and physical chemistry.

The *scientific method* consists of six interrelated processes: observation, questioning, pattern recognition, development of *theories* from *hypotheses*, experimentation, and summarizing information. A *law* summarizes a large quantity of information.

The development of the scientific method has played a major role in civilization's rapid growth during the past two centuries.

1.2 Matter and Properties

Properties (characteristics) of matter may be classified as either physical or chemical. *Physical properties* can be observed without changing the chemical composition of the sample. *Chemical properties* result in a change in composition and can be observed only through *chemical reactions*. *Intensive properties* are independent of the quantity of the substance. *Extensive properties* depend on the quantity of a substance.

Three states of matter exist (*solid, liquid,* and *gas*); these states of matter are distinguishable by differences in physical properties.

All matter is classified as either a *pure substance* or a *mixture*. A pure substance is a substance that has only one component. A mixture is a combination of two or more pure substances in which the combined substances retain their identity.

A *homogeneous mixture* has uniform composition. Its particles are well mixed. A *heterogeneous mixture* has a nonuniform composition.

An *element* is a pure substance that cannot be converted into a simpler form of matter by any chemical reaction. A *compound* is a substance produced from the combination of two or more elements in a definite, reproducible fashion.

1.3 Measurement in Chemistry

Science is the study of humans and their environment. Its tool is experimentation. A scientific experiment produces *data*. Each piece of data is the individual result of a single measurement. Mass, length, volume, time, temperature, and energy are the most common types of data obtained from chemical experiments.

Results are the outcome of an experiment. Usually, several pieces of data are combined, using a mathematical equation, to produce a result.

A *unit* defines the basic quantity of mass, volume, time, and so on. A number that is not followed by the correct unit usually conveys no useful information.

The metric system is a decimal-based system in contrast to the English system. In the metric system, mass is represented as the gram, length as the meter, and volume as the liter. Any subunit or multiple unit contains one of these units preceded by a prefix indicating the power of ten by which the base unit is to be multiplied to form the subunit or multiple unit. Scientists favor this system over the not-so-systematic English units of measurement.

To convert one unit to another, we must set up a *conversion factor* or series of conversion factors that relate two units. The proper use of these conversion factors is referred to as the factor-label method. This method is used either to convert from one unit to another within the same system or to convert units from one system to another. It is a very useful problem-solving tool.

1.4 Significant Figures and Scientific Notation

Significant figures are all digits in a number representing data or results that are known with certainty plus the first uncertain digit. The number of significant figures associated with a measurement is determined by the measuring device. Results should be rounded off to the proper number of significant figures.

Error is defined as the difference between the true value and our estimation, or measurement, of the value. *Accuracy* is the degree of agreement between the true and measured values. *Uncertainty* is the degree of doubt in a single measurement. The number of meaningful digits in a measurement is determined by the measuring device. *Precision* is a measure of the agreement of replicate measurements.

Very large and very small numbers may be represented with the proper number of significant figures by using *scientific notation*.

1.5 Experimental Quantities

Mass describes the quantity of matter in an object. The terms weight and mass are often used interchangeably, but they are not equivalent. *Weight* is the force of gravity on an object. The fundamental unit of mass in the metric system is the gram. One atomic mass unit (amu) is equal to 1.661×10^{-24} g.

The standard metric unit of length is the meter. Large distances are measured in kilometers; smaller distances are measured in millimeters or centimeters. Very small distances (on the atomic scale) are measured in nanometers (nm). The standard metric unit of volume is the liter. A liter is the volume occupied by 1000 grams of water at 4 degrees Celsius. The standard metric unit of time is the second, a unit that is used in the English system as well.

Temperature is the degree of "hotness" of an object. Many substances, such as liquid mercury, expand as their temperature increases, and this expansion provides us with a way to measure temperature and temperature changes. Three common temperature scales are Fahrenheit (°F), Celsius (°C), and Kelvin (K).

Energy, the ability to do work, may be categorized as either *kinetic energy*, the energy of motion, or *potential energy*, the energy of position. The principal forms of energy are light, heat, mechanical, electrical, nuclear, and chemical energy.

Energy absorbed or liberated in chemical reactions is most often in the form of heat energy. Heat energy may be represented in units of calories or joules: 1 calorie (cal) = 4.18 joules (J). One calorie is defined as the amount of heat energy required to change the temperature of 1 gram of water 1°C.

Concentration is a measure of the number of particles of a substance, or the mass of those particles, that are contained in a specified volume. Concentration is a widely used way of representing relative quantities of different substances in a mixture of those substances.

Density is the ratio of mass to volume and is a useful way of characterizing a substance. Values of density are often related to a standard reference, the density of pure water at 4°C. This "referenced" density is the *specific gravity*, the ratio of the density of the object in question to the density of pure water at 4°C.

KEY TERMS

accuracy (1.4)
chemical property (1.2)
chemical reaction (1.2)
chemistry (1.1)
compound (1.2)
concentration (1.5)
data (1.3)
density (1.5)
element (1.2)
energy (1.1)
error (1.4)
extensive property (1.2)
gaseous state (1.2)
heterogeneous mixture (1.2)
homogeneous mixture (1.2)
hypothesis (1.1)
intensive property (1.2)
kinetic energy (1.5)
law (1.1)
liquid state (1.2)
mass (1.5)
matter (1.1)
mixture (1.2)
physical change (1.2)
physical property (1.2)
potential energy (1.5)
precision (1.4)
properties (1.2)
pure substance (1.2)
results (1.3)
scientific method (1.1)
scientific notation (1.4)
significant figures (1.4)
solid state (1.2)
specific gravity (1.5)
temperature (1.5)
theory (1.1)
uncertainty (1.4)
unit (1.3)
weight (1.5)

QUESTIONS AND PROBLEMS

The Discovery Process

Foundations

1.23 Define each of the following terms:
   a. chemistry
   b. matter
   c. energy
1.24 Define each of the following terms:
   a. hypothesis
   b. theory
   c. law
1.25 Define each of the following terms:
   a. potential energy
   b. kinetic energy
   c. data
1.26 Define each of the following terms:
   a. results
   b. mass
   c. weight
1.27 Give the base unit for each of the following in the metric system:
   a. mass
   b. volume
   c. length
1.28 Give the base unit for each of the following in the metric system:
   a. time
   b. temperature
   c. energy

Applications

1.29 Discuss the difference between the terms *mass* and *weight*.
1.30 Discuss the difference between the terms *data* and *results*.
1.31 Distinguish between specific gravity and density.
1.32 Distinguish between kinetic energy and potential energy.
1.33 Discuss the meaning of the term *scientific method*.
1.34 Describe an application of reasoning involving the scientific method that has occurred in your day-to-day life.
1.35 Stem-cell research has the potential to provide replacement "parts" for the human body. Is this statement a hypothesis, theory, or law? Explain your reasoning.
1.36 Observed increases in global temperatures are caused by elevated levels of carbon dioxide. Is this statement a hypothesis, theory, or law? Explain your reasoning.

Matter and Properties

Foundations

1.37 Describe what is meant by a physical property.
1.38 Describe what is meant by a physical change.
1.39 Describe several chemical properties of matter.
1.40 Describe what is meant by a chemical reaction.
1.41 Distinguish between a pure substance and a mixture.
1.42 Give examples of pure substances and mixtures.
1.43 Distinguish between a homogeneous mixture and a heterogeneous mixture.
1.44 Distinguish between an intensive property and an extensive property.

Applications

1.45 Label each of the following as either a physical change or a chemical reaction:
  a. An iron nail rusts. — chem
  b. An ice cube melts. — physical
  c. A limb falls from a tree. — physical

1.46 Label each of the following as either a physical change or a chemical reaction:
  a. A puddle of water evaporates. — Phy
  b. Food is digested. — chem
  c. Wood is burned. — chem

1.47 Label each of the following properties of sodium as either a physical property or a chemical property:
  a. Sodium is a soft metal (can be cut with a knife).
  b. Sodium reacts violently with water to produce hydrogen gas and sodium hydroxide.

1.48 Label each of the following properties of sodium as either a physical property or a chemical property:
  a. When exposed to air, sodium forms a white oxide. — chem
  b. Sodium melts at 98°C. — Physical
  c. The density of sodium metal at 25°C is 0.97 g/cm³. — Phys

1.49 Label each of the following as either a pure substance or a mixture:
  a. water
  b. table salt (sodium chloride)
  c. blood

1.50 Label each of the following as either a pure substance or a mixture:
  a. sucrose (table sugar)
  b. orange juice
  c. urine

1.51 Label each of the following as either a homogeneous mixture or a heterogeneous mixture:
  a. a soft drink
  b. a saline solution
  c. gelatin

1.52 Label each of the following as either a homogeneous mixture or a heterogeneous mixture:
  a. gasoline — Homo
  b. vegetable soup — Hetero
  c. concrete — Hetero

1.53 Describe the general properties of the gaseous state.

1.54 Contrast the physical properties of the gaseous and solid states.

1.55 Label each of the following as either an intensive property or an extensive property:
  a. mass
  b. volume
  c. density

1.56 Label each of the following as either an intensive property or an extensive property:
  a. specific gravity
  b. temperature
  c. heat content

1.57 Describe the difference between the terms *atom* and *element*.

1.58 Describe the difference between the terms *atom* and *compound*.

1.59 Give at least one example of each of the following:
  a. an element
  b. a pure substance

1.60 Give at least one example of each of the following:
  a. a homogeneous mixture
  b. a heterogeneous mixture

Measurement in Chemistry

Foundations

1.61 Convert 2.0 pounds to:
  a. ounces 32   d. milligrams 909,000   9.1 × 10⁵
  b. tons .001   e. dekagrams 90.8
  c. grams 908

1.62 Convert 5.0 quarts to:
  a. gallons   d. milliliters
  b. pints   e. microliters
  c. liters

1.63 Convert 3.0 grams to:
  a. pounds   d. centigrams
  b. ounces   e. milligrams
  c. kilograms

1.64 Convert 3.0 meters to:
  a. yards 3.3   d. centimeters 300 cm
  b. inches 118   e. millimeters 3000
  c. feet 9.84

1.65 Convert 50.0°F to:
  a. °C
  b. K

1.66 Convert –10.0°F to:
  a. °C –23.3
  b. K   250

1.67 Convert 20.0°C to:
  a. K   293
  b. °F 68

1.68 Convert 300.0 K to:
  a. °C
  b. °F

Applications Pints to liter

1.69 A 150-lb adult has approximately 9 pints of blood. How many liters of blood does the individual have?

1.70 If a drop of blood has a volume of 0.05 mL, how many drops of blood are in the adult described in Problem 1.69?

1.71 A patient's temperature is found to be 38.5°C. To what Fahrenheit temperature does this correspond?

1.72 A newborn is 21 inches in length and weighs 6 lb 9 oz. Describe the baby in metric units.

1.73 Which distance is shorter: 5.0 cm or 5.0 in.?

1.74 Which volume is smaller: 50.0 mL or 0.500 L?

1.75 Which mass is smaller: 5.0 mg or 5.0 µg?

1.76 Which volume is smaller: 1.0 L or 1.0 qt?

Significant Figures and Scientific Notation

Foundations

1.77 How many significant figures are contained in each of the following numbers?
  a. 10.0   3   d. 2.062   4
  b. 0.214 3   e. 10.50   4
  c. 0.120 3   f. 1050   3

1.78 How many significant figures are contained in each of the following numbers?
  a. 3.8×10^{-3} d. 24
  b. 5.20×10^{2} e. 240
  c. 0.00261   f. 2.40

1.79 Round the following numbers to three significant figures:
  a. 3.873×10^{-3} d. 24.3387
  b. 5.202×10^{-2} e. 240.1
  c. 0.002616   f. 2.407

1.80 Round the following numbers to three significant figures:
  a. 123700   d. 53.2995
  b. 0.00285792   e. 16.96
  c. 1.421×10^{-3} f. 507.5

1.81 Define each of the following terms:
  a. precision
  b. accuracy

1.82 Define each of the following terms:
  a. error
  b. uncertainty

Applications

1.83 Perform each of the following arithmetic operations, reporting the answer with the proper number of significant figures:
  a. $(23)(657)$
  b. $0.00521 + 0.236$
  c. $\dfrac{18.3}{3.0576}$
  d. $1157.23 - 17.812$
  e. $\dfrac{(1.987)(298)}{0.0821}$

1.84 Perform each of the following arithmetic operations, reporting the answer with the proper number of significant figures:
  a. $\dfrac{(16.0)(0.1879)}{45.3}$
  b. $\dfrac{(76.32)(1.53)}{0.052}$
  c. $(0.0063)(57.8)$
  d. $18 + 52.1$
  e. $58.17 - 57.79$

1.85 Express the following numbers in scientific notation (use the proper number of significant figures):
  a. 12.3
  b. 0.0569
  c. −1527
  d. 0.000000789
  e. 92,000,000
  f. 0.005280
  g. 1.279
  h. −531.77

1.86 Using scientific notation, express the number two thousand in terms of:
  a. one significant figure
  b. two significant figures
  c. three significant figures
  d. four significant figures
  e. five significant figures

1.87 Express each of the following numbers in decimal notation:
  a. 3.24×10^3
  b. 1.50×10^{-4}
  c. 4.579×10^{-1}
  d. -6.83×10^5
  e. -8.21×10^{-2}
  f. 2.9979×10^8
  g. 1.50×10^0
  h. 6.02×10^{23}

1.88 Which of the following numbers have two significant figures? Three significant figures? Four significant figures?
  a. 327
  b. 1.049×10^4
  c. 1.70
  d. 0.000570
  e. 7.8×10^3
  f. 1507
  g. 4.8×10^2
  h. 7.389×10^{15}

Experimental Quantities

Foundations

1.89 Calculate the density of a 3.00×10^2-g object that has a volume of 50.0 mL.

1.90 What volume, in liters, will 8.00×10^2 g of air occupy if the density of air is 1.29 g/L?

1.91 What is the mass, in grams, of a piece of iron that has a volume of 1.50×10^2 mL and a density of 7.20 g/mL?

1.92 What is the mass of a femur (leg bone) having a volume of 118 cm³? The density of bone is 1.8 g/cm³.

Applications

1.93 You are given a piece of wood that is maple, teak, or oak. The piece of wood has a volume of 1.00×10^2 cm³ and a mass of 98 g. The densities of maple, teak, and oak are as follows:

Wood	Density (g/cm³)
Maple	0.70
Teak	0.98
Oak	0.85

What is the identity of the piece of wood?

1.94 The specific gravity of a patient's urine sample was measured to be 1.008. Given that the density of water is 1.000 g/mL at 4°C, what is the density of the urine sample?

1.95 The density of grain alcohol is 0.789 g/mL. Given that the density of water at 4°C is 1.00 g/mL, what is the specific gravity of grain alcohol?

1.96 The density of mercury is 13.6 g/mL. If a sample of mercury weighs 272 g, what is the volume of the sample in milliliters?

1.97 You are given three bars of metal. Each is labeled with its identity (lead, uranium, platinum). The lead bar has a mass of 5.0×10^1 g and a volume of 6.36 cm³. The uranium bar has a mass of 75 g and a volume of 3.97 cm³. The platinum bar has a mass of 2140 g and a volume of 1.00×10^2 cm³. Which of these metals has the lowest density? Which has the greatest density?

1.98 Refer to Problem 1.97. Suppose that each of the bars had the same mass. How could you determine which bar had the lowest density or highest density?

1.99 The density of methanol at 20°C is 0.791 g/mL. What is the volume of a 10.0-g sample of methanol?

1.100 The density of methanol at 20°C is 0.791 g/mL. What is the mass of a 50.0-mL sample of methanol?

CRITICAL THINKING PROBLEMS

1. An instrument used to detect metals in drinking water can detect as little as one microgram of mercury in one liter of water. Mercury is a toxic metal; it accumulates in the body and is responsible for the deterioration of brain cells. Calculate the number of mercury atoms you would consume if you drank one liter of water that contained only one microgram of mercury. (The mass of one mercury atom is 3.3×10^{-22} grams.)

2. Yesterday's temperature was 40°F. Today it is 80°F. Bill tells Sue that it is twice as hot today. Sue disagrees. Do you think Sue is correct or incorrect? Why or why not?

3. Aspirin has been recommended to minimize the chance of heart attacks in persons who have already had one or more occurrences. If a patient takes one aspirin tablet per day for ten years, how many pounds of aspirin will the patient consume? (Assume that each tablet is approximately 325 mg.)

4. Design an experiment that will allow you to measure the density of your favorite piece of jewelry.

5. The diameter of an aluminum atom is 250 picometers (1 picometer = 10^{-12} meters). How many aluminum atoms must be placed end to end to make a "chain" of aluminum atoms one foot long?

GENERAL CHEMISTRY

The Structure of the Atom and the Periodic Table

2

Learning Goals

1. Describe the important properties of protons, neutrons, and electrons.
2. Calculate the number of protons, neutrons, and electrons in any atom.
3. Distinguish among atoms, ions, and isotopes.
4. Trace the history of the development of atomic theory, beginning with Dalton.
5. Explain the critical role of spectroscopy in the development of atomic theory and in our everyday lives.
6. State the basic postulates of Bohr's theory, its utility, and its limitations.
7. Recognize the important subdivisions of the periodic table: periods, groups (families), metals, and nonmetals.
8. Use the periodic table to obtain information about an element.
9. Describe the relationship between the electronic structure of an element and its position in the periodic table.
10. Write electron configurations for atoms of the most commonly occurring elements.
11. Use the octet rule to predict the charge of common cations and anions.
12. Utilize the periodic table and its predictive power to estimate the relative sizes of atoms and ions, as well as relative magnitudes of ionization energy and electron affinity.

Organization and understanding go hand-in-hand.

Outline

Chemistry Connection: *Managing Mountains of Information*

2.1 Composition of the Atom
2.2 Development of Atomic Theory
2.3 Light, Atomic Structure, and the Bohr Atom

An Environmental Perspective: *Electromagnetic Radiation and Its Effects on Our Everyday Lives*

A Human Perspective: *Atomic Spectra and the Fourth of July*

2.4 The Periodic Law and the Periodic Table

A Medical Perspective: *Copper Deficiency and Wilson's Disease*

2.5 Electron Arrangement and the Periodic Table
2.6 The Octet Rule

A Medical Perspective: *Dietary Calcium*

2.7 Trends in the Periodic Table

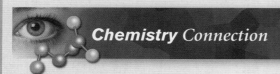

Chemistry Connection

Managing Mountains of Information

Recall for a moment the first time that you sat down in front of a computer. Perhaps it was connected to the Internet; somewhere in its memory was a word processor program, a spreadsheet, a few games, and many other features with strange-sounding names. Your challenge, very simply, was to use this device to access and organize information. Several manuals, all containing hundreds of pages of bewilderment, were your only help. How did you overcome this seemingly impossible task?

We are quite sure that you did not succeed without doing some reading and talking to people who had experience with computers. Also, you did not attempt to memorize every single word in each manual.

Success with a computer or any other storehouse of information results from developing an overall understanding of the way in which the system is organized. Certain facts must be memorized, but seeing patterns and using these relationships allows us to accomplish a wide variety of tasks that involve similar logic.

The study of chemistry is much like "real life." Just as it is impossible to memorize every single fact that will allow you to run a computer or drive an automobile in traffic, it is equally impossible to learn every fact in chemistry. Knowing the organization and logic of a process, along with a few key facts, makes a task manageable.

One powerful organizational device in chemistry is the periodic table. Its use in organizing and predicting the behavior of all of the known elements (and many of the compounds formed from these elements) is the subject of this chapter.

Introduction

Why does ice float on water? Why don't oil and water mix? Why does blood transport oxygen to our cells, whereas carbon monoxide inhibits this process? Questions such as these are best explained by understanding the behavior of substances at the atomic level.

In this chapter we will learn some of the properties of the major particles that make up the atom and look at early experiments that enabled us to develop theories of atomic structure. These theories, in turn, help us to explain the behavior of atoms themselves, as well as the compounds that result from their combination.

The structure of atoms of each element is unique, so it is useful to consider relationships and differences among the elements themselves. The unifying concept is called the periodic law, and it gives rise to an organized "map" of the elements that relates their structure to their chemical and physical properties. This "map" is the periodic table.

As we study the periodic law and periodic table, we shall see that the chemical and physical properties of elements follow directly from the electronic structure of the atoms that make up these elements. A thorough familiarity with the arrangement of the periodic table is vital to the study of chemistry. It not only allows us to predict the structure and properties of the various elements, but it also serves as the basis for developing an understanding of chemical bonding, or the process of forming molecules. Additionally, the properties and behavior of these larger units (bulk properties) are fundamentally related to the properties of the atoms that compose them.

2.1 Composition of the Atom

The basic structural unit of an element is the **atom**, which is the smallest unit of an element that retains the chemical properties of that element. A tiny sample of the element copper, too small to be seen by the naked eye, is composed of billions of copper atoms arranged in some orderly fashion. Each atom is incredibly small. Only recently have we been able to "see" atoms using modern instruments such as the scanning tunneling microscope (Figure 2.1).

Electrons, Protons, and Neutrons

We know from experience that certain kinds of atoms can "split" into smaller particles and release large amounts of energy; this process is *radioactive decay*. We also know that the atom is composed of three primary particles: the *electron*, the *proton*, and the *neutron*. Although other subatomic fragments with unusual names (neutrinos, gluons, quarks, and so forth) have also been discovered, we shall concern ourselves only with the primary particles: the protons, neutrons, and electrons.

We can consider the atom to be composed of two distinct regions:

1. The **nucleus** is a small, dense, positively charged region in the center of the atom. The nucleus is composed of positively charged **protons** and uncharged **neutrons**.
2. Surrounding the nucleus is a diffuse region of negative charge populated by **electrons**, the source of the negative charge. Electrons are very low in mass in contrast to the protons and neutrons.

The properties of these particles are summarized in Table 2.1.

Atoms of various types differ in their number of protons, neutrons, and electrons. The number of protons determines the identity of the atom. As such, the number of protons is *characteristic* of the element. When the number of protons is equal to the number of electrons, the atom is neutral because the charges are balanced and effectively cancel one another.

We may represent an element symbolically as follows:

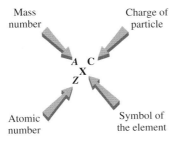

The **atomic number** (Z) is equal to the number of protons in the atom, and the **mass number** (A) is equal to the *sum* of the number of protons and neutrons (the mass of the electrons is so small as to be insignificant in comparison to that of the nucleus).

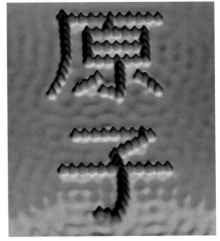

Figure 2.1
Sophisticated techniques, such as scanning tunneling electron microscopy, provide visual evidence for the structure of atoms and molecules. Each dot represents the image of a single iron atom. Even more amazing, the iron atoms have been arranged on a copper surface in the form of the Chinese characters representing the word *atom*.

Radioactivity and radioactive decay are discussed in Chapter 9.

Recall from Chapter 1 (p. 24) that 1 atomic mass unit (amu) is equivalent to 1.661×10^{-24} g.

TABLE 2.1 Selected Properties of the Three Basic Subatomic Particles

Name	Charge	Mass (amu)	Mass (grams)
Electron (e)	−1	5.4×10^{-4} $1/1800$	9.1095×10^{-28}
Proton (p)	+1	1.00	1.6725×10^{-24}
Neutron (n)	0	1.00	1.6750×10^{-24}

If

$$\text{number of protons} + \text{number of neutrons} = \text{mass number}$$

then, if the number of protons is subtracted from each side,

$$\text{number of neutrons} = \text{mass number} - \text{number of protons}$$

or, because the number of protons equals the atomic number,

$$\text{number of neutrons} = \text{mass number} - \text{atomic number}$$

For an atom, in which positive and negative charges cancel, the number of protons and electrons must be equal and identical to the atomic number.

EXAMPLE 2.1 Determining the Composition of an Atom

Calculate the numbers of protons, neutrons, and electrons in an atom of fluorine. The atomic symbol for the fluorine atom is $^{19}_{9}F$.

Solution

The mass number 19 tells us that the total number of protons + neutrons is 19. The atomic number, 9, represents the number of protons. The difference, $19 - 9$, or 10, is the number of neutrons. The number of electrons must be the same as the number of protons, hence 9, for a neutral fluorine atom.

Question 2.1

Calculate the number of protons, neutrons, and electrons in each of the following atoms:

a. $^{32}_{16}S$
b. $^{23}_{11}Na$

Question 2.2

Calculate the number of protons, neutrons, and electrons in each of the following atoms:

a. $^{1}_{1}H$
b. $^{244}_{94}Pu$

Isotopes

LEARNING GOAL

Isotopes are atoms of the same element having different masses *because they contain different numbers of neutrons*. In other words, isotopes have different mass numbers. For example, all of the following are isotopes of hydrogen:

$^{1}_{1}H$ $^{2}_{1}H$ $^{3}_{1}H$

Hydrogen   Deuterium   Tritium

(Hydrogen-1)   (Hydrogen-2)   (Hydrogen-3)

Isotopes are often written with the name of the element followed by the mass number. For example, the isotopes $^{12}_{6}C$ and $^{14}_{6}C$ may be written as carbon-12 (or C-12) and carbon-14 (or C-14), respectively.

Certain isotopes (radioactive isotopes) of elements emit particles and energy that can be used to trace the behavior of biochemical systems. These isotopes otherwise behave identically to any other isotope of the same element. Their chemical

A detailed discussion of the use of radioactive isotopes in the diagnosis and treatment of diseases is found in Chapter 9.

behavior is identical; it is their nuclear behavior that is unique. As a result, a radioactive isotope can be substituted for the "nonradioactive" isotope, and its biochemical activity can be followed by monitoring the particles or energy emitted by the isotope as it passes through the body.

The existence of isotopes explains why the average masses, measured in atomic mass units (amu), of the various elements are not whole numbers. This is contrary to what we would expect from proton and neutron masses, which are very close to unity.

Consider, for example, the mass of one chlorine atom, containing 17 protons (atomic number) and 18 neutrons:

$$17 \text{ protons} \times \frac{1.00 \text{ amu}}{\text{proton}} = 17.00 \text{ amu}$$

$$18 \text{ neutrons} \times \frac{1.00 \text{ amu}}{\text{neutron}} = 18.00 \text{ amu}$$

$$17.00 \text{ amu} + 18.00 \text{ amu} = 35.00 \text{ amu (mass of chlorine atom)}$$

Inspection of the periodic table reveals that the mass of chlorine is actually 35.45 amu, *not* 35.00 amu. The existence of isotopes accounts for this difference. A natural sample of chlorine is composed principally of two isotopes, chlorine-35 and chlorine-37, in approximately a 3:1 ratio, and the tabulated mass is the *weighted average* of the two isotopes. In our calculation the chlorine atom referred to was the isotope that has a mass number of 35 amu.

The weighted average of the masses of all of the isotopes of an element is the **atomic mass** and should be distinguished from the mass number, which is the sum of the number of protons and neutrons in a single isotope of the element.

Example 2.2 demonstrates the calculation of the atomic mass of chlorine.

Atomic mass units are convenient for representing the mass of very small particles, such as individual atoms. Refer to the discussion of units in Chapter 1, p. 24.

The weighted average is not a true average but is corrected by the relative amounts (the weighting factor) of each isotope present in nature.

EXAMPLE 2.2

Determining Atomic Mass

Calculate the atomic mass of naturally occurring chlorine if 75.77% of chlorine atoms are $^{35}_{17}\text{Cl}$ (chlorine-35) and 24.23% of chlorine atoms are $^{35}_{17}\text{Cl}$ (chlorine-37).

Solution

Step 1. Convert each percentage to a decimal fraction.

$$75.77\% \text{ chlorine-35} \times \frac{1}{100\%} = 0.7577 \text{ chlorine-35}$$

$$24.23\% \text{ chlorine-37} \times \frac{1}{100\%} = 0.2423 \text{ chlorine-37}$$

Step 2. Multiply the decimal fraction of each isotope by the mass of that isotope to determine the isotopic contribution to the average atomic mass.

$$\begin{array}{c} \text{contribution to} \\ \text{atomic mass} \\ \text{by chlorine-35} \end{array} = \begin{array}{c} \text{fraction of all} \\ \text{Cl atoms that} \\ \text{are chlorine-35} \end{array} \times \begin{array}{c} \text{mass of a} \\ \text{chlorine-35} \\ \text{atom} \end{array}$$

$$= 0.7577 \times 35.00 \text{ amu}$$

$$= 26.52 \text{ amu}$$

Continued—

EXAMPLE 2.2 —Continued

$$\begin{array}{rcl}
\text{contribution to} & & \text{fraction of all} & & \text{mass of a} \\
\text{atomic mass} & = & \text{Cl atoms that} & \times & \text{chlorine-37} \\
\text{by chlorine-37} & & \text{are chlorine-37} & & \text{atom} \\
& = & 0.2423 & \times & 37.00 \text{ amu} \\
& = & 8.965 \text{ amu}
\end{array}$$

Step 3. The weighted average is the sum of the isotopic contributions:

$$\begin{array}{rcl}
\text{atomic mass} & & \text{contribution} & & \text{contribution} \\
\text{of naturally} & = & \text{of} & + & \text{of} \\
\text{occurring Cl} & & \text{chlorine-35} & & \text{chlorine-37} \\
& = & 26.52 \text{ amu} & + & 8.965 \text{ amu} \\
& = & 35.49 \text{ amu}
\end{array}$$

which is very close to the tabulated value of 35.45 amu. An even more exact value would be obtained by using a more exact value of the mass of the proton and neutron (experimentally known to a greater number of significant figures).

A hint for numerical problem solving: Estimate (at least to an order of magnitude) your answer before beginning the calculation using your calculator.

Whenever you do calculations such as those in Example 2.2, before even beginning the calculation you should look for an approximation of the value sought. Then do the calculation and see whether you obtain a reasonable number (similar to your anticipated value). In the preceding problem, if the two isotopes have masses of 35 and 37, the atomic mass must lie somewhere between the two extremes. Furthermore, because the majority of a naturally occurring sample is chlorine-35 (about 75%), the value should be closer to 35 than to 37. An analysis of the results often avoids problems stemming from untimely events such as pushing the wrong button on a calculator.

EXAMPLE 2.3 Determining Atomic Mass

Calculate the atomic mass of naturally occurring carbon if 98.90% of carbon atoms are $^{12}_{6}C$ (carbon-12) with a mass of 12.00 amu and 1.11% are $^{13}_{6}C$ (carbon-13) with a mass of 13.00 amu. (Note that a small amount of $^{14}_{6}C$ is also present but is small enough to ignore in a calculation involving three or four significant figures.)

Solution

Step 1. Convert each percentage to a decimal fraction.

$$98.90\% \text{ carbon-12} \times \frac{1}{100\%} = 0.9890 \text{ carbon-12}$$

$$1.11\% \text{ carbon-13} \times \frac{1}{100\%} = 0.0111 \text{ carbon-13}$$

Continued—

EXAMPLE 2.3 —Continued

Step 2.

contribution to atomic mass by carbon-12 = fraction of all C atoms that are carbon-12 × (mass of a carbon-12 atom)

= 0.9890 × 12.00 amu

= 11.87 amu

contribution to atomic mass by carbon-13 = (fraction of all C atoms that are carbon-13) × (mass of a carbon-13 atom)

= 0.0111 × 13.00 amu

= 0.144 amu

Step 3. The weighted average is:

atomic mass of naturally occurring carbon = (contribution of carbon-12) + (contribution of carbon-13)

= 11.87 amu + 0.144 amu

= 12.01 amu

Helpful Hint: Because most of the carbon is carbon-12, with very little carbon-13 present, the atomic mass should be very close to that of carbon-12. Approximations, before performing the calculation, provide another check on the accuracy of the final result.

Question 2.3

The element neon has three naturally occurring isotopes. One of these has a mass of 19.99 amu and a natural abundance of 90.48%. A second isotope has a mass of 20.99 amu and a natural abundance of 0.27%. A third has a mass of 21.99 amu and a natural abundance of 9.25%. Calculate the atomic mass of neon.

Question 2.4

The element nitrogen has two naturally occurring isotopes. One of these has a mass of 14.003 amu and a natural abundance of 99.63%; the other isotope has a mass of 15.000 amu and a natural abundance of 0.37%. Calculate the atomic mass of nitrogen.

Ions

Ions are electrically charged particles that result from a gain of one or more electrons by the parent atom (forming negative ions, or **anions**) or a loss of one or more electrons from the parent atom (forming positive ions, or **cations**).

Formation of an anion may occur as follows:

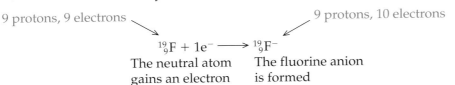

> Ions are often formed in chemical reactions, when one or more electrons are transferred from one substance to another.

Alternatively, formation of a cation of sodium may proceed as follows:

11 protons, 11 electrons → ... ← 11 protons, 10 electrons

$$^{23}_{11}Na \longrightarrow 1e^- + {}^{23}_{11}Na^+$$

The neutral atom loses an electron → The sodium cation is formed

Note that the electrons gained are written to the left of the reaction arrow (they are reactants), whereas the electrons lost are written as products to the right of the reaction arrow. For simplification, the atomic and mass numbers are often omitted, because they do not change during ion formation. For example, the sodium cation would be written as Na^+ and the anion of fluorine as F^-.

2.2 Development of Atomic Theory

LEARNING GOAL With this overview of our current understanding of the structure of the atom, we now look at a few of the most important scientific discoveries that led to modern atomic theory.

Dalton's Theory

The first experimentally based theory of atomic structure was proposed by John Dalton, an English schoolteacher, in the early 1800s. Dalton proposed the following description of atoms:

1. All matter consists of tiny particles called atoms.
2. An atom cannot be created, divided, destroyed, or converted to any other type of atom.
3. Atoms of a particular element have identical properties.
4. Atoms of different elements have different properties.
5. Atoms of different elements combine in simple whole-number ratios to produce compounds (stable aggregates of atoms).
6. Chemical change involves joining, separating, or rearranging atoms.

Although Dalton's theory was founded on meager and primitive experimental information, we regard much of it as correct today. Postulates 1, 4, 5, and 6 are currently regarded as true. The discovery of the processes of nuclear fusion, fission ("splitting" of atoms), and radioactivity has disproved the postulate that atoms cannot be created or destroyed. Postulate 3, that all the atoms of a particular element are identical, was disproved by the discovery of isotopes.

Fusion, fission, radioactivity, and isotopes are discussed in some detail in Chapter 9. Figure 2.2 uses a simple model to illustrate Dalton's theory.

Figure 2.2
An illustration of John Dalton's atomic theory. (a) Atoms of the same element are identical but different from atoms of any other element. (b) Atoms combine in whole-number ratios to form compounds.

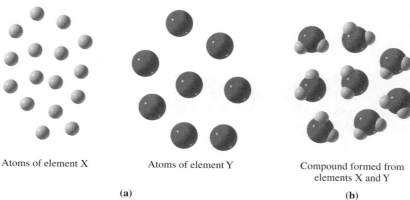

Atoms of element X | Atoms of element Y | Compound formed from elements X and Y

(a) (b)

2.2 Development of Atomic Theory

Evidence for Subatomic Particles: Electrons, Protons, and Neutrons

The next major discoveries occurred almost a century later (1879–1897). Although Dalton pictured atoms as indivisible, various experiments, particularly those of William Crookes and Eugene Goldstein, indicated that the atom is composed of charged (+ and –) particles.

Crookes connected two metal electrodes (metal discs connected to a source of electricity) at opposite ends of a sealed glass vacuum tube. When the electricity was turned on, rays of light were observed to travel between the two electrodes. They were called **cathode rays** because they traveled from the *cathode* (the negative electrode) to the *anode* (the positive electrode).

Later experiments by J. J. Thomson, an English scientist, demonstrated the electrical and magnetic properties of cathode rays (Figure 2.3). The rays were deflected toward the positive electrode of an external electric field. Because opposite charges attract, this indicates the negative character of the rays. Similar experiments with an external magnetic field showed a deflection as well; hence these cathode rays also have magnetic properties.

A change in the material used to fabricate the electrode discs brought about no change in the experimental results. This suggested that the ability to produce cathode rays is a characteristic of all materials.

In 1897, Thomson announced that cathode rays are streams of negative particles of energy. These particles are *electrons*. Similar experiments, conducted by Goldstein, led to the discovery of particles that are equal in charge to the electron but opposite in sign. These particles, much heavier than electrons (actually 1837 times as heavy), are called *protons*.

As we have seen, the third fundamental atomic particle is the *neutron*. It has a mass virtually identical (it is less than 1% heavier) to that of the proton and has zero charge. The neutron was first postulated in the early 1920s, but it was not until 1932 that James Chadwick demonstrated its existence with a series of experiments involving the use of small particle bombardment of nuclei.

LEARNING GOAL

Crookes's cathode ray tube was the forerunner of the computer screen (often called CRT) and the television.

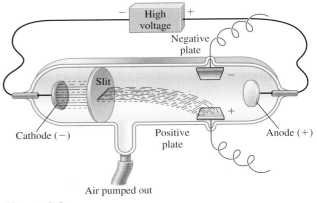

Figure 2.3
Illustration of an experiment demonstrating the charge of cathode rays. The application of an external electric field causes the electron beam to deflect toward a positive charge, implying that the cathode ray is negative.

Evidence for the Nucleus

In the early 1900s it was believed that protons and electrons were uniformly distributed throughout the atom. However, an experiment by Hans Geiger led Ernest Rutherford (in 1911) to propose that the majority of the mass and positive charge of the atom was actually located in a small, dense region, the *nucleus*, with small, negatively charged electrons occupying a much larger volume outside of the nucleus.

To understand how Rutherford's theory resulted from the experimental observations of Geiger, let us examine this experiment in greater detail. Rutherford and others had earlier demonstrated that some atoms spontaneously "decay" to produce three types of radiation: alpha (α), beta (β), and gamma (γ) radiation. This process is known as *natural radioactivity* (Figure 2.4). Geiger used radioactive

LEARNING GOAL

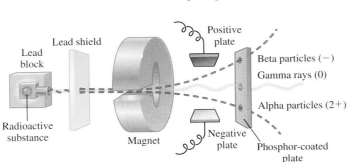

Figure 2.4
Types and characteristics of radioactive emissions. The direction taken by the radioactive emissions indicates the presence of three types of emissions: positive, negative, and neutral components.

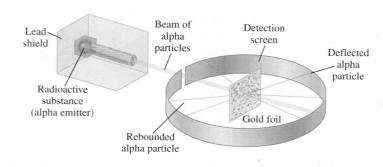

Figure 2.5
The alpha particle scattering experiment. Most alpha particles passed through the foil without being deflected; a few were deflected from their path by nuclei in the gold atoms.

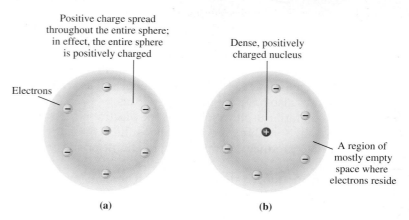

Figure 2.6
(a) A model of the atom (credited to Thomson) prior to the work of Geiger and Rutherford. (b) A model of the atom supported by the alpha-particle scattering experiments of Geiger and Rutherford.

materials, such as *radium*, as projectile sources, "firing" alpha particles at a thin metal foil target (gold leaf). He then observed the interaction of the metal and alpha particles with a detection screen (Figure 2.5) and found that:

a. Most alpha particles pass through the foil without being deflected.
b. A small fraction of the particles were deflected, some even *directly back to the source.*

Rutherford interpreted this to mean that most of the atom is empty space, because most alpha particles were not deflected. Further, most of the mass and positive charge must be located in a small, dense region; collision of the heavy and positively charged alpha particle with this small dense and positive region (the nucleus) caused the great deflections. Rutherford summarized his astonishment at observing the deflected particles: "It was almost as incredible as if you fired a 15-inch shell at a piece of tissue and it came back and hit you."

The significance of Rutherford's contribution cannot be overstated. It caused a revolutionary change in the way that scientists pictured the atom (Figure 2.6). His discovery of the nucleus is fundamental to our understanding of chemistry. Chapter 9 will provide much more information on a special branch of chemistry: nuclear chemistry.

2.3 Light, Atomic Structure, and the Bohr Atom

Light and Atomic Structure

LEARNING GOAL

The Rutherford atom leaves us with a picture of a tiny, dense, positively charged nucleus containing protons and surrounded by electrons. The electron arrangement, or configuration, is not clearly detailed. More information is needed regarding the relationship of the electrons to each other and to the nucleus. In dealing with dimensions on the order of 10^{-9} m (the atomic level), conventional methods for measurement of location and distance of separation become impossible. An alternative approach involves the measurement of *energy* rather than the *position* of the atomic particles to determine structure. For example, information obtained from the absorption or emission of *light* by atoms (energy changes) can yield valuable insight into structure. Such studies are referred to as **spectroscopy.**

In a general sense we refer to light as **electromagnetic radiation.** Electromagnetic radiation travels in *waves* from a source. The most recognizable source of this radiation is the sun. We are aware of a rainbow, in which visible white light from the sun is broken up into several characteristic bands of different colors. Similarly, visible white light, when passed through a glass prism, is separated into its various component colors (Figure 2.7). These various colors are simply light (electromagnetic radiation) of differing *wavelengths*. Light is propagated as a collection of sine waves, and the wavelength is the distance between identical points on successive waves:

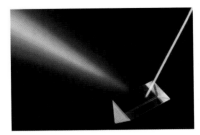

Figure 2.7
The visible spectrum of light. Light passes through a prism, producing a continuous spectrum. Color results from the way in which our eyes interpret the various wavelengths.

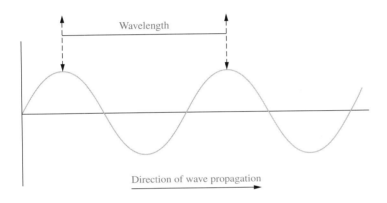

All electromagnetic radiation travels at a speed of 3.0×10^8 m/s, the **speed of light.** However, each wavelength of light, although traveling with identical velocity, has its own characteristic energy. A collection of all electromagnetic radiation, including each of these wavelengths, is referred to as the **electromagnetic spectrum.** For convenience in discussing this type of radiation we subdivide electromagnetic radiation into various spectral regions, which are characterized by physical properties of the radiation, such as its *wavelength* or its *energy* (Figure 2.8). Some of these regions are quite familiar to us from our everyday experiences; the visible and microwave regions are two common examples.

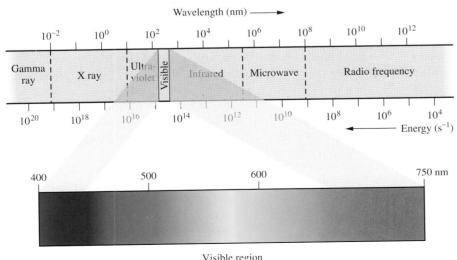

Figure 2.8
The electromagnetic spectrum. Note that the visible spectrum is only a small part of the total electromagnetic spectrum.

Higher wave length = lower the energy

An Environmental Perspective

Electromagnetic Radiation and Its Effects on Our Everyday Lives

From the preceding discussion of the interaction of electromagnetic radiation with matter—spectroscopy—you might be left with the impression that the utility of such radiation is limited to theoretical studies of atomic structure. Although this is a useful application that has enabled us to learn a great deal about the structure and properties of matter, it is by no means the only application. Useful, everyday applications of the theories of light energy and transmission are all around us. Let's look at just a few examples.

Transmission of sound and pictures is conducted at radio frequencies or radio wavelengths. We are immersed in radio waves from the day we are born. A radio or television is our "detector" of these waves. Radio waves are believed to cause no physical harm because of their very low energy, although some concern for people who live very close to transmission towers has resulted from recent research.

X-rays are electromagnetic radiation, and they travel at the speed of light just like radio waves. However, because of their higher energy, they can pass through the human body and leave an image of the body's interior on a photographic film. X-ray photographs are invaluable for medical diagnosis. However, caution is advised in exposing oneself to X-rays, because the high energy can remove electrons from biological molecules, causing subtle and potentially harmful changes in their chemistry.

The sunlight that passes through our atmosphere provides the basis for a potentially useful technology for providing heat and electricity: *solar energy*. Light is captured by absorbers, referred to as *solar collectors*, which convert the light energy into heat energy. This heat can be transferred to water circulating beneath the collectors to provide heat and hot water for homes or industry. Wafers of a silicon-based material can convert light energy to electrical energy; many believe that if the efficiency of these processes can be improved, such approaches may provide at least a partial solution to the problems of rising energy costs and pollution associated with our fossil fuel-based energy economy.

The intensity of infrared radiation from a solid or liquid is an indicator of relative temperature. This has been used to advantage in the design of infrared cameras, which can obtain images without the benefit of the visible light that is necessary for conventional cameras. The infrared photograph shows the coastline surrounding the city of San Francisco.

Microwave radiation for cooking, *infrared* lamps for heating and remote sensing, *ultraviolet* lamps used to kill microorganisms on environmental surfaces, *gamma radiation* from nuclear waste, the *visible* light from the lamp you are using to read this chapter—all are forms of the same type of energy that, for better or worse, plays such a large part in our twenty-first century technological society.

Electromagnetic radiation and spectroscopy also play a vital role in the field of diagnostic medicine. They are routinely

---

A spectrophotometer, an instrument that utilizes a prism (or similar device) and a light-sensitive detector, is capable of very accurate and precise wavelength measurement.

Light of shorter wavelength has higher energy; this means that the magnitude of the energy and wavelength is inversely proportional. The wavelength of a particular type of light can be measured, and from this the energy may be calculated.

If we take a sample of some element, such as hydrogen, in the gas phase, place it in an evacuated glass tube containing a pair of electrodes, and pass an electrical charge (cathode ray) through the hydrogen gas, light is emitted. Not all wavelengths (or energies) of light are emitted—only certain wavelengths that are characteristic of the gas under study. This is referred to as an *emission spectrum* (Figure 2.9). If a different gas, such as helium, is used, a different spectrum (different wavelengths of light) is observed. The reason for this behavior was explained by Niels Bohr.

2.3 Light, Atomic Structure, and the Bohr Atom

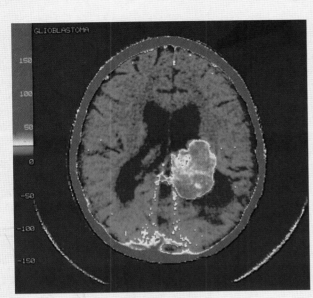

An image of a tumor detected by a CT scan.

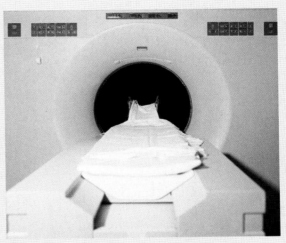

The CT scanner is a device used for diagnostic purposes.

used as diagnostic and therapeutic tools in the detection and treatment of disease.

The radiation therapy used in the treatment of many types of cancer has been responsible for saving many lives and extending the span of many others. When radiation is used as a treatment, it destroys cancer cells. This topic will be discussed in detail in Chapter 9.

As a diagnostic tool, spectroscopy has the benefit of providing data quickly and reliably; it can also provide information that might not be available through any other means. Additionally, spectroscopic procedures are often nonsurgical, outpatient procedures. Such procedures involve less risk, can be more routinely performed, and are more acceptable to the general public than surgical procedures. The potential cost savings because of the elimination of many unnecessary surgical procedures is an added benefit.

The most commonly practiced technique uses the CT scanner, an acronym for *computer-accentuated tomography*. In this technique, X-rays are directed at the tissue of interest. As the X-rays pass through the tissue, detectors surrounding the tissue gather the signal, compare it to the original X-ray beam, and, using the computer, produce a three-dimensional image of the tissue.

For Further Understanding

Diane says that a medical X-ray is risky, but a CT scan is risk free. Is Diane correct? Explain your answer.

Why would the sensor (detector) for a conventional camera and an IR camera have to be designed differently?

The Bohr Atom

Niels Bohr hypothesized that surrounding each atomic nucleus were certain fixed **energy levels** that could be occupied by electrons. He also believed that each level was defined by a spherical *orbit* around the nucleus, located at a specific distance from the nucleus. The concept of certain fixed energy levels is referred to as the **quantization** of energy. The implication is that only these orbits, or *quantum levels*, as described by Max Planck, are allowed locations for electrons. If an atom *absorbs* energy, an electron undergoes *promotion* from an orbit closer to the nucleus (lower energy) to one farther from the nucleus (higher energy), creating an *excited state*. Similarly, the release of energy by an atom, or *relaxation*, results from an electron falling into an orbit closer to the nucleus (lower energy level).

Figure 2.9

(a) The emission spectrum of hydrogen. Certain wavelengths of light, characteristic of the atom, are emitted upon electrical excitation. (b) The line spectrum of hydrogen is compared with (c and d) the line spectrum of helium and sodium and the spectrum of visible light (e).

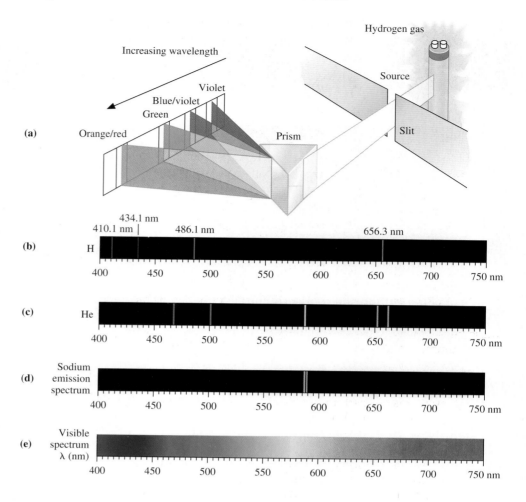

Promotion and relaxation processes are referred to as *electronic transitions*. The amount of energy absorbed in jumping from one energy level to a higher energy level is a precise quantity (hence, quantum), and that energy corresponds exactly to the energy differences between the orbits involved. Electron promotion resulting from absorption of energy results in an *excited state atom*; the process of relaxation allows the atom to return to the ground state (Figure 2.10) with the simultaneous release of light energy. The *ground state* is the lowest possible energy state. This emission process, such as the release of energy after excitation of hydrogen atoms by an

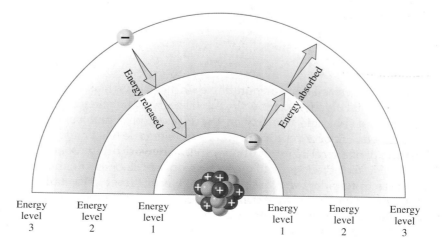

Figure 2.10

The Bohr representation of atoms. Excitation involves promotion of an electron to a higher energy level when energy is absorbed. Relaxation is the reverse process; atoms return to the ground state when the electron relaxes to a lower energy level, releasing energy.

2.3 Light, Atomic Structure, and the Bohr Atom

TABLE 2.2 Electronic Transitions Responsible for the Hydrogen Spectrum

Line Color	Wavelength Emitted (nm)	Electronic Transition $n=$	to	$n=$
Red	656.3	3		2
Green	486.1	4		2
Blue	434.1	5		2
Violet	410.1	6		2

electric arc, produces the series of emission lines (emission spectrum). Measurement of the wavelengths of these lines enables the calculation of energy levels in the atom. These energy levels represent the location of the atom's electrons.

We may picture the Bohr atom as a series of concentric orbits surrounding the nucleus. The orbits are identified using numbers ($n = 1, 2, 3, \ldots$, etc.). The number n is referred to as a *quantum number*.

The hydrogen spectrum consists of four lines in the visible region of the spectrum. Electronic transitions, calculated from the Bohr theory, account for each of these lines. Table 2.2 gives a summary of the hydrogen spectrum.

A summary of the major features of the Bohr theory is as follows:

- Atoms can absorb and emit energy via *promotion* of electrons to higher energy levels and *relaxation* to lower levels.
- Energy that is emitted upon relaxation is observed as a single wavelength of light, a collection of photons.
- These *spectral lines* are a result of electron transitions between *allowed levels* in the atom.
- The allowed levels are quantized energy levels, or orbits.
- Electrons are found only in these energy levels.
- The highest-energy orbits are located farthest from the nucleus.
- Atoms absorb energy by excitation of electrons to higher energy levels.
- Atoms release energy by relaxation of electrons to lower energy levels.
- Energy differences may be calculated from the wavelengths of light emitted.

> Each electronic transition produces "bundles," or quanta, of energy. These quanta are termed "photons." Photons resulting from a certain electronic transition have their own unique wavelength, frequency, and energy.

Modern Atomic Theory

The Bohr model was an immensely important contribution to the understanding of atomic structure. The idea that electrons exist in specific energy states and that transitions between states involve quanta of energy provided the linkage between atomic structure and atomic spectra. However, some limitations of this model quickly became apparent. Although it explained the hydrogen spectrum, it provided only a crude approximation of the spectra for more complex atoms. Subsequent development of more sophisticated experimental techniques demonstrated that there are problems with the Bohr theory even in the case of hydrogen.

Although Bohr's concept of principal energy levels is still valid, restriction of electrons to fixed orbits is too rigorous. All current evidence shows that electrons do *not*, in fact, orbit the nucleus. We now speak of the *probability* of finding an electron in a *region* of space within the principal energy level, referred to as an **atomic orbital**. The rapid movement of the electron spreads the charge into a *cloud* of charge. This cloud is more dense in certain regions, the **electron density** being proportional to the probability of finding the electron at any point in time. Insofar as these atomic orbitals are part of the principal energy levels, they are referred to as sublevels. In Chapter 3 we will see that the orbital model of the atom can be used to predict how atoms can bond together to form compounds. Furthermore, electron arrangement in orbitals enables us to predict various chemical and physical properties of these compounds.

A Human Perspective

Atomic Spectra and the Fourth of July

At one time or another we have all marveled at the bright, multicolored display of light and sound that is a fireworks display. These sights and sounds are produced by a chemical reaction that generates the energy necessary to excite a variety of elements to their higher-energy electronic states. Light emission results from relaxation of the excited atoms to the ground state. Each atom releases light of specific wavelengths. The visible wavelengths are seen as colored light.

Fireworks need a chemical reaction to produce energy. We know from common experience that oxygen and a fuel will release energy. The fuel in most fireworks preparations is sulfur or aluminum. Each reacts slowly with oxygen; a more potent solid-state source of oxygen is potassium perchlorate ($KClO_4$). The potassium perchlorate reacts with the fuel (an oxidation-reduction reaction, Chapter 8), producing a bright white flash of light. The heat produced excites the various elements packaged with the fuel and oxidant.

Sodium salts, such as sodium chloride, furnish sodium ions, which, when excited, produce yellow light. Red colors arise from salts of strontium, which emit several shades of red corresponding to wavelengths in the 600- to 700-nm region of the visible spectrum. Copper salts produce blue radiation, because copper emits in the 400- to 500-nm spectral region.

The beauty of fireworks is a direct result of the skill of the manufacturer. Selection of the proper oxidant, fuel, and color-producing elements is critical to the production of a spectacular display. Packaging these chemicals in proper quantities so that they can be stored and used safely is an equally important consideration.

A fireworks display is a dramatic illustration of light emission by excited atoms.

For Further Understanding

Explain why excited sodium emits a yellow color. (Refer to Figure 2.9.)

How does this story illustrate the interconversion of potential and kinetic energy?

Question 2.5 What is meant by the term *electron density*?

Question 2.6 How do *orbits* and *orbitals* differ?

The theory of atomic structure has progressed rapidly, from a very primitive level to its present point of sophistication, in a relatively short time. Before we proceed, let us insert a note of caution. We must not think of the present picture of the atom as final. Scientific inquiry continues, and we should view the present theory as a step in an evolutionary process. *Theories are subject to constant refinement*, as was noted in our discussion of the scientific method.

2.4 The Periodic Law and the Periodic Table

LEARNING GOAL

In 1869, Dmitri Mendeleev, a Russian, and Lothar Meyer, a German, working independently, found ways of arranging elements in order of increasing atomic mass such that elements with similar properties were grouped together in a

2.4 The Periodic Law and the Periodic Table

table of elements. The **periodic law** is embodied by Mendeleev's statement, "the elements if arranged according to their atomic weights (masses), show a distinct *periodicity* (regular variation) of their properties." The *periodic table* (Figure 2.11) is a visual representation of the periodic law.

Chemical and physical properties of elements correlate with the electronic structure of the atoms that make up these elements. In turn, the electronic structure correlates with position on the periodic table.

A thorough familiarity with the arrangement of the periodic table allows us to predict electronic structure and physical and chemical properties of the various elements. It also serves as the basis for understanding chemical bonding.

The concept of "periodicity" may be illustrated by examining a portion of the modern periodic table. The elements in the second row (beginning with lithium, Li, and proceeding to the right) show a marked difference in properties. However, sodium (Na) has properties similar to those of lithium, and sodium is therefore placed below lithium; once sodium is fixed in this position, the elements Mg through Ar have properties remarkably similar (though not identical) to those of the elements just above them. The same is true throughout the complete periodic table.

Mendeleev arranged the elements in his original periodic table in order of increasing atomic mass. However, as our knowledge of atomic structure increased,

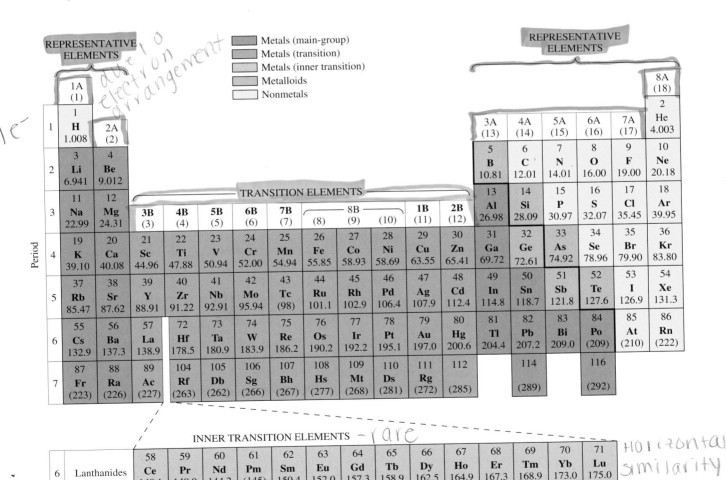

Figure 2.11
Classification of the elements: the periodic table.

TABLE 2.3	Summary of the Most Important Elements in Biological Systems	
Element	Symbol	Significance
Hydrogen	H	Components of major biological molecules
Carbon	C	
Oxygen	O	
Nitrogen	N	
Phosphorus	P	
Sulfur	S	
Potassium	K	Produce electrolytes responsible for fluid balance and nerve transmission
Sodium	Na	
Chlorine	Cl	
Calcium	Ca	Bones, nerve function
Magnesium	Mg	
Zinc	Zn	Essential trace metals in human metabolism
Strontium	Sr	
Iron	Fe	
Copper	Cu	
Cobalt	Co	
Manganese	Mn	
Cadmium	Cd	"Heavy metals" toxic to living systems
Mercury	Hg	
Lead	Pb	

atomic numbers became the basis for the organization of the table. Remarkably, his table was able to predict the existence of elements not known at the time.

The modern periodic law states that *the physical and chemical properties of the elements are periodic functions of their atomic numbers.* If we arrange the elements in order of increasing number of protons, the properties of the elements repeat at regular intervals.

Not all of the elements are of equal importance to an introductory study of chemistry. Table 2.3 lists twenty of the elements that are most important to biological systems, along with their symbols and a brief description of their functions.

We will use the periodic table as our "map," just as a traveler would use a road map. A short time spent learning how to read the map (and remembering to carry it along on your trip!) is much easier than memorizing every highway and intersection. The information learned about one element relates to an entire family of elements grouped as a recognizable unit within the table.

Numbering Groups in the Periodic Table

The periodic table created by Mendeleev has undergone numerous changes over the years. These modifications occurred as more was learned about the chemical and physical properties of the elements. The labeling of groups with Roman numerals followed by the letter *A* (representative elements) or *B* (transition elements) was standard, until 1983, in North America and Russia. However, in other parts of the world, the letters *A* and *B* were used in a different way. Consequently, two different periodic tables were in widespread use. This certainly created some confusion.

Mendeleev's original periodic table included only the elements known at the time, less than half of the current total.

The International Union of Pure and Applied Chemistry (IUPAC), in 1983, recommended that a third system, using numbers 1–18 to label the groups, replace both of the older systems. Unfortunately, multiple systems now exist and this can cause confusion for both students and experienced chemists.

The periodic tables in this textbook are "double labeled." Both the old (Roman numeral) and new (1–18) systems are used to label the groups. The label that you use is simply a guide to reading the table; the real source of information is in the structure of the table itself. The following sections will show you how to extract useful information from this structure.

Periods and Groups

 LEARNING GOAL

A **period** is a horizontal row of elements in the periodic table. The periodic table consists of six periods containing 2, 8, 8, 18, 18, and 32 elements. The seventh period is still incomplete but potentially holds 32 elements. Note that the *lanthanide series*, a collection of 14 elements that are chemically and physically similar to the element lanthanum, is a part of period six. It is written separately for convenience of presentation and is inserted between lanthanum (La), atomic number 57, and hafnium (Hf), atomic number 72. Similarly, the *actinide series*, consisting of 14 elements similar to the element actinium, is inserted between actinium, atomic number 89, and rutherfordium, atomic number 104.

Groups or *families* are columns of elements in the periodic table. The elements of a particular group or family share many similarities, as in a human family. The similarities extend to physical and chemical properties that are related to similarities in electronic structure (that is, the way in which electrons are arranged in an atom).

Group A elements are called **representative elements,** and Group B elements are **transition elements.** Certain families also have common names. For example, Group IA (or 1) elements are also known as the **alkali metals;** Group IIA (or 2), the **alkaline earth metals;** Group VIIA (or 17), the **halogens;** and Group VIIIA (or 18), the **noble gases.**

Representative elements are also known as *main-group elements*. These terms are synonymous.

Metals and Nonmetals

 LEARNING GOAL

A **metal** is a substance whose atoms tend to lose electrons during chemical change, forming positive ions. A **nonmetal,** on the other hand, is a substance whose atoms may gain electrons, forming negative ions.

A closer inspection of the periodic table reveals a bold zigzag line running from top to bottom, beginning to the left of boron (B) and ending between polonium (Po) and astatine (At). This line acts as the boundary between *metals*, to the left, and *nonmetals*, to the right. Elements straddling the boundary have properties intermediate between those of metals and nonmetals. These elements are referred to as **metalloids.** The metalloids include boron (B), silicon (Si), germanium (Ge), arsenic (As), atimony (Sb), tellurium (Te), polonium (Po), and astatine (At).

Note that aluminum (Al) is classified as a metal, not a metalloid.

Metals and nonmetals may be distinguished by differences in their physical properties in addition to their chemical tendency to lose or gain electrons. Metals have a characteristic luster and generally conduct heat and electricity well. Most (except mercury) are solids at room temperature. Nonmetals, on the other hand, are poor conductors, and several are gases at room temperature.

Atomic Number and Atomic Mass

 LEARNING GOAL

The atomic number is the number of protons in the nucleus of an atom of an element. It also corresponds to the nuclear charge, the positive charge from the nucleus. Both the atomic number and the average atomic mass of each element are readily available from the periodic table. For example,

Chapter 2 The Structure of the Atom and the Periodic Table

Not on Test

20 ← atomic number
Ca ← symbol
calcium ← name
40.08 ← atomic mass

More detailed periodic tables may also include such information as the electron arrangement, relative sizes of atoms and ions, and most probable ion charges.

Question 2.7

Refer to the periodic table (Figure 2.11) and find the following information:

a. the symbol of the element with an atomic number of 40
b. the mass of the element sodium (Na)
c. the element whose atoms contain 24 protons
d. the known element that should most resemble the as-yet undiscovered element with an atomic number of 117

Question 2.8

Refer to the periodic table (Figure 2.11) and find the following information:

a. the symbol of the noble gas in period 3
b. the element in Group IVA with the smallest mass
c. the only metalloid in Group IIIA
d. the element whose atoms contain 18 protons

Question 2.9

For each of the following element symbols, give the name of the element, its atomic number, and its atomic mass.

a. He
b. F
c. Mn

Question 2.10

For each of the following element symbols, give the name of the element, its atomic number, and its atomic mass:

a. Mg
b. Ne
c. Se

2.5 Electron Arrangement and the Periodic Table

LEARNING GOAL

A primary objective of studying chemistry is to understand the way in which atoms join together to form chemical compounds. The most important factor in this *bonding process* is the arrangement of the electrons in the atoms that are combining. The **electron configuration** describes the arrangement of electrons in atoms. The periodic table is helpful because it provides us with a great deal of information about the electron arrangement or electronic configuration of atoms.

A Medical Perspective

Copper Deficiency and Wilson's Disease

An old adage tells us that we should consume all things in moderation. This is very true of many of the trace minerals, such as copper. Too much copper in the diet causes toxicity and too little copper results in a serious deficiency disease.

Copper is extremely important for the proper functioning of the body. It aids in the absorption of iron from the intestine and facilitates iron metabolism. It is critical for the formation of hemoglobin and red blood cells in the bone marrow. Copper is also necessary for the synthesis of collagen, a protein that is a major component of the connective tissue. It is essential to the central nervous system in two important ways. First, copper is needed for the synthesis of norepinephrine and dopamine, two chemicals that are necessary for the transmission of nerve signals. Second, it is required for the deposition of the myelin sheath (a layer of insulation) around nerve cells. Release of cholesterol from the liver depends on copper, as does bone development and proper function of the immune and blood clotting systems.

The estimated safe and adequate daily dietary intake (ESADDI) for adults is 1.5–3.0 mg. Meats, cocoa, nuts, legumes, and whole grains provide significant amounts of copper. The accompanying table shows the amount of copper in some common foods.

Although getting enough copper in the diet would appear to be relatively simple, it is estimated that Americans often ingest only marginal levels of copper, and we absorb only 25–40% of that dietary copper. Despite these facts, it appears that copper deficiency is not a serious problem in the United States.

Individuals who are at risk for copper deficiency include people who are recovering from abdominal surgery, which causes decreased absorption of copper from the intestine. Others at risk are premature babies and people who are sustained solely by intravenous feedings that are deficient in copper. In addition, people who ingest high doses of antacids or take excessive supplements of zinc, iron, or vitamin C can develop copper deficiency because of reduced copper absorption. Because copper is involved in so many processes in the body, it is not surprising that the symptoms of copper deficiency are many and diverse. They include anemia; decreased red and white blood cell counts; heart disease; increased levels of serum cholesterol; loss of bone; defects in the nervous system, immune system, and connective tissue; and abnormal hair.

Some of these symptoms are seen among people who suffer from the rare genetic disease known as Menkes' kinky hair syndrome. The symptoms of this disease, which is caused by a defect in the ability to absorb copper from the intestine, include very low copper levels in the serum, kinky white hair, slowed growth, and degeneration of the brain.

Just as too little copper causes serious problems, so does an excess of copper. At doses greater than about 15 mg, copper causes toxicity that results in vomiting. The effects of extended exposure to excess copper are apparent when we look at Wilson's disease. This is a genetic disorder in which excess copper cannot be removed from the body and accumulates in the cornea of the eye, liver, kidneys, and brain. The symptoms include a greenish ring around the cornea, cirrhosis of the liver, copper in the urine, dementia and paranoia, drooling, and progressive tremors. As a result of the condition, the victim generally dies in early adolescence. Wilson's disease can be treated with moderate success if it is recognized early, before permanent damage has occurred to any tissues. The diet is modified to reduce the intake of copper; for instance, such foods as chocolate are avoided. In addition, the drug penicillamine is administered. This compound is related to the antibiotic penicillin but has no antibacterial properties; rather it has the ability to bind to copper in the blood and enhance its excretion by the kidneys into the urine. In this way the brain degeneration and tissue damage that are normally seen with the disease can be lessened.

Copper in One-Cup Portions of Food

Food	Mass of copper (mg)
Sesame seeds	5.88
Cashews	3.04
Oysters	2.88
Sunflower seeds	2.52
Peanuts, roasted	1.85
Crabmeat	1.71
Walnuts	1.28
Almonds	1.22
Cereal, All Bran	0.98
Tuna fish	0.93
Wheat germ	0.70
Prunes	0.69
Kidney beans	0.56
Dried apricots	0.56
Lentils, cooked	0.54
Sweet potato, cooked	0.53
Dates	0.51
Whole milk	0.50
Raisins	0.45
Cereal, C. W. Post, Raisins	0.40
Grape Nuts	0.38
Whole-wheat bread	0.34
Cooked cereal, Roman Meal	0.32

Source: David C. Nieman, Diane E. Butterworth, and Catherine N. Nieman, *Nutrition*, Revised First Edition. Copyright 1992 Wm. C. Brown Communications, Inc., Dubuque, Iowa. All Rights Reserved. Reprinted by permission.

For Further Understanding

Why is there an upper limit on the recommended daily amount of copper?

Iron is another essential trace metal in our diet. Go to the Web and find out if upper limits exist for daily iron consumption.

We have seen (p. 55) that elements in the periodic table are classified as either representative or transition. Representative elements consist of all group 1, 2, and 13–18 elements (IA–VIIIA). All others are transition elements. The guidelines that we will develop for writing electron configurations are intended for representative elements. Electron configurations for transition elements include several exceptions to the rule.

Valence Electrons

If we picture two spherical objects that we wish to join together, perhaps with glue, the glue can be applied to the surface and the two objects can then be brought into contact. We can extend this analogy to two atoms that are modeled as spherical objects. Although this is not a perfect analogy, it is apparent that the surface interaction is of primary importance. Although the positively charged nucleus and "interior" electrons certainly play a role in bonding, we can most easily understand the process by considering only the outermost electrons. We refer to these as *valence electrons*. **Valence electrons** are the outermost electrons in an atom, which are involved, or have the potential to become involved, in the bonding process.

For representative elements the number of valence electrons in an atom corresponds to the number of the *group* or *family* in which the atom is found. For example, elements such as hydrogen and sodium (in fact, all alkali metals, Group IA or 1) have one valence electron. From left to right in period 2, beryllium, Be (Group IIA or 2), has two valence electrons; boron, B (Group IIIA or 3), has three; carbon, C (Group IVA or 4), has four; and so forth.

We have seen that an atom may have electrons in several different energy levels. These energy levels are symbolized by n, the lowest energy level being assigned a value of $n = 1$. Each energy level may contain up to a fixed maximum number of electrons. For example, the $n = 1$ energy level may contain a maximum of two electrons. Thus hydrogen (atomic number = 1) has one electron and helium

> Metals tend to have fewer valence electrons, and nonmetals tend to have more valence electrons.

TABLE 2.4 The Electron Distribution for the First Twenty Elements of the Periodic Table

Element Symbol and Name	Total Number of Electrons	Total Number of Valence Electrons	Electrons in $n = 1$	Electrons in $n = 2$	Electrons in $n = 3$	Electrons in $n = 4$
H, hydrogen	1	1	1	0	0	0
He, helium	2	2	2	0	0	0
Li, lithium	3	1	2	1	0	0
Be, beryllium	4	2	2	2	0	0
B, boron	5	3	2	3	0	0
C, carbon	6	4	2	4	0	0
N, nitrogen	7	5	2	5	0	0
O, oxygen	8	6	2	6	0	0
F, fluorine	9	7	2	7	0	0
Ne, neon	10	8	2	8	0	0
Na, sodium	11	1	2	8	1	0
Mg, magnesium	12	2	2	8	2	0
Al, aluminum	13	3	2	8	3	0
Si, silicon	14	4	2	8	4	0
P, phosphorus	15	5	2	8	5	0
S, sulfur	16	6	2	8	6	0
Cl, chlorine	17	7	2	8	7	0
Ar, argon	18	8	2	8	8	0
K, potassium	19	1	2	8	8	1
Ca, calcium	20	2	2	8	8	2

(atomic number = 2) has two electrons in the $n = 1$ level. Only these elements have electrons *exclusively* in the first energy level:

Hydrogen: one-electron atom

Helium: two-electron atom

These two elements make up the first period of the periodic table. Period 1 contains all elements whose *maximum* energy level is $n = 1$. In other words, the $n = 1$ level is the *outermost* electron region for hydrogen and helium. Hydrogen has one electron and helium has two electrons in the $n = 1$ level.

The valence electrons of elements in the second period are in the $n = 2$ energy level. (Remember that you must fill the $n = 1$ level with two electrons before adding electrons to the next level). The third electron of lithium (Li) and the remaining electrons of the second period elements must be in the $n = 2$ level and are considered the valence electrons for lithium and remaining second period elements.

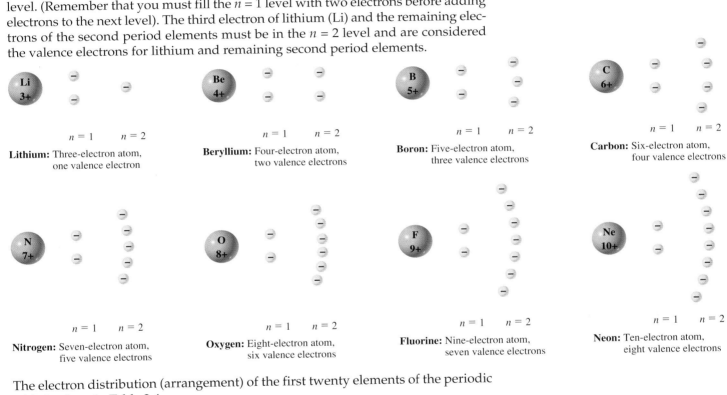

Lithium: Three-electron atom, one valence electron

Beryllium: Four-electron atom, two valence electrons

Boron: Five-electron atom, three valence electrons

Carbon: Six-electron atom, four valence electrons

Nitrogen: Seven-electron atom, five valence electrons

Oxygen: Eight-electron atom, six valence electrons

Fluorine: Nine-electron atom, seven valence electrons

Neon: Ten-electron atom, eight valence electrons

The electron distribution (arrangement) of the first twenty elements of the periodic table is given in Table 2.4.

Two general rules of electron distribution are based on the periodic law:

RULE 1: The number of valence electrons in an atom equals the *group* number for all representative (A group) elements. ■

RULE 2: The energy level ($n = 1, 2$, etc.) in which the valence electrons are located corresponds to the *period* in which the element may be found. ■

For example,

Group IA	*Group IIA*	*Group IIIA*	*Group VIIA*
Li	Ca	Al	Br
one valence electron in $n = 2$ energy level; period 2	two valence electrons in $n = 4$ energy level; period 4	three valence electrons in $n = 3$ energy level; period 3	seven valence electrons in $n = 4$ energy level; period 4

EXAMPLE 2.4 Determining Electron Arrangement

Provide the total number of electrons, total number of valence electrons, and energy level in which the valence electrons are found for the silicon (Si) atom.

Solution

Step 1. Determine the position of silicon in the periodic table. Silicon is found in Group IVA and period 3 of the table. Silicon has an atomic number of 14.

Step 2. The atomic number provides the number of electrons in an atom. Silicon therefore has 14 electrons.

Step 3. Because silicon is in Group IV, only 4 of the 14 electrons are valence electrons.

Step 4. Silicon has 2 electrons in $n = 1$, 8 electrons in $n = 2$, and 4 electrons in the $n = 3$ level.

Question 2.11

For each of the following elements, provide the *total* number of electrons and *valence* electrons in its atom:

a. Na
b. Mg
c. S
d. Cl
e. Ar

Question 2.12

For each of the following elements, provide the *total* number of electrons and *valence* electrons in its atom:

a. K
b. F
c. P
d. O
e. Ca

The Quantum Mechanical Atom

As we noted in Section 2.3, the success of Bohr's theory was short-lived. Emission spectra of multi-electron atoms (recall that the hydrogen atom has only one electron) could not be explained by Bohr's theory. Evidence that electrons have wave properties served to intensify the problem. Bohr stated that electrons in atoms had very specific locations, now termed *principal energy levels.* The very nature of waves, spread out in space, defies such an exact model of electrons in atoms. Furthermore, the exact model is contradictory to theory and subsequent experiments.

The basic concept of the Bohr theory, that the energy of an electron in an atom is quantized, was refined and expanded by an Austrian physicist, Erwin Schröedinger. He described electrons in atoms in probability terms, developing equations that emphasize the wavelike character of electrons. Although Schröedinger's approach was founded on complex mathematics, we can readily use models of electron probability regions to enable us to gain a reasonable insight into atomic structure without the need to understand the underlying mathematics.

Schröedinger's theory, often described as quantum mechanics, incorporates Bohr's principal energy levels ($n = 1, 2$, and so forth); however, it is proposed that each of these levels is made up of one or more sublevels. Each sublevel, in turn, contains one or more atomic orbitals. In the following section we shall look at each of these regions in more detail and learn how to predict the way that electrons are arranged in stable atoms.

Energy Levels and Sublevels

Principal Energy Levels

The principal energy levels are designated $n = 1, 2, 3$, and so forth. The number of possible sublevels in a principal energy level is also equal to n. When $n = 1$, there can be only one sublevel; $n = 2$ allows two sublevels, and so forth.

The total electron capacity of a principal level is $2(n)^2$. For example:

$n = 1$ $2(1)^2$ Capacity $= 2e^-$

$n = 2$ $2(2)^2$ Capacity $= 8e^-$

$n = 3$ $2(3)^2$ Capacity $= 18e^-$

Sublevels

A **sublevel** is a set of equal-energy orbitals within a principal energy level. The sublevels, or subshells, are symbolized as s, p, d, f, and so forth; they increase in energy in the following order:

$$s < p < d < f$$

We specify both the principal energy level and type of sublevel when describing the location of an electron—for example, 1s, 2s, 2p. Energy level designations for the first four principal energy levels follow:

- The first principal energy level ($n = 1$) has one possible sublevel: 1s.
- The second principal energy level ($n = 2$) has two possible sublevels: 2s and 2p.
- The third principal energy level ($n = 3$) has three possible sublevels: 3s, 3p, and 3d.
- The fourth principal energy level ($n = 4$) has four possible sublevels: 4s, 4p, 4d, and 4f.

Orbitals

An **atomic orbital** is a specific region of a sublevel containing a maximum of two electrons.

Figure 2.12 depicts a model of an s orbital. It is spherically symmetrical, much like a Ping-Pong ball. Its volume represents a region where there is a high probability of finding electrons of similar energy. This probability decreases as we approach the outer region of the atom. The nucleus is at the center of the s orbital. At that point the probability of finding the electron is zero; electrons cannot reside in the nucleus. Only one s orbital can be found in any n level. Atoms with many electrons, occupying a number of n levels, have an s orbital in each n level. Consequently 1s, 2s, 3s, and so forth are possible orbitals.

Figure 2.13 describes the shapes of the three possible p orbitals within a given level. Each has the same shape, and that shape appears much like a dumbbell; these three orbitals differ only in the direction they extend into space. Imaginary coordinates x, y, and z are superimposed on these models to emphasize this fact. These three orbitals, termed p_x, p_y, and p_z, may coexist in a single atom.

Figure 2.12
Representation of s orbitals.

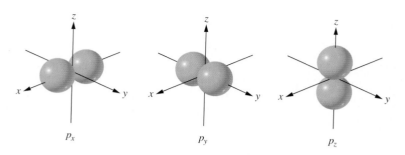

Figure 2.13
Representation of the three p orbitals, p_x, p_y, and p_z.

In a similar fashion, five possible d orbitals and seven possible f orbitals exist. The d orbitals exist only in $n = 3$ and higher principal energy levels; f orbitals exist only in $n = 4$ and higher principal energy levels. Because of their complexity, we will not consider the shapes of d and f orbitals.

Electrons in Sublevels

We can deduce the maximum electron capacity of each sublevel based on the information just given.

For the s sublevel:

$$1 \text{ orbital} \times \frac{2e^- \text{ capacity}}{\text{orbital}} = 2e^- \text{ capacity}$$

For the p sublevel:

$$3 \text{ orbitals} \times \frac{2e^- \text{ capacity}}{\text{orbital}} = 6e^- \text{ capacity}$$

For the d sublevel:

$$5 \text{ orbitals} \times \frac{2e^- \text{ capacity}}{\text{orbital}} = 10e^- \text{ capacity}$$

For the f sublevel:

$$7 \text{ orbitals} \times \frac{2e^- \text{ capacity}}{\text{orbital}} = 14e^- \text{ capacity}$$

Electron Spin

Section 2.2 discusses the properties of electrons demonstrated by Thomson.

As we have noted, each atomic orbital has a maximum capacity of two electrons. The electrons are perceived to *spin* on an imaginary axis, and the two electrons in the same orbital must have opposite spins: clockwise and counterclockwise. Their behavior is analogous to two ends of a magnet. Remember, electrons have magnetic properties. The electrons exhibit sufficient magnetic attraction to hold themselves together despite the natural repulsion that they "feel" for each other, owing to their similar charge (remember, like charges repel). Electrons must therefore have opposite spins to coexist in an orbital. A pair of electrons in one orbital that possess opposite spins are referred to as *paired* electrons.

Electron Configuration and the Aufbau Principle

LEARNING GOAL 10

The arrangement of electrons in atomic orbitals is referred to as the atom's electron configuration. The *aufbau*, or building up, *principle* helps us to represent the electron configuration of atoms of various elements. According to this principle, electrons fill the lowest-energy orbital that is available first. We should also recall that the maximum capacity of an s level is two, that of a p level is six, that of a d level is ten, and that of an f level is fourteen electrons. Consider the following guidelines for writing electron configurations:

2.5 Electron Arrangement and the Periodic Table

Guidelines for Writing Electron Configurations
- Obtain the total number of electrons in the atoms from the atomic number found on the periodic table.
- Electrons in atoms occupy the lowest energy orbitals that are available, beginning with 1s.
- Each principal energy level, n, can contain only n subshells.
- Each sublevel is composed of one (s) or more (three p, five d, seven f) orbitals.
- No more than two electrons can be placed in any orbital.
- The maximum number of electrons in any principal energy level is $2(n)^2$.
- The theoretical order of orbital filling is depicted in Figure 2.14.

Now let us look at several elements:

Figure 2.14
A useful way to remember the filling order for electrons in atoms. Begin adding electrons at the bottom (lowest energy) and follow the arrows. Remember: no more than two electrons in each orbital.

Hydrogen
Hydrogen is the simplest atom; it has only one electron. That electron must be in the lowest principal energy level ($n = 1$) and the lowest orbital (s). We indicate the number of electrons in a region with a *superscript*, so we write $1s^1$.

Helium
Helium has two electrons, which will fill the lowest energy level. The ground state (lowest energy) electron configuration for helium is $1s^2$.

Lithium
Lithium has three electrons. The first two are configured as helium. The third must go into the orbital of the lowest energy in the second principal energy level; therefore the configuration is $1s^2\,2s^1$.

Beryllium Through Neon
The second principal energy level can contain eight electrons [$2(2)^2$], two in the s level and six in the p level. The "building up" process results in

Be	$1s^2\,2s^2$
B	$1s^2\,2s^2\,2p^1$
C	$1s^2\,2s^2\,2p^2$
N	$1s^2\,2s^2\,2p^3$
O	$1s^2\,2s^2\,2p^4$
F	$1s^2\,2s^2\,2p^5$
Ne	$1s^2\,2s^2\,2p^6$

Sodium Through Argon
Electrons in these elements retain the basic $1s^2\,2s^2\,2p^6$ arrangement of the preceding element, neon; new electrons enter the third principal energy level:

Na	$1s^2\,2s^2\,2p^6\,3s^1$
Mg	$1s^2\,2s^2\,2p^6\,3s^2$
Al	$1s^2\,2s^2\,2p^6\,3s^2\,3p^1$
Si	$1s^2\,2s^2\,2p^6\,3s^2\,3p^2$
P	$1s^2\,2s^2\,2p^6\,3s^2\,3p^3$
S	$1s^2\,2s^2\,2p^6\,3s^2\,3p^4$
Cl	$1s^2\,2s^2\,2p^6\,3s^2\,3p^5$
Ar	$1s^2\,2s^2\,2p^6\,3s^2\,3p^6$

By knowing the order of filling of atomic orbitals, lowest to highest energy, you may write the electron configuration for any element. The order of orbital filling can be represented by the diagram in Figure 2.14. Such a diagram provides an easy way of predicting the electron configuration of the elements. Remember that the diagram is based on an energy scale, with the lowest energy orbital at the beginning of the "path" and the highest energy orbital at the end of the "path." An alternative way of representing orbital energies is through the use of an energy level diagram, such as the one in Figure 2.15.

EXAMPLE 2.5 Writing the Electron Configuration of Tin

Tin, Sn, has an atomic number of 50; thus we must place fifty electrons in atomic orbitals. We must also remember the total electron capacities of orbital types: s, 2; p, 6; d, 10; and f, 14. The first principal energy level has one sublevel, the second has two sublevels, and so on. The order of filling (Figure 2.14) is 1s, 2s, 2p, 3s, 3p, 4s, 3d, 4p. The electron configuration is as follows:

$$1s^2\ 2s^2\ 2p^6\ 3s^2\ 3p^6\ 4s^2\ 3d^{10}\ 4p^6\ 5s^2\ 4d^{10}\ 5p^2$$

As a check, count electrons in the electron configuration (add all of the superscripted numbers) to see that we have accounted for all fifty electrons of the Sn atom.

Question 2.13

Give the electron configuration for an atom of:

a. sulfur
b. calcium

Question 2.14

Give the electron configuration for an atom of:

a. potassium
b. phosphorus

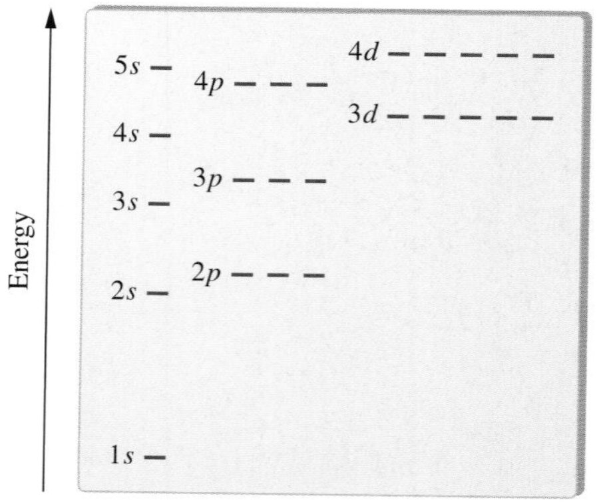

Figure 2.15
An orbital energy-level diagram. Electrons fill orbitals in the order of increasing energy.

Shorthand Electron Configurations

As we noted earlier the electron configuration for the sodium atom (Na, atomic number 11) is

$$1s^2\ 2s^2\ 2p^6\ 3s^1$$

The electron configuration for the preceding noble gas, neon (Ne, atomic number 10), is

$$1s^2\ 2s^2\ 2p^6$$

The electron configuration for sodium is really the electron configuration of Ne, with $3s^1$ added to represent one additional electron. So it is permissible to write

[Ne] $3s^1$ as equivalent to $1s^2\ 2s^2\ 2p^6\ 3s^1$

[Ne] $3s^1$ is the *shorthand electron configuration* for sodium.

Similarly,

$$[Ne]\ 3s^2 \text{ representing Mg}$$

$$[Ne]\ 3s^2\ 3p^5 \text{ representing Cl}$$

$$[Ar]\ 4s^1 \text{ representing K}$$

are valid electron configurations.

The use of abbreviated electron configurations, in addition to being faster and easier to write, serves to highlight the valence electrons, those electrons involved in bonding. The symbol of the noble gas represents the *core*, nonvalence electrons and the valence electron configuration follows the noble gas symbol.

Question 2.15

Give the *shorthand* electron configuration for each atom in Question 2.13.

Question 2.16

Give the *shorthand* electron configuration for each atom in Question 2.14.

2.6 The Octet Rule

Elements in the last family, the noble gases, have either two valence electrons (helium) or eight valence electrons (neon, argon, krypton, xenon, and radon). These elements are extremely stable and were often termed *inert gases* because they do not readily bond to other elements, although they can be made to do so under extreme experimental conditions. A full $n = 1$ energy level (as in helium) or an outer *octet* of electrons (eight valence electrons, as in all of the other noble gases) is responsible for this unique stability.

Atoms of elements in other groups are more reactive than the noble gases because in the process of chemical reaction they are trying to achieve a more stable "noble gas" configuration by gaining or losing electrons. This is the basis of the **octet rule** which states that elements usually react in such a way as to attain the electron configuration of the noble gas closest to them in the periodic table (a stable octet of electrons). In chemical reactions they will gain, lose, or share the minimum number of electrons necessary to attain this more stable energy state. The octet rule, although simple in concept, is a remarkably reliable predictor of chemical change, especially for representative elements.

> We may think of stability as a type of contentment; a noble gas atom does not need to rearrange its electrons or lose or gain any electrons to get to a more stable, lower energy, or more "contented" configuration.

Ion Formation and the Octet Rule

Metals and nonmetals differ in the way in which they form ions. Metallic elements (located at the left of the periodic table) tend to form positively charged ions called *cations*. Positive ions are formed when an atom loses one or more electrons, for example,

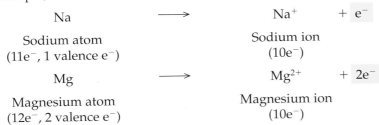

Na $\longrightarrow$ Na$^+$ + e$^-$

Sodium atom (11e$^-$, 1 valence e$^-$) → Sodium ion (10e$^-$)

Mg $\longrightarrow$ Mg^{2+} + 2e$^-$

Magnesium atom (12e$^-$, 2 valence e$^-$) → Magnesium ion (10e$^-$)

11 LEARNING GOAL

Chapter 2 The Structure of the Atom and the Periodic Table

$$\text{Al} \longrightarrow \text{Al}^{3+} + 3e^-$$

Aluminum atom   Aluminum ion
($13e^-$, 3 valence e^-)   ($10e^-$)

In each of these cases the atom has lost *all* of its valence electrons. The resulting ion has the same number of electrons as the nearest noble gas atom:

Na^+ ($10e^-$) and Mg^{2+} ($10e^-$) and Al^{3+} ($10e^-$) are all isoelectronic with Ne ($10e^-$).

> Recall that the prefix *iso* (Greek *isos*) means equal.

These ions are particularly stable. Each ion is **isoelectronic** (that is, it has the same number of electrons) with its nearest noble gas neighbor and has an octet of electrons in its outermost energy level.

Sodium is typical of each element in its group. Knowing that sodium forms a 1+ ion leads to the prediction that H, Li, K, Rb, Cs, and Fr also will form 1+ ions. Furthermore, magnesium, which forms a 2+ ion, is typical of each element in its group; Be^{2+}, Ca^{2+}, Sr^{2+}, and so forth are the resulting ions.

Nonmetallic elements, located at the right of the periodic table, tend to gain electrons to become isoelectronic with the nearest noble gas element, forming negative ions called *anions*.

> *Section 3.2 discusses the naming of ions.*

Consider:

$$\text{F} + 1e^- \longrightarrow \text{F}^- \quad \text{(isoelectronic with Ne, } 10e^-\text{)}$$

Fluorine atom   Fluoride ion
($9e^-$, 7 valence e^-)   ($10e^-$)

> The ion of fluorine is the *fluoride ion*; the ion of oxygen is the *oxide ion*; and the ion of nitrogen is the *nitride ion*.

$$\text{O} + 2e^- \longrightarrow \text{O}^{2-} \quad \text{(isoelectronic with Ne, } 10e^-\text{)}$$

Oxygen atom   Oxide ion
($8e^-$, 6 valence e^-)   ($10e^-$)

$$\text{N} + 3e^- \longrightarrow \text{N}^{3-} \quad \text{(isoelectronic with Ne, } 10e^-\text{)}$$

Nitrogen atom   Nitride ion
($7e^-$, 5 valence e^-)   ($10e^-$)

As in the case of positive ion formation, each of these negative ions has an octet of electrons in its outermost energy level.

The element fluorine, forming F^-, indicates that the other halogens, Cl, Br, and I, behave as a true family and form Cl^-, Br^-, and I^- ions. Also, oxygen and the other nonmetals in its group form 2- ions; nitrogen and phosphorus form 3- ions. It is important to recognize that ions are formed by gain or loss of *electrons*. No change occurs in the nucleus; the number of protons remains the same.

Question 2.17

Give the charge of the most probable ion resulting from each of the following elements. With what element is the ion isoelectronic?

a. Ca
b. Sr
c. S
d. Mg
e. P

Question 2.18

Which of the following pairs of atoms and ions are isoelectronic?

a. Cl^-, Ar
b. Na^+, Ne
c. Mg^{2+}, Na^+
d. Li^+, Ne
e. O^{2-}, F^-

A Medical Perspective

Dietary Calcium

"Drink your milk!" "Eat all of your vegetables!" These imperatives are almost universal memories from our childhood. Our parents knew that calcium, present in abundance in these foods, was an essential element for the development of strong bones and healthy teeth.

Recent studies, spanning the fields of biology, chemistry, and nutrition science indicate that the benefits of calcium go far beyond bones and teeth. This element has been found to play a role in the prevention of disease throughout our bodies.

Calcium is the most abundant mineral (metal) in the body. It is ingested as the calcium ion (Ca^{2+}) either in its "free" state or "combined," as a part of a larger compound; calcium dietary supplements often contain ions in the form of calcium carbonate. The acid naturally present in the stomach produces the calcium ion:

$$CaCO_3 + 2H^+ \longrightarrow Ca^{2+} + H_2O + CO_2$$
calcium carbonate + stomach acid → calcium ion + water + carbon dioxide

Vitamin D serves as the body's regulator of calcium ion uptake, release, and transport in the body. (You can find further information online at www.mhhe.com/denniston5e, in "Lipid-Soluble Vitamins.")

Calcium is responsible for a variety of body functions including:

- transmission of nerve impulses
- release of "messenger compounds" that enable communication among nerves
- blood clotting
- hormone secretion
- growth of living cells throughout the body

The body's storehouse of calcium is bone tissue. When the supply of calcium from external sources, the diet, is insufficient, the body uses a mechanism to compensate for this shortage. With vitamin D in a critical role, this mechanism removes calcium from bone to enable other functions to continue to take place. It is evident then that prolonged dietary calcium deficiency can weaken the bone structure. Unfortunately, current studies show that as many as 75% of the American population may not be consuming sufficient amounts of calcium. Developing an understanding of the role of calcium in premenstrual syndrome, cancer, and blood pressure regulation is the goal of three current research areas.

Calcium and premenstrual syndrome (PMS). Dr. Susan Thys-Jacobs, a gynecologist at St. Luke's-Roosevelt Hospital Center in New York City, and colleagues at eleven other medical centers are conducting a study of calcium's ability to relieve the discomfort of PMS. They believe that women with chronic PMS have calcium blood levels that are normal only because calcium is continually being removed from the bone to maintain an adequate supply in the blood. To complicate the situation, vitamin D levels in many young women are very low (as much as 80% of a person's vitamin D is made in the skin, upon exposure to sunlight; many of us now minimize our exposure to the sun because of concerns about ultraviolet radiation and skin cancer). Because vitamin D plays an essential role in calcium metabolism, even if sufficient calcium is consumed, it may not be used efficiently in the body.

Colon cancer. The colon is lined with a type of cell (epithelial cell) that is similar to those that form the outer layers of skin. Various studies have indicated that by-products of a high-fat diet are irritants to these epithelial cells and produce abnormal cell growth in the colon. Dr. Martin Lipkin, Rockefeller University in New York, and his colleagues have shown that calcium ions may bind with these irritants, reducing their undesirable effects. It is believed that a calcium-rich diet, low in fat, and perhaps use of a calcium supplement can prevent or reverse this abnormal colon cell growth, delaying or preventing the onset of colon cancer.

Blood pressure regulation. Dr. David McCarron, a blood pressure specialist at the Oregon Health Sciences University, believes that dietary calcium levels may have a significant influence on hypertension (high blood pressure). Preliminary studies show that a diet rich in low-fat dairy products, fruits, and vegetables, all high in calcium, may produce a significant lowering of blood pressure in adults with mild hypertension.

The take-home lesson appears clear: a high calcium, low fat diet promotes good health in many ways. Once again, our parents were right!

For Further Understanding

Distinguish between "free" and "combined" calcium in the diet.

Why might calcium supplements be ineffective in treating all cases of osteoporosis?

The transition metals tend to form positive ions by losing electrons, just like the representative metals. Metals, whether representative or transition, share this characteristic. However, the transition elements are characterized as "variable valence" elements; depending on the type of substance with which they react, they may form more than one stable ion. For example, iron has two stable ionic forms:

$$Fe^{2+} \text{ and } Fe^{3+}$$

68   Chapter 2 The Structure of the Atom and the Periodic Table

Copper can exist as

$$Cu^+ \text{ and } Cu^{2+}$$

and elements such as vanadium, V, and manganese, Mn, each can form four different stable ions.

Predicting the charge of an ion or the various possible ions for a given transition metal is not an easy task. Energy differences between valence electrons of transition metals are small and not easily predicted from the position of the element in the periodic table. In fact, in contrast to representative metals, the transition metals show great similarities within a *period* as well as within a *group*.

2.7 Trends in the Periodic Table

Atomic Size

LEARNING GOAL

Many atomic properties correlate with electronic structure, hence, with their position in the periodic table. Given the fact that interactions among multiple charged particles are very complex, we would not expect the correlation to be perfect. Nonetheless, the periodic table remains an excellent guide to the prediction of properties.

If our model of the atom is a tiny sphere whose radius is determined by the distance between the center of the nucleus and the boundary of the region where the valence electrons have a probability of being located, the size of the atom will be determined principally by two factors.

The radius of an atom is traditionally defined as one-half of the distance between atoms in a covalent bond. The covalent bond is discussed in Section 3.1.

1. The energy level (n level) in which the outermost electron(s) is (are) found increases as we go *down* a group. (Recall that the outermost n level correlates with period number.) Thus the size of atoms should increase from top to bottom of the periodic table as we fill successive energy levels of the atoms with electrons (Figure 2.16).
2. As the magnitude of the positive charge of the nucleus increases, its "pull" on all of the electrons increases, and the electrons are drawn closer to the nucleus. This results in a contraction of the atomic radius and therefore a decrease in atomic size. This effect is apparent as we go *across* the periodic table within a period. Atomic size decreases from left to right in the periodic table. See how many exceptions you can find in Figure 2.16.

Ion Size

Positive ions (cations) are smaller than the parent atom. The cation has more protons than electrons (an increased nuclear charge). The excess nuclear charge pulls

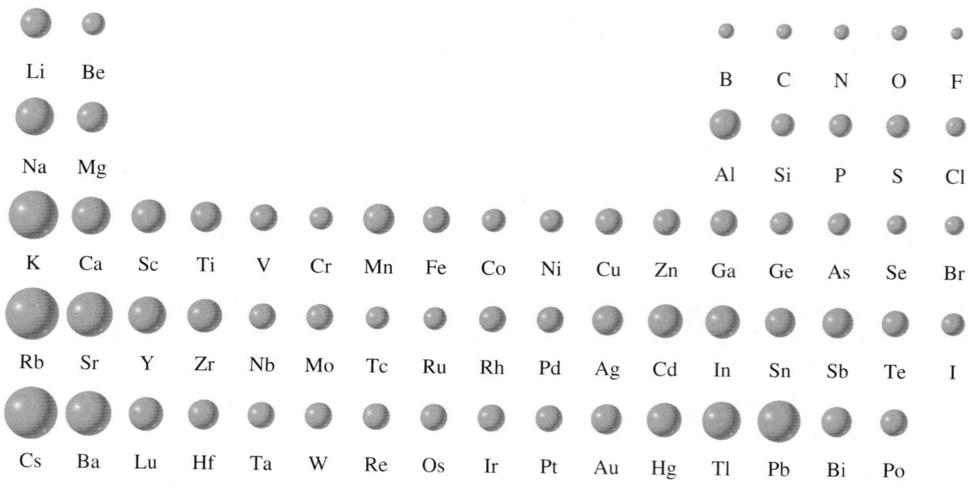

Figure 2.16
Variation in the size of atoms as a function of their position in the periodic table. Note particularly the decrease in size from left to right in the periodic table and the increase in size as we proceed down the table, although some exceptions do exist. (Lanthanide and actinide elements are not included here.)

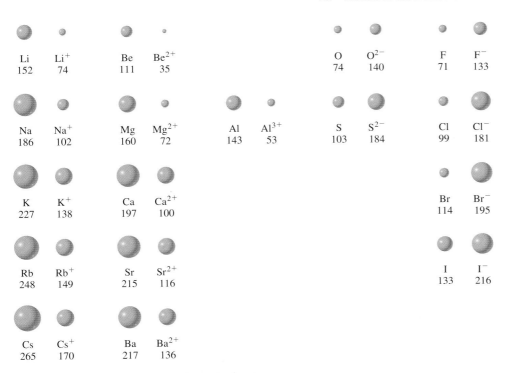

Figure 2.17
Relative size of ions and their parent atoms. Atomic radii are provided in units of picometers.

the remaining electrons closer to the nucleus. Also, cation formation often results in the loss of all outer-shell electrons, resulting in a significant decrease in radius.

Negative ions (anions) are larger than the parent atom. The anion has more electrons than protons. Owing to the excess negative charge, the nuclear "pull" on each individual electron is reduced. The electrons are held less tightly, resulting in a larger anion radius in contrast to the neutral atom.

Ions with multiple positive charge (such as Cu^{2+}) are even *smaller* than their corresponding monopositive ion (Cu^+); ions with multiple negative charge (such as O^{2-}) are *larger* than their corresponding less negative ion.

Figure 2.17 depicts the relative sizes of several atoms and their corresponding ions.

Ionization Energy

The energy required to remove an electron from an isolated atom is the **ionization energy**. The process for sodium is represented as follows:

$$\text{ionization energy} + Na \longrightarrow Na^+ + e^-$$

The magnitude of the ionization energy should correlate with the strength of the attractive force between the nucleus and the outermost electron.

- Reading *down* a group, note that the ionization energy decreases, because the atom's size is increasing. The outermost electron is progressively farther from the nuclear charge, hence easier to remove.
- Reading *across* a period, note that atomic size decreases, because the outermost electrons are closer to the nucleus, more tightly held, and more difficult to remove. Therefore the ionization energy generally increases.

A correlation does indeed exist between trends in atomic size and ionization energy. Atomic size generally *decreases* from the bottom to top of a group and from left to right in a period. Ionization energies generally *increase* in the same periodic way. Note also that ionization energies are highest for the noble gases (Figure 2.18a). A high value for ionization energy means that it is difficult to remove electrons from the atom, and this, in part, accounts for the extreme stability and nonreactivity of the noble gases.

Remember: ionization energy and electron affinity (below) are predictable from *trends* in the periodic table. As with most trends, exceptions occur.

Chapter 2 The Structure of the Atom and the Periodic Table

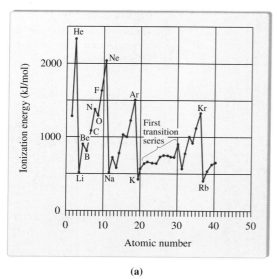

(a)

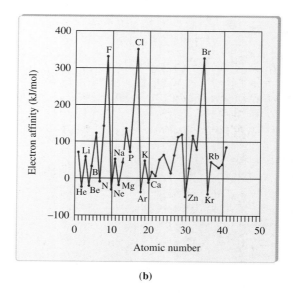

(b)

Figure 2.18
(a) The ionization energies of the first forty elements versus their atomic numbers. Note the very high values for elements located on the right in the periodic table, and low values for those on the left. Some exceptions to the trends are evident. (b) The periodic variation of electron affinity. Note the very low values for the noble gases and the elements on the far left of the periodic table. These elements do not form negative ions. In contrast, F, Cl, and Br readily form negative ions.

Electron Affinity

The energy released when a single electron is added to an isolated atom is the **electron affinity.** If we consider ionization energy in relation to positive ion formation (remember that the magnitude of the ionization energy tells us the ease of *removal* of an electron, hence the ease of forming positive ions), then electron affinity provides a measure of the ease of forming negative ions. A large electron affinity (energy released) indicates that the atom becomes more stable as it becomes a negative ion (through gaining an electron). Consider the gain of an electron by a bromine atom:

$$Br + e^- \longrightarrow Br^- + energy$$

Electron affinity

Periodic trends for electron affinity are as follows:

- Electron affinities generally decrease down a group.
- Electron affinities generally increase across a period.

Remember these trends are not absolute. Exceptions exist, as seen in the irregularities in Figure 2.18b.

Question 2.19

Rank Be, N, and F in order of increasing

a. atomic size
b. ionization energy
c. electron affinity

Question 2.20

Rank Cl, Br, I, and F in order of increasing

a. atomic size
b. ionization energy
c. electron affinity

SUMMARY

2.1 Composition of the Atom

The basic structural unit of an element is the *atom,* which is the smallest unit of an element that retains the chemical properties of that element.

The atom has two distinct regions. The *nucleus* is a small, dense, positively charged region in the center of the atom composed of positively charged *protons* and uncharged *neutrons*. Surrounding the nucleus is a diffuse region of negative charge occupied by *electrons,* the source of the negative charge. Electrons are very low in mass in comparison to protons and neutrons.

The *atomic number* (Z) is equal to the number of protons in the atom. The *mass number* (A) is equal to the sum of the protons and neutrons (the mass of the electrons is insignificant).

Isotopes are atoms of the same element that have different masses because they have different numbers of neutrons (different mass numbers). Isotopes have chemical behavior identical to that of any other isotope of the same element.

Ions are electrically charged particles that result from a gain or loss of one or more electrons by the parent atom. *Anions,* negative ions, are formed by a gain of one or more electrons by the parent atom. *Cations,* positive ions, are formed by a loss of one or more electrons from the parent atom.

2.2 Development of Atomic Theory

The first experimentally based theory of atomic structure was proposed by John Dalton. Although Dalton pictured atoms as indivisible, the experiments of William Crookes, Eugene Goldstein, and J. J. Thomson indicated that the atom is composed of charged particles: protons and electrons. The third fundamental atomic particle is the neutron. An experiment conducted by Hans Geiger led Ernest Rutherford to propose that the majority of the mass and positive charge of the atom is located in a small, dense region, the *nucleus,* with small, negatively charged electrons occupying a much larger, diffuse space outside of the nucleus.

2.3 Light, Atomic Structure, and the Bohr Atom

The study of the interaction of *light* and *matter* is termed *spectroscopy*. Light, *electromagnetic radiation,* travels at a speed of 3.0×10^8 m/s, the *speed of light*. Light is made up of many wavelengths. Collectively, they make up the *electromagnetic spectrum*. Samples of elements emit certain wavelengths of light when an electrical current is passed through the sample. Different elements emit a different pattern (different wavelengths) of light.

Niels Bohr proposed an atomic model that described the atom as a nucleus surrounded by fixed *energy levels* (or *quantum levels*) that can be occupied by electrons. He believed that each level was defined by a spherical *orbit* located at a specific distance from the nucleus.

Promotion and relaxation processes are referred to as *electronic transitions*. Electron promotion resulting from absorption of energy results in an excited state atom; the process of relaxation allows the atom to return to the ground state by emitting a certain wavelength of light.

The modern view of the atom describes the probability of finding an electron in a region of space within the principal energy level, referred to as an *atomic orbital*. The rapid movement of the electrons spreads them into a cloud of charge. This cloud is more dense in certain regions, the density being proportional to the probability of finding the electron at any point in time. The orbital is strikingly different from Bohr's orbit. The electron does not orbit the nucleus; rather, its behavior is best described as that of a wave.

2.4 The Periodic Law and the Periodic Table

The *periodic law* is an organized "map" of the elements that relates their structure to their chemical and physical properties. It states that the elements, when arranged according to their atomic numbers, show a distinct periodicity (regular variation) of their properties. The periodic table is the result of the periodic law.

The modern periodic table exists in several forms. The most important variation is in group numbering. The tables in this text use the two most commonly accepted numbering systems.

A horizontal row of elements in the periodic table is referred to as a *period*. The periodic table consists of seven periods. The lanthanide series is a part of period 6; the actinide series is a part of period 7.

The columns of elements in the periodic table are called *groups* or *families*. The elements of a particular family share many similarities in physical and chemical properties because of the similarities in electronic structure. Some of the most important groups are named; for example, the *alkali metals* (IA or 1), *alkaline earth metals* (IIA or 2), the *halogens* (VIIA or 17), and the *noble gases* (VIII or 18).

Group A elements are called *representative elements;* Group B elements are *transition elements*. A bold zigzag line runs from top to bottom of the table, beginning to the left of boron (B) and ending between polonium (Po) and astatine (At). This line acts as the boundary between *metals* to the left and *nonmetals* to the right. Elements straddling the boundary, *metalloids,* have properties intermediate between those of metals and nonmetals.

2.5 Electron Arrangement and the Periodic Table

The outermost electrons in an atom are *valence electrons*. For representative elements the number of valence electrons in an atom corresponds to the group or family number (old numbering system using Roman numerals). Metals tend to have fewer valence electrons than nonmetals.

Electron configuration of the elements is predictable, using the aufbau principle. Knowing the electron configuration, we can identify valence electrons and begin to predict the kinds of reactions that the elements will undergo.

Elements in the last family, the noble gases, have either two valence electrons (helium) or either valence electrons (neon, argon, krypton, xenon, and radon). Their most important properties are their extreme stability and lack of reactivity. A full valence level is responsible for this unique stability.

2.6 The Octet Rule

The *octet rule* tells us that in chemical reactions, elements will gain, lose, or share the minimum number of electrons necessary to achieve the electron configuration of the nearest noble gas.

Metallic elements tend to form cations. The ion is *isoelectronic* with its nearest noble gas neighbor and has a stable octet of electrons in its outermost energy level. Nonmetallic elements tend to gain electrons to become isoelectronic with the nearest noble gas element, forming anions.

2.7 Trends in the Periodic Table

Atomic size decreases from left to right and from bottom to top in the periodic table. *Cations* are smaller than the parent atom. *Anions* are larger than the parent atom. Ions with multiple positive charge are even smaller than their corresponding monopositive ion; ions with multiple negative charge are larger than their corresponding less negative ion.

The energy required to remove an electron from the atom is the *ionization energy*. Down a group, the ionization energy generally decreases. Across a period, the ionization energy generally increases.

The energy released when a single electron is added to a neutral atom in the gaseous state is known as the *electron affinity*. Electron affinities generally decrease proceeding down a group and increase proceeding across a period.

KEY TERMS

alkali metal (2.4)
alkaline earth metal (2.4)
anion (2.1)
atom (2.1)
atomic mass (2.1)
atomic number (2.1)
atomic orbital (2.3 and 2.5)
cathode rays (2.2)
cation (2.1)
electromagnetic radiation (2.3)
electromagnetic spectrum (2.3)
electron (2.1)
electron affinity (2.7)

electron configuration (2.5)
electron density (2.3)
energy level (2.3)
group (2.4)
halogen (2.4)
ion (2.1)
ionization energy (2.7)
isoelectronic (2.6)
isotope (2.1)
mass number (2.1)
metal (2.4)
metalloid (2.4)
neutron (2.1)

noble gas (2.4)
nonmetal (2.4)
nucleus (2.1)
octet rule (2.6)
period (2.4)
periodic law (2.4)
proton (2.1)

quantization (2.3)
representative element (2.4)
spectroscopy (2.3)
speed of light (2.3)
sublevel (2.5)
transition element (2.4)
valence electron (2.5)

QUESTIONS AND PROBLEMS

Composition of the Atom

Foundations

2.21 Calculate the number of protons, neutrons, and electrons in:
  a. $^{16}_{8}O$
  b. $^{31}_{15}P$

2.22 Calculate the number of protons, neutrons, and electrons in:
  a. $^{136}_{56}Ba$
  b. $^{209}_{84}Po$

2.23 State the mass and charge of the:
  a. electron
  b. proton
  c. neutron

2.24 Calculate the number of protons, neutrons, and electrons in:
  a. $^{37}_{17}Cl$ 17 P - 17 E - 20 N
  b. $^{23}_{11}Na$ 11 P - 11 E - 12 N
  c. $^{84}_{36}Kr$ 36 P - 36 E - 48 N

2.25 a. What is an ion? — charged atom or group of atoms
  b. What process results in the formation of a cation? — loss of electron
  c. What process results in the formation of an anion? — gain electron

2.26 a. What are isotopes?
  b. What is the major difference among isotopes of an element?
  c. What is the major similarity among isotopes of an element?

Applications

2.27 How many protons are in the nucleus of the isotope Rn-220?
2.28 How many neutrons are in the nucleus of the isotope Rn-220?
2.29 Selenium-80 is a naturally occurring isotope. It is found in over-the-counter supplements.
  a. How many protons are found in one atom of selenium-80? — 34
  b. How many neutrons are found in one atom of selenium-80? — 46
2.30 Iodine-131 is an isotope used in thyroid therapy.
  a. How many protons are found in one atom of iodine-131?
  b. How many neutrons are found in one atom of iodine-131?
2.31 Write symbols for each isotope:
  a. Each atom contains 1 proton and 0 neutrons.
  b. Each atom contains 6 protons and 8 neutrons.
2.32 Write symbols for each isotope:
  a. Each atom contains 1 proton and 2 neutrons. $^{3}_{1}X = ^{3}_{1}H$
  b. Each atom contains 92 protons and 146 neutrons. $^{238}_{92}U$
2.33 Fill in the blanks:

Symbol	No. of Protons	No. of Neutrons	No. of Electrons	Charge
Example: $^{40}_{20}Ca$	20	20	20	0
$^{23}_{11}Na$	11	___	11	0
$^{32}_{16}S^{2-}$	16	16	___	2−
	8	8	8	0
$^{24}_{12}Mg^{2+}$	___	12	___	2+
___	19	20	18	___

2.34 Fill in the blanks:

Atomic Symbol	No. of Protons	No. of Neutrons	No. of Electrons	Charge
Example: $^{27}_{13}Al$	13	14	13	0
$^{39}_{19}K$	19		19	0
$^{31}_{15}P^{3-}$	15	16		
	29	34	27	2+
$^{55}_{26}Fe^{2+}$		29		2+
	8	8	10	

2.35 Fill in the blanks:
  a. An isotope of an element differs in mass because the atom has a different number of _Neutrons_.
  b. The atomic number gives the number of _protons_ in the nucleus.
  c. The mass number of an atom is due to the number of _P_ and _N_ in the nucleus.
  d. A charged atom is called a(n) _ion_.
  e. Electrons surround the _nucleus_ and have a _−_ charge.

2.36 Label each of the following statements as true or false:
  a. An atom with an atomic number of 7 and a mass of 14 is identical to an atom with an atomic number of 6 and a mass of 14.
  b. Neutral atoms have the same number of electrons as protons.
  c. The mass of an atom is due to the sum of the number of protons, neutrons, and electrons.

Development of Atomic Theory
Foundations
2.37 What are the major postulates of Dalton's atomic theory?
2.38 What points of Dalton's theory are no longer current?
2.39 Note the major accomplishment of each of the following:
  a. Chadwick
  b. Goldstein
2.40 Note the major accomplishment of each of the following:
  a. Geiger
  b. Bohr
2.41 Note the major accomplishment of each of the following:
  a. Dalton
  b. Crookes
2.42 Note the major accomplishment of each of the following:
  a. Thomson
  b. Rutherford

Applications
2.43 Describe the experiment that provided the basis for our understanding of the nucleus.
2.44 Describe the series of experiments that characterized the electron.
2.45 List at least three properties of the electron.
2.46 Describe the process that occurs when electrical energy is applied to a sample of hydrogen gas. _electrons energized + release light_
2.47 What is a cathode ray? Which subatomic particle is detected?
2.48 Pictured is a cathode ray tube. Show the path that an electron would follow in the tube.

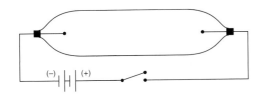

Light, Atomic Structure, and the Bohr Atom
Foundations
2.49 Rank the various regions of the electromagnetic spectrum in order of increasing wavelength.
2.50 Rank the various regions of the electromagnetic spectrum in order of increasing energy.
2.51 Which form of radiation has greater energy, microwave or infrared? _infrared - shorter wavelength_
2.52 Which form of radiation has the longer wavelength, ultraviolet or infrared?
2.53 What is meant by the term *spectroscopy*?
2.54 What is meant by the term *electromagnetic spectrum*?

Applications
2.55 Critique this statement: Electrons can exist in any position outside of the nucleus.
2.56 Critique this statement: Promotion of electrons is accompanied by a release of energy.
2.57 What are the most important points of the Bohr theory?
2.58 Give two reasons why the Bohr theory did not stand the test of time.
2.59 What was the major contribution of Bohr's atomic model?
2.60 What was the major deficiency of Bohr's atomic model?

The Periodic Law and the Periodic Table
Foundations
2.61 Provide the name of the element represented by each of the following symbols:
  a. Na
  b. K
  c. Mg
  d. B
2.62 Provide the name of the element represented by each of the following symbols:
  a. Ca
  b. Cu
  c. Co
  d. Si
2.63 Which group of the periodic table is known as the alkali metals? List them.
2.64 Which group of the periodic table is known as the alkaline earth metals? List them.
2.65 Which group of the periodic table is known as the halogens? List them.
2.66 Which group of the periodic table is known as the noble gases? List them.

Applications
2.67 Label each of the following statements as true or false:
  a. Elements of the same group have similar properties. — _True_
  b. Atomic size decreases from left to right across a period. — _True_
2.68 Label each of the following statements as true or false:
  a. Ionization energy increases from top to bottom within a group.
  b. Representative metals are located on the left in the periodic table.
2.69 For each of the elements Na, Ni, Al, P, Cl, and Ar, provide the following information:
  a. Which are metals? — _Na, Ni, Al_
  b. Which are representative metals? — _Na, Al_
  c. Which tend to form positive ions? — _Na, Ni, Al_
  d. Which are inert or noble gases? — _Ar_

2.70 For each of the elements Ca, K, Cu, Zn, Br, and Kr provide the following information:
  a. Which are metals?
  b. Which are representative metals?
  c. Which tend to form positive ions?
  d. Which are inert or noble gases?

Electron Arrangement and the Periodic Table
Foundations
2.71 How many valence electrons are found in an atom of each of the following elements?
  a. H – 1
  b. Na – 1
  c. B – 3
  d. F – 7
  e. Ne – 8
  f. He – 2

2.72 How many valence electrons are found in an atom of each of the following elements?
  a. Mg
  b. K
  c. C
  d. Br
  e. Ar
  f. Xe

2.73 Distinguish between a principal energy level and a sublevel.
2.74 Distinguish between a sublevel and an orbital.
2.75 Sketch a diagram and describe our current model of an s orbital.
2.76 How is a $2s$ orbital different from a $1s$ orbital?
2.77 How many p orbitals can exist in a given principal energy level?
2.78 Sketch diagrams of a set of p orbitals. How does a p_x orbital differ from a p_y orbital? From a p_z orbital?
2.79 How does a $3p$ orbital differ from a $2p$ orbital?
2.80 What is the maximum number of electrons that an orbital can hold?

Applications
2.81 What is the maximum number of electrons in each of the following energy levels?
  a. $n = 1$ – 2
  b. $n = 2$ – 8
  c. $n = 3$ – 18

2.82 a. What is the maximum number of s electrons that can exist in any one principal energy level? – 2
  b. How many p electrons? – 6
  c. How many d electrons? – 10
  d. How many f electrons? – 14

2.83 In which orbital is the highest-energy electron located in each of the following elements?
  a. Al
  b. Na
  c. Sc
  d. Ca
  e. Fe
  f. Cl

2.84 Using only the periodic table or list of elements, write the electron configuration of each of the following atoms:
  a. B
  b. S
  c. Ar
  d. V
  e. Cd
  f. Te

2.85 Which of the following electron configurations are not possible? Why?
  a. $1s^2 1p^2$ – NO
  b. $1s^2 2s^2 2p^2$ – Yes (carbon)
  c. $2s^2, 2s^2, 2p^6, 2d^1$ – NO
  d. $1s^2, 2s^3$ – NO

2.86 For each incorrect electron configuration in Question 2.85, assume that the number of electrons is correct, identify the element, and write the correct electron configuration.

The Octet Rule
Foundations
2.87 Give the most probable ion formed from each of the following elements:
  a. Li
  b. O
  c. Ca
  d. Br
  e. S
  f. Al

2.88 Using only the periodic table or list of elements, write the electron configuration of each of the following ions:
  a. I^-
  b. Ba^{2+}
  c. Se^{2-}
  d. Al^{3+}

2.89 Which of the following pairs of atoms and/or ions are isoelectronic with one another?
  a. O^{2-}, Ne
  b. S^{2-}, Cl^-

2.90 Which of the following pairs of atoms and/or ions are isoelectronic with one another?
  a. F^-, Cl^-
  b. K^+, Ar

Applications
2.91 Which species in each of the following groups would you expect to find in nature?
  a. Na, (Na$^+$), Na$^-$
  b. (S^{2-}), S$^-$, S$^+$
  c. Cl, (Cl$^-$), Cl$^+$

2.92 Which atom or ion in each of the following groups would you expect to find in nature?
  a. K, K^+, K^-
  b. O^{2-}, O, O^{2+}
  c. Br, Br^-, Br^+

2.93 Write the electron configuration of each of the following biologically important ions:
  a. Ca^{2+}
  b. Mg^{2+}

2.94 Write the electron configuration of each of the following biologically important ions:
  a. K^+
  b. Cl^-

Trends in the Periodic Table
Foundations
2.95 Arrange each of the following lists of elements in order of increasing atomic size:
  a. N, O, F – F, O, N
  b. Li, K, Cs – Li, K, Cs
  c. Cl, Br, I – Cl, Br, I

2.96 Arrange each of the following lists of elements in order of increasing atomic size:
   a. Al, Si, P, Cl, S
   b. In, Ga, Al, B, Tl
   c. Sr, Ca, Ba, Mg, Be
   d. P, N, Sb, Bi, As

2.97 Arrange each of the following lists of elements in order of increasing ionization energy:
   a. N, O, F
   b. Li, K, Cs
   c. Cl, Br, I

2.98 Arrange each of the following lists of elements in order of decreasing electron affinity:
   a. Na, Li, K
   b. Br, F, Cl
   c. S, O, Se

Applications

[handwritten note: loses an electron, more P]

2.99 Explain why a positive ion is always smaller than its parent atom.

2.100 Explain why a negative ion is always larger than its parent atom.

2.101 Explain why a fluoride ion is commonly found in nature but a fluorine atom is not.

2.102 Explain why a sodium ion is commonly found in nature but a sodium atom is not.

CRITICAL THINKING PROBLEMS

1. A natural sample of chromium, taken from the ground, will contain four isotopes: Cr-50, Cr-52, Cr-53, and Cr-54. Predict which isotope is in greatest abundance. Explain your reasoning.

2. Crookes's cathode ray tube experiment inadvertently supplied the basic science for a number of modern high-tech devices. List a few of these devices and describe how they involve one or more aspects of this historic experiment.

3. Name five elements that you came in contact with today. Were they in combined form or did they exist in the form of atoms? Were they present in pure form or in mixtures? If mixtures, were they heterogeneous or homogeneous? Locate each in the periodic table by providing the group and period designation, for example: Group IIA (2), period 3.

4. The periodic table is incomplete. It is possible that new elements will be discovered from experiments using high-energy particle accelerators. Predict as many properties as you can that might characterize the element that would have an atomic number of 118. Can you suggest an appropriate name for this element?

5. The element titanium is now being used as a structural material for bone and socket replacement (shoulders, knees). Predict properties that you would expect for such applications; go to the library or internet and look up the properties of titanium and evaluate your answer.

6. Imagine that you have undertaken a voyage to an alternate universe. Using your chemical skills, you find a collection of elements quite different than those found here on earth. After measuring their properties and assigning symbols for each, you wish to organize them as Mendeleev did for our elements. Design a periodic table using the information you have gathered:

Symbol	Mass (amu)	Reactivity	Electrical Conductivity
A	2.0	High	High
B	4.0	High	High
C	6.0	Moderate	Trace
D	8.0	Low	0
E	10.0	Low	0
F	12.0	High	High
G	14.0	High	High
H	16.0	Moderate	Trace
I	18.0	Low	0
J	20.0	None	0
K	22.0	High	High
L	24.0	High	High

Predict the reactivity and conductivity of an element with a mass of 30.0 amu. What element in our universe does this element most closely resemble?

GENERAL CHEMISTRY

Structure and Properties of Ionic and Covalent Compounds

3

Pattern formation and communication in nature.

Learning Goals

1. Classify compounds as having ionic, covalent, or polar covalent bonds.

2. Write the formula of a compound when provided with the name of the compound.

3. Name common inorganic compounds using standard conventions and recognize the common names of frequently used substances.

4. Predict differences in physical state, melting and boiling points, solid-state structure, and solution chemistry that result from differences in bonding.

5. Draw Lewis structures for covalent compounds and polyatomic ions.

6. Describe the relationship between stability and bond energy.

7. Predict the geometry of molecules and ions using the octet rule and Lewis structures.

8. Understand the role that molecular geometry plays in determining the solubility and melting and boiling points of compounds.

9. Use the principles of VSEPR theory and molecular geometry to predict relative melting points, boiling points, and solubilities of compounds.

Outline

Chemistry Connection: *Magnets and Migration*

3.1 Chemical Bonding

3.2 Naming Compounds and Writing Formulas of Compounds

A Human Perspective: *Origin of the Elements*

3.3 Properties of Ionic and Covalent Compounds

3.4 Drawing Lewis Structures of Molecules and Polyatomic Ions

A Medical Perspective: *Blood Pressure and the Sodium Ion/Potassium Ion Ratio*

3.5 Properties Based on Electronic Structure and Molecular Geometry

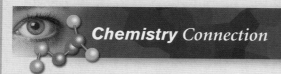

Chemistry Connection

Magnets and Migration

All of us, at one time or another, have wondered at the magnificent sight of thousands of migrating birds, flying in formation, heading south for the winter and returning each spring.

Less visible, but no less impressive, are the schools of fish that travel thousands of miles, returning to the same location year after year. Almost instantly, when faced with some external stimulus such as a predator, they snap into a formation that rivals an army drill team for precision.

The questions of how these life-forms know when and where they are going and how they establish their formations have perplexed scientists for many years. The explanations so far are really just hypotheses.

Some clues to the mystery may be hidden in very tiny particles of magnetite, Fe_3O_4. Magnetite contains iron that is naturally magnetic, and collections of these particles behave like a compass needle; they line up in formation aligned with the earth's magnetic field.

Magnetotactic bacteria contain magnetite in the form of magnetosomes, small particles of Fe_3O_4. Fe_3O_4 is a compound whose atoms are joined by chemical bonds. Electrons in the iron atoms have an electron configuration that results in single electrons (not pairs of electrons) occupying orbitals. These unpaired electrons impart magnetic properties to the compound.

The normal habitat of magnetotactic bacteria is either fresh water or the ocean; the bacteria orient themselves to the earth's magnetic field and swim to the nearest pole (north or south). This causes them to swim into regions of nutrient-rich sediment.

Could the directional device, the simple F_3O_4 unit, also be responsible for direction finding in higher organisms in much the same way that an explorer uses a compass? Perhaps so! Recent studies have shown evidence of magnetosomes in the brains of birds, tuna, green turtles, and dolphins.

Most remarkably, at least one study has shown evidence that magnetite is present in the human brain.

These preliminary studies offer hope of unraveling some of the myth and mystery of guidance and communication in living systems. The answers may involve a very basic compound that is like those we will study in this chapter.

Introduction

A chemical compound is formed when two or more atoms of different elements are joined by attractive forces called chemical bonds. These bonds result from either a transfer of electrons from one atom to another (the ionic bond) or a sharing of electrons between two atoms (the covalent bond). The elements, once converted to a compound, cannot be recovered by any physical process. A chemical reaction must take place to regenerate the individual elements. The chemical and physical properties of a compound are related to the structure of the compound, and this structure is, in turn, determined by the arrangement of electrons in the atoms that produced the compounds. Properties such as solubility, boiling point, and melting point correlate well with the shape and charge distribution in the individual units of the compound.

We need to learn how to properly name and write formulas for ionic and covalent compounds. We should become familiar with some of their properties and be able to relate these properties to the structure and bonding of the compounds.

3.1 Chemical Bonding

When two or more atoms form a chemical compound, the atoms are held together in a characteristic arrangement by attractive forces. The **chemical bond** is the force of attraction between any two atoms in a compound. The attraction is the force that overcomes the repulsion of the positively charged nuclei of the two atoms.

3.1 Chemical Bonding

Interactions involving valence electrons are responsible for the chemical bond. We shall focus our attention on these electrons and the electron arrangement of atoms both before and after bond formation.

Lewis Symbols

The **Lewis symbol,** or Lewis structure, developed by G. N. Lewis early in the twentieth century, is a convenient way of representing atoms singly or in combination. Its principal advantage is that *only* valence electrons (those that may participate in bonding) are shown. Lewis symbolism is based on the octet rule that was described in Chapter 2.

To draw Lewis structures, we first write the chemical symbol of the atom; this symbol represents the nucleus and all of the lower energy nonvalence electrons. The valence electrons are indicated by dots arranged around the atomic symbol. For example:

H·
Hydrogen

He:
Helium

Li·
Lithium

·Be·
Beryllium

·B·
Boron

·C·
Carbon

·N·
Nitrogen

·O·
Oxygen

:F·
Fluorine

:Ne:
Neon

Recall that the number of valence electrons can be determined from the position of the element in the periodic table (see Figure 2.11).

Note particularly that the number of dots corresponds to the number of valence electrons in the outermost shell of the atoms of the element. The four "sides" of the chemical symbol represent an atomic orbital capable of holding one or two valence electrons. Because each atomic orbital can hold no more than two electrons, we can show a maximum of two dots on each side of the element's symbol. Using the same logic employed in writing electron configurations in Chapter 2, we place one dot on each side then sequentially add a second dot, filling each side in turn. This process is limited by the total number of available valence electrons. Each unpaired dot (representing an unpaired electron) is available to form a chemical bond with another element, producing a compound. Figure 3.1 depicts the Lewis dot structures for the representative elements.

Principal Types of Chemical Bonds: Ionic and Covalent

Two principal classes of chemical bonds exist: ionic and covalent. Both involve valence electrons.

Ionic bonding involves a transfer of one or more electrons from one atom to another, leading to the formation of an ionic bond. **Covalent bonding** involves a sharing of electrons resulting in the covalent bond.

Before discussing each type, we should recognize that the distinction between ionic and covalent bonding is not always clear-cut. Some compounds are clearly ionic, and some are clearly covalent, but many others possess both ionic and covalent characteristics.

Ionic Bonding

Representative elements form ions that obey the octet rule. Ions of opposite charge attract each other and this attraction is the essence of the ionic bond. Consider the reaction of a sodium atom and a chlorine atom to produce sodium chloride:

$$Na + Cl \longrightarrow NaCl$$

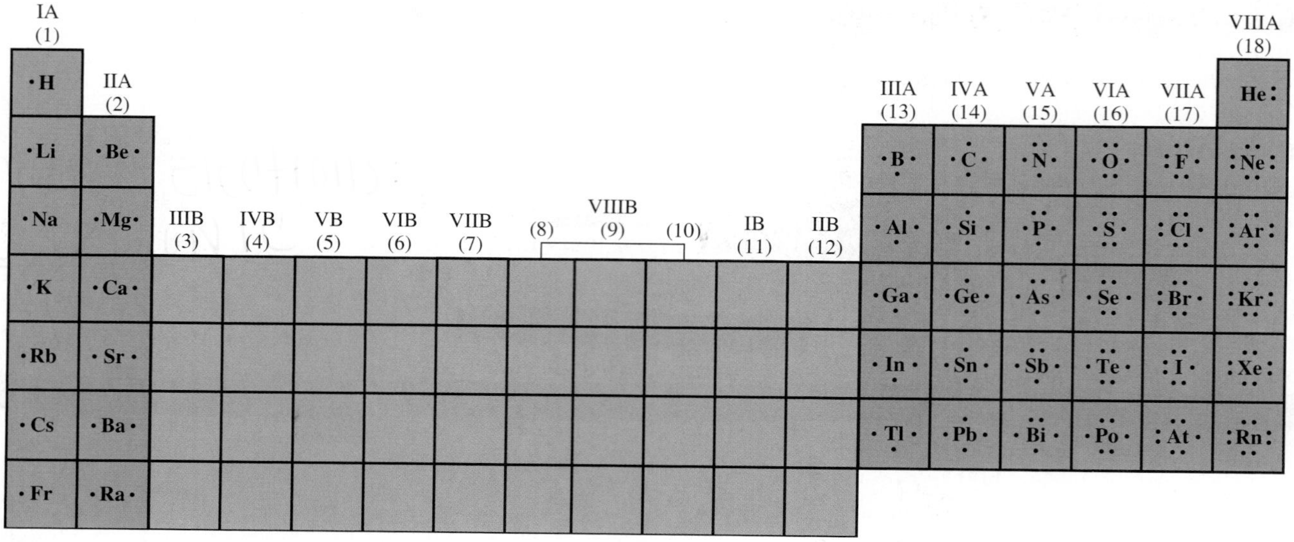

Chapter 3 Structure and Properties of Ionic and Covalent Compounds

Figure 3.1
Lewis dot symbols for representative elements. Each unpaired electron is a potential bond.

Refer to Section 2.7 for a discussion of ionization energy and electron affinity.

Na + Cl → NaCl

Recall that the sodium atom has

- a low ionization energy (it readily loses an electron) and
- a low electron affinity (it does not want more electrons).

If sodium loses its valence electron, it will become isoelectronic (same number of electrons) with neon, a very stable noble gas atom. This tells us that the sodium atom would be a good electron donor, forming the sodium ion:

$$Na \cdot \longrightarrow Na^+ + e^-$$

Recall that the chlorine atom has

- a high ionization energy (it will not easily give up an electron) and
- a high electron affinity (it readily accepts another electron).

Chlorine will gain one more electron. By doing so, it will complete an octet (eight outermost electrons) and be isoelectronic with argon, a stable noble gas. Therefore, chlorine behaves as a willing electron acceptor, forming a chloride ion:

$$:\ddot{Cl}\cdot + e^- \longrightarrow [:\ddot{Cl}:]^-$$

The electron released by sodium (*electron donor*) is the electron received by chlorine (*electron acceptor*):

$$Na \cdot \longrightarrow Na^+ + e^-$$

$$e^- + \cdot \ddot{Cl}: \longrightarrow [:\ddot{Cl}:]^-$$

The resulting ions of opposite charge, Na^+ and Cl^-, are attracted to each other (opposite charges attract) and held together by this *electrostatic force* as an **ion pair**: Na^+Cl^-.

This electrostatic force, the attraction of opposite charges, is quite strong and holds the ions together. It is the ionic bond.

The essential features of ionic bonding are the following:

- Atoms of elements with low ionization energy and low electron affinity tend to form positive ions.
- Atoms of elements with high ionization energy and high electron affinity tend to form negative ions.

3.1 Chemical Bonding   81

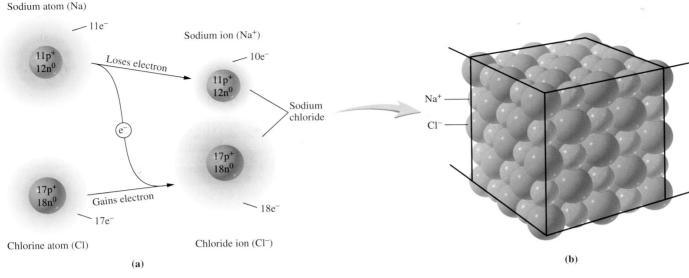

Figure 3.2
The arrangement of ions in a crystal of NaCl (sodium chloride, table salt). (a) A sodium atom loses one electron to become a smaller sodium ion, and a chlorine atom gains that electron, becoming a larger chloride ion. (b) Attraction of Na⁺ and Cl⁻ forms NaCl ion pairs that aggregate in a three-dimensional crystal lattice structure. (c) A microscopic view of NaCl crystals shows their cubic geometry. Each tiny crystal contains billions of sodium and chloride ions.

- Ion formation takes place by an electron transfer process.
- The positive and negative ions are held together by the electrostatic force between ions of opposite charge in an ionic bond.
- Reactions between representative metals and nonmetals (elements far to the left and right, respectively, in the periodic table) tend to result in ionic bonds.

Although ionic compounds are sometimes referred to as ion pairs, in the solid state these ion pairs do not actually exist as individual units. The positive ions exert attractive forces on several negative ions, and the negative ions are attracted to several positive centers. Positive and negative ions arrange themselves in a regular three-dimensional repeating array to produce a stable arrangement known as a **crystal lattice.** The lattice structure for sodium chloride is shown in Figure 3.2.

Covalent Bonding

The octet rule is not just for ionic compounds. Covalently bonded compounds share electrons to complete the octet of electrons for each of the atoms participating in the bond. Consider the bond formed between two hydrogen atoms, producing the diatomic form of hydrogen: H_2. Individual hydrogen atoms are not stable, and two hydrogen atoms readily combine to produce diatomic hydrogen:

$$H + H \longrightarrow H_2$$

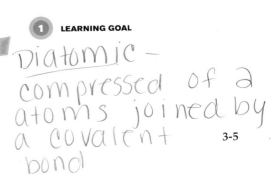

82 Chapter 3 Structure and Properties of Ionic and Covalent Compounds

> A diatomic compound is one that is composed of two atoms joined by a covalent bond.

If a hydrogen atom were to gain a second electron, it would be isoelectronic with the stable electron configuration of helium. However, because two identical hydrogen atoms have an equal tendency to gain or lose electrons, an electron transfer from one atom to the other is unlikely to occur under normal conditions. Each atom may attain a noble gas structure only by *sharing* its electron with the other, as shown with Lewis symbols:

$$H\cdot + \cdot H \longrightarrow H:H$$

When electrons are shared rather than transferred, the *shared electron pair* is referred to as a *covalent bond* (Figure 3.3). Compounds characterized by covalent bonding are called *covalent compounds*. Covalent bonds tend to form between atoms with similar tendencies to gain or lose electrons. The most obvious examples are the diatomic molecules H_2, N_2, O_2, F_2, Cl_2, Br_2, and I_2. Bonding in these molecules is *totally covalent* because there can be no net tendency for electron transfer between identical atoms. The formation of F_2, for example, may be represented as

$$:\!\ddot{F}\!\cdot\; +\; \cdot\!\ddot{F}\!:\; \longrightarrow\; :\!\ddot{F}\!:\!\ddot{F}\!:$$

> Fourteen valence electrons are arranged in such a way that each fluorine atom is surrounded by eight electrons. The octet rule is satisfied for each fluorine atom.

As in H_2, a single covalent bond is formed. The bonding electron pair is said to be *localized*, or largely confined to the region between the two fluorine nuclei.

Two atoms do not have to be identical to form a covalent bond. Consider compounds such as the following:

$H:\ddot{F}:$	$H:\ddot{O}:H$	$H:\overset{H}{\underset{H}{C}}:H$	$H:\overset{H}{N}:H$
Hydrogen fluoride	Water	Methane	Ammonia
7e⁻ from F	6e⁻ from O	4e⁻ from C	5e⁻ from N
1e⁻ from H	2e⁻ from 2H	4e⁻ from 4H	3e⁻ from 3H
8e⁻ for F	8e⁻ for O	8e⁻ for C	8e⁻ for N
2e⁻ for H	2e⁻ for H	2e⁻ for H	2e⁻ for H

In each of these cases, bond formation satisfies the octet rule. A total of eight electrons surround each atom other than hydrogen. Hydrogen has only two electrons (corresponding to the electronic structure of helium).

Polar Covalent Bonding and Electronegativity
The Polar Covalent Bond

Covalent bonding is the sharing of an electron pair by two atoms. However, just as we may observe in our day-to-day activities, sharing is not always equal. In a molecule like H_2 (or N_2, or any other diatomic molecule composed of only one element), the electrons, on average, spend the same amount of time in the vicinity of each atom; the electrons have no preference because both atoms are identical.

Now consider a diatomic molecule composed of two different elements; HF is a common example. It has been experimentally shown that the electrons in the H—F bond are not equally shared; the electrons spend more time in the vicinity of the fluorine atom. This unequal sharing can be described in various ways:

Partial electron transfer: This describes the bond as having both covalent and ionic properties.

Unequal electron density: The density of electrons around F is greater than the density of electrons around H.

Hydrogen atoms approach at high velocity

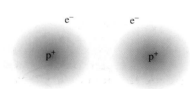

Hydrogen nuclei begin to attract each other's electrons.

Hydrogen atoms form the hydrogen molecule; atoms are held together by the shared electrons, the covalent bond.

Figure 3.3
Covalent bonding in hydrogen.

3.1 Chemical Bonding

Polar covalent bond is the preferred term for a bond made up of unequally shared electron pairs. One end of the bond (in this case, the F atom) is more electron rich (higher electron density), hence, more negative. The other end of the bond (in this case, the H atom) is less electron rich (lower electron density), hence, more positive. These two ends, one somewhat positive (δ^+) and the other somewhat negative (δ^-) may be described as electronic poles, hence the term polar covalent bonds.

The water molecule is perhaps the best-known example of a molecule that exhibits polar covalent bonding (Figure 3.4). In Section 3.4 we will see that the water molecule itself is polar and this fact is the basis for many of water's unique properties.

Once again, we can use the predictive power of the periodic table to help us determine whether a particular bond is polar or nonpolar covalent. We already know that elements that tend to form negative ions (by gaining electrons) are found to the right of the table whereas positive ion formers (that may lose electrons) are located on the left side of the table. Elements whose atoms strongly attract electrons are described as electronegative elements. Linus Pauling, a chemist noted for his theories on chemical bonding, developed a scale of relative electronegativities that correlates reasonably well with the positions of the elements in the periodic table.

Electronegativity

Electronegativity (E_n) is a measure of the ability of an atom to attract electrons in a chemical bond. Elements with high electronegativity have a greater ability to attract electrons than do elements with low electronegativity. Pauling developed a method to assign values of electronegativity to many of the elements in the periodic table. These values range from a low of 0.7 to a high of 4.0, 4.0 being the most electronegative element.

Figure 3.5 shows that the most electronegative elements (excluding the nonreactive noble gas elements) are located in the upper right corner of the periodic table, whereas the least electronegative elements are found in the lower left corner of the table. In general, electronegativity values increase as we proceed left to right and bottom to top of the table. Like other periodic trends, numerous exceptions occur.

If we picture the covalent bond as a competition for electrons between two positive centers, it is the difference in electronegativity, ΔE_n, that determines the extent of polarity. Consider: H_2 or H—H

$$\Delta E_n = \begin{bmatrix} \text{Electronegativity} \\ \text{of hydrogen} \end{bmatrix} - \begin{bmatrix} \text{Electronegativity} \\ \text{of hydrogen} \end{bmatrix}$$

$$\Delta E_n = 2.1 - 2.1 = 0$$

The bond in H_2 is nonpolar covalent. Bonds between *identical* atoms are *always* nonpolar covalent. Also, Cl_2 or Cl—Cl

$$\Delta E_n = \begin{bmatrix} \text{Electronegativity} \\ \text{of chlorine} \end{bmatrix} - \begin{bmatrix} \text{Electronegativity} \\ \text{of chlorine} \end{bmatrix}$$

$$\Delta E_n = 3.0 - 3.0 = 0$$

The bond in Cl_2 is nonpolar covalent. Now consider HCl or H—Cl

$$\Delta E_n = \begin{bmatrix} \text{Electronegativity} \\ \text{of chlorine} \end{bmatrix} - \begin{bmatrix} \text{Electronegativity} \\ \text{of hydrogen} \end{bmatrix}$$

$$\Delta E_n = 3.0 - 2.1 = 0.9$$

The bond in HCl is polar covalent.

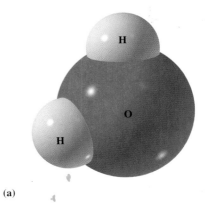

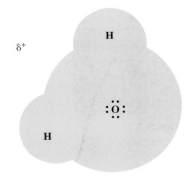

Figure 3.4
Polar covalent bonding in water. Oxygen is electron-rich (δ^-) and hydrogen is electron deficient (δ^+) due to unequal electron sharing. Water has two polar covalent bonds.

Linus Pauling is the only person to receive two Nobel Prizes in very unrelated fields; the chemistry award in 1954 and eight years later, the Nobel Peace Prize. His career is a model of interdisciplinary science, with important contributions ranging from chemical physics to molecular biology.

By convention, the electronegativity difference is calculated by subtracting the less electronegative element's value from the value for the more electronegative element. In this way, negative numbers are avoided.

Figure 3.5
Electronegativities of the elements.

Legend:
- Less than 1.0
- Between 1.0–3.0
- Greater than or equal to 3.0

Group																	
IA (1)	IIA (2)	IIIB (3)	IVB (4)	VB (5)	VIB (6)	VIIB (7)	VIIIB (8)	VIIIB (9)	VIIIB (10)	IB (11)	IIA (12)	IIIA (13)	IVA (14)	VA (15)	VIA (16)	VIIA (17)
H 2.1																
Li 1.0	Be 1.5											B 2.0	C 2.5	N 3.0	O 3.5	F 4.0
Na 0.9	Mg 1.2											Al 1.5	Si 1.8	P 2.1	S 2.5	Cl 3.0
K 0.8	Ca 1.0	Sc 1.3	Ti 1.5	V 1.6	Cr 1.6	Mn 1.5	Fe 1.8	Co 1.9	Ni 1.9	Cu 1.9	Zn 1.6	Ga 1.6	Ge 1.8	As 2.0	Se 2.4	Br 2.8
Rb 0.8	Sr 1.0	Y 1.2	Zr 1.4	Nb 1.6	Mo 1.8	Tc 1.9	Ru 2.2	Rh 2.2	Pd 2.2	Ag 1.9	Cd 1.7	In 1.7	Sn 1.8	Sb 1.9	Te 2.1	I 2.5
Cs 0.7	Ba 0.9	La* 1.1	Hf 1.3	Ta 1.5	W 1.7	Re 1.9	Os 2.2	Ir 2.2	Pt 2.2	Au 2.4	Hg 1.9	Tl 1.8	Pb 1.9	Bi 1.9	Po 2.0	At 2.2
Fr 0.7	Ra 0.9	Ac† 1.1														

*Lanthanides: 1.1 – 1.3
†Actinides: 1.1 – 1.5

3.2 Naming Compounds and Writing Formulas of Compounds

Nomenclature is the assignment of a correct and unambiguous name to each and every chemical compound. Assignment of a name to a structure or deducing the structure from a name is a necessary first step in any discussion of these compounds.

Ionic Compounds

The **formula** is the representation of the fundamental compound using chemical symbols and numerical subscripts. It is the "shorthand" symbol for a compound—for example,

$$\text{NaCl} \quad \text{and} \quad \text{MgBr}_2$$

The formula identifies the number and type of the various atoms that make up the compound unit. The number of like atoms in the unit is shown by the use of a subscript. The presence of only one atom is understood when no subscript is present.

The formula NaCl indicates that each ion pair consists of one sodium cation (Na^+) and one chloride anion (Cl^-). Similarly, the formula $MgBr_2$ indicates that one magnesium ion and two bromide ions combine to form the compound.

In Chapter 2 we learned that positive ions are formed from elements that

- are located at the left of the periodic table,
- are referred to as *metals*, and
- have low ionization energies, low electron affinities, and hence easily *lose* electrons.

Elements that form negative ions, on the other hand,

- are located at the right of the periodic table (but exclude the noble gases),
- are referred to as *nonmetals*, and
- have high ionization energies, high electron affinities, and hence easily *gain* electrons.

3.2 Naming Compounds and Writing Formulas of Compounds

In short, metals and nonmetals usually react to produce ionic compounds resulting from the transfer of one or more electrons from the metal to the nonmetal.

An electronegativity difference of 1.9 is generally accepted as the boundary between polar covalent and ionic bonds. Although, strictly speaking, any electronegativity difference, no matter how small, produces a polar bond, the degree of polarity for bonds with electronegativity differences less than 0.5 is minimal. Consequently, we shall classify these bonds as nonpolar. The *formula* of an ionic compound is the smallest whole-number ratio of ions in the substance.

Writing Formulas of Ionic Compounds from the Identities of the Component Ions

It is important to be able to write the formula of an ionic compound when provided with the identities of the ions that make up the compound. The charge of each ion can usually be determined from the group (family) of the periodic table in which the parent element is found. The cations and anions must combine in such a way that the resulting formula unit has a net charge of zero.

Consider the following examples.

Predicting the Formula of an Ionic Compound EXAMPLE 3.1

Predict the formula of the ionic compound formed from the reaction of sodium and oxygen atoms.

Solution

Sodium is in group IA (or 1); it has *one* valence electron. Loss of this electron produces Na^+. Oxygen is in group VIA (or 16); it has *six* valence electrons. A gain of two electrons (to create a stable octet) produces O^{2-}. Two positive charges are necessary to counterbalance two negative charges on the oxygen anion. Because each sodium ion carries a 1+ charge, two sodium ions are needed for each O^{2-}. The subscript 2 is used to indicate that the formula unit contains two sodium ions. Thus the formula of the compound is Na_2O.

Predicting the Formula of an Ionic Compound EXAMPLE 3.2

Predict the formula of the compound formed by the reaction of aluminum and oxygen atoms.

Solution

Aluminum is in group IIIA (or 13) of the periodic table; we predict that it has three valence electrons. Loss of these electrons produces Al^{3+}. Oxygen is in group VIA (or 16) of the periodic table and has six valence electrons. A gain of two electrons (to create a stable octet) produces O^{2-}. How can we combine Al^{3+} and O^{2-} to yield a unit of zero charge? It is necessary that *both* the cation and anion be multiplied by factors that will result in a zero net charge:

$$2 \times (+3) = +6 \quad \text{and} \quad 3 \times (-2) = -6$$
$$2 \times Al^{3+} = +6 \quad \text{and} \quad 3 \times O^{2-} = -6$$

Hence the formula is Al_2O_3.

Question 3.1

Predict the formulas of the compounds formed from the combination of ions of the following elements:

a. lithium and bromine
b. calcium and bromine
c. calcium and nitrogen

Question 3.2

Predict the formulas of the compounds formed from the combination of ions of the following elements:

a. potassium and chlorine
b. magnesium and bromine
c. magnesium and nitrogen

Writing Names of Ionic Compounds from the Formula of the Compound

LEARNING GOAL

Nomenclature, the way in which compounds are named, is based on their formulas. The name of the cation appears first, followed by the name of the anion. The positive ion has the name of the element; the negative ion is named by using the *stem* of the name of the element joined to the suffix *-ide*. Some examples follow.

Formula	cation	and	anion stem	+	ide	=	Compound name
NaCl	sodium		chlor	+	ide		sodium chloride
Na$_2$O	sodium		ox	+	ide		sodium oxide
Li$_2$S	lithium		sulf	+	ide		lithium sulfide
AlBr$_3$	aluminum		brom	+	ide		aluminum bromide
CaO	calcium		ox	+	ide		calcium oxide

If the cation and anion exist in only one common charged form, there is no ambiguity between formula and name. Sodium chloride *must be* NaCl, and lithium sulfide *must be* Li$_2$S, so that the sum of positive and negative charges is zero. With many elements, such as the transition metals, several ions of different charge may exist. Fe^{2+}, Fe^{3+} and Cu^+, Cu^{2+} are two common examples. Clearly, an ambiguity exists if we use the name iron for both Fe^{2+} and Fe^{3+} or copper for both Cu^+ and Cu^{2+}. Two systems have been developed to avoid this problem: the *Stock system* and the *common nomenclature system*.

In the Stock system (systematic name), a Roman numeral placed immediately after the name of the ion indicates the magnitude of the cation's charge. In the older common nomenclature system, the suffix *-ous* indicates the lower ionic charge, and the suffix *-ic* indicates the higher ionic charge. Consider the examples in Table 3.1.

Systematic names are easier and less ambiguous than common names. Whenever possible, we will use this system of nomenclature. The older, common names (-ous, -ic) are less specific; furthermore, they often use the Latin names of the elements (for example, iron compounds use *ferr-*, from *ferrum*, the Latin word for iron).

Monatomic ions are ions consisting of a single atom. Common monatomic ions are listed in Table 3.2. The ions that are particularly important in biological systems are highlighted in blue.

Polyatomic ions, such as the hydroxide ion, OH^-, are composed of two or more atoms bonded together. These ions, although bonded to other ions with ionic bonds, are themselves held together by covalent bonds.

3.2 Naming Compounds and Writing Formulas of Compounds

TABLE 3.1 Systemic (Stock) and Common Names for Iron and Copper Ions

For systematic name:

Formula	+ Ion Charge	Cation Name	Compound Name
$FeCl_2$	2+	Iron(II)	Iron(II) chloride
$FeCl_3$	3+	Iron(III)	Iron(III) chloride
Cu_2O	1+	Copper(I)	Copper(I) oxide
CuO	2+	Copper(II)	Copper(II) oxide

For common nomenclature:

Formula	+ Ion Charge	Cation Name	Common -ous/ic Name
$FeCl_2$	2+	Ferr*ous*	Ferrous chloride
$FeCl_3$	3+	Ferr*ic*	Ferric chloride
Cu_2O	1+	Cupr*ous*	Cuprous oxide
CuO	2+	Cupr*ic*	Cupric oxide

TABLE 3.2 Common Monatomic Cations and Anions

Cation	Name	Anion	Name
H^+	Hydrogen ion	H^-	Hydride ion
Li^+	Lithium ion	F^-	Fluoride ion
Na^+	Sodium ion	Cl^-	Chloride ion
K^+	Potassium ion	Br^-	Bromide ion
Cs^+	Cesium ion	I^-	Iodide ion
Be^{2+}	Beryllium ion	O^{2-}	Oxide ion
Mg^{2+}	Magnesium ion	S^{2-}	Sulfide ion
Ca^{2+}	Calcium ion	N^{3-}	Nitride ion
Ba^{2+}	Barium ion	P^{3-}	Phosphide ion
Al^{3+}	Aluminum ion		
Ag^+	Silver ion		

Note: The ions of principal importance are highlighted in magenta.

The polyatomic ion has an *overall* positive or negative charge. Some common polyatomic ions are listed in Table 3.3. The formulas, charges, and names of these polyatomic ions, especially those highlighted in blue, should be memorized.

The following examples are formulas of several compounds containing polyatomic ions.

Formula	Cation	Anion	Name
NH_4Cl	NH_4^+	Cl^-	ammonium chloride
$Ca(OH)_2$	Ca^{2+}	OH^-	calcium hydroxide
Na_2SO_4	Na^+	SO_4^{2-}	sodium sulfate
$NaHCO_3$	Na^+	HCO_3^-	sodium bicarbonate

Sodium bicarbonate may also be named sodium hydrogen carbonate, a preferred and less ambiguous name. Likewise, Na_2HPO_4 is named sodium hydrogen phosphate, and other ionic compounds are named similarly.

memorize

TABLE 3.3 Common Polyatomic Cations and Anions

Ion	Name
✓ NH_4^+	Ammonium
NO_2^-	Nitrite
✓ NO_3^-	Nitrate
SO_3^{2-}	Sulfite
✓ SO_4^{2-}	Sulfate
HSO_4^-	Hydrogen sulfate
✓ OH^-	Hydroxide
✓ CN^-	Cyanide
✓ PO_4^{3-}	Phosphate
HPO_4^{2-}	Hydrogen phosphate
$H_2PO_4^-$	Dihydrogen phosphate
✓ CO_3^{2-}	Carbonate
✓ HCO_3^-	Bicarbonate
ClO^-	Hypochlorite
ClO_2^-	Chlorite
✓ ClO_3^-	Chlorate
ClO_4^-	Perchlorate
✓ CH_3COO^- (or $C_2H_3O_2^-$)	Acetate
MnO_4^-	Permanganate
$Cr_2O_7^{2-}$	Dichromate
CrO_4^{2-}	Chromate
O_2^{2-}	Peroxide

Note: The most commonly encountered ions are highlighted in magenta.

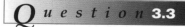

Name each of the following compounds:

a. KCN
b. MgS
c. $Mg(CH_3COO)_2$

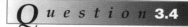

Name each of the following compounds:

a. Li_2CO_3
b. $FeBr_2$
c. $CuSO_4$

Writing Formulas of Ionic Compounds from the Name of the Compound

LEARNING GOAL

It is also important to be able to write the correct formula when given the compound name. To do this, we must be able to predict the charge of monatomic ions and remember the charge and formula of polyatomic ions. Equally important, the relative number of positive and negative ions in the unit must result in a net (compound) charge of zero. The compounds are electrically neutral. Two examples follow.

3.2 Naming Compounds and Writing Formulas of Compounds

EXAMPLE 3.3 Writing a Formula When Given the Name of the Compound

Write the formula of sodium sulfate.

Solution

Step 1. The sodium ion is Na^+, a group I (or 1) element. The sulfate ion is SO_4^{2-} (from Table 3.3).

Step 2. Two positive charges, two sodium ions, are needed to cancel the charge on one sulfate ion (two negative charges).

Hence the formula is Na_2SO_4.

EXAMPLE 3.4 Writing a Formula When Given the Name of the Compound

Write the formula of ammonium sulfide.

Solution

Step 1. The ammonium ion is NH_4^+ (from Table 3.3). The sulfide ion is S^{2-} (from its position on the periodic table).

Step 2. Two positive charges are necessary to cancel the charge on one sulfide ion (two negative charges).

Hence the formula is $(NH_4)_2S$.
   Note that parentheses must be used whenever a subscript accompanies a polyatomic ion.

Question 3.5

Write the formula for each of the following compounds:

a. calcium carbonate
b. sodium bicarbonate
c. copper(I) sulfate

Question 3.6

Write the formula for each of the following compounds:

a. sodium phosphate
b. potassium bromide
c. iron(II) nitrate

Covalent Compounds
Naming Covalent Compounds

Most covalent compounds are formed by the reaction of nonmetals. **Molecules** are compounds characterized by covalent bonding. We saw earlier that ionic compounds are not composed of single units but are a part of a massive three-dimensional crystal structure in the solid state. Covalent compounds exist as discrete molecules in the solid, liquid, and gas states. This is a distinctive feature of covalently bonded substances.

TABLE 3.4 Prefixes Used to Denote Numbers of Atoms in a Compound

Prefix	Number of Atoms
Mono-	1
Di-	2
Tri-	3
Tetra-	4
Penta-	5
Hexa-	6
Hepta-	7
Octa-	8
Nona-	9
Deca-	10

By convention the prefix *mono-* is often omitted from the second element as well (dinitrogen oxide, not dinitrogen monoxide). In other cases, common usage retains the prefix (carbon monoxide, not carbon oxide).

The conventions for naming covalent compounds follow:

1. The names of the elements are written in the order in which they appear in the formula.
2. A prefix (Table 3.4) indicating the number of each kind of atom found in the unit is placed before the name of the element.
3. If only one atom of a particular kind is present in the molecule, the prefix mono- is usually omitted from the first element.
4. The stem of the name of the last element is used with the suffix -ide.
5. The final vowel in a prefix is often dropped before a vowel in the stem name.

EXAMPLE 3.5 Naming a Covalent Compound

Name the covalent compound N_2O_4.

Solution

Step 1. two nitrogen atoms   four oxygen atoms
Step 2. di-   tetra-
Step 3. dinitrogen   tetr(a)oxide

The name is dinitrogen tetroxide.

The following are examples of other covalent compounds.

Formula	Name
N_2O	dinitrogen monoxide
NO_2	nitrogen dioxide
SiO_2	silicon dioxide
CO_2	carbon dioxide
CO	carbon monoxide

3.2 Naming Compounds and Writing Formulas of Compounds

Question 3.7

Name each of the following compounds:

a. B_2O_3 c. ICl
b. NO   d. PCl_3

Question 3.8

Name each of the following compounds:

a. H_2S c. PCl_5
b. CS_2 d. P_2O_5

Writing Formulas of Covalent Compounds

Many compounds are so familiar to us that their *common names* are generally used. For example, H_2O is water, NH_3 is ammonia, C_2H_5OH (ethanol) is ethyl alcohol, and $C_6H_{12}O_6$ is glucose. It is useful to be able to correlate both systematic and common names with the corresponding molecular formula and vice versa.

When common names are used, formulas of covalent compounds can be written *only* from memory. You *must* remember that water is H_2O, ammonia is NH_3, and so forth. This is the major disadvantage of common names. Because of their widespread use, however, they cannot be avoided and must be memorized.

Compounds named by using Greek prefixes are easily converted to formulas. Consider the following examples.

LEARNING GOAL 2

EXAMPLE 3.6

Writing the Formula of a Covalent Compound

Write the formula of nitrogen monoxide.

Solution

Nitrogen has no prefix; one is understood. Oxide has the prefix *mono*—one oxygen. Hence the formula is NO.

EXAMPLE 3.7

Writing the Formula of a Covalent Compound

Write the formula of dinitrogen tetroxide.

Solution

Nitrogen has the prefix *di*—two nitrogen atoms. Oxygen has the prefix *tetr(a)*—four oxygen atoms. Hence the formula is N_2O_4.

Question 3.9

Write the formula of each of the following compounds:

a. diphosphorus pentoxide
b. silicon dioxide

A Human Perspective

Origin of the Elements

The current, most widely held theory of the origin of the universe is the "big bang" theory. An explosion of very dense matter was followed by expansion into space of the fragments resulting from this explosion. This is one of the scenarios that have been created by scientists fascinated by the origins of matter, the stars and planets, and life as we know it today.

The first fragments, or particles, were protons and neutrons moving with tremendous velocity and possessing large amounts of energy. Collisions involving these high-energy protons and neutrons formed deuterium atoms (^{2}H), which are isotopes of hydrogen. As the universe expanded and cooled, tritium (^{3}H), another hydrogen isotope, formed as a result of collisions of neutrons with deuterium atoms. Subsequent capture of a proton produced helium (He). Scientists theorize that a universe that was principally composed of hydrogen and helium persisted for perhaps 100,000 years until the temperature decreased sufficiently to allow the formation of a simple molecule, hydrogen, two atoms of hydrogen bonded together (H_2).

Many millions of years later, the effect of gravity caused these small units to coalesce, first into clouds and eventually into stars, with temperatures of millions of degrees. In this setting, these small collections of protons and neutrons combined to form larger atoms such as carbon (C) and oxygen (O), then sodium (Na), neon (Ne), magnesium (Mg), silicon (Si), and so forth. Subsequent explosions of stars provided the conditions that formed many larger atoms. These fragments, gathered together by the force of gravity, are the most probable origin of the planets in our own solar system.

The reactions that formed the elements as we know them today were a result of a series of *fusion reactions,* the joining of nuclei to produce larger atoms at very high temperatures (millions of degrees Celsius). These fusion reactions are similar to processes that are currently being studied as a possible alternative source of nuclear power. We shall study such nuclear processes in more detail in Chapter 9.

Nuclear reactions of this type do not naturally occur on the earth today. The temperature is simply too low. As a result we have, for the most part, a collection of stable elements existing as chemical compounds, atoms joined together by chemical bonds while retaining their identity even in the combined state. Silicon exists all around us as sand and soil in a combined form, silicon dioxide; most metals exist as a part of a chemical compound, such as iron ore. We are learning more about the structure and properties of these compounds in this chapter.

For Further Understanding

How does tritium differ from "normal" hydrogen?

Would you expect to find similar atoms on other planets?

Question 3.10

Write the formula of each of the following compounds:

a. nitrogen trifluoride
b. carbon monoxide

3.3 Properties of Ionic and Covalent Compounds

LEARNING GOAL

The differences in ionic and covalent bonding result in markedly different properties for ionic and covalent compounds. Because covalent molecules are distinct units, they have less tendency to form an extended structure in the solid state. Ionic compounds, with ions joined by electrostatic attraction, do not have definable units but form a crystal lattice composed of enormous numbers of positive and negative ions in an extended three-dimensional network. The effects of this basic structural difference are summarized in this section.

Physical State

All ionic compounds (for example, NaCl, KCl, and $NaNO_3$) are solids at room temperature; covalent compounds may be solids (sugar), liquids (H_2O, ethanol), or gases (carbon monoxide, carbon dioxide). The three-dimensional crystal structure that is characteristic of ionic compounds holds them in a rigid, solid arrangement, whereas molecules of covalent compounds may be fixed, as in a solid, or more mobile, a characteristic of liquids and gases.

Melting and Boiling Points

The **melting point** is the temperature at which a solid is converted to a liquid, and the **boiling point** is the temperature at which a liquid is converted to a gas at a specified pressure. Considerable energy is required to break apart an ionic crystal lattice with uncountable numbers of ionic interactions and convert the ionic substance to a liquid or a gas. As a result, the melting and boiling temperatures for ionic compounds are generally higher than those of covalent compounds, whose molecules interact less strongly in the solid state. A typical ionic compound, sodium chloride, has a melting point of 801°C; methane, a covalent compound, melts at −182°C. Exceptions to this general rule do exist; diamond, a covalent solid with an extremely high melting point, is a well-known example.

Structure of Compounds in the Solid State

Ionic solids are *crystalline*, characterized by a regular structure, whereas covalent solids may either be crystalline or have no regular structure. In the latter case they are said to be *amorphous*.

Solutions of Ionic and Covalent Compounds

In Chapter 1 we saw that mixtures are either heterogeneous or homogeneous. A homogeneous mixture is a solution. Many ionic solids dissolve in solvents, such as water. An ionic solid, if soluble, will form positive and negative ions in solution by **dissociation.**

Because ions in water are capable of carrying (conducting) a current of electricity, we refer to these compounds as **electrolytes,** and the solution is termed an **electrolytic solution.** Covalent solids dissolved in solution usually retain their neutral (molecular) character and are **nonelectrolytes.** The solution is not an electrical conductor.

The role of the solvent in the dissolution of solids is discussed in Section 3.5.

3.4 Drawing Lewis Structures of Molecules and Polyatomic Ions

Lewis Structures of Molecules

In Section 3.1, we used Lewis structures of individual atoms to help us understand the bonding process. To begin to explain the relationship between molecular structure and molecular properties, we will first need a set of guidelines to help us write Lewis structures for more complex molecules.

 **LEARNING GOAL**

1. *Use chemical symbols for the various elements to write the skeletal structure of the compound.* To accomplish this, place the bonded atoms next to one another. This is relatively easy for simple compounds; however, as the number of atoms in the compound increases, the possible number of arrangements increases dramatically. We may be told the pattern of arrangement of the atoms in advance; if not, we can make an intelligent guess and see if a reasonable Lewis structure can be constructed. Three considerations are very important here:
   • the least electronegative atom will be placed in the central position (the central atom),

The skeletal structure indicates only the relative positions of atoms in the molecule or ion. Bonding information results from the Lewis structure.

A Medical Perspective

Blood Pressure and the Sodium Ion/Potassium Ion Ratio

When you have a physical exam, the physician measures your blood pressure. This indicates the pressure of blood against the walls of the blood vessels each time the heart pumps. A blood pressure reading is always characterized by two numbers. With every heartbeat there is an increase in pressure; this is the systolic blood pressure. When the heart relaxes between contractions, the pressure drops; this is the diastolic pressure. Thus the blood pressure is expressed as two values—for instance, 117/72—measured in millimeters of mercury. Hypertension is simply defined as high blood pressure. To the body it means that the heart must work too hard to pump blood, and this can lead to heart failure or heart disease.

Heart disease accounts for 50% of all deaths in the United States. Epidemiological studies correlate the following major risk factors with heart disease: heredity, sex, race, age, diabetes, cigarette smoking, high blood cholesterol, and hypertension. Obviously, we can do little about our age, sex, and genetic heritage, but we can stop smoking, limit dietary cholesterol, and maintain a normal blood pressure.

The number of Americans with hypertension is alarmingly high: 60 million adults and children. More than 10 million of these individuals take medication to control blood pressure, at a cost of nearly $2.5 billion each year. In many cases, blood pressure can be controlled without medication by increasing physical activity, losing weight, decreasing consumption of alcohol, and limiting intake of sodium.

It has been estimated that the average American ingests 7.5–10 g of salt (NaCl) each day. Because NaCl is about 40% (by mass) sodium ions, this amounts to 3–4 g of sodium daily. Until 1989 the Food and Nutrition Board of the National Academy of Sciences National Research Council's defined *estimated safe and adequate daily dietary intake* (ESADDI) of sodium ion was 1.1–3.3 g. Clearly, Americans exceed this recommendation.

Recently, studies have shown that excess sodium is not the sole consideration in the control of blood pressure. More important is the sodium ion/potassium ion (Na^+/K^+) ratio. That ratio should be about 0.6; in other words, our diet should contain about 67% more potassium than sodium. Does the typical American diet fall within this limit? Definitely not! Young American males (25–30 years old) consume a diet with a $Na^+/K^+ = 1.07$, and the diet of females of the same age range has a $Na^+/K^+ = 1.04$. It is little wonder that so many Americans suffer from hypertension.

How can we restrict sodium in the diet, while increasing the potassium? The following table lists a variety of foods that are low in sodium and high in potassium. These include fresh fruits and vegetables and fruit juices, a variety of cereals, unsalted nuts, and cooked dried beans (legumes). The table also notes some high-sodium, low-potassium foods. Notice that most of these are processed or prepared foods. This points out how difficult it can be to control sodium in the diet. The majority of the sodium that we ingest comes from commercially prepared foods. The consumer must read the nutritional information printed on cans and packages to determine whether the sodium levels are within acceptable limits.

Low Sodium Ion, High Potassium Ion Foods

Food Category	Examples
Fruit and fruit juices	Pineapple, grapefruit, pears, strawberries, watermelon, raisins, bananas, apricots, oranges
Low-sodium cereals	Oatmeal (unsalted), Roman Meal Hot Cereal, shredded wheat
Nuts (unsalted)	Hazelnuts, macadamia nuts, almonds, peanuts, cashews, coconut
Vegetables	Summer squash, zucchini, eggplant, cucumber, onions, lettuce, green beans, broccoli
Beans (dry, cooked)	Great Northern beans, lentils, lima beans, red kidney beans

High Sodium Ion, Low Potassium Ion Foods

Food Category	Examples
Fats	Butter, margarine, salad dressings
Soups	Onion, mushroom, chicken noodle, tomato, split pea
Breakfast cereals	Many varieties; consult the label for specific nutritional information.
Breads	Most varieties
Processed meats	Most varieties
Cheese	Most varieties

For Further Understanding

Find several commercial food products on the shelves of your local grocery store; read the label and calculate the sodium ion–potassium ion ratio.

Describe each product that you have chosen in terms of its suitability for inclusion in the diet of a person with moderately elevated blood pressure.

3.4 Drawing Lewis Structures of Molecules and Polyatomic Ions

- hydrogen and fluorine (and the other halogens) often occupy terminal positions,
- carbon often forms chains of carbon-carbon covalent bonds.

2. **Determine the number of valence electrons associated with each atom; combine them to determine the total number of valence electrons in the compound.** However, if we are representing polyatomic cations or anions, we must account for the charge on the ion. Specifically:
   - for polyatomic cations, subtract one electron for each unit of positive charge. This accounts for the fact that the positive charge arises from electron loss.
   - for polyatomic anions, add one electron for each unit of negative charge. This accounts for excess negative charge resulting from electron gain.

3. **Connect the central atom to each of the surrounding atoms using electron pairs.** Then complete the octets of all of the atoms bonded to the central atom. Recall that hydrogen needs only two electrons to complete its valence shell. Electrons not involved in bonding must be represented as lone pairs and the total number of electrons in the structure must equal the number of valence electrons computed in our second step.

4. **If the octet rule is not satisfied for the central atom, move one or more electron pairs from the surrounding atoms.** Use these electrons to create double or triple bonds until all atoms have an octet.

5. **After you are satisfied with the Lewis structure that you have constructed, perform a final electron count.** This allows you to verify that the total number of electrons and the number around each atom are correct.

> The central atom is often the element farthest to the left and/or lowest in the periodic table.
> The central atom is often the element in the compound for which there is only one atom.
> Hydrogen is *never* the central atom.

Now, let us see how these guidelines are applied in the examples that follow.

Drawing Lewis Structures of Covalent Compounds   **EXAMPLE 3.8**

Draw the Lewis structure of carbon dioxide, CO_2.

Solution

Draw a skeletal structure of the molecule, arranging the atoms in their most probable order.

For CO_2, two possibilities exist:

$$C-O-O \quad \text{and} \quad O-C-O$$

Referring to Figure 3.5, we find that the electronegativity of oxygen is 3.5 whereas that of carbon is 2.5. Our strategy dictates that the least electronegative atom, in this case carbon, is the central atom. Hence the skeletal structure $O-C-O$ may be presumed correct.

Next, we want to determine the number of valence electrons on each atom and add them to arrive at the total for the compound.

For CO_2,

$$\begin{aligned} 1\text{ C atom} \times 4 \text{ valence electrons} &= 4\text{ e}^- \\ 2\text{ O atoms} \times 6 \text{ valence electrons} &= 12\text{ e}^- \\ \hline &16\text{ e}^- \text{ total} \end{aligned}$$

Now, use electron pairs to connect the central atom, C, to each oxygen with a single bond.

$$O:C:O$$

Continued—

EXAMPLE 3.8 —Continued

Distribute the electrons around the atoms (in pairs if possible) in an attempt to satisfy the octet rule, eight electrons around each element.

$$:\overset{..}{\underset{..}{O}}:C:\overset{..}{\underset{..}{O}}:$$

This structure satisfies the octet rule for each oxygen atom, but not the carbon atom (only four electrons surround the carbon).

However, when this structure is modified by moving two electrons from each oxygen atom to a position between C and O, each oxygen and carbon atom is surrounded by eight electrons. The octet rule is satisfied, and the structure below is the most probable Lewis structure for CO_2.

$$\overset{..}{\underset{..}{O}}::C::\overset{..}{\underset{..}{O}}$$

In this structure, four electrons (two electron pairs) are located between C and each O, and these electrons are shared in covalent bonds. Because a **single bond** is composed of two electrons (one electron pair) and because four electrons "bond" the carbon atom to each oxygen atom in this structure, there must be two bonds between each oxygen atom and the carbon atom, a **double bond:**

The notation for a single bond : is equivalent to — (one pair of electrons).

The notation for a double bond : : is equivalent to = (two pairs of electrons).

We may write CO_2 as shown above or, replacing dots with dashes to indicate bonding electron pairs,

$$\overset{..}{\underset{..}{O}}=C=\overset{..}{\underset{..}{O}}$$

As a final step, let us do some "electron accounting." There are eight electron pairs, and they correspond to sixteen valence electrons (8 pair × 2e⁻/pair). Furthermore, there are eight electrons around each atom and the octet rule is satisfied. Therefore

$$\overset{..}{\underset{..}{O}}=C=\overset{..}{\underset{..}{O}}$$

is a satisfactory way to depict the structure of CO_2.

EXAMPLE 3.9 Drawing Lewis Structures of Covalent Compounds

Draw the Lewis structure of ammonia, NH_3.

Solution

When trying to implement the first step in our strategy we may be tempted to make H our central atom because it is less electronegative than N. But, remember the margin note in this section:

"Hydrogen is *never* the central atom"

Continued—

3.4 Drawing Lewis Structures of Molecules and Polyatomic Ions

EXAMPLE 3.9 —*Continued*

Hence:

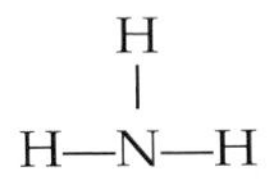

$$\begin{array}{c} H \\ | \\ H-N-H \end{array}$$

is our skeletal structure.

Applying our strategy to determine the total valence electrons for the molecule, we find that there are five valence electrons in nitrogen and one in each of the three hydrogens, for a total of eight valence electrons.

Applying our strategy for distribution of valence electrons results in the following Lewis diagram:

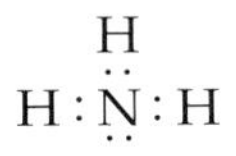

$$\begin{array}{c} H \\ \ddot{} \\ H:\ddot{N}:H \end{array}$$

This satisfies the octet rule for nitrogen (eight electrons around N) and hydrogen (two electrons around each H) and is an acceptable structure for ammonia. Ammonia may also be written:

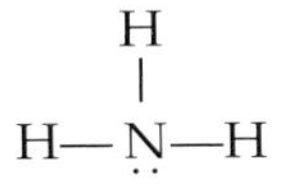

$$\begin{array}{c} H \\ | \\ H-\ddot{N}-H \end{array}$$

Note the pair of nonbonding electrons on the nitrogen atom. These are often called a **lone pair,** or *unshared* pair, of electrons. As we will see later in this section, lone pair electrons have a profound effect on molecular geometry. The geometry, in turn, affects the reactivity of the molecule.

Question 3.11

Draw a Lewis structure for each of the following covalent compounds:

a. H_2O (water)
b. CH_4 (methane)

Question 3.12

Draw a Lewis structure for each of the following covalent compounds:

a. C_2H_6 (ethane)
b. N_2 (nitrogen gas)

Lewis Structures of Polyatomic Ions

The strategies for writing the Lewis structures of polyatomic ions are similar to those for neutral compounds. There is, however, one major difference: the charge on the ion must be accounted for when computing the total number of valence electrons.

5 LEARNING GOAL

EXAMPLE 3.10 Drawing Lewis Structures of Polyatomic Cations

Draw the Lewis structure of the ammonium ion, NH_4^+.

Solution

The ammonium ion has the following skeletal structure and charge:

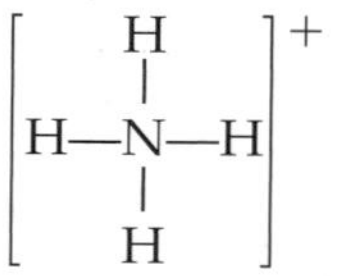

$$\left[\begin{array}{c} H \\ | \\ H-N-H \\ | \\ H \end{array} \right]^+$$

The total number of valence electrons is determined by subtracting one electron for each unit of positive charge.

$$\begin{aligned} 1 \text{ N atom} \times 5 \text{ valence electrons} &= 5 \text{ e}^- \\ 4 \text{ H atoms} \times 1 \text{ valence electron} &= 4 \text{ e}^- \\ \underline{-1 \text{ electron for } +1 \text{ charge}} &\underline{= -1 \text{ e}^-} \\ & 8 \text{ e}^- \text{ total} \end{aligned}$$

Distribute these eight electrons around our skeletal structure:

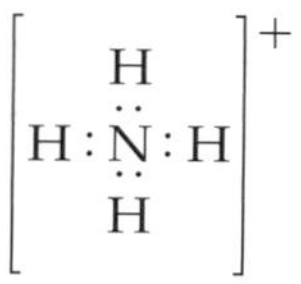

$$\left[\begin{array}{c} H \\ \cdot\cdot \\ H:N:H \\ \cdot\cdot \\ H \end{array} \right]^+ \quad \text{or} \quad \left[\begin{array}{c} H \\ | \\ H-N-H \\ | \\ H \end{array} \right]^+$$

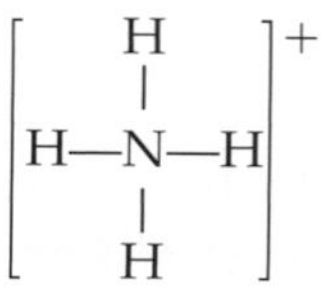

A final check shows eight total electrons, eight around the central atom, nitrogen, and two electrons associated with each hydrogen. Hence the structure is satisfactory.

EXAMPLE 3.11 Drawing Lewis Structures of Polyatomic Anions

Draw the Lewis structure of the carbonate ion, CO_3^{2-}.

Solution

Carbon is less electronegative than oxygen. Therefore carbon is the central atom. The carbonate ion has the following skeletal structure and charge:

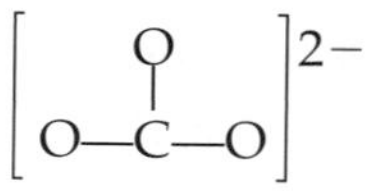

$$\left[\begin{array}{c} O \\ | \\ O-C-O \end{array} \right]^{2-}$$

The total number of valence electrons is determined by adding one electron for each unit of negative charge:

$$\begin{aligned} 1 \text{ C atom} \times 4 \text{ valence electrons} &= 4 \text{ e}^- \\ 3 \text{ O atoms} \times 6 \text{ valence electrons} &= 18 \text{ e}^- \\ \underline{+2 \text{ negative charges}} &\underline{= 2 \text{ e}^-} \\ & 24 \text{ e}^- \text{ total} \end{aligned}$$

Continued—

3.4 Drawing Lewis Structures of Molecules and Polyatomic Ions

EXAMPLE 3.11 —Continued

Distributing the electron dots around the central carbon atom (forming four bonds) and around the surrounding oxygen atoms in an attempt to satisfy the octet rule results in the structure:

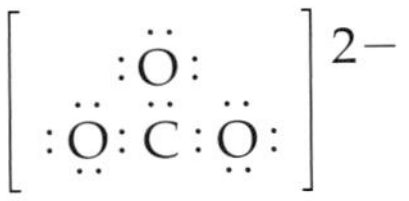

$$\left[\begin{array}{c} :\ddot{O}: \\ :\ddot{O}:C:\ddot{O}: \end{array} \right]^{2-} \quad \text{or} \quad \left[\begin{array}{c} :\ddot{O}: \\ | \\ :\ddot{O}-C-\ddot{O}: \end{array} \right]^{2-}$$

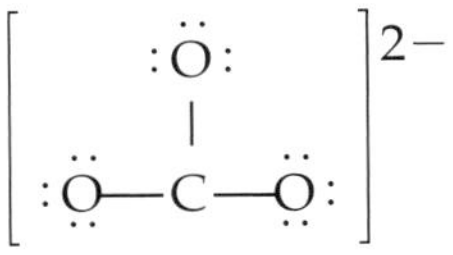

Although the octet rule is satisfied for the three O atoms, it is not for the C atom. We must move a lone pair from one of the O atoms to form another bond with C:

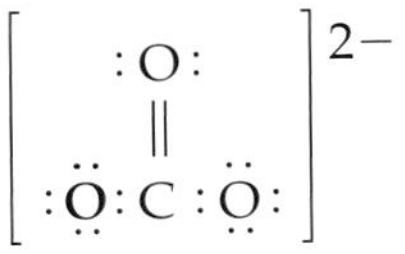

$$\left[\begin{array}{c} :O: \\ \| \\ :\ddot{O}:C:\ddot{O}: \end{array} \right]^{2-} \quad \text{or} \quad \left[\begin{array}{c} |O| \\ \| \\ |\overline{O}-C-\overline{O}| \end{array} \right]^{2-}$$

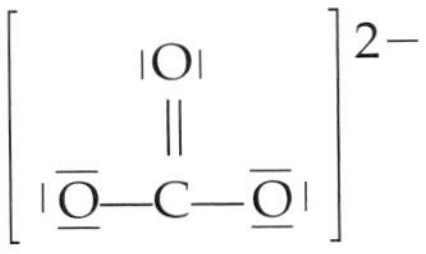

Now the octet rule is also satisfied for the C atom.

A final check shows twenty-four electrons and eight electrons around each atom. Hence the structure is a satisfactory representation of the carbonate ion.

Drawing Lewis Structures of Polyatomic Anions

EXAMPLE 3.12

Draw the Lewis structure of the acetate ion, CH_3COO^-.

Solution

A commonly encountered anion, the acetate ion has a skeletal structure that is more complex than any of the examples that we have studied thus far. Which element should we choose as the central atom? We have three choices: H, O, and C. H is eliminated because hydrogen can never be the central atom. Oxygen is more electronegative than carbon, so carbon must be the central atom. There are two carbon atoms; often they are joined. Further clues are obtained from the formula itself; CH_3COO^- implies three hydrogen atoms attached to the first carbon atom and two oxygen atoms joined to the second carbon. A plausible skeletal structure is:

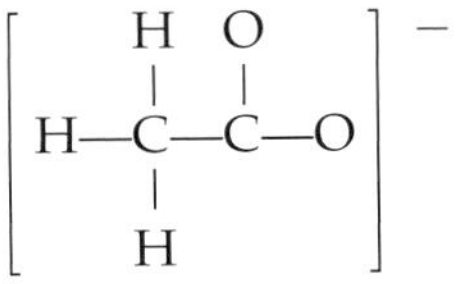

$$\left[\begin{array}{c} H \quad O \\ | \quad\ | \\ H-C-C-O \\ | \\ H \end{array} \right]^{-}$$

The pool of valence electrons for anions is determined by adding one electron for each unit of negative charge:

$$\begin{array}{lr} 2\ \text{C atoms} \times 4\ \text{valence electrons} & =\ 8\ e^- \\ 3\ \text{H atoms} \times 1\ \text{valence electron} & =\ 3\ e^- \\ 2\ \text{O atoms} \times 6\ \text{valence electrons} & =\ 12\ e^- \\ +\ 1\ \text{negative charge} & =\ 1\ e^- \\ \hline & 24\ e^-\ \text{total} \end{array}$$

Continued—

EXAMPLE 3.12 —Continued

Distributing these twenty-four electrons around our skeletal structure gives

$$\begin{bmatrix} H:\ddot{O}: \\ H:\ddot{C}:\ddot{C}:\ddot{O}: \\ H \end{bmatrix}^{-} \quad \text{or} \quad \begin{bmatrix} H & |\ddot{O}| \\ | & \| \\ H-C-C-\bar{O}| \\ | \\ H \end{bmatrix}^{-}$$

This Lewis structure satisfies the octet rule for carbon and oxygen and surrounds each hydrogen with two electrons. All twenty-four electrons are used in this process.

Question 3.13

Draw the Lewis structure for each of the following ions:

a. H_3O^+ (the hydronium ion)
b. OH^- (the hydroxide ion)

Question 3.14

Draw the Lewis structure for each of the following ions:

a. CN^- (the cyanide ion)
b. CO_3^{2-} (the carbonate ion)

Question 3.15

Write a Lewis structure describing the bonding in each of the following polyatomic ions:

a. the bicarbonate ion, HCO_3^-
b. the phosphate ion, PO_4^{3-}

Question 3.16

Write a Lewis structure describing the bonding in each of the following polyatomic ions:

a. the hydrogen sulfide ion, HS^-
b. the peroxide ion, O_2^{2-}

Lewis Structure, Stability, Multiple Bonds, and Bond Energies

LEARNING GOAL

Hydrogen, oxygen, and nitrogen are present in the atmosphere as diatomic gases, H_2, O_2, and N_2. All are covalent molecules. Their stability and reactivity, however, are quite different. Hydrogen is an explosive material, sometimes used as a fuel. Oxygen, although more stable than hydrogen, reacts with fuels in combustion. The explosion of the space shuttle *Challenger* resulted from the reaction of massive amounts of hydrogen and oxygen. Nitrogen, on the other hand, is extremely nonreactive. Because nitrogen makes up about 80% of the atmosphere, it dilutes the oxygen, which accounts for only about 20% of the atmosphere.

3.4 Drawing Lewis Structures of Molecules and Polyatomic Ions

Breathing pure oxygen for long periods, although necessary in some medical situations, causes the breakdown of nasal and lung tissue over time. Oxygen diluted with nonreactive nitrogen is an ideal mixture for humans and animals to breathe.

Why is there such a great difference in reactivity among these three gases? We can explain this, in part, in terms of their bonding characteristics. The Lewis structure for H_2 (two valence electrons) is

$$H:H \quad \text{or} \quad H\text{—}H$$

For oxygen (twelve valence electrons, six on each atom), the only Lewis structure that satisfies the octet rule is

$$\ddot{\text{O}}::\ddot{\text{O}} \quad \text{or} \quad \ddot{\text{O}}=\ddot{\text{O}}$$

The Lewis structure of N_2 (ten total valence electrons) must be

$$:N:::N: \quad \text{or} \quad :N\equiv N:$$

Therefore

- N_2 has a *triple bond* (six bonding electrons).
- O_2 has a *double bond* (four bonding electrons).
- H_2 has a *single bond* (two bonding electrons).

A **triple bond**, in which three pairs of electrons are shared by two atoms, is very stable. More energy is required to break a triple bond than a double bond, and a double bond is stronger than a single bond. Stability is related to the bond energy. The **bond energy** is the amount of energy, in units of kilocalories or kilojoules, required to break a bond holding two atoms together. Bond energy is therefore a *measure* of stability. The values of bond energies decrease in the order *triple bond > double bond > single bond*.

The bond length is related to the presence or absence of multiple bonding. The distance of separation of two nuclei is greatest for a single bond, less for a double bond, and still less for a triple bond. The bond length decreases in the order *single bond > double bond > triple bond*.

> The term *bond order* is sometimes used to distinguish among single, double, and triple bonds. A bond order of 1 corresponds to a single bond, 2 corresponds to a double bond, and 3 corresponds to a triple bond.

Question **3.17**

Contrast a single and double bond with regard to:

a. distance of separation of the bonded nuclei
b. strength of the bond

How are these two properties related?

Question **3.18**

Two nitrogen atoms in a nitrogen molecule are held together more strongly than the two chlorine atoms in a chlorine molecule. Explain this fact by comparing their respective Lewis structures.

Lewis Structures and Resonance

In some cases we find that it is possible to write more than one Lewis structure that satisfies the octet rule for a particular compound. Consider sulfur dioxide, SO_2. Its skeletal structure is

$$O\text{—}S\text{—}O$$

Total valence electrons may be calculated as follows:

$$1 \text{ sulfur atom} \times 6 \text{ valence } e^-/\text{atom} = 6 \text{ } e^-$$
$$+ \text{ 2 oxygen atoms} \times 6 \text{ valence } e^-/\text{atom} = 12 \text{ } e^-$$
$$18 \text{ } e^- \text{ total}$$

The resulting Lewis structures are

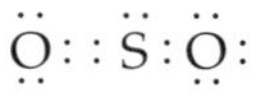

$$\ddot{\underset{..}{O}}::\ddot{S}:\ddot{\underset{..}{O}}: \quad \text{and} \quad :\ddot{\underset{..}{O}}:\ddot{S}::\ddot{\underset{..}{O}}$$

Both satisfy the octet rule. However, experimental evidence shows no double bond in SO$_2$. The two sulfur-oxygen bonds are equivalent. Apparently, neither structure accurately represents the structure of SO$_2$, and neither actually exists. The actual structure is said to be an average or *hybrid* of these two Lewis structures. When a compound has two or more Lewis structures that contribute to the real structure, we say that the compound displays the property of **resonance.** The contributing Lewis structures are **resonance forms.** The true structure, a hybrid of the resonance forms, is known as a **resonance hybrid** and may be represented as:

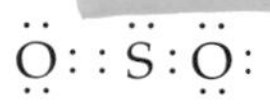

$$\ddot{\underset{..}{O}}::\ddot{S}:\ddot{\underset{..}{O}}: \quad \leftrightarrow \quad :\ddot{\underset{..}{O}}:\ddot{S}::\ddot{\underset{..}{O}}$$

A common analogy might help to clarify this concept. A horse and a donkey may be crossbred to produce a hybrid, the mule. The mule doesn't look or behave exactly like either parent, yet it has attributes of both. The resonance hybrid of a molecule has properties of each resonance form but is not identical to any one form. Unlike the mule, resonance hybrids *do not actually exist*. Rather, they comprise a model that results from the failure of any one Lewis structure to agree with experimentally obtained structural information.

The presence of resonance enhances molecular stability. The more resonance forms that exist, the greater is the stability of the molecule they represent. This concept is important in understanding the chemical reactions of many complex organic molecules and is used extensively in organic chemistry.

EXAMPLE 3.13 Drawing Resonance Hybrids of Covalently Bonded Compounds

Draw the possible resonance structures of the nitrate ion, NO$_3^-$, and represent them as a resonance hybrid.

Solution

Nitrogen is less electronegative than oxygen; therefore, nitrogen is the central atom and the skeletal structure is:

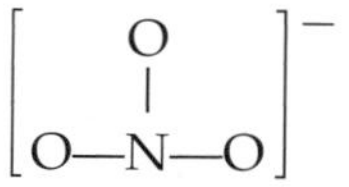

$$\begin{bmatrix} & \text{O} & \\ & | & \\ \text{O}-&\text{N}-&\text{O} \end{bmatrix}^-$$

The pool of valence electrons for anions is determined by adding one electron for each unit of negative charge:

$$1 \text{ N atom} \times 5 \text{ valence electrons} = 5 \text{ } e^-$$
$$3 \text{ O atoms} \times 6 \text{ valence electrons} = 18 \text{ } e^-$$
$$+ 1 \text{ negative charge} = 1 \text{ } e^-$$
$$24 \text{ } e^- \text{ total}$$

Continued—

EXAMPLE 3.13 —Continued

Distributing the electrons throughout the structure results in the legitimate Lewis structures:

$$\left[\begin{array}{c} :\ddot{O}: \\ \ddot{O}::N:\ddot{O}: \end{array}\right]^{-} \text{ and } \left[\begin{array}{c} :O: \\ :\ddot{O}:N:\ddot{O}: \end{array}\right]^{-} \text{ and } \left[\begin{array}{c} :\ddot{O}: \\ :\ddot{O}:N::\ddot{O} \end{array}\right]^{-}$$

All contribute to the true structure of the nitrate ion, represented as a resonance hybrid.

$$\left[\begin{array}{c} :\ddot{O}: \\ \ddot{O}::N:\ddot{O}: \end{array}\right]^{-} \leftrightarrow \left[\begin{array}{c} :O: \\ :\ddot{O}:N:\ddot{O}: \end{array}\right]^{-} \leftrightarrow \left[\begin{array}{c} :\ddot{O}: \\ :\ddot{O}:N::\ddot{O} \end{array}\right]^{-}$$

Question 3.19

SeO_2, like SO_2, has two resonance forms. Draw their Lewis structures.

Question 3.20

Explain any similarities between the structures for SeO_2 and SO_2 (in Question 3.19) in light of periodic relationships.

Lewis Structures and Exceptions to the Octet Rule

The octet rule is remarkable in its ability to realistically model bonding and structure in covalent compounds. But, like any model, it does not adequately describe all systems. Beryllium, boron, and aluminum, in particular, tend to form compounds in which they are surrounded by fewer than eight electrons. This situation is termed an *incomplete octet*. Other molecules, such as nitric oxide:

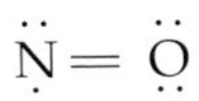

$$\dot{\ddot{N}} = \ddot{O}$$

are termed *odd electron* molecules. Note that it is impossible to pair all electrons to achieve an octet simply because the compound contains an odd number of valence electrons. Elements in the third period and beyond may involve *d* orbitals and form an *expanded octet*, with ten or even twelve electrons surrounding the central atom. Examples 3.14 and 3.15 illustrate common exceptions to the octet rule.

EXAMPLE 3.14

Drawing Lewis Structures of Covalently Bonded Compounds That Are Exceptions to the Octet Rule

Draw the Lewis structure of beryllium hydride, BeH_2.

Solution

A reasonable skeletal structure of BeH_2 is:

H—Be—H

Continued—

EXAMPLE 3.14 —Continued

The total number of valence electrons in BeH_2 is:

$$1 \text{ beryllium atom} \times 2 \text{ valence } e^-/\text{atom} = 2\ e^-$$
$$2 \text{ hydrogen atoms} \times 1 \text{ valence } e^-/\text{atom} = 2\ e^-$$
$$4\ e^- \text{ total}$$

The resulting Lewis structure must be:

$$H:Be:H \quad \text{or} \quad H\text{—}Be\text{—}H$$

It is apparent that there is no way to satisfy the octet rule for Be in this compound. Consequently, BeH_2 is an exception to the octet rule. It contains an incomplete octet.

EXAMPLE 3.15 Drawing Lewis Structures of Covalently Bonded Compounds That Are Exceptions to the Octet Rule

Draw the Lewis structure of phosphorus pentafluoride.

Solution

A reasonable skeletal structure of PF_5 is:

Phosphorus is a third-period element; it may have an expanded octet. The total number of valence electrons is:

$$1 \text{ phosphorus atom} \times 5 \text{ valence } e^-/\text{atom} = 5\ e^-$$
$$5 \text{ fluorine atoms} \times 7 \text{ valence } e^-/\text{atom} = 35\ e^-$$
$$40\ e^- \text{ total}$$

Distributing the electrons around each F in the skeletal structure results in the Lewis structure:

PF_5 is an example of a compound with an expanded octet.

Lewis Structures and Molecular Geometry; VSEPR Theory

LEARNING GOAL

The shape of a molecule plays a large part in determining its properties and reactivity. We may predict the shapes of various molecules by inspecting their Lewis structures for the orientation of their electron pairs. The covalent bond, for instance, in which bonding electrons are localized between the nuclear centers of the

3.4 Drawing Lewis Structures of Molecules and Polyatomic Ions

atoms, is *directional*; the bond has a specific orientation in space between the bonded atoms. Electrostatic forces in ionic bonds, in contrast, are *nondirectional*; they have no specific orientation in space. The specific orientation of electron pairs in covalent molecules imparts a characteristic shape to the molecules. Consider the following series of molecules whose Lewis structures are shown.

BeH_2 H : Be : H

BF_3 :F̈:
                :F̈ : B̈ : F̈:

CH_4 H
                H : C̈ : H
                H

NH_3 H : N̈ : H
                H

H_2O H : Ö : H

The electron pairs around the central atom of the molecule arrange themselves to minimize electronic repulsion. This means that the electron pairs arrange themselves so that they can be as far as possible from each other. We may use this fact to predict molecular shape. This approach is termed the **valence shell electron pair repulsion (VSEPR) theory.**

Let's see how the VSEPR theory can be used to describe the bonding and structure of each of the preceding molecules.

BeH_2

As we saw in Example 3.14, beryllium hydride has two shared electron pairs around the beryllium atom. These electron pairs have minimum repulsion if they are located as far apart as possible while still bonding the hydrogen to the central atom. This condition is met if the electron pairs are located on opposite sides of the molecule, resulting in a **linear structure,** 180° apart:

$$H : Be : H \quad \text{or} \quad H—Be—H$$

The *bond angle,* the angle between H—Be and Be—H bonds, formed by the two bonding pairs is 180° (Figure 3.6).

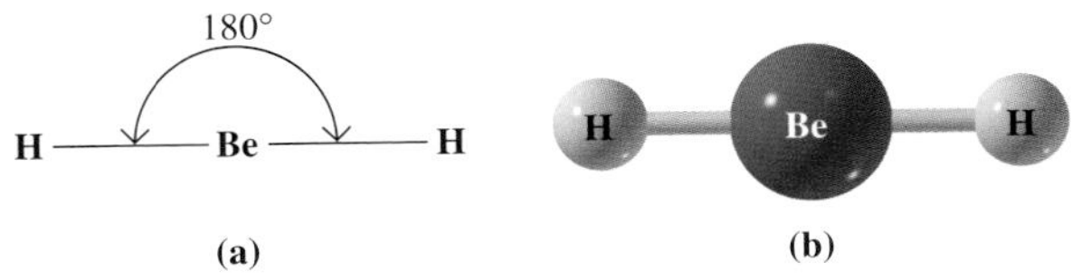

Only four electrons surround the beryllium atom in BeH_2. Consequently, BeH_2 is a stable exception to the octet rule.

Figure 3.6
Bonding and geometry in beryllium hydride, BeH_2. (a) Linear geometry in BeH_2. (b) Computer-generated model of linear BeH_2.

BF_3

Boron trifluoride has three shared electron pairs around the central atom. Placing the electron pairs in a plane, forming a triangle, minimizes the electron pair repulsion in this molecule, as depicted in Figure 3.7 and the following sketches:

BF_3 has only six electrons around the central atom, B. It is one of a number of stable compunds that are exceptions to the octet rule.

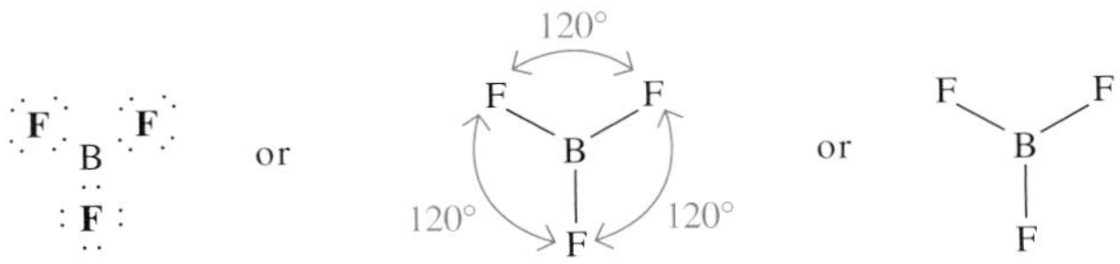

Figure 3.7
Bonding and geometry in boron trifluoride, BF$_3$: (a) trigonal planar geometry in BF$_3$; (b) computer-generated model of trigonal planar BF$_3$.

Such a structure is *trigonal planar,* and each F—B—F bond angle is 120°. We also find that compounds with central atoms in the same group of the periodic table have similar geometry. Aluminum, in the same group as boron, produces compounds such as AlH$_3$, which is also trigonal planar.

CH$_4$

Methane has four shared pairs of electrons. Here, minimum electron repulsion is achieved by arranging the electrons at the corners of a tetrahedron (Figure 3.8). Each H—C—H bond angle is 109.5°. Methane has a three-dimensional **tetrahedral structure.** Silicon, in the same group as carbon, forms compounds such as SiCl$_4$ and SiH$_4$ that also have tetrahedral structures.

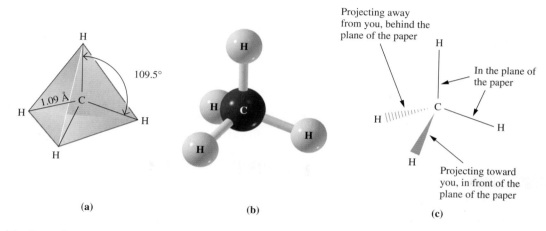

Figure 3.8
Representations of the three-dimensional structure of methane, CH$_4$. (a) Tetrahedral methane structure. (b) Computer-generated model of tetrahedral methane. (c) Three-dimensional representation of structure (b).

CH$_4$, NH$_3$, and H$_2$O all have eight electrons around their central atoms; all obey the octet rule.

NH$_3$

Ammonia also has four electron pairs about the central atom. In contrast to methane, in which all four pairs are bonding, ammonia has three pairs of bonding electrons and one nonbonding lone pair of electrons. We might expect CH$_4$ and NH$_3$ to have electron pair arrangements that are similar but not identical. The lone pair in ammonia is more negative than the bonding pairs; some of the negative charge on the bonding pairs is offset by the presence of the hydrogen atoms with their positive nuclei. Thus the arrangement of electron pairs in ammonia is distorted.

The hydrogen atoms in ammonia are pushed closer together than in methane (Figure 3.9). The bond angle is 107° because lone pair–bond pair repulsions are greater than bond pair–bond pair repulsions. The structure or shape is termed *trigonal pyramidal,* and the molecule is termed a **trigonal pyramidal molecule.**

Figure 3.9
The structure of the ammonia molecule. (a) Pyramidal ammonia structure. (b) Computer-generated model of pyramidal ammonia. (c) A three-dimensional sketch. (d) The H–N–H bond angle in ammonia.

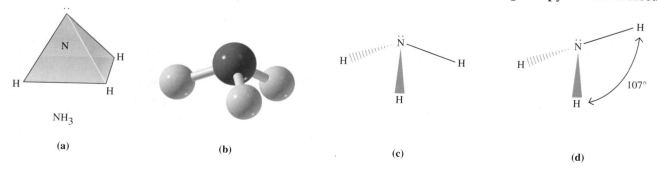

3.4 Drawing Lewis Structures of Molecules and Polyatomic Ions

H₂O

Water also has four electron pairs around the central atom; two pairs are bonding, and two pairs are nonbonding. These four electron pairs are approximately tetrahedral to each other; however, because of the difference between bonding and nonbonding electrons, noted earlier, the tetrahedral relationship is only approximate.

The **angular** (or *bent*) **structure** has a bond angle of 104.5°, which is 5° smaller than the tetrahedral angle, because of the repulsive effects of the lone pairs of electrons (as shown in Figure 3.10).

The characteristics of linear, trigonal planar, angular, and tetrahedral structures are summarized in Table 3.5.

> Molecules with five and six electron pairs also exist. They may have structures that are *trigonal bipyramidal* (forming a six-sided figure) or *octahedral* (forming an eight-sided figure).

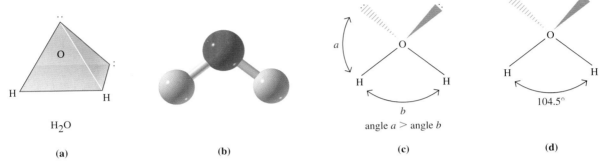

Figure 3.10
The structure of the water molecule. (a) Angular water structure. (b) Computer-generated model of angular water. (c) A three-dimensional sketch. (d) The H—O—H bond angle in water.

Periodic Structural Relationships

The molecules considered above contain the central atoms Be (Group IIA), B (Group IIIA), C (Group IVA), N (Group VA), and O (Group VIA). We may expect that a number of other compounds, containing the same central atom, will have

| TABLE 3.5 | Molecular Structure: The Geometry of a Molecule Is Affected by the Number of Nonbonded Electron Pairs Around the Central Atom and the Number of Bonded Atoms |

Bonded Atoms	Nonbonding Electron Pairs	Bond Angle	Molecular Structure	Example	Structure
2	0	180°	Linear	CO_2	
3	0	120°	Trigonal planar	SO_3	
2	1	<120°	Angular	SO_2	
4	0	~109°	Tetrahedral	CH_4	
3	1	~107°	Trigonal pyramidal	NH_3	
2	2	~104.5°	Angular	H_2O	

structures with similar geometries. This is an approximation, not always true, but still useful in expanding our ability to write reasonable, geometrically accurate structures for a large number of compounds.

The periodic similarity of group members is also useful in predictions involving bonding. Consider Group VI, oxygen, sulfur, and selenium (Se). Each has six valence electrons. Each needs two more electrons to complete its octet. Each should react with hydrogen, forming H_2O, H_2S, and H_2Se.

If we recall that H_2O is an angular molecule with the following Lewis structure,

$$H:\ddot{\underset{..}{O}}:$$
$$H$$

it follows that H_2S and H_2Se would also be angular molecules with similar Lewis structures, or

$$H:\ddot{\underset{..}{S}}: \quad \text{and} \quad H:\ddot{\underset{..}{Se}}:$$
$$H \qquad\qquad\qquad H$$

This logic applies equally well to the other representative elements.

More Complex Molecules

A molecule such as dimethyl ether, CH_3—O—CH_3, has two different central atoms: oxygen and carbon. We could picture the parts of the molecule containing the CH_3 group (commonly referred to as the *methyl group*) as exhibiting tetrahedral geometry (analogous to methane):

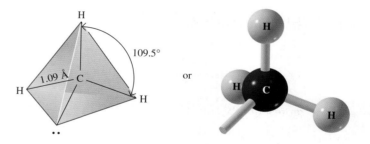

The part of the molecule connecting these two methyl groups (the oxygen) would have a bond angle similar to that of water (in which oxygen is also the central atom), approximately 104°, as seen in Figure 3.11. This is a reasonable way to represent the molecule dimethyl ether.

Trimethylamine, $(CH_3)_3N$, is a member of the amine family. As in the case of ether, two different central atoms are present. Carbon and nitrogen determine the geometry of amines. In this case the methyl group should assume the tetrahedral geometry of methane, and the nitrogen atom should have the methyl groups in a pyramidal arrangement, similar to the hydrogen atoms in ammonia, as seen in

Figure 3.11
A comparison of the bonding in water and dimethylether.

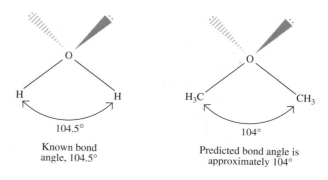

3.4 Drawing Lewis Structures of Molecules and Polyatomic Ions

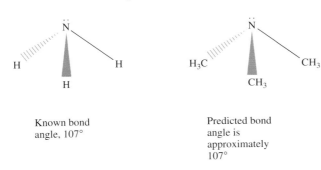

Known bond angle, 107°

Predicted bond angle is approximately 107°

Figure 3.12
A comparison of the bonding in ammonia and trimethylamine.

Figure 3.12. This creates a pyramidal geometry around nitrogen. H—N—H bond angles in ammonia are 107°; experimental information shows a very similar C—N—C bond angle in trimethylamine.

Question 3.21

Sketch the geometry of each of the following molecules (basing your structure on the Lewis electron dot representation of the molecule):

a. PH_3
b. SiH_4

Question 3.22

Sketch the geometry of each of the following molecules (basing your structure on the Lewis electron dot representation of the molecule):

a. C_2H_4
b. C_2H_2

It is essential to represent the molecule in its correct geometric form, using the Lewis and VSEPR theories, to understand its physical and chemical behavior. In the rest of this section we use these models to predict molecular behavior.

Lewis Structures and Polarity

A molecule is *polar* if its centers of positive and negative charges do not coincide. Molecules whose positive and negative charges are separated when the molecules are placed in an electric field align themselves with the field. The molecule behaves as a *dipole* (having two "poles" or ends, one pole is more negative and the other pole is more positive) and is said to be polar.

Nonpolar molecules will not align with the electric field because their positive and negative centers are the same; no dipole exists. These molecules are nonpolar.

The hydrogen molecule is the simplest nonpolar molecule:

$$H:H \quad \text{or} \quad H—H \quad -NONPOLAR$$

Both electrons, on average, are located at the center of the molecule and positively charged nuclei are on either side. The center of both positive and negative charge is at the center of the molecule; therefore the bond is nonpolar.

We may arrive at the same conclusion by considering the equality of electron sharing between the atoms being bonded. Electron sharing is related to the concept of electronegativity introduced in Section 3.1.

The atoms that comprise H_2 are identical; their electronegativity (electron attracting power) is the same. Thus the electrons remain at the center of the molecule, and the molecule is nonpolar.

Remember: Electronegativity deals with atoms in molecules, whereas electron affinity and ionization energy deal with isolated atoms.

110   Chapter 3 Structure and Properties of Ionic and Covalent Compounds

Similarly, O_2, N_2, Cl_2, and F_2 are nonpolar molecules with nonpolar bonds. Arguments analogous to those made for hydrogen explain these observations as well.

Let's next consider hydrogen fluoride, HF. Fluorine is more electronegative than hydrogen. This indicates that the electrons are more strongly attracted to a fluorine atom than they are to a hydrogen atom. This results in a bond and molecule that are polar. The symbol

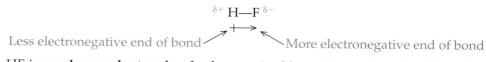

Less electronegative part of bond ⟶   ⟵ More electronegative part of bond

placed below a bond indicates the direction of polarity. The more negative end of the bond is near the head of the arrow, and the less negative end of the bond is next to the tail of the arrow. Symbols using the Greek letter *delta* may also be used to designate polarity. In this system the more negative end of the bond is designated δ^- (partial negative), and the less negative end is designated δ^+ (partial positive). The symbols are applied to the hydrogen fluoride molecule as follows:

$$\delta^+ \; H\text{—}F \; \delta^-$$

Less electronegative end of bond ⟶   ⟵ More electronegative end of bond

HF is a **polar covalent molecule** characterized by **polar covalent bonding**. This implies that the electrons are shared unequally.

A molecule containing all nonpolar bonds must also be nonpolar. In contrast, a molecule containing polar bonds may be either polar or nonpolar depending on the relative arrangement of the bonds and any lone pairs of electrons.

Let's now examine the bonding in carbon tetrachloride. All four bonds of CCl_4 are polar because of the electronegativity difference between C and Cl. However, because of the symmetrical arrangement of the four C—Cl bonds, their polarities cancel, and the molecule is nonpolar covalent:

> The word *partial* implies less than a unit charge. Thus δ^+ and δ^- do not imply overall charge on a unit (such as the + or − sign on an ion); they are meant to show only the relative distribution of charge within a unit. The HF molecule shown is *neutral* but has an unequal charge distribution *within* the molecule.

Now look at H_2O. Because of its angular (bent) structure, the polar bonds do not cancel, and the molecule is polar covalent:

The electron density is shifted away from the hydrogens toward oxygen in the water molecule. In carbon tetrachloride, equal electron "pull" in all directions results in a nonpolar covalent molecule.

Question 3.23

Predict which of the following bonds are polar, and, if polar, in which direction the electrons are pulled:

a. O—S
b. C≡N
c. Cl—Cl
d. I—Cl

Question 3.24

Predict which of the following bonds are polar, and, if polar, in which direction the electrons are pulled:

a. Si—Cl
b. S—Cl
c. H—C
d. C—C

Question 3.25

Predict whether each of the following molecules is polar:

a. BCl_3
b. NH_3
c. HCl
d. $SiCl_4$

Question 3.26

Predict whether each of the following molecules is polar:

a. CO_2
b. SCl_2
c. BrCl
d. CS_2

3.5 Properties Based on Electronic Structure and Molecular Geometry

Intramolecular forces are attractive forces *within* molecules. They are the chemical bonds that determine the shape and polarity of individual molecules. **Intermolecular forces,** on the other hand, are forces *between* molecules.

It is important to distinguish between these two kinds of forces. It is the *intermolecular* forces that determine such properties as the solubility of one substance in another and the freezing and boiling points of liquids. But, at the same time we must realize that these forces are a direct consequence of the *intramolecular* forces in the individual units, the molecules.

In the following section we will see some of the consequences of bonding that are directly attributable to differences in intermolecular forces (solubility, boiling and melting points). In Section 5.2 we will investigate, in some detail, the nature of the intermolecular forces themselves.

8 LEARNING GOAL

9 LEARNING GOAL

Solubility

Solubility is defined as the maximum amount of solute that dissolves in a given amount of solvent at a specified temperature. Polar molecules are most soluble in polar solvents, whereas nonpolar molecules are most soluble in nonpolar solvents. This is the rule of *"like dissolves like."* Substances of similar polarity are mutually soluble, and large differences in polarity lead to insolubility.

The solute is the substance that is present in lesser quantity, and the solvent is the substance that is present in the greater amount (see Section 6.3).

Case I: Ammonia and Water

Ammonia is soluble in water because both ammonia and water are polar molecules:

The interaction of water and ammonia is an example of a particularly strong intermolecular force, the hydrogen bond; this phenomenon is discussed in Chapter 5.

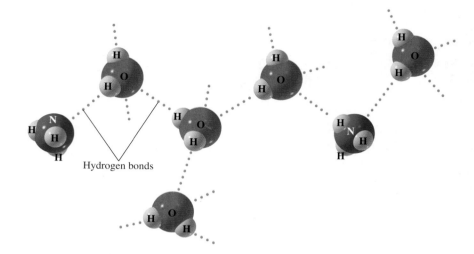

Figure 3.13
The interaction of polar covalent water molecules (the solvent) with polar covalent solute molecules such as ammonia, resulting in the formation of a solution.

Dissolution of ammonia in water is a consequence of the intermolecular forces present among the ammonia and water molecules. The δ^- end (a nitrogen) of the ammonia molecule is attracted to the δ^+ end (a hydrogen) of the water molecule; at the same time the δ^+ end (a hydrogen) of the ammonia molecule is attracted to the δ^- end (an oxygen) of the water molecule. These attractive forces thus "pull" ammonia into water (and water into ammonia), and the ammonia molecules are randomly distributed throughout the solvent, forming a homogeneous solution (Figure 3.13).

Case II: Oil and Water

Oil and water do not mix; oil is a nonpolar substance composed primarily of molecules containing carbon and hydrogen. Water molecules, on the other hand, are quite polar. The potential solvent, water molecules, have partially charged ends, whereas the molecules of oil do not. As a result, water molecules exert their attractive forces on other water molecules, not on the molecules of oil; the oil remains insoluble, and because it is less dense than water, the oil simply floats on the surface of the water. This is illustrated in Figure 3.14.

Boiling Points of Liquids and Melting Points of Solids

Boiling a liquid requires energy. The energy is used to overcome the intermolecular attractive forces in the liquid, driving the molecules into the less associated gas phase. The amount of energy required to accomplish this is related to the boiling temperature. This, in turn, depends on the strength of the intermolecular attractive forces in the liquid, which parallel the polarity. This is not the only determinant of boiling point. Molecular mass is also an important consideration. The larger the mass of the molecule, the more difficult it becomes to convert the collection of molecules to the gas phase.

A similar argument can be made for the melting points of solids. The ease of conversion of a solid to a liquid also depends on the magnitude of the attractive forces in the solid. The situation actually becomes very complex for ionic solids because of the complexity of the crystal lattice.

As a general rule, polar compounds have strong attractive (intermolecular) forces, and their boiling and melting points tend to be higher than those of nonpolar substances of similar molecular mass.

Melting and boiling points of a variety of substances are included in Table 3.6.

Figure 3.14
The interaction of polar water molecules and nonpolar oil molecules. The familiar salad dressing, oil and vinegar, forms two layers. The oil does not dissolve in vinegar, an aqueous solution of acetic acid.

See Section 5.3, the solid state.

TABLE 3.6 Melting and Boiling Points of Selected Compounds in Relation to Their Bonding Type

Formula (Name)	Bonding Type	M.P. (°C)	B.P. (°C)
N_2 (nitrogen)	Nonpolar covalent	−210	−196
O_2 (oxygen)	Nonpolar covalent	−219	−183
NH_3 (ammonia)	Polar covalent	−78	−33
H_2O (water)	Polar covalent	0	100
NaCl (sodium chloride)	Ionic	801	1413
KBr (potassium bromide)	Ionic	730	1435

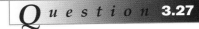

Predict which compound in each of the following groups should have the higher melting and boiling points (Hint: Write the Lewis dot structure and determine whether the molecule is polar or nonpolar.):

a. H_2O and C_2H_4
b. CO and CH_4
c. NH_3 and N_2
d. Cl_2 and ICl

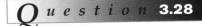

Predict which compound in each of the following groups should have the higher melting and boiling points (Hint: Write the Lewis dot structure and determine whether the molecule is polar or nonpolar.):

a. C_2H_6 and CH_4
b. CO and NO
c. F_2 and Br_2
d. $CHCl_3$ and CCl_4

[handwritten: Br₂ lewis structure Br̄—Br̄ 2×7 = 14 e⁻]

SUMMARY

3.1 Chemical Bonding

When two atoms are joined to make a chemical compound, the force of attraction between the two atoms is the *chemical bond*. *Ionic bonding* is characterized by an electron transfer process occurring before bond formation, forming an *ion pair*. In *covalent bonding*, electrons are shared between atoms in the bonding process. *Polar covalent bonding*, like covalent bonding, is based on the concept of electron sharing; however, the sharing is unequal and based on the *electronegativity* difference between joined atoms. The *Lewis symbol*, showing only valence electrons, is a convenient way of representing atoms singly or in combination.

3.2 Naming Compounds and Writing Formulas of Compounds

The "shorthand" symbol for a compound is its *formula*. The formula identifies the number and type of atoms in the compound.

An ion that consists of only a single atom is said to be *monatomic*. *Polyatomic ions*, such as the hydroxide ion, OH^-, are composed of two or more atoms bonded together.

Names of ionic compounds are derived from the names of their ions. The name of the cation appears first, followed by the name of the anion. In the Stock system for naming an ion (the systematic name), a Roman numeral indicates the charge of the cation. In the older common nomenclature system, the suffix *-ous* indicates the lower of the ionic charges, and the suffix *-ic* indicates the higher ionic charge.

Most covalent compounds are formed by the reaction of nonmetals. Covalent compounds exist as *molecules.*

The convention used for naming covalent compounds is as follows:

- The names of the elements are written in the order in which they appear in the formula.
- A prefix indicating the number of each kind of atom found in the unit is placed before the name of the element.
- The stem of the name of the last element is used with the suffix *-ide.*

Many compounds are so familiar to us that their common names are used. It is useful to be able to correlate both systematic and common names with the corresponding molecular formula.

3.3 Properties of Ionic and Covalent Compounds

Covalently bonded molecules are discrete units, and they have less tendency to form an extended structure in the solid state. Ionic compounds, with ions joined by electrostatic attraction, do not have definable units but form a *crystal lattice* composed of positive and negative ions in an extended three-dimensional network.

The *melting point* is the temperature at which a solid is converted to a liquid; the *boiling point* is the temperature at which a liquid is converted to a gas. Melting and boiling temperatures for ionic compounds are generally higher than those of covalent compounds.

Ionic solids are crystalline, whereas covalent solids may be either crystalline or amorphous.

Many ionic solids dissolve in water, *dissociating* into positive and negative ions (an *electrolytic solution*). Because these ions can carry (conduct) a current of electricity, they are called *electrolytes.* Covalent solids in solution usually retain their neutral character and are *nonelectrolytes.*

3.4 Drawing Lewis Structures of Molecules and Polyatomic Ions

The procedure for drawing Lewis structures of molecules involves writing a skeletal structure of the molecule, arranging the atoms in their most probable order, determining the number of valence electrons on each atom, and combining them to get the total for the compound. The electrons are then distributed around the atoms (in pairs if possible) to satisfy the octet rule. At this point electron pairs may be moved, creating double or triple bonds, in an effort to satisfy the octet rule for all atoms and produce the final structure.

Stability of a covalent compound is related to the *bond energy.* The magnitude of the bond energy decreases in the order *triple bond > double bond > single bond.* The bond length decreases in the order *single bond > double bond > triple bond.*

The *valence shell electron pair repulsion theory* states that electron pairs around the central atom of the molecule arrange themselves to minimize electronic repulsion; the electrons orient themselves as far as possible from each other. Two electron pairs around the central atom lead to a *linear* arrangement of the attached atoms; three indicate a *trigonal planar* arrangement, and four result in a *tetrahedral geometry.* Both lone pair and bonding pair electrons must be taken into account when predicting structure. Molecules with fewer than four and as many as five or six electron pairs around the central atom also exist. They are exceptions to the octet rule.

A molecule is polar if its centers of positive and negative charges do not coincide. A *polar covalent* molecule has at least one *polar covalent bond.* An understanding of the concept of *electronegativity,* the relative electron-attracting power of atoms in molecules, helps us to assess the polarity of a bond.

A molecule containing all nonpolar bonds must be nonpolar. A molecule containing polar bonds may be either polar or nonpolar, depending on the relative position of the bonds.

3.5 Properties Based on Electronic Structure and Molecular Geometry

Attractions between molecules are called *intermolecular forces. Intramolecular forces,* on the other hand, are the attractive forces within molecules. It is the intermolecular forces that determine such properties as the solubility of one substance in another and the freezing and boiling points of liquids.

Solubility is the maximum amount of solute that dissolves in a given amount of solvent at a specified temperature. Polar molecules are most soluble in polar solvents; nonpolar molecules are most soluble in nonpolar solvents. This is the rule of "like dissolves like."

As a general rule, polar compounds have strong intermolecular forces, and their boiling and melting points tend to be higher than nonpolar compounds of similar molecular mass.

KEY TERMS

angular structure (3.4)
boiling point (3.3)
bond energy (3.4)
chemical bond (3.1)
covalent bond (3.1)
crystal lattice (3.2)
dissociation (3.3)
double bond (3.4)
electrolyte (3.3)
electrolytic solution (3.3)
electronegativity (3.1)
formula (3.2)
intermolecular force (3.5)
intramolecular force (3.5)
ionic bonding (3.1)

ion pair (3.1)
Lewis symbol (3.1)
linear structure (3.4)
lone pair (3.4)
melting point (3.3)
molecule (3.2)
monatomic ion (3.2)
nomenclature (3.2)
nonelectrolyte (3.3)
polar covalent bonding (3.4)
polar covalent molecule (3.4)
polyatomic ion (3.2)
resonance (3.4)
resonance form (3.4)
resonance hybrid (3.4)

single bond (3.4)
solubility (3.5)
tetrahedral structure (3.4)
trigonal pyramidal
   molecule (3.4)
triple bond (3.4)
valence shell electron
   pair repulsion (VSEPR)
   theory (3.4)

QUESTIONS AND PROBLEMS

Chemical Bonding

Foundations

3.29 Classify each of the following compounds as ionic or covalent:
  a. $MgCl_2$
  b. CO_2
  c. H_2S
  d. NO_2

3.30 Classify each of the following compounds as ionic or covalent:
  a. NaCl
  b. CO
  c. ICl
  d. H_2

3.31 Classify each of the following compounds as ionic or covalent:
  a. SiO_2
  b. SO_2
  c. SO_3
  d. $CaCl_2$

3.32 Classify each of the following compounds as ionic or covalent:
  a. NF_3
  b. NaF
  c. CsF — ionic
  d. $SiCl_4$

Applications

3.33 Using Lewis symbols, write an equation predicting the product of the reaction of:
  a. Li + Br
  b. Mg + Cl

3.34 Using Lewis symbols, write an equation predicting the product of the reaction of:
  a. Na + O
  b. Na + S

3.35 Using Lewis symbols, write an equation predicting the product of the reaction of:
  a. S + H
  b. P + H

3.36 Using Lewis symbols, write an equation predicting the product of the reaction of:
  a. Si + H
  b. Ca + F

3.37 Explain, using Lewis symbols and the octet rule, why helium is so nonreactive.

3.38 Explain, using Lewis symbols and the octet rule, why neon is so nonreactive.

Naming Compounds and Writing Formulas of Compounds

3.39 Name each of the following ions:
  a. Na^+
  b. Cu^+
  c. Mg^{2+}

3.40 Name each of the following ions:
  a. Cu^{2+}
  b. Fe^{2+}
  c. Fe^{3+}

3.41 Name each of the following ions:
  a. S^{2-}
  b. Cl^-
  c. CO_3^{2-}

3.42 Name each of the following ions:
  a. ClO^-
  b. NH_4^+
  c. CH_3COO^-

3.43 Write the formula for each of the following monatomic ions:
  a. the potassium ion
  b. the bromide ion

3.44 Write the formula for each of the following monatomic ions:
  a. the calcium ion
  b. the chromium(VI) ion

3.45 Write the formula for each of the following complex ions:
  a. the sulfate ion
  b. the nitrate ion

3.46 Write the formula for each of the following complex ions:
  a. the phosphate ion
  b. the bicarbonate ion

3.47 Write the correct formula for each of the following:
  a. sodium chloride
  b. magnesium bromide

3.48 Write the correct formula for each of the following:
  a. copper(II) oxide
  b. iron(III) oxide

3.49 Write the correct formula for each of the following:
  a. silver cyanide
  b. ammonium chloride

3.50 Write the correct formula for each of the following:
  a. magnesium carbonate
  b. magnesium bicarbonate

3.51 Name each of the following compounds:
  a. $MgCl_2$
  b. $AlCl_3$

3.52 Name each of the following compounds:
  a. Na_2O
  b. $Fe(OH)_3$

3.53 Name each of the following covalent compounds:
  a. NO_2
  b. SO_3

3.54 Name each of the following covalent compounds:
  a. N_2O_4
  b. CCl_4

3.55 Predict the formula of a compound formed from:
  a. aluminum and oxygen
  b. lithium and sulfur

3.56 Predict the formula of a compound formed from:
  a. boron and hydrogen
  b. magnesium and phosphorus

3.57 Predict the formula of a compound formed from:
  a. carbon and oxygen
  b. sulfur and hydrogen

3.58 Predict the formula of a compound formed from:
  a. calcium and oxygen
  b. silicon and hydrogen

3.59 Write a suitable formula for each of the following:
  a. sodium nitrate
  b. magnesium nitrate

3.60 Write a suitable formula for each of the following:
  a. aluminum nitrate
  b. ammonium nitrate

3.61 Write a suitable formula for each of the following:
  a. ammonium iodide
  b. ammonium sulfate

3.62 Write a suitable formula for each of the following:
  a. ammonium acetate
  b. ammonium cyanide

3.63 Name each of the following:
  a. CuS
  b. $CuSO_4$

3.64 Name each of the following:
  a. $Cu(OH)_2$
  b. CuO

3.65 Name each of the following:
   a. NaClO
   b. NaClO$_2$
3.66 Name each of the following:
   a. NaClO$_3$
   b. NaClO$_4$

Properties of Ionic and Covalent Compounds

Foundations

3.67 Contrast ionic and covalent compounds with respect to their solid state structure.
3.68 Contrast ionic and covalent compunds with respect to their behavior in solution.
3.69 Contrast ionic and covalent compounds with respect to their relative boiling points.
3.70 Contrast ionic and covalent compounds with respect to their relative melting points.

Applications

3.71 Would KCl be expected to be a solid at room temperature? Why?
3.72 Would CCl$_4$ be expected to be a solid at room temperature? Why?
3.73 Would H$_2$O or CCl$_4$ be expected to have a higher boiling point? Why?
3.74 Would H$_2$O or CCl$_4$ be expected to have a higher melting point? Why?

Drawing Lewis Structures of Molecules and Polyatomic Ions

Foundations

3.75 Draw the appropriate Lewis structure for each of the following atoms:
   a. H   c. C
   b. He   d. N
3.76 Draw the appropriate Lewis structure for each of the following atoms:
   a. Be   c. F
   b. B   d. S
3.77 Draw the appropriate Lewis structure for each of the following ions:
   a. Li$^+$ c. Cl$^-$
   b. Mg^{2+} d. P^{3-}
3.78 Draw the appropriate Lewis structure for each of the following ions:
   a. Be^{2+} c. O^{2-}
   b. Al^{3+} d. S^{2-}

Applications

3.79 Give the Lewis structure for each of the following compounds:
   a. NCl$_3$ b. CH$_3$OH   c. CS$_2$
3.80 Give the Lewis structure for each of the following compounds:
   a. HNO$_3$ b. CCl$_4$ c. PBr$_3$
3.81 Using the VSEPR theory, predict the geometry, polarity, and water solubility of each compound in Question 3.79.
3.82 Using the VSEPR theory, predict the geometry, polarity, and water solubility of each compound in Question 3.80.
3.83 Discuss the concept of resonance, being certain to define the terms *resonance*, *resonance form*, and *resonance hybrid*.
3.84 Why is resonance an important concept in bonding?
3.85 The acetate ion (Example 3.12) exhibits resonance. Draw two resonance forms of the acetate ion.
3.86 The nitrate ion (Table 3.3) has three resonance forms. Draw each form.
3.87 Ethanol (ethyl alcohol or grain alcohol) has a molecular formula of C$_2$H$_5$OH. Represent the structure of ethanol using the Lewis electron dot approach.

3.88 Formaldehyde, H$_2$CO, in water solution has been used as a preservative for biological specimens. Represent the Lewis structure of formaldehyde.
3.89 Acetone, C$_3$H$_6$O, is a common solvent. It is found in such diverse materials as nail polish remover and industrial solvents. Draw its Lewis structure if its skeletal structure is

3.90 Ethylamine is an example of an important class of organic compounds. The molecular formula of ethylamine is CH$_3$CH$_2$NH$_2$. Draw its Lewis structure.
3.91 Predict whether the bond formed between each of the following pairs of atoms would be ionic, nonpolar, or polar covalent:
   a. S and O   d. Na and O
   b. Si and P   e. Ca and Br
   c. Na and Cl
3.92 Predict whether the bond formed between each of the following pairs of atoms would be ionic, nonpolar, or polar covalent:
   a. Cl and Cl   d. Li and F
   b. H and H   e. O and O
   c. C and H
3.93 Draw an appropriate covalent Lewis structure formed by the simplest combination of atoms in Problem 3.91 for each solution that involves a nonpolar or polar covalent bond.
3.94 Draw an appropriate covalent Lewis structure formed by the simplest combination of atoms in Problem 3.92 for each solution that involves a nonpolar or polar covalent bond.

Properties Based on Electronic Structure and Molecular Geometry

3.95 What is the relationship between the polarity of a bond and the polarity of the molecule?
3.96 What effect does polarity have on the solubility of a compound in water?
3.97 What effect does polarity have on the melting point of a pure compound?
3.98 What effect does polarity have on the boiling point of a pure compound?
3.99 Would you expect KCl to dissolve in water?
3.100 Would you expect ethylamine (Question 3.90) to dissolve in water?

CRITICAL THINKING PROBLEMS

1. Predict differences in our global environment that may have arisen if the freezing point and boiling point of water were 20°C higher than they are.
2. Would you expect the compound C$_2$S$_2$H$_4$ to exist? Why or why not?
3. Draw the resonance forms of the carbonate ion. What conclusions, based on this exercise, can you draw about the stability of the carbonate ion?
4. Which of the following compounds would be predicted to have the higher boiling point? Explain your reasoning.

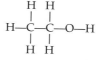

Ethanol   Ethane

5. Why does the octet rule not work well for compounds of lanthanide and actinide elements? Suggest a number other than eight that may be more suitable.

GENERAL CHEMISTRY

Calculations and the Chemical Equation

4

Careful measurements validate chemical equations.

Learning Goals

1. Know the relationship between the mole and Avogadro's number, and the usefulness of these quantities.

2. Perform calculations using Avogadro's number and the mole.

3. Write chemical formulas for common inorganic substances.

4. Calculate the formula weight and molar mass of a compound.

5. Know the major function served by the chemical equation, the basis for chemical calculations.

6. Classify chemical reactions by type: combination, decomposition, or replacement.

7. Recognize the various classes of chemical reactions: precipitation, reactions with oxygen, acid-base, and oxidation-reduction.

8. Balance chemical equations given the identity of products and reactants.

9. Calculate the number of moles or grams of product resulting from a given number of moles or grams of reactants or the number of moles or grams of reactant needed to produce a certain number of moles or grams of product.

10. Calculate theoretical and percent yield.

Outline

Chemistry Connection:
The Chemistry of Automobile Air Bags

4.1 The Mole Concept and Atoms

4.2 The Chemical Formula, Formula Weight, and Molar Mass

4.3 The Chemical Equation and the Information It Conveys

4.4 Balancing Chemical Equations

A Medical Perspective:
Carbon Monoxide Poisoning: A Case of Combining Ratios

4.5 Calculations Using the Chemical Equation

A Medical Perspective:
Pharmaceutical Chemistry: The Practical Significance of Percent Yield

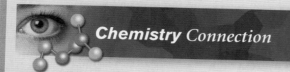

Chemistry Connection

The Chemistry of Automobile Air Bags

Each year, thousands of individuals are killed or seriously injured in automobile accidents. Perhaps most serious is the front-end collision. The car decelerates or stops virtually on impact; the momentum of the passengers, however, does not stop, and the driver and passengers are thrown forward toward the dashboard and the front window. Suddenly, passive parts of the automobile, such as control knobs, the rearview mirror, the steering wheel, the dashboard, and the windshield, become lethal weapons.

Automobile engineers have been aware of these problems for a long time. They have made a series of design improvements to lessen the potential problems associated with front-end impact. Smooth switches rather than knobs, recessed hardware, and padded dashboards are examples. These changes, coupled with the use of lap and shoulder belts, which help to immobilize occupants of the car, have decreased the frequency and severity of the impact and lowered the death rate for this type of accident.

An almost ideal protection would be a soft, fluffy pillow, providing a cushion against impact. Such a device, an air bag inflated only on impact, is now standard equipment for the protection of the driver and front-seat passenger.

How does it work? Ideally, it inflates only when severe front-end impact occurs; it inflates very rapidly (in approximately 40 milliseconds), then deflates to provide a steady deceleration, cushioning the occupants from impact. A remarkably simple chemical reaction makes this a reality.

When solid sodium azide (NaN_3) is detonated by mechanical energy produced by an electric current, it decomposes to form solid sodium and nitrogen gas:

$$2NaN_3(s) \longrightarrow 2Na(s) + 3N_2(g)$$

The nitrogen gas inflates the air bag, cushioning the driver and front-seat passenger.

The solid sodium azide has a high density (characteristic of solids) and thus occupies a small volume. It can easily be stored in the center of a steering wheel or in the dashboard. The rate of the detonation is very rapid. In milliseconds it produces three moles of N_2 gas for every two moles of NaN_3. The N_2 gas occupies a relatively large volume because its density is low. This is a general property of gases.

Figuring out how much sodium azide is needed to produce enough nitrogen to properly inflate the bag is an example of a practical application of the chemical arithmetic that we are learning in this chapter.

Introduction

The calculation of chemical quantities based on chemical equations, termed stoichiometry, is the application of logic and arithmetic to chemical systems to answer questions such as the following:

A pharmaceutical company wishes to manufacture 1000 kg of a product next year. How much of each of the starting materials must be ordered? If the starting materials cost $20/g, how much money must be budgeted for the project?

We often need to predict the quantity of a product produced from the reaction of a given amount of material. This calculation is possible. It is equally possible to calculate how much of a material would be necessary to produce a desired amount of product. One of many examples was shown in the preceding Chemistry Connection: the need to solve a very practical problem.

What is required is a recipe: a procedure to follow. The basis for our recipe is the chemical equation. A properly written chemical equation provides all of the necessary information for the chemical calculation. That critical information is the combining ratio of elements or compounds that must occur to produce a certain amount of product or products.

In this chapter we define the mole, the fundamental unit of measure of chemical arithmetic, learn to write and balance chemical equations, and use these tools to perform calculations of chemical quantities.

4.1 The Mole Concept and Atoms

Atoms are exceedingly small, yet their masses have been experimentally determined for each of the elements. The unit of measurement for these determinations is the **atomic mass unit,** abbreviated amu:

 $1 \text{ amu} = 1.661 \times 10^{-24} \text{ g}$

The Mole and Avogadro's Number

The exact value of the atomic mass unit is defined in relation to a standard, just as the units of the metric system represent defined quantities. The carbon-12 isotope has been chosen and is assigned a mass of exactly 12 atomic mass units. Hence this standard reference point defines an atomic mass unit as exactly one-twelfth the mass of a carbon-12 atom.

The periodic table provides atomic weights in atomic mass units. These atomic weights are average values, based on the contribution of all naturally occurring isotopes of the particular element. For example, the average mass of a carbon atom is 12.01 amu and

$$\frac{12.01 \text{ amu C}}{\text{C atom}} \times \frac{1.661 \times 10^{-24} \text{ g C}}{1 \text{ amu C}} = 1.995 \times \frac{10^{-23} \text{ g C}}{\text{C atom}}$$

The average mass of a helium atom is 4.003 amu and

$$\frac{4.003 \text{ amu He}}{\text{He atom}} \times \frac{1.661 \times 10^{-24} \text{ g He}}{1 \text{ amu He}} = 6.649 \times \frac{10^{-24} \text{ g He}}{\text{He atom}}$$

In everyday work, chemists use much larger quantities of matter (typically, grams or kilograms). A more practical unit for defining a "collection" of atoms is the **mole:**

 $1 \text{ mol of atoms} = 6.022 \times 10^{23} \text{ atoms of an element}$

This number is **Avogadro's number.** Amedeo Avogadro, a nineteenth-century scientist, conducted a series of experiments that provided the basis for the mole concept. This quantity is based on the number of carbon-12 atoms in one mole of carbon-12.

The practice of defining a unit for a quantity of small objects is common; a *dozen* eggs, a *ream* of paper, and a *gross* of pencils are well-known examples. Similarly, a mole is 6.022×10^{23} individual units of anything. We could, if we desired, speak of a mole of eggs or a mole of pencils. However, in chemistry we use the mole to represent a specific quantity of atoms, ions, or molecules.

The mole (mol) and the atomic mass unit (amu) are related. The atomic mass of an element corresponds to the average mass of a single atom in amu *and* the mass of a mole of atoms in grams.

The mass of 1 mol of atoms, in grams, is defined as the **molar mass.** Consider this relationship for sodium in Example 4.1.

> **1** LEARNING GOAL

The term *atomic weight* is not correct but is a fixture in common usage. Just remember that atomic weight is really "average atomic mass."

> **2** LEARNING GOAL

Relating Avogadro's Number to Molar Mass

EXAMPLE 4.1

Calculate the mass, in grams, of Avogadro's number of sodium atoms.

Continued—

EXAMPLE 4.1 —Continued

Solution

The periodic table indicates that the average mass of one sodium atom is 22.99 amu. This may be represented as:

$$\frac{22.99 \text{ amu Na}}{1 \text{ atom Na}}$$

In order to answer the question, we must calculate the molar mass in units of g/mol. We need two conversion factors; one to convert amu → grams and another to convert atoms → mol.

As previously noted, 1 amu is 1.661×10^{-24} g, and 6.022×10^{23} atoms of sodium is Avogadro's number. Similarly, these relationships may be formatted as:

$$1.661 \times 10^{-24} \frac{\text{g Na}}{\text{amu}} \quad \text{and} \quad 6.022 \times 10^{23} \frac{\text{atoms Na}}{\text{mol Na}}$$

Representing this information as a series of conversion factors, using the factor-label method, we have

$$22.99 \frac{\cancel{\text{amu Na}}}{\text{atom Na}} \times 1.661 \times 10^{-24} \frac{\text{g Na}}{\cancel{\text{amu Na}}} \times 6.022 \times 10^{23} \frac{\cancel{\text{atoms Na}}}{\text{mol Na}} = 22.99 \frac{\text{g Na}}{\text{mol Na}}$$

The average mass of one *atom* of sodium, in units of amu, is *numerically identical* to the mass of *Avogadro's number of atoms*, expressed in units of grams. Hence the molar mass of sodium is 22.99 g Na/mol.

Helpful Hint: Section 1.3 discusses the use of conversion factors.

The sodium example is not unique. The relationship holds for every element in the periodic table.

Because Avogadro's number of particles (atoms) is 1 mol, it follows that the average mass of one atom of hydrogen is 1.008 amu and the mass of 1 mol of hydrogen atoms is 1.008 g or the average mass of one atom of carbon is 12.01 amu and the mass of 1 mol of carbon atoms is 12.01 g.

In fact, one mole of atoms of *any element* contains the same number, Avogadro's number, of atoms, 6.022×10^{23} atoms.

The difference in mass of a mole of two different elements can be quite striking (Figure 4.1). For example, a mole of hydrogen atoms is 1.008 g, and a mole of lead atoms is 207.19 g.

Figure 4.1
The comparison of approximately one mole each of silver (as Morgan and Peace dollars), gold (as Canadian Maple Leaf coins), and copper (as pennies) shows the considerable difference in mass (as well as economic value) of equivalent moles of different substances.

Question 4.1

Calculate the mass, in grams, of Avogadro's number of aluminum atoms.

Question 4.2

Calculate the mass, in grams, of Avogadro's number of mercury atoms.

Calculating Atoms, Moles, and Mass

Performing calculations based on a chemical equation requires a facility for relating the number of atoms of an element to a corresponding number of moles of that element and ultimately to their mass in grams. Such calculations involve the use of conversion factors. The use of conversion factors was first described in Chapter 1. Some examples follow.

EXAMPLE 4.2

Converting Moles to Atoms

How many iron atoms are present in 3.0 mol of iron metal?

Solution

The calculation is based on choosing the appropriate conversion factor. The relationship

$$\frac{6.022 \times 10^{23} \text{ atoms Fe}}{1 \text{ mol Fe}}$$

follows directly from

$$1 \text{ mol Fe} = 6.022 \times 10^{23} \text{ atoms Fe}$$

Using this conversion factor, we have

$$\text{number of atoms of Fe} = 3.0 \text{ mol Fe} \times \frac{6.022 \times 10^{23} \text{ atoms Fe}}{1 \text{ mol Fe}}$$

$$= 18 \times 10^{23} \text{ atoms of Fe, or}$$

$$= 1.8 \times 10^{24} \text{ atoms of Fe}$$

EXAMPLE 4.3

Converting Atoms to Moles

Calculate the number of moles of sulfur represented by 1.81×10^{24} atoms of sulfur.

Solution

$$1.81 \times 10^{24} \text{ atoms S} \times \frac{1 \text{ mol S}}{6.022 \times 10^{23} \text{ atoms S}} = 3.01 \text{ mol S}$$

Note that this conversion factor is the inverse of that used in Example 4.2. Remember, the conversion factor must cancel units that should not appear in the final answer.

EXAMPLE 4.4

Converting Moles of a Substance to Mass in Grams

What is the mass, in grams, of 3.01 mol of sulfur?

Solution

We know from the periodic table that 1 mol of sulfur has a mass of 32.06 g. Setting up a suitable conversion factor between grams and moles results in

$$3.01 \text{ mol S} \times \frac{32.06 \text{ g S}}{1 \text{ mol S}} = 96.5 \text{ g S}$$

EXAMPLE 4.5 Converting Kilograms to Moles

Calculate the number of moles of sulfur in 1.00 kg of sulfur.

Solution

$$1.00 \text{ kg S} \times \frac{10^3 \text{ g S}}{1 \text{ kg S}} \times \frac{1 \text{ mol S}}{32.06 \text{ g S}} = 31.2 \text{ mol S}$$

EXAMPLE 4.6 Converting Grams to Number of Atoms

Calculate the number of atoms of sulfur in 1.00 g of sulfur.

Solution

$$1.00 \text{ g S} \times \frac{1 \text{ mol S}}{32.06 \text{ g S}} \times \frac{6.022 \times 10^{23} \text{ atoms S}}{1 \text{ mol S}} = 1.88 \times 10^{22} \text{ atoms S}$$

The preceding examples demonstrate the use of a sequence of conversion factors to proceed from the information *provided* in the problem to the information *requested* by the problem.

It is generally useful to map out a pattern for the required conversion. In Example 4.6 we are given the number of grams and need the number of atoms that correspond to that mass.

Begin by "tracing a path" to the answer:

$$\boxed{\text{grams sulfur}} \xrightarrow{\text{Step 1}} \boxed{\text{moles sulfur}} \xrightarrow{\text{Step 2}} \boxed{\text{atoms sulfur}}$$

Two transformations, or conversions, are required:

Step 1. Convert grams to moles.

Step 2. Convert moles to atoms.

To perform step 1, we could consider either

$$\frac{1 \text{ mol S}}{32.06 \text{ g S}} \quad \text{or} \quad \frac{32.06 \text{ g S}}{1 \text{ mol S}}$$

If we want grams to cancel, $\frac{1 \text{ mol S}}{32.06 \text{ g S}}$ is the correct choice, resulting in

$$\text{g S} \times \frac{1 \text{ mol S}}{32.06 \text{ g S}} = \text{value in mol S}$$

To perform step 2, the conversion of moles to atoms, the moles of S must cancel; therefore

$$\text{mol S} \times \frac{6.022 \times 10^{23} \text{ atoms S}}{1 \text{ mol S}} = \text{number of atoms S}$$

which are the units desired in the solution.

4.2 The Chemical Formula, Formula Weight, and Molar Mass

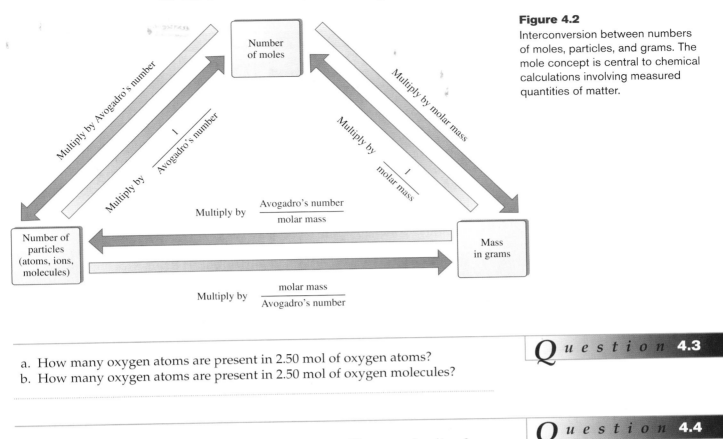

Figure 4.2
Interconversion between numbers of moles, particles, and grams. The mole concept is central to chemical calculations involving measured quantities of matter.

Question 4.3

a. How many oxygen atoms are present in 2.50 mol of oxygen atoms?
b. How many oxygen atoms are present in 2.50 mol of oxygen molecules?

Question 4.4

How many moles of sodium are represented by 9.03×10^{23} atoms of sodium?

Question 4.5

What is the mass, in grams, of 3.50 mol of the element helium?

Question 4.6

How many oxygen atoms are present in 40.0 g of oxygen molecules?

The conversion between the three principal measures of quantity of matter—the number of grams (mass), the number of moles, and the number of individual particles (atoms, ions, or molecules)—is essential to the art of problem solving in chemistry. Their interrelationship is depicted in Figure 4.2.

4.2 The Chemical Formula, Formula Weight, and Molar Mass

The Chemical Formula

Compounds are pure substances. They are composed of two or more elements that are chemically combined. A **chemical formula** is a combination of symbols of the various elements that make up the compound. It serves as a convenient way to

 LEARNING GOAL

Figure 4.3
The marked difference in color of (a) hydrated and (b) anhydrous copper sulfate is clear evidence that they are, in fact, different compounds.

represent a compound. The chemical formula is based on the formula unit. The **formula unit** is the smallest collection of atoms that provides two important pieces of information:

- the identity of the atoms present in the compound and
- the relative numbers of each type of atom.

Let's look at the following formulas:

- *Hydrogen gas,* H_2. This indicates that two atoms of hydrogen are chemically bonded forming diatomic hydrogen, hence the subscript 2.
- *Water,* H_2O. Water is composed of molecules that contain two atoms of hydrogen (subscript 2) and one atom of oxygen (lack of a subscript means *one* atom).
- *Sodium chloride,* NaCl. One atom of sodium and one atom of chlorine combine to make sodium chloride.
- *Calcium hydroxide,* $Ca(OH)_2$. Calcium hydroxide contains one atom of calcium and two atoms each of oxygen and hydrogen. The subscript outside the parentheses applies to *all* atoms inside the parentheses.
- *Ammonium sulfate,* $(NH_4)_2SO_4$. Ammonium sulfate contains two ammonium ions (NH_4^+) and one sulfate ion (SO_4^{2-}). Each ammonium ion contains one nitrogen and four hydrogen atoms. The formula shows that ammonium sulfate contains two nitrogen atoms, eight hydrogen atoms, one sulfur atom, and four oxygen atoms.
- *Copper(II) sulfate pentahydrate,* $CuSO_4 \cdot 5H_2O$. This is an example of a compound that has water in its structure. Compounds containing one or more water molecules as an integral part of their structure are termed **hydrates.** Copper sulfate pentahydrate has five units of water (or ten H atoms and five O atoms) in addition to one copper atom, one sulfur atom, and four oxygen atoms for a total atomic composition of:

  1 copper atom
  1 sulfur atom
  9 oxygen atoms
  10 hydrogen atoms

It is possible to determine the correct molecular formula of a compound from experimental data. You can find further information online at www.mhhe.com/denniston5e in "Composition and Formulas of Compounds."

Note that the symbol for water is preceded by a dot, indicating that, although the water is a formula unit capable of standing alone, in this case it is a part of a larger structure. Copper sulfate also exists as a structure free of water, $CuSO_4$. This form is described as anhydrous (no water) copper sulfate. The physical and chemical properties of a hydrate often differ markedly from the anhydrous form (Figure 4.3).

Formula Weight and Molar Mass

LEARNING GOAL

Just as the atomic weight of an element is the average atomic mass for one atom of the naturally occurring element, expressed in atomic mass units, the **formula weight** of a compound is the sum of the atomic weights of all atoms in the compound, as represented by its formula. To calculate the formula weight of a compound we *must* know the correct formula. The formula weight is expressed in atomic mass units.

When working in the laboratory, we do not deal with individual molecules; instead, we use units of moles or grams. Eighteen grams of water (less than one ounce) contain approximately Avogadro's number of molecules (6.022×10^{23} molecules). Defining our working units as moles and grams makes good chemical sense.

We earlier concluded that the atomic mass of an element in amu from the periodic table corresponds to the mass of a mole of atoms of that element in units of grams/mol. It follows that **molar mass,** the mass of a mole of compound, is numerically equal to the formula weight in atomic mass units.

4.2 The Chemical Formula, Formula Weight, and Molar Mass

EXAMPLE 4.7
Calculating Formula Weight and Molar Mass

Calculate the formula weight and molar mass of water, H_2O.

Solution

Each water molecule contains two hydrogen atoms and one oxygen atom. The formula weight is

$$\begin{aligned} \text{2 atoms of hydrogen} \times 1.008 \text{ amu/atom} &= 2.016 \text{ amu} \\ \text{1 atom of oxygen} \times 16.00 \text{ amu/atom} &= \underline{16.00 \text{ amu}} \\ &\ 18.02 \text{ amu} \end{aligned}$$

The average mass of a single molecule of H_2O is 18.02 amu and is the formula weight. Therefore the mass of a mole of H_2O is 18.02 g or 18.02 g/mol.

Helpful Hint: Adding 2.016 and 16.00 shows a result of 18.016 on your calculator. Proper use of significant figures (Chapter 1) dictates rounding that result to 18.02.

EXAMPLE 4.8
Calculating Formula Weight and Molar Mass

Calculate the formula weight and molar mass of sodium sulfate.

Solution

The sodium ion is Na^+, and the sulfate ion is SO_4^{2-}. Two sodium ions must be present to neutralize the negative charges on the sulfate ion. The formula is Na_2SO_4. Sodium sulfate contains two sodium atoms, one sulfur atom, and four oxygen atoms. The formula weight is

$$\begin{aligned} \text{2 atoms of sodium} \times 22.99 \text{ amu/atom} &= 45.98 \text{ amu} \\ \text{1 atom of sulfur} \times 32.06 \text{ amu/atom} &= 32.06 \text{ amu} \\ \text{4 atoms of oxygen} \times 16.00 \text{ amu/atom} &= \underline{64.00 \text{ amu}} \\ &\ 142.04 \text{ amu} \end{aligned}$$

The average mass of a single unit of Na_2SO_4 is 142.04 amu and is the formula weight. Therefore the mass of a mole of Na_2SO_4 is 142.04 g, or 142.04 g/mol.

In Example 4.8, Na_2SO_4 is an ionic compound. As we have seen, it is not technically correct to describe ionic compounds as molecules; similarly, the term *molecular weight* is not appropriate for Na_2SO_4. The term *formula weight* may be used to describe the formula unit of a substance, whether it is made up of ions, ion pairs, or molecules. We shall use the term *formula weight* in a general way to represent each of these species.

Figure 4.4 illustrates the difference between molecules and ion pairs.

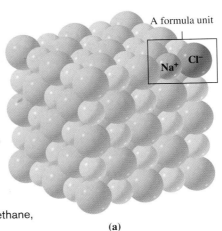

Figure 4.4
Formula units of (a) sodium chloride, an ionic compound, and (b) methane, a covalent compound.

EXAMPLE 4.9 Calculating Formula Weight and Molar Mass

Calculate the formula weight and molar mass of calcium phosphate.

Solution

The calcium ion is Ca^{2+}, and the phosphate ion is PO_4^{3-}. To form a neutral unit, three Ca^{2+} must combine with two PO_4^{3-}; $[3 \times (+2)]$ calcium ion charges are balanced by $[2 \times (-3)]$, the phosphate ion charge. Thus, for calcium phosphate, $Ca_3(PO_4)_2$, the subscript 2 for phosphate dictates that there are two phosphorus atoms and eight oxygen atoms (2×4) in the formula unit. Therefore

$$\begin{aligned}
3 \text{ atoms of Ca} \times 40.08 \text{ amu/atom} &= 120.24 \text{ amu} \\
2 \text{ atoms of P} \times 30.97 \text{ amu/atom} &= 61.94 \text{ amu} \\
8 \text{ atoms of O} \times 16.00 \text{ amu/atom} &= 128.00 \text{ amu} \\
\hline
& 310.18 \text{ amu}
\end{aligned}$$

The formula weight of calcium phosphate is 310.18 amu, and the molar mass is 310.18 g/mol.

Question 4.7

Calculate the formula weight and molar mass of each of the following compounds:

a. NH_3 (ammonia)
b. $C_6H_{12}O_6$ (a sugar, glucose)
c. $CoCl_2 \cdot 6H_2O$ (cobalt chloride hexahydrate)

Question 4.8

Calculate the formula weight and molar mass of each of the following compounds:

a. $C_2F_2Cl_4$ (a Freon gas)
b. C_3H_7OH (isopropyl alcohol, rubbing alcohol)
c. CH_3Br (bromomethane, a pesticide)

4.3 The Chemical Equation and the Information It Conveys

A Recipe for Chemical Change

LEARNING GOAL

The **chemical equation** is the shorthand notation for a chemical reaction. It describes all of the substances that react and all the products that form. **Reactants**, or starting materials, are all substances that undergo change in a chemical reaction; **products** are substances produced by a chemical reaction.

The chemical equation also describes the physical state of the reactants and products as solid, liquid, or gas. It tells us whether the reaction occurs and identifies the solvent and experimental conditions employed, such as heat, light, or electrical energy added to the system.

Most important, the relative number of moles of reactants and products appears in the equation. According to the **law of conservation of mass,** matter cannot be either gained or lost in the process of a chemical reaction. The total mass of the products must be equal to the total mass of the reactants. In other words, the law of conservation of mass tells us that we must have a balanced chemical equation.

4.3 The Chemical Equation and the Information It Conveys

Features of a Chemical Equation

Consider the decomposition of calcium carbonate:

$$CaCO_3(s) \xrightarrow{\Delta} CaO(s) + CO_2(g)$$

Calcium carbonate   Calcium oxide   Carbon dioxide

> This equation reads: One mole of solid calcium carbonate decomposes upon heating to produce one mole of solid calcium oxide and one mole of gaseous carbon dioxide.

The factors involved in writing equations of this type are described as follows:

1. *The identity of products and reactants must be specified using chemical symbols.* In some cases it is possible to predict the products of a reaction. More often, the reactants and products must be verified by chemical analysis. (Generally, you will be given information regarding the identity of the reactants and products.)
2. *Reactants are written to the left of the reaction arrow (→), and products are written to the right.* The direction in which the arrow points indicates the direction in which the reaction proceeds. In the decomposition of calcium carbonate, the reactant on the left ($CaCO_3$) is converted to products on the right ($CaO + CO_2$) during the course of the reaction.
3. *The physical states of reactants and products may be shown in parentheses.* For example:
   - $Cl_2(g)$ means that chlorine is in the gaseous state.
   - $Mg(s)$ indicates that magnesium is a solid.
   - $Br_2(l)$ indicates that bromine is present as a liquid.
   - $NH_3(aq)$ tells us that ammonia is present as an aqueous solution (dissolved in water).
4. *The symbol Δ over the reaction arrow means that energy is necessary for the reaction to occur.* Often, this and other special conditions are noted above or below the reaction arrow. For example, "light" means that a light source provides energy necessary for the reaction. Such reactions are termed photochemical reactions.
5. *The equation must be balanced.* All of the atoms of every reactant must also appear in the products, although in different compounds. We will treat this topic in detail later in this chapter.

According to the factors outlined, the equation for the decomposition of calcium carbonate may now be written as

$$CaCO_3(s) \xrightarrow{\Delta} CaO(s) + CO_2(g)$$

The Experimental Basis of a Chemical Equation

The chemical equation must represent a chemical change: One or more substances are changed into new substances, with different chemical and physical properties. Evidence for the reaction may be based on observations such as

- the release of carbon dioxide gas when an acid is added to a carbonate,
- the formation of a solid (or precipitate) when solutions of iron ions and hydroxide ions are mixed,
- the production of heat when using hot packs for treatment of injury, and
- the change in color of a solution upon addition of a second substance.

See discussion of acid-base reactions in Chapter 8.

See A Medical Perspective: Hot and Cold Packs in Chapter 7.

Many reactions are not so obvious. Sophisticated instruments are now available to the chemist. These instruments allow the detection of subtle changes in chemical systems that would otherwise go unnoticed. Such instruments may measure

- heat or light absorbed or emitted as the result of a reaction,
- changes in the way the sample behaves in an electric or magnetic field before and after a reaction, and
- changes in electrical properties before and after a reaction.

Whether we use our senses or a million dollar computerized instrument, the "bottom line" is the same: We are measuring a change in one or more chemical or physical properties in an effort to understand the changes taking place in a chemical system.

Disease can be described as a chemical system (actually a biochemical system) gone awry. Here, too, the underlying changes may not be obvious. Just as technology has helped chemists see subtle chemical changes in the laboratory, medical diagnosis has been revolutionized in our lifetimes using very similar technology. Some of these techniques are described in the Medical Perspective: Magnetic Resonance Imaging, in Chapter 9.

Writing Chemical Reactions

LEARNING GOAL

Chemical reactions, whether they involve the formation of precipitate, reaction with oxygen, acids and bases, or oxidation-reduction, generally follow one of a few simple patterns: combination, decomposition, and single- or double-replacement. Recognizing the underlying pattern will improve your ability to write and understand chemical reactions.

Combination Reactions

Combination reactions involve the joining of two or more elements or compounds, producing a product of different composition. The general form of a combination reaction is

$$A + B \longrightarrow AB$$

in which A and B represent reactant elements or compounds and AB is the product. Examples include

1. combination of a metal and a nonmetal to form a salt,

$$Ca(s) + Cl_2(g) \longrightarrow CaCl_2(s)$$

2. combination of hydrogen and chlorine molecules to produce hydrogen chloride,

$$H_2(g) + Cl_2(g) \longrightarrow 2HCl(g)$$

3. formation of water from hydrogen and oxygen molecules,

$$2H_2(g) + O_2(g) \longrightarrow 2H_2O(g)$$

4. reaction of magnesium oxide and carbon dioxide to produce magnesium carbonate,

$$MgO(s) + CO_2(g) \longrightarrow MgCO_3(s)$$

Decomposition Reactions

Decomposition reactions produce two or more products from a single reactant. The general form of these reactions is the reverse of a combination reaction:

$$AB \longrightarrow A + B$$

Some examples are

1. the heating of calcium carbonate to produce calcium oxide and carbon dioxide,

$$CaCO_3(s) \longrightarrow CaO(s) + CO_2(g)$$

Hydrated compounds are described on page 124.

2. the removal of water from a hydrated material (a *hydrate* is a substance that has water molecules incorporated in its structure),

$$CuSO_4 \cdot 5H_2O(s) \longrightarrow CuSO_4(s) + 5H_2O(g)$$

Replacement Reactions

Replacement reactions include both *single-replacement* and *double-replacement*. In a **single-replacement reaction,** one atom replaces another in the compound, producing a new compound:

$$A + BC \longrightarrow AC + B$$

Examples include

1. the replacement of copper by zinc in copper sulfate,

$$Zn(s) + CuSO_4(aq) \longrightarrow ZnSO_4(aq) + Cu(s)$$

2. the replacement of aluminum by sodium in aluminum nitrate,

$$3Na(s) + Al(NO_3)_3(aq) \longrightarrow 3NaNO_3(aq) + Al(s)$$

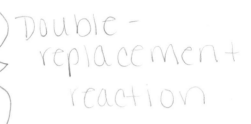
single-replacement reaction

A **double-replacement reaction,** on the other hand, involves *two compounds* undergoing a "change of partners." Two compounds react by exchanging atoms to produce two new compounds:

$$AB + CD \longrightarrow AD + CB$$

Examples include

1. the reaction of an acid (hydrochloric acid) and a base (sodium hydroxide) to produce water and salt, sodium chloride,

$$HCl(aq) + NaOH(aq) \longrightarrow H_2O(l) + NaCl(aq)$$

2. the formation of solid barium sulfate from barium chloride and potassium sulfate,

$$BaCl_2(aq) + K_2SO_4(aq) \longrightarrow BaSO_4(s) + 2KCl(aq)$$

Double-replacement reaction

Question 4.9

Classify each of the following reactions as decomposition (D), combination (C), single-replacement (SR), or double-replacement (DR):

a. $HNO_3(aq) + KOH(aq) \longrightarrow KNO_3(aq) + H_2O(aq)$
b. $Al(s) + 3NiNO_3(aq) \longrightarrow Al(NO_3)_3(aq) + 3Ni(s)$
c. $KCN(aq) + HCl(aq) \longrightarrow HCN(aq) + KCl(aq)$
d. $MgCO_3(s) \longrightarrow MgO(s) + CO_2(g)$

Question 4.10

Classify each of the following reactions as decomposition (D), combination (C), single-replacement (SR), or double-replacement (DR):

a. $2Al(OH)_3(s) \xrightarrow{\Delta} Al_2O_3(s) + 3H_2O(g)$
b. $Fe_2S_3(s) \xrightarrow{\Delta} 2Fe(s) + 3S(s)$
c. $Na_2CO_3(aq) + BaCl_2(aq) \longrightarrow BaCO_3(s) + 2NaCl(aq)$
d. $C(s) + O_2(g) \xrightarrow{\Delta} CO_2(g)$

Types of Chemical Reactions
Precipitation Reactions — forming a solid

Precipitation reactions include any chemical change in solution that results in one or more insoluble product(s). For aqueous solution reactions the product is insoluble in water.

An understanding of precipitation reactions is useful in many ways. They may explain natural phenomena, such as the formation of stalagmites and stalactites in

TABLE 4.1 Solubilities of Some Common Ionic Compounds

Solubility Predictions

Sodium, potassium, and ammonium compounds are generally *soluble*.
Nitrates and acetates are generally *soluble*.
Chlorides, bromides, and iodides (halides) are generally *soluble*. However, halide compounds containing lead(II), silver(I), and mercury(I) are *insoluble*.
Carbonates and phosphates are generally *insoluble*. Sodium, potassium, and ammonium carbonates and phosphates are, however, *soluble*.
Hydroxides and sulfides are generally *insoluble*. Sodium, potassium, calcium, and ammonium compounds are, however, *soluble*.

Precipitation reactions may be written as net ionic equations. You can find further information online at www.mhhe.com/denniston5e in "Writing Net Ionic Equations."

caves; they are simply precipitates in rocklike form. Kidney stones may result from the precipitation of calcium oxalate (CaC_2O_4). The routine act of preparing a solution requires that none of the solutes will react to form a precipitate.

How do you know whether a precipitate will form? Readily available solubility tables, such as Table 4.1, make prediction rather easy.

The following example illustrates the process.

EXAMPLE 4.10 Predicting Whether Precipitation Will Occur

Will a precipitate form if two solutions of the soluble salts NaCl and $AgNO_3$ are mixed?

Solution

Two soluble salts, if they react to form a precipitate, will probably "exchange partners":

$$NaCl(aq) + AgNO_3(aq) \longrightarrow AgCl(?) + NaNO_3(?)$$

Next, refer to Table 4.1 to determine the solubility of AgCl and $NaNO_3$. We predict that $NaNO_3$ is soluble and AgCl is not:

$$NaCl(aq) + AgNO_3(aq) \longrightarrow AgCl(s) + NaNO_3(aq)$$

The fact that the solid AgCl is predicted to form classifies this double-replacement reaction as a precipitation reaction.

Helpful Hints: (aq) indicates a soluble species; (s) indicates an insoluble species.

Question 4.11

Predict whether the following reactants, when mixed in aqueous solution, undergo a precipitation reaction. Write a balanced equation for each precipitation reaction.

a. potassium chloride and silver nitrate
b. potassium acetate and silver nitrate

Question 4.12

Predict whether the following reactants, when mixed in aqueous solution, undergo a precipitation reaction. Write a balanced equation for each precipitation reaction.

a. sodium hydroxide and ammonium chloride
b. sodium hydroxide and iron(II) chloride

4.3 The Chemical Equation and the Information It Conveys

Reactions with Oxygen

Many substances react with oxygen. These reactions generally release energy. The combustion of gasoline is used for transportation. Fossil fuel combustion is used to heat homes and provide energy for industry. Reactions involving oxygen provide energy for all sorts of biochemical processes.

When organic (carbon-containing) compounds react with the oxygen in air (burning), carbon dioxide is usually produced. If the compound contains hydrogen, water is the other product.

The reaction between oxygen and methane, CH_4, the major component of natural gas, is

$$CH_4(g) + 2O_2(g) \longrightarrow CO_2(g) + 2H_2O(g)$$

Energetics of reactions is discussed in Section 7.1.

CO_2 and H_2O are waste products, and CO_2 may contribute to the greenhouse effect and global warming. The really important, unseen product is heat energy. That is why we use this reaction in our furnaces!

Inorganic substances also react with oxygen and produce heat, but these reactions usually proceed more slowly. *Corrosion* (rusting iron) is a familiar example:

$$4Fe(s) + 3O_2(g) \longrightarrow \underset{\text{Rust}}{2Fe_2O_3(s)}$$

See An Environmental Perspective: The Greenhouse Effect and Global Warming, Chapter 5.

Some reactions of metals with oxygen are very rapid. A dramatic example is the reaction of magnesium with oxygen (see page 207):

$$2Mg(s) + O_2(g) \longrightarrow 2MgO(s)$$

Acid-Base Reactions

Another approach to the classification of chemical reactions is based on the gain or loss of hydrogen ions. **Acid-base reactions** involve the transfer of a *hydrogen ion*, H^+, from one reactant (the acid) to another (the base).

A common example of an acid-base reaction involves hydrochloric acid and sodium hydroxide:

$$\underset{\text{Acid}}{HCl(aq)} + \underset{\text{Base}}{NaOH(aq)} \longrightarrow \underset{\text{Water}}{H_2O(l)} + \underset{\text{Salt}}{Na^+(aq) + Cl^-(aq)}$$

See discussion of acid-base reactions in Chapter 8.

A hydrogen ion is transferred from the acid to the base, producing water and a salt in solution.

Acid-base reactions may also be written as net ionic equations. You can find further information online at www.mhhe.com/denniston5e in "Writing Net Ionic Equations."

Oxidation-Reduction Reactions

Another important reaction type, **oxidation-reduction**, takes place because of the transfer of negative charge (one or more *electrons*) from one reactant to another.

The reaction of zinc metal with copper(II) ions is one example of oxidation-reduction:

$$\underset{\substack{\text{Substance} \\ \text{to be oxidized}}}{Zn(s)} + \underset{\substack{\text{Substance to be} \\ \text{reduced}}}{Cu^{2+}(aq)} \longrightarrow \underset{\substack{\text{Oxidized} \\ \text{product}}}{Zn^{2+}(aq)} + \underset{\substack{\text{Reduced} \\ \text{product}}}{Cu(s)}$$

Zinc metal atoms each donate two electrons to copper(II) ions; consequently zinc atoms become zinc(II) ions and copper(II) ions become copper atoms. Zinc is oxidized (increased positive charge) and copper is reduced (decreased positive charge) as a result of electron transfer.

The principles and applications of acid-base reactions will be discussed in Sections 8.1 through 8.4, and oxidation-reduction processes will be discussed in Section 8.5.

All of the reactions with oxygen (discussed above) are oxidation-reduction reactions as well.

4.4 Balancing Chemical Equations

LEARNING GOAL 8

The chemical equation shows the *molar quantity* of reactants needed to produce a certain *molar quantity* of products.

The relative number of moles of each product and reactant is indicated by placing a whole-number *coefficient* before the formula of each substance in the chemical equation. A coefficient of 2 (for example, 2NaCl) indicates that 2 mol of sodium chloride are involved in the reaction. Also, $3NH_3$ signifies 3 mol of ammonia; it means that 3 mol of nitrogen atoms and 3×3, or 9, mol of hydrogen atoms are involved in the reaction. The coefficient 1 is understood, not written. Therefore H_2SO_4 would be interpreted as 1 mol of sulfuric acid, or 2 mol of hydrogen atoms, 1 mol of sulfur atoms, and 4 mol of oxygen atoms.

The equation

$$CaCO_3(s) \xrightarrow{\Delta} CaO(s) + CO_2(g)$$

is balanced as written. On the reactant side we have

1 mol Ca
1 mol C
3 mol O

On the product side there are

1 mol Ca
1 mol C
3 mol O

Therefore the law of conservation of mass is obeyed, and the equation is balanced as written.

Now consider the reaction of aqueous hydrogen chloride with solid calcium metal in aqueous solution:

$$HCl(aq) + Ca(s) \longrightarrow CaCl_2(aq) + H_2(g)$$

The equation, as written, is not balanced.

Reactants	Products
1 mol H atoms	2 mol H atoms
1 mol Cl atoms	2 mol Cl atoms
1 mol Ca atoms	1 mol Ca atoms

We need 2 mol of both H and Cl on the left, or reactant, side. An *incorrect* way of balancing the equation is as follows:

$$H_2Cl_2(aq) + Ca(s) \longrightarrow CaCl_2(aq) + H_2(g)$$

NOT a correct equation

The equation satisfies the law of conservation of mass; however, we have altered one of the reacting species. Hydrogen chloride is HCl, not H_2Cl_2. We must remember that *we cannot alter any chemical substance in the process of balancing the equation.* We can *only* introduce coefficients into the equation. Changing subscripts changes the identity of the chemicals involved, and that is not permitted. The equation must represent the reaction accurately. The correct equation is

$$2HCl(aq) + Ca(s) \longrightarrow CaCl_2(aq) + H_2(g)$$

Correct equation

This process is illustrated in Figure 4.5.

The coefficients indicate *relative* numbers of moles: 10 mol of $CaCO_3$ produce 10 mol of CaO; 0.5 mol of $CaCO_3$ produce 0.5 mol of CaO; and so forth.

Coefficients placed in front of the formula indicate the relative numbers of moles of compound (represented by the formula) that are involved in the reaction. Subscripts placed to the lower right of the atomic symbol indicate the relative number of atoms in the compound.

Water (H_2O) and hydrogen peroxide (H_2O_2) illustrate the effect a subscript can have. The two compounds show marked differences in physical and chemical properties.

4.4 Balancing Chemical Equations

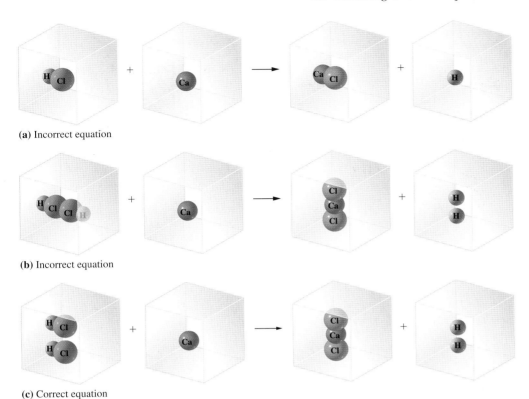

Figure 4.5
Balancing the equation HCl + Ca → CaCl$_2$ + H$_2$. (a) Neither product is the correct chemical species. (b) The reactant, HCl, is incorrectly represented as H$_2$Cl$_2$. (c) This equation is correct; all species are correct, and the law of conservation of mass is obeyed.

(a) Incorrect equation

(b) Incorrect equation

(c) Correct equation

Many equations are balanced by trial and error. After the identity of the products and reactants, the physical state, and the reaction conditions are known, the following steps provide a method for correctly balancing a chemical equation:

Step 1. Count the number of moles of atoms of each element on both product and reactant side.

Step 2. Determine which elements are not balanced.

Step 3. Balance one element at a time using coefficients.

Step 4. After you believe that you have successfully balanced the equation, check, as in Step 1, to be certain that mass conservation has been achieved.

Let us apply these steps to the reaction of calcium with aqueous hydrogen chloride:

$$HCl(aq) + Ca(s) \longrightarrow CaCl_2(aq) + H_2(g)$$

Step 1. **Reactants**   **Products**
     1 mol H atoms   2 mol H atoms
     1 mol Cl atoms   2 mol Cl atoms
     1 mol Ca atoms   1 mol Ca atoms

Step 2. The numbers of moles of H and Cl are not balanced.

Step 3. Insertion of a 2 before HCl on the reactant side should balance the equation:

$$2HCl(aq) + Ca(s) \longrightarrow CaCl_2(aq) + H_2(g)$$

Step 4. Check for mass balance:

Reactants   **Products**
2 mol H atoms   2 mol H atoms
2 mol Cl atoms   2 mol Cl atoms
1 mol Ca atoms   1 mol Ca atoms

Hence the equation is balanced.

EXAMPLE 4.11 Balancing Equations

Balance the following equation: Hydrogen gas and oxygen gas react explosively to produce gaseous water.

Solution

Recall that hydrogen and oxygen are diatomic gases; therefore

$$H_2(g) + O_2(g) \longrightarrow H_2O(g)$$

Note that the moles of hydrogen atoms are balanced but that the moles of oxygen atoms are not; therefore we must first balance the moles of oxygen atoms:

$$H_2(g) + O_2(g) \longrightarrow 2H_2O(g)$$

Balancing moles of oxygen atoms creates an imbalance in the number of moles of hydrogen atoms, so

$$2H_2(g) + O_2(g) \longrightarrow 2H_2O(g)$$

The equation is balanced, with 4 mol of hydrogen atoms and 2 mol of oxygen atoms on each side of the reaction arrow.

EXAMPLE 4.12 Balancing Equations

Balance the following equation: Propane gas, C_3H_8, a fuel, reacts with oxygen gas to produce carbon dioxide and water vapor. The reaction is

$$C_3H_8(g) + O_2(g) \longrightarrow CO_2(g) + H_2O(g)$$

Solution

First, balance the carbon atoms; there are 3 mol of carbon atoms on the left and only 1 mol of carbon atoms on the right. We need $3CO_2$ on the right side of the equation:

$$C_3H_8(g) + O_2(g) \longrightarrow 3CO_2(g) + H_2O(g)$$

Next, balance the hydrogen atoms; there are 2 mol of hydrogen atoms on the right and 8 mol of hydrogen atoms on the left. We need $4H_2O$ on the right:

$$C_3H_8(g) + O_2(g) \longrightarrow 3CO_2(g) + 4H_2O(g)$$

There are now 10 mol of oxygen atoms on the right and 2 mol of oxygen atoms on the left. To balance, we must have $5O_2$ on the left side of the equation:

$$C_3H_8(g) + 5O_2(g) \longrightarrow 3CO_2(g) + 4H_2O(g)$$

Remember: In every case, be sure to check the final equation for mass balance.

EXAMPLE 4.13 Balancing Equations

Balance the following equation: Butane gas, C_4H_{10}, a fuel used in pocket lighters, reacts with oxygen gas to produce carbon dioxide and water vapor. The reaction is

$$C_4H_{10}(g) + O_2(g) \longrightarrow CO_2(g) + H_2O(g)$$

Continued—

EXAMPLE 4.13 —Continued

Solution

First, balance the carbon atoms; there are 4 mol of carbon atoms on the left and only 1 mol of carbon atoms on the right:

$$C_4H_{10}(g) + O_2(g) \longrightarrow 4CO_2(g) + H_2O(g)$$

Next, balance hydrogen atoms; there are 10 mol of hydrogen atoms on the left and only 2 mol of hydrogen atoms on the right:

$$C_4H_{10}(g) + O_2(g) \longrightarrow 4CO_2(g) + 5H_2O(g)$$

There are now 13 mol of oxygen atoms on the right and only 2 mol of oxygen atoms on the left. Therefore a coefficient of 6.5 is necessary for O_2.

$$C_4H_{10}(g) + 6.5O_2(g) \longrightarrow 4CO_2(g) + 5H_2O(g)$$

Fractional or decimal coefficients are often needed and used. However, the preferred form requires all integer coefficients. Multiplying each term in the equation by a suitable integer (2, in this case) satisfies this requirement. Hence

$$2C_4H_{10}(g) + 13O_2(g) \longrightarrow 8CO_2(g) + 10H_2O(g)$$

The equation is balanced, with 8 mol of carbon atoms, 20 mol of hydrogen atoms, and 26 mol of oxygen atoms on each side of the reaction arrow.

Helpful Hint: When balancing equations, we find that it is often most efficient to begin by balancing the atoms in the most complicated formulas.

EXAMPLE 4.14 Balancing Equations

Balance the following equation: Aqueous ammonium sulfate reacts with aqueous lead nitrate to produce aqueous ammonium nitrate and solid lead sulfate. The reaction is

$$(NH_4)_2SO_4(aq) + Pb(NO_3)_2(aq) \longrightarrow NH_4NO_3(aq) + PbSO_4(s)$$

Solution

In this case the polyatomic ions remain as intact units. Therefore we can balance them as we would balance molecules rather than as atoms.

There are two ammonium ions on the left and only one ammonium ion on the right. Hence

$$(NH_4)_2SO_4(aq) + Pb(NO_3)_2(aq) \longrightarrow 2NH_4NO_3(aq) + PbSO_4(s)$$

No further steps are necessary. The equation is now balanced. There are two ammonium ions, two nitrate ions, one lead ion, and one sulfate ion on each side of the reaction arrow.

Question 4.13

Balance each of the following chemical equations:

a. $Fe(s) + O_2(g) \longrightarrow Fe_2O_3(s)$
b. $C_6H_6(l) + O_2(g) \longrightarrow CO_2(g) + H_2O(g)$

A Medical Perspective

Carbon Monoxide Poisoning: A Case of Combining Ratios

A fuel, such as methane, CH_4, burned in an excess of oxygen produces carbon dioxide and water:

$$CH_4(g) + 2O_2(g) \longrightarrow CO_2(g) + 2H_2O(g)$$

The same combustion in the presence of insufficient oxygen produces carbon monoxide and water:

$$2CH_4(g) + 3O_2(g) \longrightarrow 2CO(g) + 4H_2O(g)$$

The combustion of methane, repeated over and over in millions of gas furnaces, is responsible for heating many of our homes in the winter. The furnace is designed to operate under conditions that favor the first reaction and minimize the second; excess oxygen is available from the surrounding atmosphere. Furthermore, the vast majority of exhaust gases (containing principally CO, CO_2, H_2O, and unburned fuel) are removed from the home through the chimney. However, if the chimney becomes obstructed, or the burner malfunctions, carbon monoxide levels within the home can rapidly reach hazardous levels.

Why is exposure to carbon monoxide hazardous? Hemoglobin, an iron-containing compound, binds with O_2 and transports it throughout the body. Carbon monoxide also combines with hemoglobin, thereby blocking oxygen transport. The binding affinity of hemoglobin for carbon monoxide is about two hundred times as great as for O_2. Therefore, to maintain O_2 binding and transport capability, our exposure to carbon monoxide must be minimal. Proper ventilation and suitable oxygen-to-fuel ratio are essential for any combustion process in the home, automobile, or workplace. In recent years carbon monoxide sensors have been developed. These sensors sound an alarm when toxic levels of CO are reached. These warning devices have helped to create a safer indoor environment.

The example we have chosen is an illustration of what is termed the *law of multiple proportions*. This law states that identical reactants may produce different products, depending on their combining ratio. The experimental conditions (in this case, the quantity of available oxygen) determine the preferred path of the chemical reaction. In Section 4.5 we will learn how to use a properly balanced equation, representing the chemical change occurring, to calculate quantities of reactants consumed or products produced.

For Further Understanding

Why may new, more strict insulation standards for homes and businesses inadvertently increase the risk of carbon monoxide poisoning?

Explain the link between smoking and carbon monoxide that has motivated many states and municipalities to ban smoking in restaurants, offices, and other indoor spaces.

Balance each of the following chemical equations:

a. $S_2Cl_2(s) + NH_3(g) \longrightarrow N_4S_4(s) + NH_4Cl(s) + S_8(s)$
b. $C_2H_5OH(l) + O_2(g) \longrightarrow CO_2(g) + H_2O(g)$

4.5 Calculations Using the Chemical Equation

General Principles

LEARNING GOAL

The calculation of quantities of products and reactants based on a balanced chemical equation is important in many fields. The synthesis of drugs and other complex molecules on a large scale is conducted on the basis of a balanced equation. This minimizes the waste of expensive chemical compounds used in these reactions. Similarly, the ratio of fuel and air in a home furnace or automobile must be adjusted carefully, according to their combining ratio, to maximize energy conversion, minimize fuel consumption, and minimize pollution.

In carrying out chemical calculations we apply the following guidelines.

1. The chemical formulas of all reactants and products must be known.

2. The basis for the calculations is a balanced equation because the conservation of mass must be obeyed. If the equation is not properly balanced, the calculation is meaningless.
3. The calculations are performed in terms of moles. The coefficients in the balanced equation represent the relative number of moles of products and reactants.

We have seen that the number of moles of products and reactants often differs in a balanced equation. For example,

$$C(s) + O_2(g) \longrightarrow CO_2(g)$$

is a balanced equation. Two moles of reactants combine to produce one mole of product:

$$1 \text{ mol C} + 1 \text{ mol O}_2 \longrightarrow 1 \text{ mol CO}_2$$

However, 1 mol of C *atoms* and 2 mol of O *atoms* produce 1 mol of C *atoms* and 2 mol of O *atoms*. In other words, the number of moles of reactants and products may differ, but the number of moles of atoms cannot. The formation of CO_2 from C and O_2 may be described as follows:

$$C(s) + O_2(g) \longrightarrow CO_2(g)$$
$$1 \text{ mol C} + 1 \text{ mol O}_2 \longrightarrow 1 \text{ mol CO}_2$$
$$12.0 \text{ g C} + 32.0 \text{ g O}_2 \longrightarrow 44.0 \text{ g CO}_2$$

The mole is the basis of our calculations. However, moles are generally measured in grams (or kilograms). A facility for interconversion of moles and grams is fundamental to chemical arithmetic (see Figure 4.2). These calculations, discussed earlier in this chapter, are reviewed in Example 4.15.

Use of Conversion Factors
Conversion Between Moles and Grams

Conversion from moles to grams, and vice versa, requires only the formula weight of the compound of interest. Consider the following examples.

EXAMPLE 4.15

Converting Between Moles and Grams

a. Convert 1.00 mol of oxygen gas, O_2, to grams.

Solution

Use the following path:

moles of oxygen $\longrightarrow$ grams of oxygen

The molar mass of oxygen (O_2) is

$$\frac{32.0 \text{ g O}_2}{1 \text{ mol O}_2}$$

Therefore

$$1.00 \text{ mol O}_2 \times \frac{32.0 \text{ g O}_2}{1 \text{ mol O}_2} = 32.0 \text{ g O}_2$$

b. How many grams of carbon dioxide are contained in 10.0 mol of carbon dioxide?

Continued—

EXAMPLE 4.15 —Continued

Solution

Use the following path:

moles of carbon dioxide → grams of carbon dioxide

The formula weight of CO_2 is

$$\frac{44.0 \text{ g } CO_2}{1 \text{ mol } CO_2}$$

and

$$10.0 \text{ mol } CO_2 \times \frac{44.0 \text{ g } CO_2}{1 \text{ mol } CO_2} = 4.40 \times 10^2 \text{ g } CO_2$$

c. How many moles of sodium are contained in 1 lb (454 g) of sodium metal?

Solution

Use the following path:

grams of sodium → moles of sodium

The number of moles of sodium atoms is

$$454 \text{ g Na} \times \frac{1 \text{ mol Na}}{22.99 \text{ g Na}} = 19.7 \text{ mol Na}$$

Helpful Hint: Note that each factor can be inverted producing a second possible factor. Only one will allow the appropriate unit cancellation.

Question 4.15

Perform each of the following conversions:

a. 5.00 mol of water to grams of water
b. 25.0 g of LiCl to moles of LiCl

Question 4.16

Perform each of the following conversions:

a. 1.00×10^{-5} mol of $C_6H_{12}O_6$ to micrograms of $C_6H_{12}O_6$
b. 35.0 g of $MgCl_2$ to moles of $MgCl_2$

Conversion of Moles of Reactants to Moles of Products

In Example 4.12 we balanced the equation for the reaction of propane and oxygen as follows:

$$C_3H_8(g) + 5O_2(g) \rightarrow 3CO_2(g) + 4H_2O(g)$$

4.5 Calculations Using the Chemical Equation

In this reaction, 1 mol of C_3H_8 corresponds to, or results in,

- 5 mol of O_2 being consumed and
- 3 mol of CO_2 being formed and
- 4 mol of H_2O being formed.

This information may be written in the form of a conversion factor or ratio:

$$1 \text{ mol } C_3H_8 / 5 \text{ mol } O_2$$

Translated: One mole of C_3H_8 reacts with *five* moles of O_2.

$$1 \text{ mol } C_3H_8 / 3 \text{ mol } CO_2$$

Translated: One mole of C_3H_8 produces *three* moles of CO_2.

$$1 \text{ mol } C_3H_8 / 4 \text{ mol } H_2O$$

Translated: One mole of C_3H_8 produces *four* moles of H_2O.

Conversion factors, based on the chemical equation, permit us to perform a variety of calculations.

Let us look at a few examples, based on the combustion of propane and the equation that we balanced in Example 4.12.

EXAMPLE 4.16 Calculating Reacting Quantities

Calculate the number of grams of O_2 that will react with 1.00 mol of C_3H_8.

Solution

Two conversion factors are necessary to solve this problem:

1. conversion from moles of C_3H_8 to moles of O_2 and
2. conversion of moles of O_2 to grams of O_2.

Therefore our path is

moles C_3H_8 $\longrightarrow$ moles O_2 $\longrightarrow$ grams O_2

and

$$1.00 \text{ mol } C_3H_8 \times \frac{5 \text{ mol } O_2}{1 \text{ mol } C_3H_8} \times \frac{32.0 \text{ g } O_2}{1 \text{ mol } O_2} = 1.60 \times 10^2 \text{ g } O_2$$

EXAMPLE 4.17 Calculating Grams of Product from Moles of Reactant

Calculate the number of grams of CO_2 produced from the combustion of 1.00 mol of C_3H_8.

Solution

Employ logic similar to that used in Example 4.16 and use the following path:

moles C_3H_8 $\longrightarrow$ moles CO_2 $\longrightarrow$ grams CO_2

Continued—

EXAMPLE 4.17 —Continued

Then

$$1.00 \text{ mol } C_3H_8 \times \frac{3 \text{ mol } CO_2}{1 \text{ mol } C_3H_8} \times \frac{44.0 \text{ g } CO_2}{1 \text{ mol } CO_2} = 132 \text{ g } CO_2$$

EXAMPLE 4.18 Relating Masses of Reactants and Products

Calculate the number of grams of C_3H_8 required to produce 36.0 g of H_2O.

Solution

It is necessary to convert

1. grams of H_2O to moles of H_2O,
2. moles of H_2O to moles of C_3H_8, and
3. moles of C_3H_8 to grams of C_3H_8.

Use the following path:

grams H_2O $\longrightarrow$ moles H_2O $\longrightarrow$ moles C_3H_8 $\longrightarrow$ grams C_3H_8

Then

$$36.0 \text{ g } H_2O \times \frac{1 \text{ mol } H_2O}{18.0 \text{ g } H_2O} \times \frac{1 \text{ mol } C_3H_8}{4 \text{ mol } H_2O} \times \frac{44.0 \text{ g } C_3H_8}{1 \text{ mol } C_3H_8} = 22.0 \text{ g } C_3H_8$$

Question 4.17

The balanced equation for the combustion of ethanol (ethyl alcohol) is:

$$C_2H_5OH(l) + 3O_2(g) \longrightarrow 2CO_2(g) + 3H_2O(g)$$

a. How many moles of O_2 will react with 1 mol of ethanol?
b. How many grams of O_2 will react with 1 mol of ethanol?

Question 4.18

How many grams of CO_2 will be produced by the combustion of 1 mol of ethanol? (See Question 4.17.)

Let's consider an example that requires us to write and balance the chemical equation, use conversion factors, and calculate the amount of a reactant consumed in the chemical reaction.

EXAMPLE 4.19 Calculating a Quantity of Reactant

Calcium hydroxide may be used to neutralize (completely react with) aqueous hydrochloric acid. Calculate the number of grams of hydrochloric acid that would be neutralized by 0.500 mol of solid calcium hydroxide.

Continued—

4.5 Calculations Using the Chemical Equation

EXAMPLE 4.19 —Continued

Solution

The formula for calcium hydroxide is Ca(OH)$_2$ and that for hydrochloric acid is HCl. The unbalanced equation produces calcium chloride and water as products:

$$Ca(OH)_2(s) + HCl(aq) \longrightarrow CaCl_2(aq) + H_2O(l)$$

First, balance the equation:

$$Ca(OH)_2(s) + 2HCl(aq) \longrightarrow CaCl_2(aq) + 2H_2O(l)$$

Next, determine the necessary conversion:

1. moles of Ca(OH)$_2$ to moles of HCl and
2. moles of HCl to grams of HCl.

Use the following path:

moles Ca(OH)$_2$ $\longrightarrow$ moles HCl $\longrightarrow$ grams HCl

$$0.500 \text{ mol Ca(OH)}_2 \times \frac{2 \text{ mol HCl}}{1 \text{ mol Ca(OH)}_2} \times \frac{36.5 \text{ g HCl}}{1 \text{ mol HCl}} = 36.5 \text{ g HCl}$$

This reaction is illustrated in Figure 4.6.

Helpful Hints:
1. The reaction between an acid and a base produces a salt and water (Chapter 8).
2. Remember to balance the chemical equation; the proper coefficients are essential parts of the subsequent calculations.

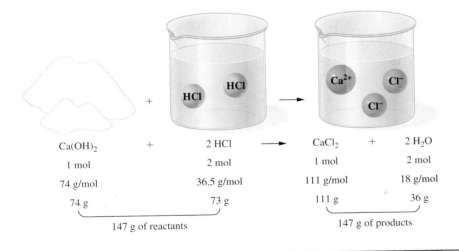

Figure 4.6
An illustration of the law of conservation of mass. In this example, 1 mol of calcium hydroxide and 2 mol of hydrogen chloride react to produce 3 mol of product (2 mol of water and 1 mol of calcium chloride). The total mass, in grams, of reactant(s) consumed is equal to the total mass, in grams, of product(s) formed. *Note:* In reality, HCl does not exist as discrete molecules in water. The HCl separates to form H$^+$ and Cl$^-$. Ionization in water will be discussed with the chemistry of acids and bases in Chapter 8.

EXAMPLE 4.20

Calculating Reactant Quantities

What mass of sodium hydroxide, NaOH, would be required to produce 8.00 g of the antacid milk of magnesia, Mg(OH)$_2$, by the reaction of MgCl$_2$ with NaOH?

Continued—

EXAMPLE 4.20 —Continued

Solution

$$MgCl_2(aq) + 2NaOH(aq) \longrightarrow Mg(OH)_2(s) + 2NaCl(aq)$$

The equation tells us that 2 mol of NaOH form 1 mol of $Mg(OH)_2$. If we calculate the number of moles of $Mg(OH)_2$ in 8.00 g of $Mg(OH)_2$, we can determine the number of moles of NaOH necessary and then the mass of NaOH required:

mass $Mg(OH)_2$ $\longrightarrow$ moles $Mg(OH)_2$ $\longrightarrow$ moles NaOH $\longrightarrow$ mass NaOH

$$58.3 \text{ g } Mg(OH)_2 = 1 \text{ mol } Mg(OH)_2$$

Therefore

$$8.00 \text{ g } Mg(OH)_2 \times \frac{1 \text{ mol } Mg(OH)_2}{58.3 \text{ g } Mg(OH)_2} = 0.137 \text{ mol } Mg(OH)_2$$

Two moles of NaOH react to give one mole of $Mg(OH)_2$. Therefore

$$0.137 \text{ mol } Mg(OH)_2 \times \frac{2 \text{ mol NaOH}}{1 \text{ mol } Mg(OH)_2} = 0.274 \text{ mol NaOH}$$

40.0 g of NaOH = 1 mol of NaOH. Therefore

$$0.274 \text{ mol NaOH} \times \frac{40.0 \text{ g NaOH}}{1 \text{ mol NaOH}} = 11.0 \text{ g NaOH}$$

The calculation may be done in a single step:

$$8.00 \text{ g } Mg(OH)_2 \times \frac{1 \text{ mol } Mg(OH)_2}{58.3 \text{ g } Mg(OH)_2} \times \frac{2 \text{ mol NaOH}}{1 \text{ mol } Mg(OH)_2} \times \frac{40.0 \text{ g NaOH}}{1 \text{ mol NaOH}} = 11.0 \text{ g NaOH}$$

Note once again that we have followed a logical and predictable path to the solution:

grams $Mg(OH)_2$ $\longrightarrow$ moles $Mg(OH)_2$ $\longrightarrow$ moles NaOH $\longrightarrow$ grams NaOH

Helpful Hint: Mass is a laboratory unit, whereas moles is a calculation unit. The laboratory balance is calibrated in units of mass (grams). Although moles are essential for calculation, often the starting point and objective are in mass units. As a result, our path is often grams → moles → grams.

A general problem-solving strategy is summarized in Figure 4.7. By systematically applying this strategy, you will be able to solve virtually any problem requiring calculations based on the chemical equation.

Question 4.19

Metallic iron reacts with O_2 gas to produce iron(III) oxide.

a. Write and balance the equation.
b. Calculate the number of grams of iron needed to produce 5.00 g of product.

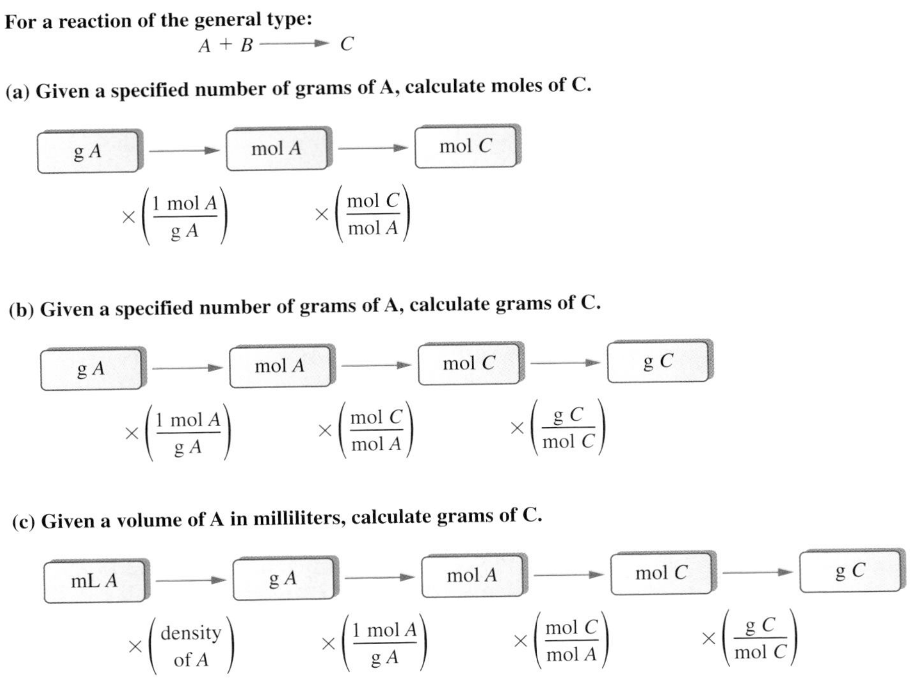

4.5 Calculations Using the Chemical Equation

For a reaction of the general type:
$$A + B \longrightarrow C$$

(a) Given a specified number of grams of A, calculate moles of C.

$$\boxed{g\ A} \longrightarrow \boxed{\text{mol } A} \longrightarrow \boxed{\text{mol } C}$$

$$\times \left(\frac{1 \text{ mol } A}{g\ A}\right) \qquad \times \left(\frac{\text{mol } C}{\text{mol } A}\right)$$

(b) Given a specified number of grams of A, calculate grams of C.

$$\boxed{g\ A} \longrightarrow \boxed{\text{mol } A} \longrightarrow \boxed{\text{mol } C} \longrightarrow \boxed{g\ C}$$

$$\times \left(\frac{1 \text{ mol } A}{g\ A}\right) \qquad \times \left(\frac{\text{mol } C}{\text{mol } A}\right) \qquad \times \left(\frac{g\ C}{\text{mol } C}\right)$$

(c) Given a volume of A in milliliters, calculate grams of C.

$$\boxed{\text{mL } A} \longrightarrow \boxed{g\ A} \longrightarrow \boxed{\text{mol } A} \longrightarrow \boxed{\text{mol } C} \longrightarrow \boxed{g\ C}$$

$$\times \left(\frac{\text{density}}{\text{of } A}\right) \quad \times \left(\frac{1 \text{ mol } A}{g\ A}\right) \quad \times \left(\frac{\text{mol } C}{\text{mol } A}\right) \quad \times \left(\frac{g\ C}{\text{mol } C}\right)$$

Figure 4.7
A general problem-solving strategy, using molar quantities.

Question 4.20

Barium carbonate decomposes upon heating to barium oxide and carbon dioxide.

a. Write and balance the equation.
b. Calculate the number of grams of carbon dioxide produced by heating 50.0 g of barium carbonate.

Theoretical and Percent Yield

The **theoretical yield** is the *maximum* amount of product that can be produced (in an ideal world). In the "real" world it is difficult to produce the amount calculated as the theoretical yield. This is true for a variety of reasons. Some experimental error is unavoidable. Moreover, many reactions simply are not complete; some amount of reactant remains at the end of the reaction. We will study these processes, termed *equilibrium reactions* in Chapter 7.

A **percent yield**, the ratio of the actual and theoretical yields multiplied by 100%, is often used to show the relationship between predicted and experimental quantities. Thus

$$\% \text{ yield} = \frac{\text{actual yield}}{\text{theoretical yield}} \times 100\%$$

In Example 4.17, the theoretical yield of CO_2 is 132 g. For this reaction let's assume that a chemist actually obtained 125 g CO_2. This is the actual yield and would normally be provided as a part of the data in the problem.
Calculate the percent yield as follows:

$$\% \text{ yield} = \frac{\text{actual yield}}{\text{theoretical yield}} \times 100\%$$

$$= \frac{125 \text{ g } CO_2 \text{ actual}}{132 \text{ g } CO_2 \text{ theoretical}} \times 100\% = 94.7\%$$

10 LEARNING GOAL

A Medical Perspective

Pharmaceutical Chemistry: The Practical Significance of Percent Yield

In recent years the major pharmaceutical industries have introduced a wide variety of new drugs targeted to cure or alleviate the symptoms of a host of diseases that afflict humanity.

The vast majority of these drugs are synthetic; they are made in a laboratory or by an industrial process. These substances are complex molecules that are patiently designed and constructed from relatively simple molecules in a series of chemical reactions. A series of ten to twenty "steps," or sequential reactions, is not unusual to put together a final product that has the proper structure, geometry, and reactivity for efficacy against a particular disease.

Although a great deal of research occurs to ensure that each of these steps in the overall process is efficient (having a large percent yield), the overall process is still very inefficient (low percent yield). This inefficiency, and the research needed to minimize it, at least in part determines the cost and availability of both prescription and over-the-counter preparations.

Consider a hypothetical five-step sequential synthesis. If each step has a percent yield of 80% our initial impression might be that this synthesis is quite efficient. However, on closer inspection we find quite the contrary to be true.

The overall yield of the five-step reaction is the product of the decimal fraction of the percent yield of each of the sequential reactions. So, if the decimal fraction corresponding to 80% is 0.80:

$$0.80 \times 0.80 \times 0.80 \times 0.80 \times 0.80 = 0.33$$

Converting the decimal fraction to percentage:

$$0.33 \times 100\% = 33\% \text{ yield}$$

Many reactions are considerably less than 80% efficient, especially those that are used to prepare large molecules with complex arrangements of atoms. Imagine a more realistic scenario in which one step is only 20% efficient (a 20% yield) and the other four steps are 50%, 60%, 70%, and 80% efficient. Repeating the calculation with these numbers (after conversion to a decimal fraction):

$$0.20 \times 0.50 \times 0.60 \times 0.70 \times 0.80 = 0.0336$$

Converting the decimal fraction to a percentage:

$$0.0336 \times 100\% = 3.36\% \text{ yield}$$

a very inefficient process.

If we apply this logic to a fifteen- or twenty-step synthesis we gain some appreciation of the difficulty of producing modern pharmaceutical products. Add to this the challenge of predicting the most appropriate molecular structure that will have the desired biological effect and be relatively free of side effects. All these considerations give new meaning to the term *wonder drug* that has been attached to some of the more successful synthetic products.

We will study some of the elementary steps essential to the synthesis of a wide range of pharmaceutical compounds in later chapters, beginning with Chapter 10.

For Further Understanding

Explain the possible connection of this perspective to escalating costs of pharmaceutical products.

Can you describe other situations, not necessarily in the field of chemistry, where multiple-step processes contribute to inefficiency?

EXAMPLE 4.21 Calculation of Percent Yield

Assume that the theoretical yield of iron in the process

$$2Al(s) + Fe_2O_3(s) \longrightarrow Al_2O_3(l) + 2Fe(l)$$

was 30.0 g.

Continued—

EXAMPLE 4.21 —Continued

If the actual yield of iron were 25.0 g in the process, calculate the percent yield.

Solution

$$\% \text{ yield} = \frac{\text{actual yield}}{\text{theoretical yield}} \times 100\%$$

$$= \frac{25.0 \text{ g}}{30.0 \text{ g}} \times 100\%$$

$$= 83.3\%$$

Question 4.21

Given the reaction represented by the balanced equation

$$Sn(s) + 2HF(aq) \longrightarrow SnF_2(s) + H_2(g)$$

a. Calculate the number of grams of SnF_2 produced by mixing 100.0 g Sn with excess HF.
b. If only 5.00 g SnF_2 were produced, calculate the % yield.

Question 4.22

Given the reaction represented by the balanced equation

$$CH_4(g) + 3Cl_2(g) \longrightarrow 3HCl(g) + CHCl_3(g)$$

a. Calculate the number of grams of $CHCl_3$ produced by mixing 105 g Cl_2 with excess CH_4.
b. If 10.0 g $CHCl_3$ were produced, calculate the % yield.

SUMMARY

4.1 The Mole Concept and Atoms

Atoms are exceedingly small, yet their masses have been experimentally determined for each of the elements. The unit of measurement for these determinations is the *atomic mass unit*, abbreviated amu:

$$1 \text{ amu} = 1.661 \times 10^{-24} \text{ g}$$

The periodic table provides atomic masses in atomic mass units.

A more practical unit for defining a "collection" of atoms is the *mole*:

$$1 \text{ mol of atoms} = 6.022 \times 10^{23} \text{ atoms of an element}$$

This number is referred to as *Avogadro's number.*

The mole and the atomic mass unit are related. The atomic mass of a given element corresponds to the average mass of a single atom in atomic mass units and the mass of a mole of atoms in grams. The mass of one mole of atoms is termed the *molar mass* of the element. One mole of atoms of any element contains the same number, Avogadro's number, of atoms.

4.2 The Chemical Formula, Formula Weight, and Molar Mass

Compounds are pure substances that are composed of two or more elements that are chemically combined. They are represented by their *chemical formula*, a combination of symbols of the various elements that make up the compounds. The chemical formula is based on the *formula unit*. This is the smallest collection of atoms that provides the identity of the atoms present in the compound and the relative numbers of each type of atom.

Just as a mole of atoms is based on the atomic mass, a mole of a compound is based on the formula mass or *formula weight*. The formula weight is calculated by addition of the masses of all the atoms or ions of which the unit is composed. To calculate the formula weight, the formula unit must be known. The formula weight of one mole of a compound is its *molar mass* in units of g/mol.

4.3 The Chemical Equation and the Information It Conveys

The *chemical equation* is the shorthand notation for a chemical reaction. It describes all of the substances that react to produce the product(s). *Reactants*, or starting materials, are all substances that undergo change in a chemical reaction; *products* are substances produced by a chemical reaction.

According to the *law of conservation of mass*, matter can neither be gained nor lost in the process of a chemical reaction. The law of conservation of mass states that we must have a balanced chemical equation.

Features of a suitable equation include the following:

- The identity of products and reactants must be specified.
- Reactants are written to the left of the reaction arrow (→) and products to the right.
- The physical states of reactants and products are shown in parentheses.
- The symbol Δ over the reaction arrow means that heat energy is necessary for the reaction to occur.
- The equation must be balanced.

Chemical reactions involve the *combination* of reactants to produce products, the *decomposition* of reactant(s) into products, or the *replacement* of one or more elements in a compound to yield products. Replacement reactions are subclassified as either *single-* or *double-*replacement.

Reactions that produce products with similar characteristics are often classified as a single group. The formation of an insoluble solid, a *precipitate*, is very common. Such reactions are precipitation reactions.

Chemical reactions that have a common reactant may be grouped together. Reactions involving oxygen, *combustion reactions*, are such a class.

Another approach to the classification of chemical reactions is based on transfer of hydrogen ions (+ charge) or electrons (– charge). *Acid-base reactions* involve the transfer of a hydrogen ion, H^+, from one reactant to another. Another important reaction type, *oxidation-reduction*, takes place because of the transfer of negative charge, one or more electrons, from one reactant to another.

4.4 Balancing Chemical Equations

The chemical equation enables us to determine the quantity of reactants needed to produce a certain molar quantity of products. The chemical equation expresses these quantities in terms of moles.

The relative number of moles of each product and reactant is indicated by placing a whole-number coefficient before the formula of each substance in the chemical equation.

Many equations are balanced by trial and error. If the identity of the products and reactants, the physical state, and the reaction conditions are known, the following steps provide a method for correctly balancing a chemical equation:

- Count the number of atoms of each element on both product and reactant sides.
- Determine which atoms are not balanced.
- Balance one element at a time using coefficients.
- After you believe that you have successfully balanced the equation, check to be certain that mass conservation has been achieved.

4.5 Calculations Using the Chemical Equation

Calculations involving chemical quantities are based on the following requirements:

- The basis for the calculations is a balanced equation.
- The calculations are performed in terms of moles.
- The conservation of mass must be obeyed.

The mole is the basis for calculations. However, masses are generally measured in grams (or kilograms). Therefore you must be able to interconvert moles and grams to perform chemical arithmetic.

KEY TERMS

acid-base reaction (4.3)
atomic mass unit (4.1)
Avogadro's number (4.1)
chemical equation (4.3)
chemical formula (4.2)
combination reaction (4.3)
decomposition reaction (4.3)
double-replacement reaction (4.3)
formula unit (4.2)
formula weight (4.2)
hydrate (4.2)
law of conservation of mass (4.3)
molar mass (4.1)
mole (4.1)
oxidation-reduction reaction (4.3)
percent yield (4.5)
product (4.3)
reactant (4.3)
single-replacement reaction (4.3)
theoretical yield (4.5)

QUESTIONS AND PROBLEMS

The Mole Concept and Atoms

Foundations

4.23 We purchase eggs by the dozen. Name several other familiar packaging units.
4.24 One dozen eggs is a convenient consumer unit. Explain why the mole is a convenient chemist's unit.
4.25 What is the average molar mass of:
   a. Si
   b. Ag
4.26 What is the average molar mass of:
   a. S
   b. Na
4.27 What is the mass of Avogadro's number of argon atoms?
4.28 What is the mass of Avogadro's number of iron atoms?

Applications

4.29 How many grams are contained in 2.00 mol of neon atoms?
4.30 How many grams are contained in 3.00 mol of carbon atoms?
4.31 What is the mass in grams of 1.00 mol of helium atoms?
4.32 What is the mass in grams of 1.00 mol of nitrogen atoms?
4.33 Calculate the number of moles corresponding to:
   a. 20.0 g He
   b. 0.040 kg Na
   c. 3.0 g Cl_2
4.34 Calculate the number of moles corresponding to:
   a. 0.10 g Ca
   b. 4.00 g Fe
   c. 2.00 kg N_2
4.35 What is the mass, in grams, of 15.0 mol of silver?
4.36 What is the mass, in grams, of 15.0 mol of carbon?

The Chemical Formula, Formula Weight, and Molar Mass

Foundations

4.37 Distinguish between the terms *molecule* and *ion pair*.
4.38 Distinguish between the terms *formula weight* and *molecular weight*.
4.39 Calculate the molar mass, in grams per mole, of each of the following formula units:
   a. NaCl
   b. Na_2SO_4
   c. $Fe_3(PO_4)_2$
4.40 Calculate the molar mass, in grams per mole, of each of the following formula units:
   a. S_8
   b. $(NH_4)_2SO_4$
   c. CO_2
4.41 Calculate the molar mass, in grams per mole, of oxygen gas, O_2.
4.42 Calculate the molar mass, in grams per mole, of ozone, O_3.
4.43 Calculate the molar mass of $CuSO_4 \cdot 5H_2O$.
4.44 Calculate the molar mass of $CaCl_2 \cdot 2H_2O$.

Applications

4.45 Calculate the number of moles corresponding to:
   a. 15.0 g NaCl
   b. 15.0 g Na_2SO_4
4.46 Calculate the number of moles corresponding to:
   a. 15.0 g NH_3
   b. 16.0 g O_2
4.47 Calculate the mass in grams corresponding to:
   a. 1.000 mol H_2O
   b. 2.000 mol NaCl
4.48 Calculate the mass in grams corresponding to:
   a. 0.400 mol NH_3
   b. 0.800 mol $BaCO_3$
4.49 Calculate the mass in grams corresponding to:
   a. 10.0 mol He
   b. 1.00×10^2 mol H_2
4.50 Calculate the mass in grams corresponding to:
   a. 2.00 mol CH_4
   b. 0.400 mol $Ca(NO_3)_2$
4.51 How many grams are required to have 0.100 mol of each of the following?
   a. Mg
   b. $CaCO_3$
4.52 How many grams are required to have 0.100 mol of each of the following?
   a. $C_6H_{12}O_6$ (glucose)
   b. NaCl
4.53 How many grams are required to have 0.100 mol of each of the following compounds?
   a. NaOH
   b. H_2SO_4
4.54 How many grams are required to have 0.100 mol of each of the following compounds?
   a. C_2H_5OH (ethanol)
   b. $Ca_3(PO_4)_2$
4.55 How many moles are in 50.0 g of each of the following substances?
   a. KBr
   b. $MgSO_4$
4.56 How many moles are in 50.0 g of each of the following substances?
   a. Br_2
   b. NH_4Cl
4.57 How many moles are in 50.0 g of each of the following substances?
   a. CS_2
   b. $Al_2(CO_3)_3$
4.58 How many moles are in 50.0 g of each of the following substances?
   a. $Sr(OH)_2$
   b. $LiNO_3$

The Chemical Equation and the Information It Conveys

4.59 What law is the ultimate basis for a correct chemical equation?
4.60 List the general types of information that a chemical equation provides.
4.61 What is the meaning of the subscript in a chemical formula?
4.62 What is the meaning of the coefficient in a chemical equation?
4.63 Give an example of:
   a. a decomposition reaction
   b. a single-replacement reaction
4.64 Give an example of:
   a. a combination reaction
   b. a double-replacement reaction
4.65 Give an example of a precipitate-forming reaction.
4.66 Give an example of a reaction in which oxygen is a reactant.
4.67 What is the meaning of Δ over the reaction arrow?
4.68 What is the meaning of (s), (l), or (g) immediately following the symbol for a chemical substance?

Balancing Chemical Equations

Foundations

4.69 When you are balancing an equation, why must the subscripts in the chemical formula remain unchanged?
4.70 Describe the process of checking to ensure that an equation is properly balanced.
4.71 What is a reactant?
4.72 On which side of the reaction arrow are reactants found?
4.73 What is a product?
4.74 On which side of the reaction arrow are products found?

Applications

4.75 Balance each of the following equations:
   a. $C_2H_6(g) + O_2(g) \longrightarrow CO_2(g) + H_2O(g)$
   b. $K_2O(s) + P_4O_{10}(s) \longrightarrow K_3PO_4(s)$
   c. $MgBr_2(aq) + H_2SO_4(aq) \longrightarrow HBr(g) + MgSO_4(aq)$
4.76 Balance each of the following equations:
   a. $C_6H_{12}O_6(s) + O_2(g) \longrightarrow CO_2(g) + H_2O(g)$
   b. $H_2O(l) + P_4O_{10}(s) \longrightarrow H_3PO_4(aq)$
   c. $PCl_5(g) + H_2O(l) \longrightarrow HCl(aq) + H_3PO_4(aq)$
4.77 Complete, then balance, each of the following equations:
   a. $Ca(s) + F_2(g) \longrightarrow$
   b. $Mg(s) + O_2(g) \longrightarrow$
   c. $H_2(g) + N_2(g) \longrightarrow$

4.78 Complete, then balance, each of the following equations:
  a. Li(s) + O_2(g) ⟶
  b. Ca(s) + N_2(g) ⟶
  c. Al(s) + S(s) ⟶

4.79 Balance each of the following equations:
  a. C_4H_{10}(g) + O_2(g) ⟶ H_2O(g) + CO_2(g)
  b. Au_2S_3(s) + H_2(g) ⟶ Au(s) + H_2S(g)
  c. $Al(OH)_3$(s) + HCl(aq) ⟶ $AlCl_3$(aq) + H_2O(l)
  d. $(NH_4)_2Cr_2O_7$(s) ⟶ Cr_2O_3(s) + N_2(g) + H_2O(g)
  e. C_2H_5OH(l) + O_2(g) ⟶ CO_2(g) + H_2O(g)

4.80 Balance each of the following equations:
  a. Fe_2O_3(s) + CO(g) ⟶ Fe_3O_4(s) + CO_2(g)
  b. C_6H_6(l) + O_2(g) ⟶ CO_2(g) + H_2O(g)
  c. I_4O_9(s) + I_2O_6(s) ⟶ I_2(s) + O_2(g)
  d. $KClO_3$(s) ⟶ KCl(s) + O_2(g)
  e. $C_6H_{12}O_6$(s) ⟶ C_2H_6O(l) + CO_2(g)

4.81 Write a balanced equation for each of the following reactions:
  a. Ammonia is formed by the reaction of nitrogen and hydrogen.
  b. Hydrochloric acid reacts with sodium hydroxide to produce water and sodium chloride.

4.82 Write a balanced equation for each of the following reactions:
  a. Nitric acid reacts with calcium hydroxide to produce water and calcium nitrate.
  b. Butane (C_4H_{10}) reacts with oxygen to produce water and carbon dioxide.

4.83 Write a balanced equation for each of the following reactions:
  a. Glucose, a sugar, $C_6H_{12}O_6$, is oxidized in the body to produce water and carbon dioxide.
  b. Sodium carbonate, upon heating, produces sodium oxide and carbon dioxide.

4.84 Write a balanced equation for each of the following reactions:
  a. Sulfur, present as an impurity in coal, is burned in oxygen to produce sulfur dioxide.
  b. Hydrofluoric acid (HF) reacts with glass (SiO_2) in the process of etching to produce silicon tetrafluoride and water.

Calculations Using the Chemical Equation

4.85 How many grams of boron oxide, B_2O_3, can be produced from 20.0 g diborane (B_2H_6)?
$$B_2H_6(l) + 3O_2(g) \longrightarrow B_2O_3(s) + 3H_2O(l)$$

4.86 How many grams of Al_2O_3 can be produced from 15.0 g Al?
$$4Al(s) + 3O_2(g) \longrightarrow 2Al_2O_3(s)$$

4.87 Calculate the amount of $CrCl_3$ that could be produced from 50.0 g Cr_2O_3 according to the equation
$$Cr_2O_3(s) + 3CCl_4(l) \longrightarrow 2CrCl_3(s) + 3COCl_2(aq)$$

4.88 A 3.5-g sample of water reacts with PCl_3 according to the following equation:
$$3H_2O(l) + PCl_3(g) \longrightarrow H_3PO_3(aq) + 3HCl(aq)$$
How many grams of H_3PO_3 are produced?

4.89 For the reaction
$$N_2(g) + H_2(g) \longrightarrow NH_3(g)$$
  a. Balance the equation.
  b. How many moles of H_2 would react with 1 mol of N_2?
  c. How many moles of product would form from 1 mol of N_2?
  d. If 14.0 g of N_2 were initially present, calculate the number of moles of H_2 required to react with all of the N_2.
  e. For conditions outlined in part (d), how many grams of product would form?

4.90 Aspirin (acetylsalicylic acid) may be formed from salicylic acid and acetic acid as follows:
$$\underset{\text{Salicylic acid}}{C_7H_6O_3(aq)} + \underset{\text{Acetic acid}}{CH_3COOH(aq)} \longrightarrow \underset{\text{Aspirin}}{C_9H_8O_4(s)} + H_2O(l)$$
  a. Is this equation balanced? If not, complete the balancing.
  b. How many moles of aspirin may be produced from 1.00×10^2 mol salicylic acid?
  c. How many grams of aspirin may be produced from 1.00×10^2 mol salicylic acid?
  d. How many grams of acetic acid would be required to react completely with the 1.00×10^2 mol salicylic acid?

4.91 The proteins in our bodies are composed of molecules called amino acids. One amino acid is methionine; its molecular formula is $C_5H_{11}NO_2S$. Calculate:
  a. the formula weight of methionine
  b. the number of oxygen atoms in a mole of this compound
  c. the mass of oxygen in a mole of the compound
  d. the mass of oxygen in 50.0 g of the compound

4.92 Triglycerides (Chapters 17 and 23) are used in biochemical systems to store energy; they can be formed from glycerol and fatty acids. The molecular formula of glycerol is $C_3H_8O_3$. Calculate:
  a. the formula weight of glycerol
  b. the number of oxygen atoms in a mole of this compound
  c. the mass of oxygen in a mole of the compound
  d. the mass of oxygen in 50.0 g of the compound

4.93 Joseph Priestley discovered oxygen in the eighteenth century by using heat to decompose mercury(II) oxide:
$$2HgO(s) \xrightarrow{\Delta} 2Hg(l) + O_2(g)$$
How much oxygen is produced from 1.00×10^2 g HgO?

4.94 Dinitrogen monoxide (also known as nitrous oxide and used as an anesthetic) can be made by heating ammonium nitrate:
$$NH_4NO_3(s) \longrightarrow N_2O(g) + 2H_2O(g)$$
How much dinitrogen monoxide can be made from 1.00×10^2 g of ammonium nitrate?

4.95 The burning of acetylene (C_2H_2) in oxygen is the reaction in the oxyacetylene torch. How much oxygen is needed to burn 20.0 kg of acetylene? The unbalanced equation is
$$C_2H_2(g) + O_2(g) \longrightarrow CO_2(g) + H_2O(g)$$

4.96 The reaction of calcium hydride with water can be used to prepare hydrogen gas:
$$CaH_2(s) + 2H_2O(l) \longrightarrow Ca(OH)_2(aq) + 2H_2(g)$$
How many moles of hydrogen gas are produced in the reaction of 1.00×10^2 g calcium hydride with water?

4.97 Various members of a class of compounds, alkenes (Chapter 11), react with hydrogen to produce a corresponding alkane (Chapter 10). Termed hydrogenation, this type of reaction is used to produce products such as margarine. A typical hydrogenation reaction is
$$\underset{\text{Decene}}{C_{10}H_{20}(l)} + H_2(g) \longrightarrow \underset{\text{Decane}}{C_{10}H_{22}(s)}$$
How much decane can be produced in a reaction of excess decene with 1.00 g hydrogen?

4.98 The Human Perspective: Alcohol Consumption and the Breathalyzer Test (Chapter 12), describes the reaction between the dichromate ion and ethanol to produce acetic acid. How much acetic acid can be produced from a mixture containing excess of dichromate ion and 1.00×10^{-1} g of ethanol?

4.99 A rocket can be powered by the reaction between dinitrogen tetroxide and hydrazine:

$$N_2O_4(l) + 2N_2H_4(l) \longrightarrow 3N_2(g) + 4H_2O(g)$$

An engineer designed the rocket to hold 1.00 kg N_2O_4 and excess N_2H_4. How much N_2 would be produced according to the engineer's design?

4.100 A 4.00-g sample of Fe_3O_4 reacts with O_2 to produce Fe_2O_3:

$$4Fe_3O_4(s) + O_2(g) \longrightarrow 6Fe_2O_3(s)$$

Determine the number of grams of Fe_2O_3 produced.

4.101 If the actual yield of decane in Problem 4.97 is 65.4 g, what is the % yield?

4.102 If the actual yield of acetic acid in Problem 4.98 is 0.110 g, what is the % yield?

4.103 If the % yield of nitrogen gas in Problem 4.99 is 75.0%, what is the actual yield of nitrogen?

4.104 If the % yield of Fe_2O_3 in Problem 4.100 is 90.0%, what is the actual yield of Fe_2O_3?

CRITICAL THINKING PROBLEMS

1. Which of the following has fewer moles of carbon: 100 g of $CaCO_3$ or 0.5 mol of CCl_4?
2. Which of the following has fewer moles of carbon: 6.02×10^{22} molecules of C_2H_6 or 88 g of CO_2?
3. How many molecules are found in each of the following?
   a. 1.0 lb of sucrose, $C_{12}H_{22}O_{11}$ (table sugar)
   b. 1.57 kg of N_2O (anesthetic)
4. How many molecules are found in each of the following?
   a. 4×10^5 tons of SO_2 (produced by the 1980 eruption of the Mount St. Helens volcano)
   b. 25.0 lb of SiO_2 (major constituent of sand)

GENERAL CHEMISTRY

States of Matter 5

Gases, Liquids, and Solids

Learning Goals

1. Describe the major points of the kinetic molecular theory of gases.
2. Explain the relationship between the kinetic molecular theory and the physical properties of macroscopic quantities of gases.
3. Describe the behavior of gases expressed by the gas laws: Boyle's law, Charles's law, combined gas law, Avogadro's law, the ideal gas law, and Dalton's law.
4. Use gas law equations to calculate conditions and changes in conditions of gases.
5. Describe properties of the liquid state in terms of the properties of the individual molecules that comprise the liquid.
6. Describe the processes of melting, boiling, evaporation, and condensation.
7. Describe the dipolar attractions known collectively as van der Waals forces.
8. Describe hydrogen bonding and its relationship to boiling and melting temperatures.
9. Relate the properties of the various classes of solids (ionic, covalent, molecular, and metallic) to the structure of these solids.

An exciting application of the gas laws.

Outline

Chemistry Connection:
The Demise of the Hindenburg

5.1 The Gaseous State

A Environmental Perspective:
The Greenhouse Effect and Global Warming

5.2 The Liquid State

A Medical Perspective:
Blood Gases and Respiration

5.3 The Solid State

Chemistry Connection

The Demise of the Hindenburg

One of the largest and most luxurious airships of the 1930s, the Hindenburg, completed thirty-six transatlantic flights within a year after its construction. It was the flagship of a new era of air travel. But, on May 6, 1937, while making a landing approach near Lakehurst, New Jersey, the hydrogen-filled airship exploded and burst into flames. In this tragedy, thirty-seven of the ninety-six passengers were killed and many others were injured.

We may never know the exact cause. Many believe that the massive ship (it was more than 800 feet long) struck an overhead power line. Others speculate that lightning ignited the hydrogen and some believe that sabotage may have been involved.

In retrospect, such an accident was inevitable. Hydrogen gas is very reactive, it combines with oxygen readily and rapidly, and this reaction liberates a large amount of energy. An explosion is the result of rapid, energy-releasing reactions.

Why was hydrogen chosen? Hydrogen is the lightest element. One mole of hydrogen has a mass of 2 grams. Hydrogen can be easily prepared in pure form, an essential requirement; more than seven million cubic feet of hydrogen were needed for each airship. Hydrogen has a low density; hence it provides great lift. The lifting power of a gas is based on the difference in density of the gas and the surrounding air (air is composed of gases with much greater molar masses; N_2 is 28 g and O_2 is 32 g). Engineers believed that the hydrogen would be safe when enclosed by the hull of the airship.

Hindenburg

Today, airships are filled with helium (its molar mass is 4 g) and are used principally for advertising and television. A Goodyear blimp can be seen hovering over almost every significant outdoor sporting event.

In this chapter we will study the relationships that predict the behavior of gases in a wide variety of applications from airships to pressurized oxygen for respiration therapy.

Introduction

Section 1.2 introduces the properties of the three states of matter.

We have learned that the major differences between solids, liquids, and gases are due to the relationships among particles. These relationships include:

- the average distance of separation of particles in each state,
- the kinds of interactions between the particles, and
- the degree of organization of particles.

We have already discovered that the solid state is the most organized, with particles close together, allowing significant interactions among the particles. This results in high melting and boiling points for solid substances. Large amounts of energy are needed to overcome the attractive forces and disrupt the orderly structure.

Substances that are gases, on the other hand, are disordered, with particles widely separated and weak interactions between particles. Their melting and boiling points are relatively low. Gases at room temperature must be cooled a great deal for them to liquefy or solidify. For example, the melting and boiling points of N_2 are $-210°C$ and $-196°C$, respectively.

Liquids are intermediate in character. The molecules of a liquid are close together, like those of solids. However, the molecules of a liquid are disordered, like those of a gas.

TABLE 5.1 A Comparison of Physical Properties of Gases, Liquids, and Solids

	Gas	Liquid	Solid
Volume and Shape	Expands to fill the volume of its container; consequently, it takes the shape of the container	Has a fixed volume at a given mass and temperature; volume principally dependent on its mass and secondarily on temperature; it assumes the shape of its container	Has a fixed volume; volume principally dependent on its mass and secondarily on temperature; it has a definite shape
Density	Low (typically ~10^{-3} g/mL)	High (typically ~1 g/mL)	High (typically 1–10 g/mL)
Compressibility	High	Very low	Virtually incompressible
Particle Motion	Virtually free	Molecules or atoms "slide" past each other	Vibrate about a fixed position
Intermolecular Distance	Very large	Molecules or atoms are close to each other	Molecules, ions, or atoms are close to each other

Changes in state are described as physical changes. When a substance undergoes a change in state, many of its physical properties change. For example, when ice forms from liquid water, changes occur in density and hardness, but it is still water. Table 5.1 summarizes the important differences in physical properties among gases, liquids, and solids.

5.1 The Gaseous State

Ideal Gas Concept

An **ideal gas** is simply a model of the way that particles (molecules or atoms) behave at the microscopic level. The behavior of the individual particles can be inferred from the macroscopic behavior of samples of real gases. We can easily measure temperature, volume, pressure, and quantity (mass) of real gases. Similarly, when we systematically change one of these properties, we can determine the effect on each of the others. For example, putting more molecules in a balloon (the act of blowing up a balloon) causes its volume to increase in a predictable way. In fact, careful measurements show a direct proportionality between the quantity of molecules and the volume of the balloon, an observation made by Amadeo Avogadro more than 200 years ago.

We owe a great deal of credit to the efforts of scientists Boyle, Charles, Avogadro, Dalton, and Gay-Lussac, whose careful work elucidated the relationships among the gas properties. Their efforts are summarized in the ideal gas law and are the subject of the first section of this chapter.

Measurement of Gases

The most important gas laws (Boyle's law, Charles's law, Avogadro's law, Dalton's law, and the ideal gas law) involve the relationships between pressure (P), volume (V), temperature (T), and number of moles (n) of gas. We are already familiar with the measurement of temperature and quantity from our laboratory experience. Measurement of pressure is perhaps not as obvious.

Gas pressure is a result of the force exerted by the collision of particles with the walls of the container. **Pressure** is force per unit area. The pressure of a gas may be measured with a **barometer,** invented by Evangelista Torricelli in the mid-1600s.

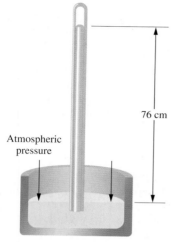

Figure 5.1
A mercury barometer of the type invented by Torricelli. The height of the column of mercury (*h*) is a function of the magnitude of the surrounding atmospheric pressure. The mercury in the tube is supported by atmospheric pressure.

The most common type of barometer is the mercury barometer depicted in Figure 5.1. A tube, sealed at one end, is filled with mercury and inverted in a dish of mercury. The pressure of the atmosphere pushing down on the mercury surface in the dish supports the column of mercury. The height of the column is proportional to the atmospheric pressure. The tube can be calibrated to give a numerical reading in millimeters, centimeters, or inches of mercury. A commonly used unit of measurement is the atmosphere (atm). One standard atmosphere (1 atm) of pressure is equivalent to a height of mercury that is equal to

760 mm Hg (millimeters of mercury)

76.0 cm Hg (centimeters of mercury)

1 mm of Hg is also = 1 torr, in honor of Torricelli.

The English system equivalent is a pressure of 14.7 lb/in.² (pounds per square inch) or 29.9 in. Hg (inches of mercury). A recommended, yet less frequently used, systematic unit is the pascal (or kilopascal), named in honor of Blaise Pascal, a seventeenth-century French mathematician and scientist:

$$1 \text{ atm} = 1.01 \times 10^5 \text{ Pa (pascal)} = 101 \text{ kPa (kilopascal)}$$

Atmospheric pressure is due to the cumulative force of the air molecules (N_2 and O_2, for the most part) that are attracted to the earth's surface by gravity.

Express each of the following in units of atmospheres:

a. 725 mm Hg
b. 29.0 cm Hg
c. 555 torr

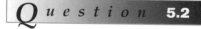

Express each of the following in units of atmospheres:

a. 10.0 torr
b. 61.0 cm Hg
c. 275 mm Hg

Kinetic Molecular Theory of Gases

The kinetic molecular theory of gases provides a reasonable explanation of the behavior of gases that we have studied in this chapter. The macroscopic properties result from the action of the individual molecules comprising the gas.

The **kinetic molecular theory** can be summarized as follows:

1. Gases are made up of small atoms or molecules that are in constant, random motion.
2. The distance of separation among these atoms or molecules is very large in comparison to the size of the individual atoms or molecules. In other words, a gas is mostly empty space.
3. All of the atoms and molecules behave independently. No attractive or repulsive forces exist between atoms or molecules in a gas.
4. Atoms and molecules collide with each other and with the walls of the container without *losing* energy. The energy is *transferred* from one atom or molecule to another.
5. The average kinetic energy of the atoms or molecules increases or decreases in proportion to absolute temperature.

(a)

(b)

Figure 5.2
Gaseous diffusion. (a) Ammonia (17.0 g/mol) and hydrogen chloride (36.5 g/mol) are introduced into the ends of a glass tube containing indicating paper. Red indicates the presence of hydrogen chloride and blue indicates ammonia. (b) Note that ammonia has diffused much farther than hydrogen chloride in the same amount of time. This is a verification of the kinetic molecular theory. Light molecules move faster than heavier molecules at a specified temperature.

We know that gases are easily *compressible*. The reason is that a gas is mostly empty space, providing space for the particles to be pushed closer together.

Gases will *expand* to fill any available volume because they move freely with sufficient energy to overcome their attractive forces.

Gases have a *low density*. Density is defined as mass per volume. Because gases are mostly empty space, they have a low mass per volume.

Gases readily *diffuse* through each other simply because they are in continuous motion and paths are readily available because of the large space between adjacent atoms or molecules. Light molecules diffuse rapidly; heavier molecules diffuse more slowly (Figure 5.2).

Gases exert *pressure* on their containers. Pressure is a force per unit area resulting from collisions of gas particles with the walls of their container.

Gases behave most *ideally at low pressures and high temperatures*. At low pressures, the average distance of separation among atoms or molecules is greatest, minimizing interactive forces. At high temperatures, the atoms and molecules are in rapid motion and are able to overcome interactive forces more easily.

Boyle's Law

The Irish scientist Robert Boyle found that the volume of a gas varies *inversely* with the pressure exerted by the gas if the number of moles and temperature of gas are held constant. This relationship is known as **Boyle's law**.

Mathematically, the *product* of pressure (P) and volume (V) is a constant:

$$PV = k_1$$

This relationship is illustrated in Figure 5.3.

LEARNING GOAL 2

Kinetic energy (K.E.) is equal to $1/2\ mv^2$, in which m = mass and v = velocity. Thus increased velocity at higher temperature correlates with an increase in kinetic energy.

LEARNING GOAL 3

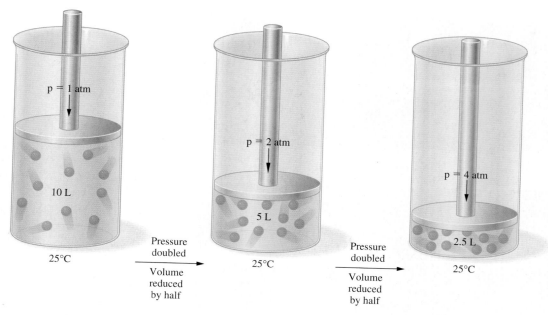

Figure 5.3
An illustration of Boyle's law. Note the inverse relationship of pressure and volume.

Boyle's law is often used to calculate the volume resulting from a pressure change or vice versa. We consider

$$P_i V_i = k_1$$

the *initial* condition and

$$P_f V_f = k_1$$

the final condition. Because PV, initial or final, is constant and is equal to k_1,

$$P_i V_i = P_f V_f$$

Consider a gas occupying a volume of 10.0 L at 1.00 atm of pressure. The product, $PV = (10.0 \text{ L})(1.00 \text{ atm})$, is a constant, k_1. Doubling the pressure, to 2.0 atm, decreases the volume to 5.0 L:

$$(2.0 \text{ atm})(V_x) = (10.0 \text{ L})(1.00 \text{ atm})$$

$$V_x = 5.0 \text{ L}$$

Tripling the pressure decreases the volume by a factor of 3:

$$(3.0 \text{ atm})(V_x) = (10.0 \text{ L})(1.00 \text{ atm})$$

$$V_x = 3.3 \text{ L}$$

You can find further information online at www.mhhe.com/denniston5e in "A Review of Mathematics."

EXAMPLE 5.1 **Calculating a Final Pressure**

LEARNING GOAL 4

A sample of oxygen, at 25°C, occupies a volume of 5.00×10^2 mL at 1.50 atm pressure. What pressure must be applied to compress the gas to a volume of 1.50×10^2 mL, with no temperature change?

Continued—

EXAMPLE 5.1 —Continued

Solution

Boyle's law applies directly, because there is no change in temperature or number of moles (no gas enters or leaves). Begin by identifying each term in the Boyle's law expression:

$$P_i = 1.50 \text{ atm}$$
$$V_i = 5.00 \times 10^2 \text{ mL}$$
$$V_f = 1.50 \times 10^2 \text{ mL}$$
$$P_i V_i = P_f V_f$$

and solve

$$P_f = \frac{P_i V_i}{V_f}$$
$$= \frac{(1.50 \text{ atm})(5.00 \times 10^2 \text{ mL})}{1.50 \times 10^2 \text{ mL}}$$
$$= 5.00 \text{ atm}$$

Helpful Hints:
1. Go to www.mhhe.com/denniston5e for a review of the mathematics used here.
2. The calculation can be done with any volume units. It is important only that the units be the *same* on both sides of the equation.

Question 5.3

Complete the following table:

	Initial Pressure (atm)	Final Pressure (atm)	Initial Volume (L)	Final Volume (L)
a.	X	5.0	1.0	7.5
b.	5.0	X	1.0	0.20

Question 5.4

Complete the following table:

	Initial Pressure (atm)	Final Pressure (atm)	Initial Volume (L)	Final Volume (L)
a.	1.0	0.50	X	0.30
b.	1.0	2.0	0.75	X

Charles's Law

3 LEARNING GOAL

Jacques Charles, a French scientist, studied the relationship between gas volume and temperature. This relationship, **Charles's law,** states that the volume of a gas varies *directly* with the absolute temperature (K) if pressure and number of moles of gas are constant.

Mathematically, the *ratio* of volume (V) and temperature (T) is a constant:

$$\frac{V}{T} = k_2$$

In a way analogous to Boyle's law, we may establish a set of initial conditions,

$$\frac{V_i}{T_i} = k_2$$

Temperature is a measure of the energy of molecular motion. The Kelvin scale is *absolute,* that is, directly proportional to molecular motion. Celsius and Fahrenheit are simply numerical scales based on the melting and boiling points of water. It is for this reason that Kelvin is used for energy-dependent relationships such as the gas laws.

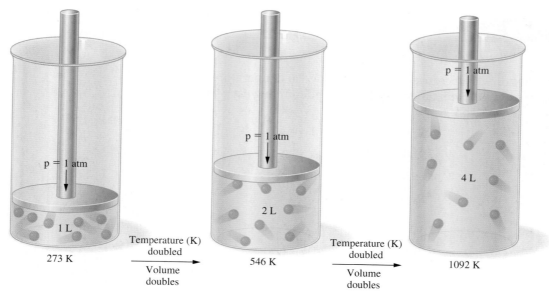

Figure 5.4
An illustration of Charles's law. Note the direct relationship between volume and temperature.

and final conditions,

$$\frac{V_f}{T_f} = k_2$$

Because k_2 is a constant, we may equate them, resulting in

$$\frac{V_i}{T_i} = \frac{V_f}{T_f}$$

and use this expression to solve some practical problems.

You can find further information online at www.mhhe.com/denniston5e in "A Review of Mathematics."

Consider a gas occupying a volume of 10.0 L at 273 K. The ratio V/T is a constant, k_2. Doubling the temperature, to 546 K, increases the volume to 20.0 L as shown here:

$$\frac{10.0 \text{ L}}{273 \text{ K}} = \frac{V_f}{546 \text{ K}}$$

$$V_f = 20.0 \text{ L}$$

Tripling the temperature, to 819 K, increases the volume by a factor of 3:

$$\frac{10.0 \text{ L}}{273 \text{ K}} = \frac{V_f}{819 \text{ K}}$$

$$V_f = 30.0 \text{ L}$$

These relationships are illustrated in Figure 5.4.

EXAMPLE 5.2 **Calculating a Final Volume**

LEARNING GOAL 4

A balloon filled with helium has a volume of 4.0×10^3 L at 25°C. What volume will the balloon occupy at 50°C if the pressure surrounding the balloon remains constant?

Continued—

EXAMPLE 5.2 —Continued

Solution

Remember, the temperature must be converted to Kelvin before Charles's law is applied:

$$T_i = 25°C + 273 = 298 \text{ K}$$
$$T_f = 50°C + 273 = 323 \text{ K}$$
$$V_i = 4.0 \times 10^3 \text{ L}$$
$$V_f = ?$$

Using

$$\frac{V_i}{T_i} = \frac{V_f}{T_f}$$

and substituting our data, we get

$$V_f = \frac{(V_i)(T_f)}{T_i} = \frac{(4.0 \times 10^3 \text{ L})(323 \text{ K})}{298 \text{ K}} = 4.3 \times 10^3 \text{ L}$$

Question 5.5

A sample of nitrogen gas has a volume of 3.00 L at 25°C. What volume will it occupy at each of the following temperatures if the pressure and number of moles are constant?

a. 100°C
b. 150°F
c. 273 K

Question 5.6

A sample of nitrogen gas has a volume of 3.00 L at 25°C. What volume will it occupy at each of the following temperatures if the pressure and number of moles are constant?

a. 546 K
b. 0°C
c. 373 K

The behavior of a hot-air balloon is a commonplace consequence of Charles's law. The balloon rises because air expands when heated (Figure 5.5). The volume of the balloon is fixed because the balloon is made of an inelastic material; as a result, when the air expands some of the air must be forced out. Hence the density of the remaining air is less (less mass contained in the same volume), and the balloon rises. Turning down the heat reverses the process, and the balloon descends.

Figure 5.5
Charles's law predicts that the volume of air in the balloon will increase when heated. We assume that the volume of the balloon is fixed; consequently, some air will be pushed out. The air remaining in the balloon is less dense (same volume, less mass) and the balloon will rise. When the heater is turned off the air cools, the density increases, and the balloon returns to earth.

Chapter 5 States of Matter: Gases, Liquids, and Solids

Combined Gas Law

Boyle's law describes the inverse proportional relationship between volume and pressure; Charles's law shows the direct proportional relationship between volume and temperature. Often, a sample of gas (a fixed number of moles of gas) undergoes change involving volume, pressure, and temperature simultaneously. It would be useful to have one equation that describes such processes.

LEARNING GOAL 3

The **combined gas law** is such an equation. It can be derived from Boyle's law and Charles's law and takes the form:

$$\frac{P_i V_i}{T_i} = \frac{P_f V_f}{T_f}$$

Let's look at two examples that use this expression.

EXAMPLE 5.3 Using the Combined Gas Law

LEARNING GOAL 4

Calculate the volume of N_2 that results when 0.100 L of the gas is heated from 300. K to 350. K at 1.00 atm.

Solution

Summarize the data:

$P_i = 1.00$ atm $P_f = 1.00$ atm

$V_i = 0.100$ L $V_f = ?$ L

$T_i = 300.$ K $T_f = 350.$ K

$$\frac{P_i V_i}{T_i} = \frac{P_f V_f}{T_f}$$

that can be rearranged as

$$P_f V_f T_i = P_i V_i T_f$$

and

$$V_f = \frac{P_i V_i T_f}{P_f T_i}$$

Because $P_i = P_f$

$$V_f = \frac{V_i T_f}{T_i}$$

Substituting gives

$$V_f = \frac{(0.100 \text{ L})(350.\text{ K})}{300.\text{ K}}$$

$$= 0.117 \text{ L}$$

Helpful Hints:

1. Go to www.mhhe.com/denniston5e for a review of the mathematics used here.
2. In this case, because the pressure is constant, the combined gas law reduces to Charles's law.

EXAMPLE 5.4

Using the Combined Gas Law

LEARNING GOAL 4

A sample of helium gas has a volume of 1.27 L at 149 K and 5.00 atm. When the gas is compressed to 0.320 L at 50.0 atm, the temperature increases markedly. What is the final temperature?

Solution

Summarize the data:

$P_i = 5.00$ atm $P_f = 50.0$ atm

$V_i = 1.27$ L $V_f = 0.320$ L

$T_i = 149$ K $T_f = ?$ K

The combined gas law expression is

$$\frac{P_i V_i}{T_i} = \frac{P_f V_f}{T_f}$$

which we rearrange as

$$P_f V_f T_i = P_i V_i T_f$$

and

$$T_f = \frac{P_f V_f T_i}{P_i V_i}$$

Substituting yields

$$T_f = \frac{(50.0 \text{ atm})(0.320 \text{ L})(149 \text{ K})}{(5.00 \text{ atm})(1.27 \text{ L})}$$

$$= 375 \text{ K}$$

Helpful Hint: Go to www.mhhe.com/denniston5e for a review of the mathematics used here.

Question 5.7

Hydrogen sulfide, H_2S, has the characteristic odor of rotten eggs. If a sample of H_2S gas at 760. torr and 25.0°C in a 2.00-L container is allowed to expand into a 10.0-L container at 25.0°C, what is the pressure in the 10.0-L container?

Question 5.8

Cyclopropane, C_3H_6, is used as a general anesthetic. If a sample of cyclopropane stored in a 2.00-L container at 10.0 atm and 25.0°C is transferred to a 5.00-L container at 5.00 atm, what is the resulting temperature?

Avogadro's Law

LEARNING GOAL 3

The relationship between the volume and number of moles of a gas at constant temperature and pressure is known as **Avogadro's law**. It states that equal volumes of any ideal gas contain the same number of moles if measured under the same conditions of temperature and pressure.

Chapter 5 States of Matter: Gases, Liquids, and Solids

Mathematically, the *ratio* of volume (V) to number of moles (n) is a constant:

$$\frac{V}{n} = k_3$$

Consider 1 mol of gas occupying a volume of 10.0 L; using logic similar to the application of Boyle's and Charles's laws, 2 mol of the gas would occupy 20.0 L, 3 mol would occupy 30.0 L, and so forth. As we have done with the previous laws, we can formulate a useful expression relating initial and final conditions:

$$\frac{V_i}{n_i} = \frac{V_f}{n_f}$$

You can find further information online at www.mhhe.com/denniston5e in "A Review of Mathematics."

EXAMPLE 5.5 Using Avogadro's Law

LEARNING GOAL 4

If 5.50 mol of CO occupy 20.6 L, how many liters will 16.5 mol of CO occupy at the same temperature and pressure?

Solution

The quantities moles and volume are related through Avogadro's law. Summarizing the data:

$V_i = 20.6$ L $V_f = ?$ L

$n_i = 5.50$ mol $n_f = 16.5$ mol

Using the mathematical expression for Avogadro's law:

$$\frac{V_i}{n_i} = \frac{V_f}{n_f}$$

and rearranging as

$$V_f = \frac{V_i n_f}{n_i}$$

then substituting yields

$$V_f = \frac{(20.6 \text{ L})(16.5 \text{ mol})}{(5.50 \text{ mol})}$$

$$= 61.8 \text{ L of CO}$$

Question 5.9

1.00 mole of hydrogen gas occupies 22.4 L. How many moles of hydrogen are needed to fill a 100.0 L container at the same pressure and temperature?

Question 5.10

How many moles of hydrogen are needed to triple the volume occupied by 0.25 mol of hydrogen, assuming no changes in pressure or temperature?

Molar Volume of a Gas

The volume occupied by *1 mol* of any gas is referred to as its **molar volume**. At **standard temperature and pressure (STP)** the molar volume of any gas is 22.4 L. STP conditions are defined as follows:

$$T = 273 \text{ K (or } 0°C)$$

$$P = 1 \text{ atm}$$

Thus, 1 mol of N_2, O_2, H_2, or He all occupy the *same volume, 22.4 L*, at STP.

Gas Densities

It is also possible to compute the density of various gases at STP. If we recall that density is the mass/unit volume,

$$d = \frac{m}{V}$$

and that 1 mol of helium weighs 4.00 g,

$$d_{He} = \frac{4.00 \text{ g}}{22.4 \text{ L}} = 0.178 \text{ g/L at STP}$$

or, because 1 mol of nitrogen weighs 28.0 g, then

$$d_{N_2} = \frac{28.0 \text{ g}}{22.4 \text{ L}} = 1.25 \text{ g/L at STP}$$

The large difference in gas densities of helium and nitrogen (which makes up about 80% of the air) accounts for the lifting power of helium. A balloon filled with helium will rise through a predominantly nitrogen atmosphere because its gas density is less than 15% of the density of the surrounding atmosphere:

$$\frac{0.178 \text{ g/L}}{1.25 \text{ g/L}} \times 100\% = 14.2\%$$

Heating a gas, such as air, will decrease its density and have a lifting effect as well.

The Ideal Gas Law

Boyle's law (relating volume and pressure), Charles's law (relating volume and temperature), and Avogadro's law (relating volume to the number of moles) may be combined into a single expression relating all four terms. This expression is the **ideal gas law**:

$$PV = nRT$$

in which R, based on k_1, k_2, and k_3 (Boyle's, Charles's, and Avogadro's law constants), is a constant and is referred to as the *ideal gas constant*:

$$R = 0.0821 \text{ L-atm K}^{-1} \text{ mol}^{-1}$$

if the units

P in atmospheres

V in liters

n in number of moles

T in Kelvin

are used.

 LEARNING GOAL

Remember that 0.0821 L-atm/K mol is identical to 0.0821 L-atm K^{-1} mol^{-1}.

Consider some examples of the application of the ideal gas equation.

EXAMPLE 5.6 Calculating a Molar Volume

LEARNING GOAL 4

Demonstrate that the molar volume of oxygen gas at STP is 22.4 L.

Solution

$$PV = nRT$$

$$V = \frac{nRT}{P}$$

At standard temperature and pressure,

$$T = 273 \text{ K}$$
$$P = 1.00 \text{ atm}$$

The other constants are

$$n = 1.00 \text{ mol}$$
$$R = 0.0821 \text{ L-atm K}^{-1} \text{ mol}^{-1}$$

Then

$$V = \frac{(1.00 \text{ mol})(0.0821 \text{ L-atm K}^{-1} \text{ mol}^{-1})(273 \text{ K})}{(1.00 \text{ atm})}$$

$$= 22.4 \text{ L}$$

EXAMPLE 5.7 Calculating the Number of Moles of a Gas

LEARNING GOAL 4

Calculate the number of moles of helium in a 1.00-L balloon at 27°C and 1.00 atm of pressure.

Solution

$$PV = nRT$$

$$n = \frac{PV}{RT}$$

If

$$P = 1.00 \text{ atm}$$
$$V = 1.00 \text{ L}$$
$$T = 27°C + 273 = 300. \text{ K}$$
$$R = 0.0821 \text{ L-atm K}^{-1} \text{ mol}^{-1}$$

then

$$n = \frac{(1.00 \text{ atm})(1.00 \text{ L})}{(0.0821 \text{ L-atm K}^{-1} \text{ mol}^{-1})(300. \text{ K})}$$

$$n = 0.0406 \text{ or } 4.06 \times 10^{-2} \text{ mol}$$

EXAMPLE 5.8

Converting Mass to Volume

Oxygen used in hospitals and laboratories is often obtained from cylinders containing liquefied oxygen. If a cylinder contains 1.00×10^2 kg of liquid oxygen, how many liters of oxygen can be produced at 1.00 atm of pressure at room temperature (20.0°C)?

Solution

$$PV = nRT$$

$$V = \frac{nRT}{P}$$

Using conversion factors, we obtain

$$n_{O_2} = 1.00 \times 10^2 \text{ kg O}_2 \times \frac{10^3 \text{ g O}_2}{1 \text{ kg O}_2} \times \frac{1 \text{ mol O}_2}{32.0 \text{ g O}_2}$$

Then

$$n = 3.13 \times 10^3 \text{ mol O}_2$$

and

$$T = 20.0°C + 273 = 293 \text{ K}$$

$$P = 1.00 \text{ atm}$$

then

$$V = \frac{(3.13 \times 10^3 \text{ mol})(0.0821 \text{ L-atm K}^{-1} \text{ mol}^{-1})(293 \text{ K})}{1.00 \text{ atm}}$$

$$= 7.53 \times 10^4 \text{ L}$$

LEARNING GOAL 4

Question 5.11

What volume is occupied by 10.0 g N_2 at 30.0°C and a pressure of 750 torr?

Question 5.12

A 20.0-L gas cylinder contains 4.80 g H_2 at 25°C. What is the pressure of this gas?

Question 5.13

How many moles of N_2 gas will occupy a 5.00-L container at standard temperature and pressure?

Question 5.14

At what temperature will 2.00 mol He fill a 2.00-L container at standard pressure?

An Environmental Perspective

The Greenhouse Effect and Global Warming

A greenhouse is a bright, warm, and humid environment for growing plants, vegetables, and flowers even during the cold winter months. It functions as a closed system in which the concentration of water vapor is elevated and visible light streams through the windows; this creates an ideal climate for plant growth.

Some of the visible light is absorbed by plants and soil in the greenhouse and radiated as infrared radiation. This radiated energy is blocked by the glass or absorbed by water vapor and carbon dioxide (CO_2). This trapped energy warms the greenhouse and is a form of solar heating: light energy is converted to heat energy.

On a global scale, the same process takes place. Although more than half of the sunlight that strikes the earth's surface is reflected back into space, the fraction of light that is absorbed produces sufficient heat to sustain life. How does this happen? Greenhouse gases, such as CO_2, trap energy radiated from the earth's surface and store it in the atmosphere. This moderates our climate. The earth's surface would be much colder and more inhospitable if the atmosphere was not able to capture some reasonable amount of solar energy.

Can we have too much of a good thing? It appears so. Since 1900 the atmospheric concentration of CO_2 has increased from 296 parts per million (ppm) to over 350 ppm (approximately 17% increase). The energy demands of technological and population growth have caused massive increases in the combustion of organic matter and carbon-based fuels (coal, oil, and natural gas), adding over 50 billion tons of CO_2 to that already present in the atmosphere. Photosynthesis naturally removes CO_2 from the atmosphere. However, the removal of forestland to create living space and cropland decreases the amount of vegetation available to consume atmospheric CO_2 through photosynthesis. The rapid destruction of the Amazon rain forest is just the latest of many examples.

If our greenhouse model is correct, an increase in CO_2 levels should produce global warming, perhaps changing our climate in unforeseen and undesirable ways.

For Further Understanding
What steps might be taken to decrease levels of CO_2 in the atmosphere over time?

In what ways might our climate and our lives change as a consequence of significant global warming?

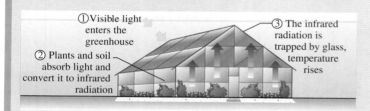

(a) A greenhouse traps solar radiation as heat. (b) Our atmosphere also acts as a solar collector. Carbon dioxide, like the windows of a greenhouse, allows the visible light to enter and traps the heat.

Dalton's Law of Partial Pressures

LEARNING GOAL 3

Our discussion of gases so far has presumed that we are working with a single pure gas. A *mixture* of gases exerts a pressure that is the *sum* of the pressures that each gas would exert if it were present alone under the same conditions. This is known as **Dalton's law** of partial pressures.

Stated another way, the total pressure of a mixture of gases is the sum of the **partial pressures.** That is,

$$P_t = p_1 + p_2 + p_3 + \ldots$$

in which P_t = total pressure and $p_1, p_2, p_3, \ldots,$ are the partial pressures of the component gases. For example, the total pressure of our atmosphere is equal to the sum of the pressures of N_2 and O_2 (the principal components of air):

$$P_{air} = p_{N_2} + p_{O_2}$$

The ideal gas law applies to mixtures of gases as well as pure gases.

Other gases, such as argon (Ar), carbon dioxide (CO_2), carbon monoxide (CO), and methane (CH_4) are present in the atmosphere at very low partial pressures. However, their presence may result in dramatic consequences; one such gas is carbon dioxide. Classified as a "greenhouse gas," it exerts a significant effect on our climate. Its role is described in An Environmental Perspective: The Greenhouse Effect and Global Warming.

Ideal Gases Versus Real Gases

To this point we have assumed, in both theory and calculations, that all gases behave as ideal gases. However, in reality there is no such thing as an ideal gas. As we noted at the beginning of this section, the ideal gas is a model (a very useful one) that describes the behavior of individual atoms and molecules; this behavior translates to the collective properties of measurable quantities of these atoms and molecules. Limitations of the model arise from the fact that interactive forces, even between the widely spaced particles of gas, are not totally absent in any sample of gas.

Attractive forces are present in gases composed of polar molecules. Nonuniform charge distribution on polar molecules creates positive and negative regions, resulting in electrostatic attraction and deviation from ideality.

Calculations involving polar gases such as HF, NO, and SO_2 based on ideal gas equations (which presume no such interactions) are approximations. However, at low pressures, such approximations certainly provide useful information. Nonpolar molecules, on the other hand, are only weakly attracted to each other and behave much more ideally in the gas phase.

See Sections 3.5 and 5.2 for a discussion of interactions of polar molecules.

5.2 The Liquid State

LEARNING GOAL 5

Molecules in the liquid state are close to one another. Attractive forces are large enough to keep the molecules together in contrast to gases, whose cohesive forces are so low that a gas expands to fill any volume. However, these attractive forces in a liquid are not large enough to restrict movement, as in solids. Let's look at the various properties of liquids in more detail.

Compressibility

Liquids are practically incompressible. In fact, the molecules are so close to one another that even the application of many atmospheres of pressure does not significantly decrease the volume. This makes liquids ideal for the transmission of force, as in the brake lines of an automobile. The force applied by the driver's foot on the brake pedal does not compress the brake fluid in the lines; rather, it transmits the force directly to the brake pads, and the friction between the brake pads and rotors (that are attached to the wheel) stops the car.

Viscosity

The **viscosity** of a liquid is a measure of its resistance to flow. Viscosity is a function of both the attractive forces between molecules and molecular geometry.

Molecules with complex structures, which do not "slide" smoothly past each other, and polar molecules, tend to have higher viscosity than less structurally complex, less polar liquids. Glycerol, which is used in a variety of skin treatments, has the structural formula:

A Medical Perspective

Blood Gases and Respiration

Respiration must deliver oxygen to cells and the waste product, carbon dioxide, to the lungs to be exhaled. Dalton's law of partial pressures helps to explain the way in which this process occurs.

Gases (such as O_2 and CO_2) move from a region of higher partial pressure to one of lower partial pressure in an effort to establish an equilibrium. At the interface of the lung, the membrane barrier between the blood and the surrounding atmosphere, the following situation exists: Atmospheric O_2 partial pressure is high, and atmospheric CO_2 partial pressure is low. The reverse is true on the other side of the membrane (blood). Thus CO_2 is efficiently removed from the blood, and O_2 is efficiently moved into the bloodstream.

At the other end of the line, capillaries are distributed in close proximity to the cells that need to expel CO_2 and gain O_2. The partial pressure of CO_2 is high in these cells, and the partial pressure of O_2 is low, having been used up by the energy-harvesting reaction, the oxidation of glucose:

$$C_6H_{12}O_6 + 6O_2 \longrightarrow 6CO_2 + 6H_2O + energy$$

The O_2 diffuses into the cells (from a region of high to low partial pressure), and the CO_2 diffuses from the cells to the blood (again from a region of high to low partial pressure).

The net result is a continuous process proceeding according to Dalton's law. With each breath we take, oxygen is distributed to the cells and used to generate energy, and the waste product, CO_2, is expelled by the lungs.

For Further Understanding

Carbon dioxide and carbon monoxide can be toxic, but for different reasons. Use the Internet to research this topic and:

Explain why carbon dioxide is toxic.

Explain why carbon monoxide is toxic.

It is quite viscous, owing to its polar nature and its significant intermolecular attractive forces. This is certainly desirable in a skin treatment because its viscosity keeps it on the area being treated. Gasoline, on the other hand, is much less viscous and readily flows through the gas lines of your auto; it is composed of nonpolar molecules.

Viscosity generally decreases with increasing temperature. The increased kinetic energy at higher temperatures overcomes some of the intermolecular attractive forces. The temperature effect is an important consideration in the design of products that must remain fluid at low temperatures, such as motor oils and transmission fluids found in automobiles.

Surface Tension

The **surface tension** of a liquid is a measure of the attractive forces exerted among molecules at the surface of a liquid. It is only the surface molecules that are not totally surrounded by other liquid molecules (the top of the molecule faces the atmosphere). These surface molecules are surrounded and attracted by fewer liquid molecules than those below and to each side. Hence the net attractive forces on surface molecules pull them downward, into the body of the liquid. As a result, the surface molecules behave as a "skin" that covers the interior.

This increased surface force is responsible for the spherical shape of drops of liquid. Drops of water "beading" on a polished surface, such as a waxed automobile, illustrate this effect.

Because surface tension is related to the attractive forces exerted among molecules, surface tension generally decreases with an increase in temperature or a decrease in the polarity of molecules that make up the liquid.

Substances known as **surfactants** can be added to a liquid to decrease surface tension. Common surfactants include soaps and detergents that reduce water's surface tension; this promotes the interaction of water with grease and dirt, making it easier to remove.

Vapor Pressure of a Liquid

Evaporation, condensation, and the meaning of the term *boiling point* are all related to the concept of liquid vapor pressure. Consider the following example. A liquid, such as water, is placed in a sealed container. After a time the contents of the container are analyzed. Both liquid water and water vapor are found at room temperature, when we might expect water to be found only as a liquid. In this closed system, some of the liquid water was converted to a gas:

$$\boxed{\text{energy}} + H_2O(l) \longrightarrow H_2O(g)$$

How did this happen? The temperature is too low for conversion of a liquid to a gas by boiling. According to the kinetic theory, liquid molecules are in continuous motion, with their *average* kinetic energy directly proportional to the Kelvin temperature. The word *average* is the key. Although the average kinetic energy is too low to allow "average" molecules to escape from the liquid phase to the gas phase, there exists a range of molecules with different energies, some low and some high, that make up the "average" (Figure 5.6). Thus some of these high-energy molecules possess sufficient energy to escape from the bulk liquid.

At the same time a fraction of these gaseous molecules lose energy (perhaps by collision with the walls of the container) and return to the liquid state:

$$H_2O(g) \longrightarrow H_2O(l) + \boxed{\text{energy}}$$

The process of conversion of liquid to gas, at a temperature too low to boil, is **evaporation**. The reverse process, conversion of the gas to the liquid state, is **condensation**. After some time the rates of evaporation and condensation become *equal*, and this sets up a dynamic equilibrium between liquid and vapor states. The **vapor pressure of a liquid** is defined as the pressure exerted by the vapor *at equilibrium*.

$$H_2O(g) \rightleftharpoons H_2O(l)$$

The equilibrium process of evaporation and condensation of water is depicted in Figure 5.7.

The boiling point of a liquid is defined as the temperature at which the vapor pressure of the liquid becomes equal to the atmospheric pressure. The "normal" atmospheric pressure is 760 torr, or 1 atm, and the **normal boiling point** is the temperature at which the vapor pressure of the liquid is equal to 1 atm.

It follows from the definition that the boiling point of a liquid is not constant. It depends on the atmospheric pressure. At high altitudes, where the atmospheric pressure is low, the boiling point of a liquid, such as water, is lower than the normal boiling point (for water, 100°C). High atmospheric pressure increases the boiling point.

LEARNING GOAL

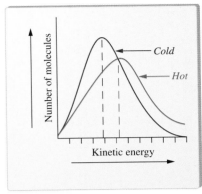

Figure 5.6
The temperature dependence of liquid vapor pressure is illustrated. The average molecular kinetic energy increases with temperature. Note that the average values are indicated by dashed lines. The small number of high-energy molecules may evaporate.

> The process of evaporation of perspiration from the skin produces a cooling effect, because heat is stored in the evaporating molecules.

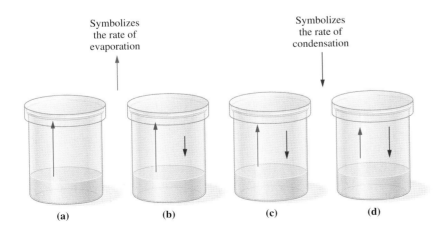

Figure 5.7
Liquid water in equilibrium with water vapor. (a) Initiation: process of evaporation exclusively. (b, c) After a time, both evaporation and condensation occur, but evaporation predominates. (d) Dynamic equilibrium established. Rates of evaporation and condensation are equal.

Apart from its dependence on the surrounding atmospheric pressure, the boiling point depends on the nature of the attractive forces between the liquid molecules. Polar liquids, such as water, with large intermolecular attractive forces have *higher* boiling points than nonpolar liquids, such as gasoline, which exhibit weak attractive forces.

Van der Waals Forces

Physical properties of liquids, such as those discussed in the previous section, can be explained in terms of their intermolecular forces. We have seen (see Section 3.5) that attractive forces between polar molecules, **dipole-dipole interactions,** significantly decrease vapor pressure and increase the boiling point. However, nonpolar substances can exist as liquids as well; many are liquids and even solids at room temperature. What is the nature of the attractive forces in these nonpolar compounds?

In 1930 Fritz London demonstrated that he could account for a weak attractive force between any two molecules, whether polar or nonpolar. He postulated that the electron distribution in molecules is not fixed; electrons are in continuous motion, relative to the nucleus. So, for a short time a nonpolar molecule could experience an *instantaneous dipole,* a short-lived polarity caused by a temporary dislocation of the electron cloud. These temporary dipoles could interact with other temporary dipoles, just as permanent dipoles interact in polar molecules. We now call these intermolecular forces **London forces.**

London forces and dipole-dipole interactions are collectively known as **van der Waals forces.** London forces exist among polar and nonpolar molecules because electrons are in constant motion in all molecules. Dipole-dipole attractions occur only among polar molecules. In addition to van der Waals forces, a special type of dipole-dipole force, the *hydrogen bond,* has a very significant effect on molecular properties, particularly in biological systems.

Hydrogen Bonding

Typical forces in polar liquids, discussed above, are only about 1–2% as strong as ionic and covalent bonds. However, certain liquids have boiling points that are much higher than we would predict from these dipolar interactions alone. This indicates the presence of some strong intermolecular force. This attractive force is due to **hydrogen bonding.** Molecules in which a hydrogen atom is bonded to a small, highly electronegative atom such as nitrogen, oxygen, or fluorine exhibit this effect. The presence of a highly electronegative atom bonded to a hydrogen atom creates a large dipole:

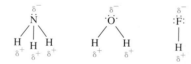

This arrangement of atoms produces a very polar bond, often resulting in a polar molecule with strong intermolecular attractive forces. Although the hydrogen bond is weaker than bonds formed *within* molecules (covalent and polar covalent *intra*molecular forces), it is the strongest attractive force *between* molecules (intermolecular force).

Consider the boiling points of four small molecules:

CH_4	NH_3	H_2O	HF
−161°C	−33°C	+100°C	+19.5°C

Clearly, ammonia, water, and hydrogen fluoride boil at significantly higher temperatures than methane. The N—H, O—H, and F—H bonds are far more polar than the C—H bond, owing to the high electronegativity of N, O, and F.

It is interesting to note that the boiling points increase as the electronegativity of the element bonded to hydrogen increases, with one exception: Fluorine, with the highest electronegativity should cause HF to have the highest boiling point. This is not the case. The order of boiling points is

water > hydrogen fluoride > ammonia > methane

not

hydrogen fluoride > water > ammonia > methane

Why? To answer this question we must look at the *number of potential bonding sites* in each molecule. Water has two partial positive sites (located at each hydrogen atom) and two partial negative sites (two lone pairs of electrons on the oxygen atom); it can form hydrogen bonds at each site. This results in a complex network of attractive forces among water molecules in the liquid state and the strength of the forces holding this network together accounts for water's unusually high boiling point. This network is depicted in Figure 5.8.

Ammonia and hydrogen fluoride can form only one hydrogen bond per molecule. Ammonia has three partial positive sites (three hydrogen atoms bonded to nitrogen) but only one partial negative site (the lone pair); the single lone pair is the limiting factor. One positive site and one negative site are needed for each hydrogen bond. Hydrogen fluoride has only one partial positive site and one partial negative site. It too can form only one hydrogen bond per molecule. Consequently, the network of attractive forces in ammonia and hydrogen fluoride is much less extensive than that found in water, and their boiling points are considerably lower than that of water.

Hydrogen bonding has an extremely important influence on the behavior of many biological systems. Molecules such as proteins and DNA require extensive hydrogen bonding to maintain their structures and hence functions. DNA (deoxyribonucleic acid, Section 20.2) is a giant among molecules with intertwined chains of atoms held together by thousands of hydrogen bonds.

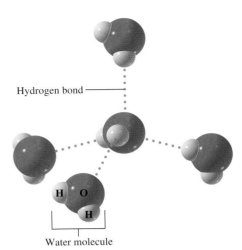

Figure 5.8
Hydrogen bonding in water. Note that the central water molecule is hydrogen bonded to four other water molecules. The attractive force between the hydrogen (δ^+) part of one water molecule and the oxygen (δ^-) part of another water molecule constitutes the hydrogen bond.

Intramolecular hydrogen bonding between polar regions helps keep proteins folded in their proper three-dimensional structure. See Chapter 18.

5.3 The Solid State

The close packing of the particles of a solid results from attractive forces that are strong enough to restrict motion. This occurs because the kinetic energy of the particles is insufficient to overcome the attractive forces among particles. The particles are "locked" together in a defined and highly organized fashion. This results in fixed shape and volume, although, at the atomic level, vibrational motion is observed.

Properties of Solids

Solids are virtually incompressible, owing to the small distance between particles. Most will convert to liquids at a higher temperature, when the increased heat energy overcomes some of the attractive forces within the solid. The temperature at which a solid is converted to the liquid phase is its **melting point.** The melting point depends on the strength of the attractive forces in the solid, hence its structure. As we might expect, polar solids have higher melting points than nonpolar solids of the same molecular weight.

A solid may be a **crystalline solid,** having a regular repeating structure, or an **amorphous solid,** having no organized structure. Diamond and sodium chloride (Figure 5.9) are examples of crystalline substances; glass, plastic, and concrete are examples of amorphous solids.

Figure 5.9
Crystalline solids.

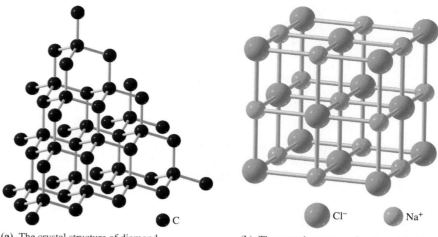

(a) The crystal structure of diamond

(b) The crystal structure of sodium chloride

(c) The crystal structure of methane, a frozen molecular solid. Only one methane molecule is shown in detail.

(d) The crystal structure of a metallic solid. The gray area represents mobile electrons around fixed metal cations.

Types of Crystalline Solids

Crystalline solids may exist in one of four general groups:

1. *Ionic solids.* The units that comprise an **ionic solid** are positive and negative ions. Electrostatic forces hold the crystal together. They generally have high melting points, and are hard and brittle. A common example of an ionic solid is sodium chloride.
2. *Covalent solids.* The units that comprise a **covalent solid** are atoms held together by covalent bonds. They have very high melting points (1200°C to 2000°C or more is not unusual) and are extremely hard. They are insoluble in most solvents. Diamond is a covalent solid composed of covalently bonded carbon atoms. Diamonds are used for industrial cutting because they are so hard and as gemstones because of their crystalline beauty.
3. *Molecular solids.* The units that make up a **molecular solid,** molecules, are held together by intermolecular attractive forces (London forces, dipole-dipole interactions, and hydrogen bonding). Molecular solids are usually soft and have low melting points. They are frequently volatile and are poor electrical conductors. A common example is ice (solid water; Figure 5.10).
4. *Metallic solids.* The units that comprise a **metallic solid** are metal atoms held together by metallic bonds. **Metallic bonds** are formed by the overlap of orbitals of metal atoms, resulting in regions of high electron density surrounding the positive metal nuclei. Electrons in these regions are extremely mobile. They are able to move freely from atom to atom through

Intermolecular forces are also discussed in Sections 3.5 and 5.2.

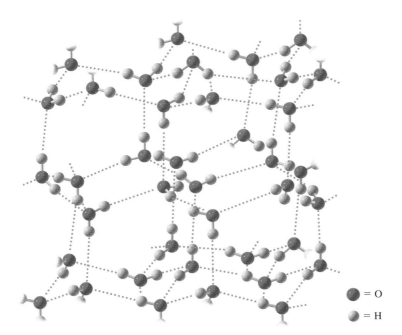

Figure 5.10
The structure of ice, a molecular solid. Hydrogen bonding among water molecules produces a regular open structure that is less dense than liquid water.

● = O
● = H

pathways that are, in reality, overlapping atomic orbitals. This results in the high *conductivity* (ability to carry electrical current) exhibited by many metallic solids. Silver and copper are common examples of metallic solids. Metals are easily shaped and are used for a variety of purposes. Most of these are practical applications such as hardware, cookware, and surgical and dental tools. Others are purely for enjoyment and decoration, such as silver and gold jewelry.

SUMMARY

5.1 The Gaseous State

The *kinetic molecular theory* describes an *ideal gas* in which gas particles exhibit no interactive or repulsive forces and the volumes of the individual gas particles are assumed to be negligible.

Boyle's law states that the volume of a gas varies inversely with the pressure exerted by the gas if the number of moles and temperature of gas are held constant ($PV = k_1$).

Charles's law states that the volume of a gas varies directly with the absolute temperature (K) if pressure and number of moles of gas are constant ($V/T = k_2$).

Avogadro's law states that equal volumes of any gas contain the same number of moles if measured at constant temperature and pressure ($V/n = k_3$).

The volume occupied by 1 mol of any gas is its *molar volume*. At *standard temperature and pressure* (STP) the molar volume of any ideal gas is 22.4 L. STP conditions are defined as 273 K (or 0°C) and 1 atm pressure.

Boyle's law, Charles's law, and Avogadro's law may be combined into a single expression relating all four terms, the *ideal gas law*: $PV = nRT$. R is the ideal gas constant (0.0821 L-atm K^{-1} mol^{-1}) if the units P (atmospheres), V (liters), n (number of moles), and T (Kelvin) are used.

The *combined gas law* provides a convenient expression for performing gas law calculations involving the most common variables: pressure, volume, and temperature.

Dalton's law of partial pressures states that a mixture of gases exerts a pressure that is the sum of the pressures that each gas would exert if it were present alone under similar conditions ($P_t = p_1 + p_2 + p_3 + \ldots$).

5.2 The Liquid State

Liquids are practically incompressible because of the closeness of the molecules. The *viscosity* of a liquid is a measure of its resistance to flow. Viscosity generally decreases with increasing temperature. The *surface tension* of a liquid is a measure of the attractive forces at the surface of a liquid. *Surfactants* decrease surface tension.

The conversion of liquid to vapor at a temperature below the boiling point of the liquid is *evaporation*. Conversion of the gas to the liquid state is *condensation*. The *vapor pressure of the liquid* is defined as the pressure exerted by the vapor at equilibrium at a specified temperature. The *normal boiling point* of a liquid is the temperature at which the vapor pressure of the liquid is equal to 1 atm.

Molecules in which a hydrogen atom is bonded to a small, highly electronegative atom such as nitrogen, oxygen, or fluorine exhibit *hydrogen bonding*. Hydrogen bonding in liquids is responsible for lower than expected vapor pressures and higher than expected boiling points. The presence of *van der Waals forces* and hydrogen bonds significantly affects the boiling points of liquids as well as the melting points of solids.

5.3 The Solid State

Solids have fixed shapes and volumes. They are *incompressible*, owing to the closeness of the particles. Solids may be *crystalline*, having a regular, repeating structure, or *amorphous*, having no organized structure.

Crystalline solids may exist as *ionic solids, covalent solids, molecular solids,* or *metallic solids*. Electrons in metallic solids are extremely mobile, resulting in the high *conductivity* (ability to carry electrical current) exhibited by many metallic solids.

KEY TERMS

amorphous solid (5.3)
Avogadro's law (5.1)
barometer (5.1)
Boyle's law (5.1)
Charles's law (5.1)
combined gas law (5.1)
condensation (5.2)
covalent solid (5.3)
crystalline solid (5.3)
Dalton's law (5.1)
dipole-dipole interactions (5.2)
evaporation (5.2)
hydrogen bonding (5.2)
ideal gas (5.1)
ideal gas law (5.1)
ionic solid (5.3)
kinetic molecular theory (5.1)
London forces (5.2)
melting point (5.3)
metallic bond (5.3)
metallic solid (5.3)
molar volume (5.1)
molecular solid (5.3)
normal boiling point (5.2)
partial pressure (5.1)
pressure (5.1)
standard temperature and pressure (STP) (5.1)
surface tension (5.2)
surfactant (5.2)
van der Waals forces (5.2)
vapor pressure of a liquid (5.2)
viscosity (5.2)

QUESTIONS AND PROBLEMS

Kinetic Molecular Theory

Foundations

5.15 Compare and contrast the gas, liquid, and solid states with regard to the average distance of particle separation.
5.16 Compare and contrast the gas, liquid, and solid states with regard to the nature of the interactions among the particles.
5.17 Describe the molecular/atomic basis of gas pressure.
5.18 Describe the measurement of gas pressure.

Applications

5.19 Why are gases easily compressible?
5.20 Why are gas densities much lower than those of liquids or solids?
5.21 Why do gases expand to fill any available volume?
5.22 Why do gases with lower molar masses diffuse more rapidly than gases with higher molar masses?
5.23 Do gases exhibit more ideal behavior at low or high pressures? Why?
5.24 Do gases exhibit more ideal behavior at low or high temperatures? Why?
5.25 Use the kinetic molecular theory to explain why dissimilar gases mix more rapidly at high temperatures than at low temperatures.
5.26 Use the kinetic molecular theory to explain why aerosol cans carry instructions warning against heating or disposing of the container in a fire.
5.27 Predict and explain any observed changes taking place when an inflated balloon is cooled (perhaps refrigerated).
5.28 Predict and explain any observed changes taking place when an inflated balloon is heated (perhaps microwaved).

Boyle's Law

Foundations

5.29 State Boyle's law in words.
5.30 State Boyle's law in equation form.
5.31 The pressure on a fixed mass of a gas is tripled at constant temperature. Will the volume increase, decrease, or remain the same?
5.32 By what factor will the volume of the gas in Question 5.31 change?

Applications

A sample of helium gas was placed in a cylinder and the volume of the gas was measured as the pressure was slowly increased. The results of this experiment are shown graphically.

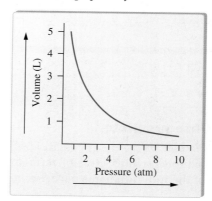

Boyle's Law

Questions 5.33–5.36 are based on this experiment.

5.33 At what pressure does the gas occupy a volume of 5 L?
5.34 What is the volume of the gas at a pressure of 5 atm?
5.35 Calculate the Boyle's law constant at a volume of 2 L.
5.36 Calculate the Boyle's law constant at a pressure of 2 atm.
5.37 Calculate the pressure, in atmospheres, required to compress a sample of helium gas from 20.9 L (at 1.00 atm) to 4.00 L.
5.38 A balloon filled with helium gas at 1.00 atm occupies 15.6 L. What volume would the balloon occupy in the upper atmosphere, at a pressure of 0.150 atm?

Charles's Law

Foundations

5.39 State Charles's law in words.
5.40 State Charles's law in equation form.
5.41 Explain why the Kelvin scale is used for gas law calculations.
5.42 The temperature on a summer day may be 90°F. Convert this value to Kelvins.

Applications

5.43 The temperature of a gas is raised from 25°C to 50°C. Will the volume double if mass and pressure do not change? Why or why not?

5.44 Verify your answer to Question 5.43 by calculating the temperature needed to double the volume of the gas.

5.45 Determine the change in volume that takes place when a 2.00-L sample of $N_2(g)$ is heated from 250°C to 500°C.

5.46 Determine the change in volume that takes place when a 2.00-L sample of $N_2(g)$ is heated from 250 K to 500 K.

5.47 A balloon containing a sample of helium gas is warmed in an oven. If the balloon measures 1.25 L at room temperature (20°C), what is its volume at 80°C?

5.48 The balloon described in Problem 5.47 was then placed in a refrigerator at 39°F. Calculate its new volume.

Combined Gas Law

Foundations

5.49 Will the volume of gas increase, decrease, or remain the same if the temperature is increased and the pressure is decreased? Explain.

5.50 Will the volume of gas increase, decrease, or remain the same if the temperature is decreased and the pressure is increased? Explain.

Applications

Use the combined gas law,

$$\frac{P_i V_i}{T_i} = \frac{P_f V_f}{T_f}$$

to answer Questions 5.51 and 5.52.

5.51 Solve the combined gas law expression for the final volume.

5.52 Solve the combined gas law expression for the final temperature.

5.53 If 2.25 L of a gas at 16°C and 1.00 atm is compressed at a pressure of 125 atm at 20°C, calculate the new volume of the gas.

5.54 A balloon filled with helium gas occupies 2.50 L at 25°C and 1.00 atm. When released, it rises to an altitude where the temperature is 20°C and the pressure is only 0.800 atm. Calculate the new volume of the balloon.

Avogadro's Law

Foundations

5.55 State Avogadro's law in words.

5.56 State Avogadro's law in equation form.

Applications

5.57 If 5.00 g helium gas is added to a 1.00 L balloon containing 1.00 g of helium gas, what is the new volume of the balloon? Assume no change in temperature or pressure.

5.58 How many grams of helium must be added to a balloon containing 8.00 g helium gas to double its volume? Assume no change in temperature or pressure.

Molar Volume and the Ideal Gas Law

Foundations

5.59 Will 1.00 mol of a gas always occupy 22.4 L?

5.60 H_2O and CH_4 are gases at 150°C. Which exhibits more ideal behavior? Why?

5.61 What are the units and numerical value of standard temperature?

5.62 What are the units and numerical value of standard pressure?

Applications

5.63 A sample of nitrogen gas, stored in a 4.0-L container at 32°C, exerts a pressure of 5.0 atm. Calculate the number of moles of nitrogen gas in the container.

5.64 Seven moles of carbon monoxide are stored in a 30.0-L container at 65°C. What is the pressure of the carbon monoxide in the container?

5.65 Calculate the volume of 44.0 g of carbon monoxide at STP.

5.66 Calculate the volume of 44.0 g of carbon dioxide at STP.

5.67 Calculate the number of moles of a gas that is present in a 7.55-L container at 45°C, if the gas exerts a pressure of 725 mm Hg.

5.68 Calculate the pressure exerted by 1.00 mol of gas, contained in a 7.55-L cylinder at 45°C.

5.69 A sample of argon (Ar) gas occupies 65.0 mL at 22°C and 750 torr. What is the volume of this Ar gas sample at STP?

5.70 A sample of O_2 gas occupies 257 mL at 20°C and 1.20 atm. What is the volume of this O_2 gas sample at STP?

5.71 Calculate the molar volume of Ar gas at STP.

5.72 Calculate the molar volume of O_2 gas at STP.

5.73 Calculate the volume of 4.00 mol Ar gas at 8.25 torr and 27°C.

5.74 Calculate the volume of 6.00 mol O_2 gas at 30 cm Hg and 72°F.

5.75 What is the temperature (°C) of 1.75 g of O_2 gas occupying 2.00 L at 1.00 atm?

5.76 How many grams of O_2 gas occupy 10.0 L at STP?

Dalton's Law

Foundations

5.77 State Dalton's law in words.

5.78 State Dalton's law in equation form.

Applications

5.79 A gas mixture has three components: N_2, F_2, and He. Their partial pressures are 0.40 atm, 0.16 atm, and 0.18 atm, respectively. What is the pressure of the gas mixture?

5.80 A gas mixture has a total pressure of 0.56 atm and consists of He and Ne. If the partial pressure of the He in the mixture is 0.27 atm, what is the partial pressure of the Ne in the mixture?

The Liquid State

Foundations

5.81 Compare the strength of intermolecular forces in liquids with those in gases.

5.82 Compare the strength of intermolecular forces in liquids with those in solids.

5.83 What is the relationship between the temperature of a liquid and the vapor pressure of that liquid?

5.84 What is the relationship between the strength of the attractive forces in a liquid and its vapor pressure?

5.85 Distinguish between the terms *evaporation* and *condensation*.

5.86 Distinguish between the terms *evaporation* and *boiling*.

5.87 Describe the process occurring at the molecular level that accounts for the property of viscosity.

5.88 Describe the process occurring at the molecular level that accounts for the property of surface tension.

Applications

Questions 5.89–5.92 are based on the following:

methane   chloromethane   methanol

5.89 Which of these molecules exhibit London forces? Why?
5.90 Which of these molecules exhibit dipole-dipole forces? Why?
5.91 Which of these molecules exhibit hydrogen bonding? Why?
5.92 Which of these molecules would you expect to have the highest boiling point? Why?

The Solid State

5.93 Explain why solids are essentially incompressible.
5.94 Distinguish between amorphous and crystalline solids.
5.95 Describe one property that is characteristic of:
    a. ionic solids
    b. covalent solids
5.96 Describe one property that is characteristic of:
    a. molecular solids
    b. metallic solids
5.97 Predict whether beryllium or carbon would be a better conductor of electricity in the solid state. Why?
5.98 Why is diamond used as an industrial cutting tool?
5.99 Mercury and chromium are toxic substances. Which element is more likely to be an air pollutant? Why?
5.100 Why is the melting point of silicon much higher than that of argon, even though argon has a greater molar mass?

CRITICAL THINKING PROBLEMS

1. An elodea plant, commonly found in tropical fish aquaria, was found to produce 5.0×10^{22} molecules of oxygen per hour. What volume of oxygen (STP) would be produced in an eight-hour period?
2. A chemist measures the volume of 1.00 mol of helium gas at STP and obtains a value of 22.4 L. After changing the temperature to 137 K, the experimental value was found to be 11.05 L. Verify the chemist's results using the ideal gas law and explain any apparent discrepancies.
3. A chemist measures the volumes of 1.00 mol of H_2 and 1.00 mol of CO and finds that they differ by 0.10 L. Which gas produced the larger volume? Do the results contradict the ideal gas law? Why or why not?
4. A 100.0-g sample of water was decomposed using an electric current (electrolysis) producing hydrogen gas and oxygen gas. Write the balanced equation for the process and calculate the volume of each gas produced (STP). Explain any relationship you may observe between the volumes obtained and the balanced equation for the process.
5. An autoclave is used to sterilize surgical equipment. It is far more effective than steam produced from boiling water in the open atmosphere because it generates steam at a pressure of 2 atm. Explain why an autoclave is such an efficient sterilization device.

GENERAL CHEMISTRY 6

Solutions

Carbonated beverages are solutions containing dissolved carbon dioxide. As they warm in an open container, carbon dioxide is released and the beverage becomes "flat."

Learning Goals

1. Distinguish among the terms *solution*, *solute*, and *solvent*.
2. Describe various kinds of solutions, and give examples of each.
3. Describe the relationship between solubility and equilibrium.
4. Calculate solution concentration in weight/volume percent, weight/weight percent, parts per thousand, and parts per million.
5. Calculate solution concentration using molarity.
6. Perform dilution calculations.
7. Interconvert molar concentration of ions and milliequivalents/liter.
8. Describe and explain concentration-dependent solution properties.
9. Describe why the chemical and physical properties of water make it a truly unique solvent.
10. Explain the role of electrolytes in blood and their relationship to the process of dialysis.

Outline

Chemistry Connection:
Seeing a Thought

6.1 Properties of Solutions
6.2 Concentration Based on Mass

A Human Perspective:
Scuba Diving: Nitrogen and the Bends

6.3 Concentration of Solutions: Moles and Equivalents

6.4 Concentration-Dependent Solution Properties

A Medical Perspective:
Oral Rehydration Therapy

6.5 Water as a Solvent
6.6 Electrolytes in Body Fluids

A Human Perspective:
An Extraordinary Molecule

A Medical Perspective:
Hemodialysis

177

Chemistry Connection

Seeing a Thought

At one time, not very long ago, mental illness was believed to be caused by some failing of the human spirit. Thoughts are nonmaterial (you can't hold a thought in your hand), and the body is quite material. No clear relationship, other than the fact that thoughts somehow come from the brain, could be shown to link the body and the spirit.

A major revolution in the diagnosis and treatment of mental illness has taken place in the last three decades. Several forms of depression, paranoia, and schizophrenia have been shown to have chemical and genetic bases. Remarkable improvement in behavior often results from altering the chemistry of the brain by using chemical therapy. Similar progress may result from the use of gene therapy (discussed in Chapter 20).

Although a treatment of mental illness, as well as of memory and logic failures, may occasionally arise by chance, a cause-and-effect relationship, based on the use of scientific methodology, certainly increases the chances of developing successful treatment. If we understand the chemical reactions involved in the thought process, we can perhaps learn to "repair" them when, for whatever reason, they go astray.

Recently, scientists at Massachusetts General Hospital in Boston have developed sophisticated versions of magnetic resonance imaging devices (MRI, discussed in Medical Perspective in Chapter 9). MRI is normally used to locate brain tumors and cerebral damage in patients. The new generation of instruments is so sensitive that it is able to detect chemical change in the brain resulting from an external stimulus. A response to a question or the observation of a flash of light produces a measurable signal. This signal is enhanced with the aid of a powerful computer that enables the location of the signal to be determined with pinpoint accuracy. So there is evidence not only for the chemical basis of thought, but for its location in the brain as well.

In this chapter and throughout your study of chemistry, you will be introduced to a wide variety of chemical reactions, some rather ordinary, some quite interesting. All are founded on the same principles that power our thoughts and actions.

Introduction

Many chemical reactions, and virtually all important organic and biochemical reactions, take place as reactants dissolved in solution. For this reason the major emphasis of this chapter will be on aqueous solution reactions.

We will see that the properties of solutions depend not only on the types of substances that make up the solution but also on the amount of each substance that is contained in a certain volume of the solution. The latter is termed the concentration of the solution.

6.1 Properties of Solutions

LEARNING GOAL

A **solution** is a homogeneous (or uniform) mixture of two or more substances. A solution is composed of one or more *solutes*, dissolved in a *solvent*. The **solute** is a compound of a solution that is present in lesser quantity than the solvent. The **solvent** is the solution component present in the largest quantity. For example, when sugar (the solute) is added to water (the solvent), the sugar dissolves in the water to produce a solution. In those instances in which the solvent is water, we refer to the homogeneous mixture as an **aqueous solution**, from the Latin *aqua*, meaning water.

The dissolution of a solid in a liquid is perhaps the most common example of solution formation. However, it is also possible to form solutions in gases and solids as well as in liquids. For example:

- Air is a gaseous mixture, but it is also a solution; oxygen and a number of trace gases are dissolved in the gaseous solvent, nitrogen.
- Alloys, such as brass and silver and the gold used to make jewelry, are also homogeneous mixtures of two or more kinds of metal atoms in the solid state.

6.1 Properties of Solutions

Although solid and gaseous solutions are important in many applications, our emphasis will be on *liquid solutions* because so many important chemical reactions take place in liquid solutions.

General Properties of Liquid Solutions

Liquid solutions are clear and transparent with no visible particles of solute. They may be colored or colorless, depending on the properties of the solute and solvent. Note that the terms *clear* and *colorless* do not mean the same thing; a clear solution has only one state of matter that can be detected; *colorless* simply means the absence of color.

Recall that solutions of **electrolytes** are formed from solutes that are soluble ionic compounds. These compounds dissociate in solution to produce ions that behave as charge carriers. Solutions of electrolytes are good conductors of electricity. For example, sodium chloride dissolving in water:

$$NaCl(s) \xrightarrow{H_2O} Na^+(aq) + Cl^-(aq)$$

Solid sodium chloride → Dissolved sodium chloride

In contrast, solutions of **nonelectrolytes** are formed from nondissociating *molecular* solutes (nonelectrolytes), and these solutions are nonconducting. For example, dissolving sugar in water:

$$C_6H_{12}O_6(s) \xrightarrow{H_2O} C_6H_{12}O_6(aq)$$

Solid glucose → Dissolved glucose

A *true solution* is a homogeneous mixture with uniform properties throughout. In a true solution the solute cannot be isolated from the solution by filtration. The particle size of the solute is about the same as that of the solvent, and solvent and solute pass directly through the filter paper. Furthermore, solute particles will not "settle out" after a time. All of the molecules of solute and solvent are intimately mixed. The continuous particle motion in solution maintains the homogeneous, random distribution of solute and solvent particles.

Volumes of solute and solvent are not additive; 1L of alcohol mixed with 1L of water does not result in exactly 2L of solution. The volume of pure liquid is determined by the way in which the individual molecules "fit together." When two or more kinds of molecules are mixed, the interactions become more complex. Solvent interacts with solvent, solute interacts with solvent, and solute may interact with other solute. This will be important to remember when we solve concentration problems later.

Solutions and Colloids

How can you recognize a solution? A beaker containing a clear liquid may be a pure substance, a true solution, or a colloid. Only chemical analysis, determining the identity of all substances in the liquid, can distinguish between a pure substance and a solution. A pure substance has *one* component, pure water being an example. A true solution will contain more than one substance, with the tiny particles homogeneously intermingled.

A **colloidal suspension** also consists of solute particles distributed throughout a solvent. However, the distribution is not completely homogeneous, owing to the size of the colloidal particles. Particles with diameters of 1×10^{-9} m (1 nm) to 2×10^{-7} m (200 nm) are colloids. Particles smaller than 1 nm are solution particles; those larger than 200 nm are precipitates (solid in contact with solvent).

To the naked eye, a colloidal suspension and a true solution appear identical; neither solute nor colloid can be seen by the naked eye. However, a simple experiment,

 LEARNING GOAL

Section 3.3 discusses properties of compounds.

Particles in electrolyte solutions are ions, making the solution an electrical conductor.

Particles in solution are individual molecules. No ions are formed in the dissolution process.

Recall that matter in solution, as in gases, is in continuous, random motion (Section 5.1).

Section 3.5 relates properties and molecular geometry.

See Section 4.3 for more information on precipitates.

Figure 6.1
The Tyndall effect. The beaker on the left contains a colloidal suspension, which scatters the light. This scattered light is visible as a haze. The beaker on the right contains a true solution; no scattered light is observed.

using only a bright light source, can readily make the distinction based upon differences in their interaction with light. Colloid particles are large enough to scatter light; solute particles are not. When a beam of light passes through a colloidal suspension, the large particles scatter light, and the liquid appears hazy. We see this effect in sunlight passing through fog. Fog is a colloidal suspension of tiny particles of liquid water dispensed throughout a gas, air. The haze is light scattered by droplets of water. You may have noticed that your automobile headlights are not very helpful in foggy weather. Visibility becomes worse rather than better because light scattering increases.

The light-scattering ability of colloidal suspensions is termed the *Tyndall effect*. True solutions, with very tiny particles, do not scatter light—no haze is observed—and true solutions are easily distinguished from colloidal suspensions by observing their light-scattering properties (Figure 6.1).

A **suspension** is a heterogeneous mixture that contains particles much larger than a colloidal suspension; over time, these particles may settle, forming a second phase. A suspension is not a true solution, nor is it a precipitate.

Question 6.1
Describe how you would distinguish experimentally between a pure substance and a true solution.

Question 6.2
Describe how you would distinguish experimentally between a true solution and a colloidal suspension.

Degree of Solubility

Section 3.5 describes solute-solvent interactions in detail.

In our discussion of the relationship of polarity and solubility, the rule *"like dissolves like"* was described as the fundamental condition for solubility. Polar solutes are soluble in polar solvents, and nonpolar solutes are soluble in nonpolar solvents. Thus, knowing a little bit about the structure of the molecule enables us to predict qualitatively the solubility of the compound.

The *degree* of **solubility,** *how much* solute can dissolve in a given volume of solvent, is a quantitative measure of solubility. It is difficult to predict the solubility of each and every compound. However, general solubility trends are based on the following considerations:

The term *qualitative* **implies identity, and the term** *quantitative* **relates to quantity.**

- *The magnitude of difference between polarity of solute and solvent.* The greater the difference, the less soluble is the solute.
- *Temperature.* An increase in temperature usually, but not always, increases solubility. Often, the effect is dramatic. For example, an increase in temperature from 0°C to 100°C increases the water solubility of KCl from 28 g/100 mL to 58 g/100 mL.
- *Pressure.* Pressure has little effect on the solubility of solids and liquids in liquids. However, the solubility of a gas in liquid is directly proportional to the applied pressure. Carbonated beverages, for example, are made by dissolving carbon dioxide in the beverage under high pressure (hence the term *carbonated*).

When a solution contains all the solute that can be dissolved at a particular temperature, it is a **saturated solution**. When solubility values are given—for example, 13.3 g of potassium nitrate in 100 mL of water at 24°C—they refer to the concentration of a saturated solution.

As we have already noted, *increasing* the temperature generally increases the amount of solute a given solution may hold. Conversely, *cooling* a saturated solution often results in a decrease in the amount of solute in solution. The excess solute falls to the bottom of the container as a **precipitate** (a solid in contact with the solution). Occasionally, on cooling, the excess solute may remain in solution for a time. Such a solution is described as a **supersaturated solution**. This type of solution is inherently unstable. With time, excess solute will precipitate, and the solution will revert to a saturated solution, which is stable.

Solubility and Equilibrium

When an excess of solute is added to a solvent, it begins to dissolve and continues until it establishes a *dynamic equilibrium* between dissolved and undissolved solute.

Initially, the rate of dissolution is large. After a time the rate of the reverse process, precipitation, increases. The rates of dissolution and precipitation eventually become equal, and there is no further change in the composition of the solution. There is, however, a continual exchange of solute particles between solid and liquid phases because particles are in constant motion. The solution is saturated. The most precise definition of a saturated solution is a solution that is in equilibrium with undissolved solute.

Solubility of Gases: Henry's Law

When a liquid and a gas are allowed to come to equilibrium, the amount of gas dissolved in the liquid reaches some maximum level. This quantity can be predicted from a very simple relationship. **Henry's law** states that the number of moles of a gas dissolved in a liquid at a given temperature is proportional to the partial pressure of the gas. In other words, the gas solubility is directly proportional to the pressure of that gas in the atmosphere that is in contact with the liquid.

Carbonated beverages are bottled at high pressures of carbon dioxide. When the cap is removed, the fizzing results from the fact that the partial pressure of carbon dioxide in the atmosphere is much less than that used in the bottling process. As a result, the equilibrium quickly shifts to one of lower gas solubility.

Gases are most soluble at low temperatures, and the gas solubility decreases markedly at higher temperatures. This explains many common observations. For example, a chilled container of carbonated beverage that is opened quickly goes flat as it warms to room temperature. As the beverage warms up, the solubility of the carbon dioxide decreases.

Henry's law helps to explain the process of respiration. Respiration depends on a rapid and efficient exchange of oxygen and carbon dioxide between the

 LEARNING GOAL

The concept of equilibrium was introduced in Section 5.2 and will be discussed in detail in Section 7.4.

The concept of partial pressure is a consequence of Dalton's law, discussed in Section 5.1.

The exchange of O_2 and CO_2 in the lungs and other tissues is a complex series of events described in greater detail in Section 18.9.

See A Medical Perspective: Blood Gases and Respiration, Chapter 5.

atmosphere and the blood. This transfer occurs through the lungs. The process, oxygen entering the blood and carbon dioxide released to the atmosphere, is accomplished in air sacs called *alveoli*, which are surrounded by an extensive capillary system. Equilibrium is quickly established between alveolar air and the capillary blood. The temperature of the blood is effectively constant. Therefore the equilibrium concentration of both oxygen and carbon dioxide are determined by the partial pressures of the gases (Henry's law). The oxygen is transported to cells, a variety of reactions takes place, and the waste product of respiration, carbon dioxide, is brought back to the lungs to be expelled into the atmosphere.

Question 6.3

Explain why, over time, a bottle of soft drink goes "flat" after it is opened.

Question 6.4

Would the soft drink in Question 6.3 go "flat" faster if the bottle warmed to room temperature? Why?

6.2 Concentration Based on Mass

LEARNING GOAL 4

Solution **concentration** is defined as the amount of solute dissolved in a given amount of solution. The concentration of a solution has a profound effect on the properties of a solution, both *physical* (melting and boiling points) and *chemical* (solution reactivity). Solution concentration may be expressed in many different units. Here we consider concentration units based on percentage.

Weight/Volume Percent

The concentration of a solution is defined as the amount of solute dissolved in a specified amount of solution,

$$\text{concentration} = \frac{\text{amount of solute}}{\text{amount of solution}}$$

If we define the amount of solute as the *mass* of solute (in grams) and the amount of solution in *volume* units (milliliters), concentration is expressed as the ratio

$$\text{concentration} = \frac{\text{grams of solute}}{\text{milliliters of solution}}$$

This concentration can then be expressed as a percentage by multiplying the ratio by the factor 100%. This results in

$$\% \text{ concentration} = \frac{\text{grams of solute}}{\text{milliliters of solution}} \times 100\%$$

The percent concentration expressed in this way is called **weight/volume percent**, or **% (W/V)**. Thus

$$\% \left(\frac{W}{V}\right) = \frac{\text{grams of solute}}{\text{milliliters of solution}} \times 100\%$$

A Human Perspective

Scuba Diving: Nitrogen and the Bends

A deep-water diver's worst fear is the interruption of the oxygen supply through equipment malfunction, forcing his or her rapid rise to the surface in search of air. If a diver must ascend too rapidly, he or she may suffer a condition known as "the bends."

Key to understanding this problem is recognition of the tremendous increase in pressure that divers withstand as they descend, because of the weight of the water above them. At the surface the pressure is approximately 1 atm. At a depth of 200 feet the pressure is approximately six times as great.

At these pressures the solubility of nitrogen in the blood increases dramatically. Oxygen solubility increases as well, although its effect is less serious (O_2 is 20% of air, N_2 is 80%). As the diver quickly rises, the pressure decreases rapidly, and the nitrogen "boils" out of the blood, stopping blood flow and impairing nerve transmission. The joints of the body lock in a bent position, hence the name of the condition: the bends.

To minimize the problem, scuba tanks are often filled with mixtures of helium and oxygen rather than nitrogen and oxygen. Helium has a much lower solubility in blood and, like nitrogen, is inert.

For Further Understanding

Why are divers who slowly rise to the surface less likely to be adversely affected?

What design features would be essential in deep-water manned exploration vessels?

Scuba diving.

Consider the following examples.

EXAMPLE 6.1

Calculating Weight/Volume Percent

Calculate the percent composition, or % (W/V), of 3.00×10^2 mL of solution containing 15.0 g of glucose.

Solution

There are 15.0 g of glucose, the solute, and 3.00×10^2 mL of total solution. Therefore, substituting in our expression for weight/volume percent:

$$\%\left(\frac{W}{V}\right) = \frac{15.0 \text{ g glucose}}{3.00 \times 10^2 \text{ mL solution}} \times 100\%$$

$$= 5.00\% \left(\frac{W}{V}\right) \text{ glucose}$$

EXAMPLE 6.2 Calculating the Weight of Solute from a Weight/Volume Percent

Calculate the number of grams of NaCl in 5.00×10^2 mL of a 10.0% solution.

Solution

Begin by substituting the data from the problem:

$$10.0\% \left(\frac{W}{V}\right) = \frac{X \text{ g NaCl}}{5.00 \times 10^2 \text{ mL solution}} \times 100\%$$

Cross-multiplying to simplify:

$$X \text{ g NaCl} \times 100\% = \left(10.0\% \frac{W}{V}\right)(5.00 \times 10^2 \text{ mL solution})$$

Dividing both sides by 100% to isolate grams NaCl on the left side of the equation:

$$X = 50.0 \text{ g NaCl}$$

Section 1.3 discusses units and unit conversion.

If the units of mass are other than grams, or if the solution volume is in units other than milliliters, the proper conversion factor must be used to arrive at the units used in the equation.

Question 6.5

Calculate the % (W/V) of 0.0600 L of solution containing 10.0 g NaCl.

Question 6.6

Calculate the volume (in milliliters) of a 25.0% (W/V) solution containing 10.0 g NaCl.

Question 6.7

Calculate the % (W/V) of 0.200 L of solution containing 15.0 g KCl.

Question 6.8

Calculate the mass (in grams) of sodium hydroxide required to make 2.00 L of a 1.00% (W/V) solution.

Question 6.9

20.0 g of oxygen gas are diluted with 80.0 g of nitrogen gas in a 78.0-L container at standard temperature and pressure. Calculate the % (W/V) of oxygen gas.

Question 6.10

50.0 g of argon gas are diluted with 80.0 g of helium gas in a 476-L container at standard temperature and pressure. Calculate the % (W/V) of argon gas.

Weight/Weight Percent

The **weight/weight percent**, or **% (W/W)**, is most useful for mixtures of solids, whose weights (masses) are easily obtained. The expression used to calculate weight/weight percentage is analogous in form to % (W/V):

$$\%\left(\frac{W}{W}\right) = \frac{\text{grams solute}}{\text{grams solution}} \times 100\%$$

EXAMPLE 6.3 — Calculating Weight/Weight Percent

Calculate the % (W/W) of platinum in a gold ring that contains 14.00 g gold and 4.500 g platinum.

Solution

Using our definition of weight/weight percent

$$\%\left(\frac{W}{W}\right) = \frac{\text{grams solute}}{\text{grams solution}} \times 100\%$$

Substituting,

$$= \frac{4.500 \text{ g platinum}}{4.500 \text{ g platinum} + 14.00 \text{ g gold}} \times 100\%$$

$$= \frac{4.500 \text{ g}}{18.50 \text{ g}} \times 100\%$$

$$= 24.32\% \text{ platinum}$$

Question 6.11

Calculate the % (W/W) of oxygen gas in Question 6.9.

Question 6.12

Calculate the % (W/W) of argon gas in Question 6.10.

Parts Per Thousand (ppt) and Parts Per Million (ppm)

The calculation of concentration in parts per thousand or parts per million is based on the same logic as weight/weight percent. Percentage is actually the number of parts of solute in 100 parts of solution. For example, a 5.00% (W/W) is made up of 5.00 g solute in 100 g solution.

$$5.00\% \text{ (W/W)} = \frac{5.00 \text{ g solute}}{100 \text{ g solution}} \times 100\%$$

It follows that a 5.00 parts per thousand (ppt) solution is made up of 5.00 g solute in 1000 g solution.

$$5.00 \text{ ppt} = \frac{5.00 \text{ g solute}}{1000 \text{ g solution}} \times 10^3 \text{ ppt}$$

Using similar logic, a 5.00 parts per million solution (ppm) is made up of 5.00 g solute in 1,000,000 g solution.

$$5.00 \text{ ppm} = \frac{5.00 \text{ g solute}}{1{,}000{,}000 \text{ g solution}} \times 10^6 \text{ ppm}$$

The general expressions are:

$$\text{ppt} = \frac{\text{grams solute}}{\text{grams solution}} \times 10^3 \text{ ppt}$$

and

$$\text{ppm} = \frac{\text{grams solute}}{\text{grams solution}} \times 10^6 \text{ ppm}$$

Ppt and ppm are most often used for expressing the concentrations of very dilute solutions.

EXAMPLE 6.4 Calculating ppt and ppm

A 1.00 g sample of stream water was found to contain 1.0×10^{-6} g lead. Calculate the concentration of lead in the stream water in units of % (W/W), ppt, and ppm. Which is the most suitable unit?

Solution

weight percent:
$$\% \text{ (W/W)} = \frac{\text{grams solute}}{\text{grams solution}} \times 100\%$$

$$\% \text{ (W/W)} = \frac{1.0 \times 10^{-6} \text{ g Pb}}{1.0 \text{ g solution}} \times 100\%$$

$$\% \text{ (W/W)} = 1.0 \times 10^{-4} \%$$

parts per thousand:
$$\text{ppt} = \frac{\text{grams solute}}{\text{grams solution}} \times 10^3 \text{ ppt}$$

$$\text{ppt} = \frac{1.0 \times 10^{-6} \text{ g Pb}}{1.0 \text{ g solution}} \times 10^3 \text{ ppt}$$

$$\text{ppt} = 1.0 \times 10^{-3} \text{ ppt}$$

parts per million:
$$\text{ppm} = \frac{\text{grams solute}}{\text{grams solution}} \times 10^6 \text{ ppm}$$

$$\text{ppm} = \frac{1.0 \times 10^{-6} \text{ g Pb}}{1.0 \text{ g solution}} \times 10^6 \text{ ppm}$$

$$\text{ppm} = 1.0 \text{ ppm}$$

Parts per million is the most reasonable unit.

> **Question 6.13**
>
> Calculate the ppt and ppm of oxygen gas in Question 6.9.

> **Question 6.14**
>
> Calculate the ppt and ppm of argon gas in Question 6.10.

6.3 Concentration of Solutions: Moles and Equivalents

In our discussion of the chemical arithmetic of reactions in Chapter 4, we saw that the chemical equation represents the relative number of *moles* of reactants producing products. When chemical reactions occur in solution, it is most useful to represent their concentrations on a *molar* basis.

Molarity

The most common mole-based concentration unit is molarity. **Molarity,** symbolized M, is defined as the number of moles of solute per liter of solution, or

  LEARNING GOAL

$$M = \frac{\text{moles solute}}{\text{L solution}}$$

> **EXAMPLE 6.5 Calculating Molarity from Moles**
>
> Calculate the molarity of 2.0 L of solution containing 5.0 mol NaOH.
>
> **Solution**
>
> Using our expression for molarity
>
> $$M = \frac{\text{moles solute}}{\text{L solution}}$$
>
> Substituting,
>
> $$M_{\text{NaOH}} = \frac{5.0 \text{ mol solute}}{2.0 \text{ L solution}}$$
>
> $$= 2.5 \, M$$

Remember the need for conversion factors to convert from mass to number of moles. Consider the following example.

Section 1.3 discussed units and unit conversion.

> **EXAMPLE 6.6 Calculating Molarity from Mass**
>
> If 5.00 g glucose are dissolved in 1.00×10^2 mL of solution, calculate the molarity, M, of the glucose solution.
>
> *Continued—*

EXAMPLE 6.6 —Continued

Solution

To use our expression for molarity it is necessary to convert from units of grams of glucose to moles of glucose. The molar mass of glucose is 1.80×10^2 g/mol. Therefore

$$5.00 \text{ g} \times \frac{1 \text{ mol}}{1.80 \times 10^2 \text{ g}} = 2.78 \times 10^{-2} \text{ mol glucose}$$

and we must convert mL to L:

$$1.00 \times 10^2 \text{ mL} \times \frac{1 \text{ L}}{10^3 \text{ mL}} = 1.00 \times 10^{-1} \text{ L}$$

Substituting these quantities:

$$M_{\text{glucose}} = \frac{2.78 \times 10^{-2} \text{ mol}}{1.00 \times 10^{-1} \text{ L}}$$

$$= 2.78 \times 10^{-1} \text{ M}$$

EXAMPLE 6.7 Calculating Volume from Molarity

Calculate the volume of a 0.750 M sulfuric acid (H_2SO_4) solution containing 0.120 mol of solute.

Solution

Substituting in our basic expression for molarity, we obtain

$$0.750 \text{ } M \text{ } H_2SO_4 = \frac{0.120 \text{ mol } H_2SO_4}{X \text{ L}}$$

$$X \text{ L} = 0.160 \text{ L}$$

Question 6.15

Calculate the number of moles of solute in 5.00×10^2 mL of 0.250 M HCl.

Question 6.16

Calculate the number of grams of silver nitrate required to prepare 2.00 L of 0.500 M AgNO$_3$.

Dilution

LEARNING GOAL

Laboratory reagents are often purchased as concentrated solutions (for example, 12 M HCl or 6 M NaOH) for reasons of safety, economy, and space limitations. We must often *dilute* such a solution to a larger volume to prepare a less concentrated solution for the experiment at hand. The approach to such a calculation is as follows.

6.3 Concentration of Solutions: Moles and Equivalents

We define

M_1 = molarity of solution *before* dilution
M_2 = molarity of solution *after* dilution
V_1 = volume of solution *before* dilution
V_2 = volume of solution *after* dilution

and

$$M = \frac{\text{moles solute}}{\text{L solution}}$$

This equation can be rearranged:

$$\text{moles solute} = (M)(\text{L solution})$$

The number of moles of solute *before* and *after* dilution is unchanged, because dilution involves only addition of extra solvent:

$$\underset{\text{Initial condition}}{\text{moles}_1 \text{ solute}} = \underset{\text{Final condition}}{\text{moles}_2 \text{ solute}}$$

or

$$(M_1)(L_1 \text{ solution}) = (M_2)(L_2 \text{ solution})$$
$$(M_1)(V_1) = (M_2)(V_2)$$

Knowing any three of these terms enables us to calculate the fourth.

EXAMPLE 6.8

Calculating Molarity After Dilution

Calculate the molarity of a solution made by diluting 0.050 L of 0.10 M HCl solution to a volume of 1.0 L.

Solution

Summarize the information provided in the problem:

$$M_1 = 0.10\ M$$
$$M_2 = X\ M$$
$$V_1 = 0.050\ L$$
$$V_2 = 1.0\ L$$

Then, using the dilution expression:

$$(M_1)(V_1) = (M_2)(V_2)$$

Solve for M_2, the final solution concentration:

$$M_2 = \frac{(M_1)(V_1)}{V_2}$$

Substituting,

$$X\ M = \frac{(0.10\ M)(0.050\ L)}{(1.0\ L)}$$

$$= 0.0050\ M \text{ or } 5.0 \times 10^{-3}\ M\ \text{HCl}$$

EXAMPLE 6.9 Calculating a Dilution Volume

Calculate the volume, in liters, of water that must be added to dilute 20.0 mL of 12.0 M HCl to 0.100 M HCl.

Solution

Summarize the information provided in the problem:

$$M_1 = 12.0\ M$$
$$M_2 = 0.100\ M$$
$$V_1 = 20.0\ \text{mL}\ (0.0200\ \text{L})$$
$$V_2 = V_{final}$$

Then, using the dilution expression:

$$(M_1)(V_1) = (M_2)(V_2)$$

Solve for V_2, the final volume:

$$V_2 = \frac{(M_1)(V_1)}{(M_2)}$$

Substituting,

$$V_{final} = \frac{(12.0\ M)(0.0200\ L)}{0.100\ M}$$

$$= 2.40\ L\ \text{solution}$$

Note that this is the *total final volume*. The amount of water added equals this volume *minus* the original solution volume, or

$$2.40\ L - 0.0200\ L = 2.38\ L\ \text{water}$$

Question 6.17

How would you prepare 1.0×10^2 mL of 2.0 M HCl, starting with concentrated (12.0 M) HCl?

Question 6.18

What volume of 0.200 M sugar solution can be prepared from 50.0 mL of 0.400 M solution?

The dilution equation is valid with any concentration units, such as % (W/V) as well as molarity, which was used in Examples 6.8 and 6.9. However, you must use the same units for both initial *and* final concentration values. Only in this way can you cancel units properly.

Representation of Concentration of Ions in Solution

LEARNING GOAL

The concentration of ions in solution may be represented in a variety of ways. The most common include moles per liter (molarity) and equivalents per liter.

When discussing solutions of ionic compounds, molarity emphasizes the number of individual ions. A one molar solution of Na^+ contains Avogadro's number, 6.022×10^{23}, of Na^+ per liter. In contrast, equivalents per liter emphasize charge; one equivalent of Na^+ contains Avogadro's number of positive charge.

We defined 1 mol as the number of grams of an atom, molecule, or ion corresponding to Avogadro's number of particles. One **equivalent** of an ion is the number of grams of the ion corresponding to Avogadro's number of electrical charges. Some examples follow:

1 mol Na^+ = 1 equivalent Na^+	(one Na^+ = 1 unit of charge/ion)
1 mol Cl^- = 1 equivalent Cl^-	(one Cl^- = 1 unit of charge/ion)
1 mol Ca^{2+} = 2 equivalents Ca^{2+}	(one Ca^{2+} = 2 units of charge/ion)
1 mol CO_3^{2-} = 2 equivalents CO_3^{2-}	(one CO_3^{2-} = 2 units of charge/ion)
1 mol PO_4^{3-} = 3 equivalents PO_4^{3-}	(one PO_4^{3-} = 3 units of charge/ion)

Changing from moles per liter to equivalents per liter (or the reverse) can be accomplished by using conversion factors.

Milliequivalents (meq) or milliequivalents/liter (meq/L) are often used when describing small amounts or low concentration of ions. These units are routinely used when describing ions in blood, urine, and blood plasma.

EXAMPLE 6.10

Calculating Ion Concentration

Calculate the number of equivalents per liter (eq/L) of phosphate ion, PO_4^{3-}, in a solution that is 5.0×10^{-3} M phosphate.

Solution

It is necessary to use two conversion factors:

$$mol\ PO_4^{3-} \longrightarrow mol\ charge$$

and

$$mol\ charge \longrightarrow eq\ PO_4^{3-}$$

Arranging these factors in sequence yields:

$$\frac{5.0 \times 10^{-3}\ mol\ PO_4^{3-}}{1\ L} \times \frac{3\ mol\ charge}{1\ mol\ PO_4^{3-}} \times \frac{1\ eq\ PO_4^{3-}}{1\ mol\ charge} = \frac{1.5 \times 10^{-2}\ eq\ PO_4^{3-}}{L}$$

6.4 Concentration-Dependent Solution Properties

 **LEARNING GOAL**

Colligative properties are solution properties that depend on the *concentration of the solute particles*, rather than the *identity of the solute*.

There are four colligative properties of solutions:

1. vapor pressure lowering
2. freezing point depression
3. boiling point elevation
4. osmotic pressure

Each of these properties has widespread practical application. We look at each in some detail in the following sections.

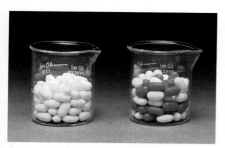

Figure 6.2
An illustration of Raoult's law: lowering of vapor pressure by addition of solute molecules. White units represent solvent molecules, and red units are solute molecules. Solute molecules present a barrier to escape of solvent molecules, thus decreasing the vapor pressure.

Recall that the concept of liquid vapor pressure was discussed in Section 5.2.

Section 7.4 discusses equilibrium.

Vapor Pressure Lowering

Raoult's law states that, when a nonvolatile solute is added to a solvent, the vapor pressure of the solvent decreases in proportion to the concentration of the solute.

Perhaps the most important consequence of Raoult's law is the effect of the solute on the freezing and boiling points of a solution.

When a nonvolatile solute is added to a solvent, the freezing point of the resulting solution decreases (a lower temperature is required to convert the liquid to a solid). The boiling point of the solution is found to increase (it requires a higher temperature to form the gaseous state).

Raoult's law may be explained in molecular terms by using the following logic: Vapor pressure of a solution results from the escape of solvent molecules from the liquid to the gas phase, thus increasing the partial pressure of the gas phase solvent molecules until the equilibrium vapor pressure is reached. Presence of solute molecules hinders the escape of solvent molecules, thus lowering the equilibrium vapor pressure (Figure 6.2).

Freezing Point Depression and Boiling Point Elevation

The freezing point depression may be explained by examining the equilibrium between solid and liquid states. At the freezing point, ice is in equilibrium with liquid water:

$$H_2O\ (l) \underset{(r)}{\overset{(f)}{\rightleftharpoons}} H_2O\ (s)$$

The solute molecules interfere with the rate at which liquid water molecules associate to form the solid state, decreasing the rate of the forward reaction. For a true equilibrium, the rate of the forward (f) and reverse (r) processes must be equal. Lowering the temperature eventually slows the rate of the reverse (r) process sufficiently to match the rate of the forward reaction. At the lower temperature, equilibrium is established, and the solution freezes.

The boiling point elevation can be explained by considering the definition of the boiling point, that is, the temperature at which the vapor pressure of the liquid equals the atmospheric pressure. Raoult's law states that the vapor pressure of a solution is decreased by the presence of a solute. Therefore a higher temperature is necessary to raise the vapor pressure to the atmospheric pressure, hence the boiling point elevation.

The extent of the freezing point depression (ΔT_f) is proportional to the solute concentration over a limited range of concentration:

$$\Delta T_f = k_f \times (\text{solute concentration})$$

The boiling point elevation (ΔT_b) is also proportional to the solute concentration:

$$\Delta T_b = k_b \times (\text{solute concentration})$$

If the value of the proportionality factor (k_f or k_b) is known for the solvent of interest, the magnitude of the freezing point depression or boiling point elevation can be calculated for a solution of known concentration.

Solute concentration must be in *mole*-based units. The number of particles (molecules or ions) is critical here, not the mass of solute. One *heavy* molecule will have exactly the same effect on the freezing or boiling point as one *light* molecule. A mole-based unit, because it is related directly to Avogadro's number, will correctly represent the number of particles in solution.

We have already worked with one mole-based unit, *molarity*, and this concentration unit can be used to calculate either the freezing point depression or the boiling point elevation.

A second mole-based concentration unit, molality, is more commonly used in these types of situations. **Molality** (symbolized m) is defined as the number of moles of solute per kilogram of solvent in a solution:

$$m = \frac{\text{moles solute}}{\text{kg solvent}}$$

6.4 Concentration-Dependent Solution Properties

Molality does not vary with temperature, whereas molarity is temperature dependent. For this reason, molality is the preferred concentration unit for studies such as freezing point depression and boiling point elevation, in which measurement of *change* in temperature is critical.

Practical applications that take advantage of freezing point depression of solutions by solutes include the following:

- Salt is spread on roads to melt ice in winter. The salt lowers the freezing point of the water, so it exists in the liquid phase below its normal freezing point, 0°C or 32°F.
- Solutes such as ethylene glycol, "antifreeze," are added to car radiators in the winter to prevent freezing by lowering the freezing point of the coolant.

> Molarity is temperature dependent simply because it is expressed as mole/volume. Volume is temperature dependent—most liquids expand measurably when heated and contract when cooled. Molality is moles/mass; both moles and mass are temperature independent.

We have referred to the concentration of *particles* in our discussion of colligative properties. Why did we stress this term? The reason is that there is a very important difference between electrolytes and nonelectrolytes. That difference is the way in which they behave when they dissolve. For example, if we dissolve 1 mol of glucose ($C_6H_{12}O_6$) in 1L of water,

$$1\ C_6H_{12}O_6(s) \xrightarrow{H_2O} 1\ C_6H_{12}O_6(aq)$$

1 mol (Avogadro's number, 6.022×10^{23} particles) of glucose is present in solution. *Glucose is a covalently bonded nonelectrolyte.* Dissolving 1 mol of sodium chloride in 1L of water,

$$1\ NaCl(s) \xrightarrow{H_2O} 1\ Na^+(aq) + 1\ Cl^-(aq)$$

produces 2 mol of particles (1 mol of sodium ions and 1 mol of chloride ions). *Sodium chloride is an ionic electrolyte.*

$$1 \text{ mol glucose} \longrightarrow 1 \text{ mol of particles in solution}$$
$$1 \text{ mol sodium chloride} \longrightarrow 2 \text{ mol of particles in solution}$$

It follows that 1 mol of sodium chloride will decrease the vapor pressure, increase the boiling point, or depress the freezing point of 1L of water *twice as much* as 1 mol of glucose in the same quantity of water.

Question 6.19

Comparing pure water and a 10% (W/V) glucose solution, which has the higher freezing point?

Question 6.20

Comparing pure water and a 10% (W/V) glucose solution, which has the higher boiling point?

Osmotic Pressure

Certain types of thin films, or *membranes*, although appearing impervious to matter, actually contain a network of small holes or pores. These pores may be large enough to allow small *solvent* molecules, such as water, to move from one side of the membrane to the other. On the other hand, *solute* molecules cannot cross the membrane because they are too large to pass through the pores. **Semipermeable membranes** are membranes that allow the solvent, but not solute, to diffuse from one side of the membrane to the other. Examples of semipermeable membranes range from synthetics, such as cellophane, to membranes of cells. When the pores are so small that only water molecules can pass through, they are called *osmotic membranes*.

Osmosis is the movement of solvent from a *dilute solution* to a more *concentrated solution* through a *semipermeable membrane*. Pressure must be applied to the

Figure 6.3
(a) Attainment of equilibrium by osmosis. Note that the solutions attain equilibrium when sufficient solvent has passed from the more dilute side (side B) to equalize the concentrations on both sides of the membrane. Side A becomes more dilute, and side B becomes more concentrated. (b) An illustration of osmosis. A semipermeable membrane separates a solution of sugar in water from the pure solvent, water. Over time, water diffuses from side B to side A in an attempt to equalize the concentration in the two compartments. The water level in side A will rise at the expense of side B because the net flow of water is from side B to side A.

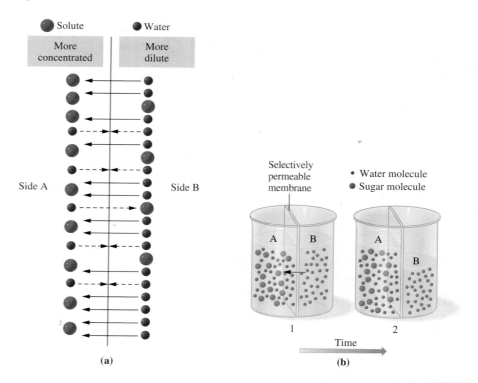

more concentrated solution to stop this flow. **Osmotic pressure is the amount of pressure required to stop the flow.**

The process of osmosis occurring between pure water and glucose (sugar) solution is illustrated in Figure 6.3. Note that the "driving force" for the osmotic process is the need to establish an equilibrium between the solutions on either side of the membrane. Pure solvent enters the more concentrated solution in an effort to dilute it. If this process is successful, and concentrations on both sides of the membrane become equal, the "driving force," or concentration difference, disappears. A dynamic equilibrium is established, and the osmotic pressure difference between the two sides is zero.

The osmotic pressure, like the pressure exerted by a gas, may be treated quantitatively. Osmotic pressure, symbolized by π, follows the same form as the ideal gas equation:

Ideal Gas	Osmotic Pressure
$PV = nRT$	$\pi V = nRT$
or	or
$P = \dfrac{n}{V}RT$	$\pi = \dfrac{n}{V}RT$
and since	and since
$M = \dfrac{n}{V}$	$M = \dfrac{n}{V}$
then	then
$P = MRT$	$\pi = MRT$

The osmotic pressure can be calculated from the solution concentration at any temperature. How do we determine "solution concentration"? Recall that osmosis is a colligative property, dependent on the concentration of solute particles. Again, it becomes necessary to distinguish between solutions of electrolytes and nonelectrolytes. For example, a 1 M glucose solution consists of 1 mol of particles per liter;

The term *selectively permeable* or *differentially permeable* is used to describe biological membranes because they restrict passage of particles based both on size and charge. Even small ions, such as H^+, cannot pass freely across a cell membrane.

glucose is a nonelectrolyte. A solution of 1 M NaCl produces 2 mol of particles per liter (1 mol of Na^+ and 1 mol of Cl^-). A 1 M $CaCl_2$ solution is 3 M in particles (1 mol of Ca^{2+} and 2 mol of Cl^- per liter).

Osmolarity, the molarity of particles in solution, and abbreviated osmol, is used for osmotic pressure calculations.

EXAMPLE 6.11

Calculating Osmolarity

Determine the osmolarity of 5.0×10^{-3} M Na_3PO_4.

Solution

Na_3PO_4 is an ionic compound and produces an electrolytic solution:

$$Na_3PO_4 \xrightarrow{H_2O} 3Na^+ + PO_4^{3-}$$

1 mol of Na_3PO_4 yields four product ions; consequently

$$5.0 \times 10^{-3} \frac{\text{mol } Na_3PO_4}{L} \times \frac{4 \text{ mol particles}}{1 \text{ mol } Na_3PO_4} = 2.0 \times 10^{-2} \frac{\text{mol particles}}{L}$$

and, using our expression for osmolarity,

$$2.0 \times 10^{-2} \frac{\text{mol particles}}{L} = 2.0 \times 10^{-2} \text{ osmol}$$

Question 6.21

Determine the osmolarity of the following solution:

$$5.0 \times 10^{-3} \text{ M } NH_4NO_3 \text{ (electrolyte)}$$

Question 6.22

Determine the osmolarity of the following solution:

$$5.0 \times 10^{-3} \text{ M } C_6H_{12}O_6 \text{ (nonelectrolyte)}$$

EXAMPLE 6.12

Calculating Osmotic Pressure

Calculate the osmotic pressure of a 5.0×10^{-2} M solution of NaCl at 25°C (298 K).

Solution

Using our definition of osmotic pressure, π:

$$\pi = MRT$$

M should be represented as osmolarity as we have shown in Example 6.11

$$M = 5.0 \times 10^{-2} \frac{\text{mol NaCl}}{L} \times \frac{2 \text{ mol particles}}{1 \text{ mol NaCl}} = 1.0 \times 10^{-1} \frac{\text{mol particles}}{L}$$

and substituting in our osmotic pressure expression:

$$\pi = 1.0 \times 10^{-1} \frac{\text{mol particles}}{L} \times 0.0821 \frac{L\text{-atm}}{K\text{-mol}} \times 298 K$$

$$= 2.4 \text{ atm}$$

Figure 6.4
The effect of hypertonic and hypotonic solutions on the cell. (a) Crenation occurs when blood cells are surrounded by a hypertonic solution (water leaving > water entering). (b) Cell rupture occurs when cells are surrounded by a hypotonic solution (water entering > water leaving). (c) Cell size remains unchanged when surrounded by an isotonic solution (water entering = water leaving).

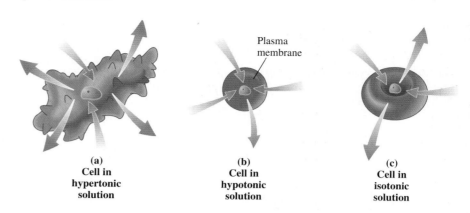

(a) Cell in hypertonic solution

(b) Cell in hypotonic solution

(c) Cell in isotonic solution

Question 6.23
Calculate the osmotic pressure of the solution described in Question 6.21. (Assume a temperature of 25°C.)

Question 6.24
Calculate the osmotic pressure of the solution described in Question 6.22. (Assume a temperature of 25°C.)

Living cells contain aqueous solution (intracellular fluid) and the cells are surrounded by aqueous solution (intercellular fluid). Cell function (and survival!) depend on maintaining approximately the same osmotic pressure inside and outside the cell. If the solute concentration of the fluid surrounding red blood cells is higher than that inside the cell (a **hypertonic solution**), water flows from the cell, causing it to collapse. This process is **crenation.** On the other hand, if the solute concentration of this fluid is too low relative to the solution within the cell (a **hypotonic solution**), water will flow into the cells, causing the cell to rupture, **hemolysis.** To prevent either of these effects from taking place when fluids are administered to a patient intravenously, aqueous fluids [0.9% (W/W) NaCl, also referred to as *physiological saline*, or 5.0% (W/W) glucose] are prepared in such a way as to be *isotonic solutions* with intracellular fluids (Figure 6.4).

Two solutions are **isotonic solutions** if they have identical osmotic pressures. In that way the osmotic pressure differential across the cell membrane is zero, and no cell disruption occurs.

Practical examples of osmosis abound, including the following:

- A sailor, lost at sea in a lifeboat, dies of dehydration while surrounded by water. Seawater, because of its high salt concentration, dehydrates the cells of the body as a result of the large osmotic pressure difference between itself and intracellular fluids.
- A cucumber, soaked in brine, shrivels into a pickle. The water in the cucumber is drawn into the brine (salt) solution because of a difference in osmotic pressure (Figure 6.5).
- Medical Perspective: Oral Rehydration Therapy describes one of the most lethal and pervasive examples of cellular fluid imbalance.

(a)

(b)

Figure 6.5
A cucumber (a) in an acidic salt solution undergoes considerable shrinkage on its way to becoming a pickle (b) because of osmosis.

A Medical Perspective

Oral Rehydration Therapy

Diarrhea kills millions of children before they reach the age of five years. This is particularly true in third world countries where sanitation, water supplies, and medical care are poor. In the case of diarrhea, death results from fluid loss, electrolyte imbalance, and hypovolemic shock (multiple organ failure due to insufficient perfusion). Cholera is one of the best-understood bacterial diarrheas. The organism *Vibrio cholera*, seen in the micrograph below, survives passage through the stomach and reproduces in the intestine, where it produces a toxin called choleragen. The toxin causes the excessive excretion of Na^+, Cl^-, and HCO_3^- from epithelial cells lining the intestine. The increased ion concentration (hypertonic solution) outside the cell results in movement of massive quantities of water into the intestinal lumen. This causes the severe, abundant, clear vomit and diarrhea that can result in the loss of 10–15 L of fluid per day. Over the four- to six-day progress of the disease, a patient may lose from one to two times his or her body mass!

The need for fluid replacement is obvious. Oral rehydration is preferred over intravenous administration of fluids and electrolytes since it is noninvasive. In many third world countries, it is the only therapy available in remote areas. The rehydration formula includes 50–80 g/L rice (or other starch), 3.5 g/L sodium chloride, 2.5 g/L sodium bicarbonate, and 1.5 g/L potassium chloride. Oral rehydration takes advantage of the cotransport of Na^+ and glucose across the cells lining the intestine. Thus, the channel protein brings glucose into the cells, and Na^+ is carried along. Movement of these materials into the cells will help alleviate the osmotic imbalance, reduce the diarrhea, and correct the fluid and electrolyte imbalance.

The disease runs its course in less than a week. In fact, antibiotics are not used to combat cholera. The only effective therapy is oral rehydration, which reduces mortality to less than 1%. A much better option is prevention. In the photo below, a woman is shown filtering water through sari cloth. This simple practice has been shown to reduce the incidence of cholera significantly.

A woman is shown filtering water through sari cloth.

For Further Understanding

Explain dehydration in terms of osmosis.

Explain why even severely dehydrated individuals continue to experience further fluid loss.

Vibrio cholera.

6.5 Water as a Solvent

LEARNING GOAL

See A Human Perspective: An Extraordinary Molecule, in this chapter.

Water is by far the most abundant substance on earth. It is an excellent solvent for most inorganic substances. In fact, it is often referred to as the "universal solvent" and is the principal biological solvent. Approximately 60% of the adult human body is water, and maintenance of this level is essential for survival. These characteristics are a direct consequence of the molecular structure of water.

Refer to Sections 3.4 and 3.5 for a more complete description of the bonding, structure, and polarity of water.

As we saw in Chapter 3, water is a bent molecule with a 104.5° bond angle. This angular structure, resulting from the effect of the two lone pairs of electrons around the oxygen atom, is responsible for the polar nature of water. The polarity, in turn, gives water its unique properties.

Because water molecules are polar, water is an excellent solvent for other polar substances ("like dissolves like"). Because much of the matter on earth is polar, hence at least somewhat water soluble, water has been described as the universal solvent. It is readily accessible and easily purified. It is nontoxic and quite nonreactive. The high boiling point of water, 100°C, compared with molecules of similar size such as N_2 (b.p. = −196°C), is also explained by water's polar character. Strong dipole-dipole interactions between a δ^+ hydrogen of one molecule and δ^- oxygen of a second, referred to as hydrogen bonding, create an interactive molecular network in the liquid phase (see Figure 5.8a). The strength of these interactions requires more energy (higher temperature) to cause water to boil. The higher than expected boiling point enhances water's value as a solvent; often, reactions are carried out at higher temperatures to increase their rate. Other solvents, with lower boiling points, would simply boil away, and the reaction would stop.

Recall the discussion of intermolecular forces in Chapters 3 and 5.

This idea is easily extended to our own chemistry—because 60% of our bodies is water, we should appreciate the polarity of water on a hot day. As a biological solvent in the human body, water is involved in the transport of ions, nutrients, and waste into and out of cells. Water is also the solvent for biochemical reactions in cells and the digestive tract. Water is a reactant or product in some biochemical processes.

EXAMPLE 6.13 Predicting Structure from Observable Properties

Sucrose is a common sugar and we know that it is used as a sweetener when dissolved in many beverages. What does this allow us to predict about the structure of sucrose?

Solution

Sucrose is used as a sweetener in teas, coffee, and a host of soft drinks. The solvent in all of these beverages is water, a polar molecule. The rule "like dissolves like" implies that sucrose must also be a polar molecule. Without even knowing the formula or structure of sucrose, we can infer this important information from a simple experiment—dissolving sugar in our morning cup of coffee.

Question 6.25

Predict whether carbon monoxide or carbon dioxide would be more soluble in water. Explain your answer. (Hint: Refer to Section 5.2, the discussion of interactions in the liquid state.)

Question 6.26

Predict whether ammonia or methane would be more soluble in water. Explain your answer. (Hint: Refer to Section 5.2, the discussion of interactions in the liquid state.)

A Human Perspective

An Extraordinary Molecule

Think for a moment. What is the only common molecule that exists in all three physical states of matter (solid, liquid, and gas) under natural conditions on earth? This molecule is absolutely essential for life; in fact, life probably arose in this substance. It is the most abundant molecule in the cells of living organisms (70–95%) and covers 75% of the earth's surface. Without it, cells quickly die, and without it the earth would not be a fit environment in which to live. By now you have guessed that we are talking about the water molecule. It is so abundant on earth that we take this deceptively simple molecule for granted.

What are some of the properties of water that cause it to be essential to life as we know it? Water has the ability to stabilize temperatures on the earth and in the body. This ability is due in part to the energy changes that occur when water changes physical state; but ultimately, this ability is due to the polar nature of the water molecule.

Life can exist only within a fairly narrow range of temperatures. Above or below that range, the chemical reactions necessary for life, and thus life itself, will cease. Water can moderate temperature fluctuation and maintain the range necessary for life, and one property that allows it to do so is its unusually high specific heat, 1 cal/g °C. This means that water can absorb or lose more heat energy than many other substances without a significant temperature change. This is because in the liquid state, every water molecule is hydrogen bonded to other water molecules. Because a temperature increase is really just a measure of increased (more rapid) molecular movement, we must get the water molecules moving more rapidly, independent of one another, to register a temperature increase. Before we can achieve this independent, increased activity, the hydrogen bonds between molecules must be broken. Much of the heat energy that water absorbs is involved in breaking hydrogen bonds and is *not* used to increase molecular movement. Thus a great deal of heat is needed to raise the temperature of water even a little bit.

Water also has a very high heat of vaporization. It takes 540 calories to change 1 g of liquid water at 100°C to a gas and even more, 603 cal/g, when the water is at 37°C, human body temperature. That is about twice the heat of vaporization of alcohol. As water molecules evaporate, the surface of the liquid cools because only the highest-energy (or "hottest") molecules leave as a gas. Only the "hottest" molecules have enough energy to break the hydrogen bonds that bind them to other water molecules. Indeed, evaporation of water molecules from the surfaces of lakes and oceans helps to maintain stable temperatures in those bodies of water. Similarly, evaporation of perspiration from body surfaces helps to prevent overheating on a hot day or during strenuous exercise.

Even the process of freezing helps stabilize and moderate temperatures. This is especially true in the fall. Water releases heat when hydrogen bonds are formed. This is an example of an exothermic process. Thus, when water freezes, solidifying into ice, additional hydrogen bonds are formed, and heat is released into the environment. As a result, the temperature change between summer and winter is more gradual, allowing organisms to adjust to the change.

Surface of a body of water.

One last feature that we take for granted is the fact that when we put ice in our iced tea on a hot summer day, the ice floats. This means that the solid state of water is actually *less* dense than the liquid state! In fact, it is about 10% less dense, having an open lattice structure with each molecule hydrogen bonded to the maximum of four other water molecules. What would happen if ice did sink? All bodies of water, including the mighty oceans would eventually freeze solid, killing all aquatic and marine plant and animal life. Even in the heat of summer, only a few inches of ice at the surface would thaw. Instead, the ice forms at the surface and provides a layer of insulation that prevents the water below from freezing.

As we continue our study of chemistry, we will refer again and again to this amazing molecule. In other Human Perspective features we will examine other properties of water that make it essential to life.

For Further Understanding

Why is the high heat of vaporization of water important to our bodies?

Why is it cooler at the ocean shore than in the desert during summer?

LEARNING GOAL

6.6 Electrolytes in Body Fluids

The concentrations of cations, anions, and other substances in biological fluids are critical to health. Consequently, the osmolarity of body fluids is carefully regulated by the kidney.

The two most important cations in body fluids are Na^+ and K^+. Sodium ion is the most abundant cation in the blood and intercellular fluids whereas potassium ion is the most abundant intracellular cation. In blood and intercellular fluid, the Na^+ concentration is 135 milliequivalents/L and the K^+ concentration is 3.5–5.0 meq/L. Inside the cell, the situation is reversed. The K^+ concentration is 125 meq/L and the Na^+ concentration is 10 meq/L.

If osmosis and simple diffusion were the only mechanisms for transporting water and ions across cell membranes, these concentration differences would not occur. One positive ion would be just as good as any other. However, the situation is more complex than this. Large protein molecules embedded in cell membranes actively pump sodium ions to the outside of the cell and potassium ions into the cell. This is termed *active transport* because cellular energy must be expended to transport those ions. Proper cell function in the regulation of muscles and the nervous system depends on the sodium ion/potassium ion ratio inside and outside of the cell.

If the Na^+ concentration in the blood becomes too low, urine output decreases, the mouth feels dry, the skin becomes flushed, and a fever may develop. The blood level of Na^+ may be elevated when large amounts of water are lost. Diabetes, certain high-protein diets, and diarrhea may cause elevated blood Na^+ level. In extreme cases, elevated Na^+ levels may cause confusion, stupor, or coma.

Concentrations of K^+ in the blood may rise to dangerously high levels following any injury that causes large numbers of cells to rupture, releasing their intracellular K^+. This may lead to death by heart failure. Similarly, very low levels of K^+ in the blood may also cause death from heart failure. This may occur following prolonged exercise that results in excessive sweating. When this happens, both body fluids and electrolytes must be replaced. Salt tablets containing both NaCl and KCl taken with water and drinks such as Gatorade effectively provide water and electrolytes and prevent serious symptoms.

The cationic charge in blood is neutralized by two major anions, Cl^- and HCO_3^-. The chloride ion plays a role in acid-base balance, maintenance of osmotic pressure within an acceptable range, and oxygen transport by hemoglobin. The bicarbonate anion is the form in which most waste CO_2 is carried in the blood.

A variety of proteins is also found in the blood. Because of their larger size, they exist in colloidal suspension. These proteins include blood clotting factors, immunoglobulins (antibodies) that help us fight infection, and albumins that act as carriers of nonpolar, hydrophobic substances (fatty acids and steroid hormones) that cannot dissolve in water.

Additionally, blood is the medium for exchange of nutrients and waste products. Nutrients, such as the polar sugar glucose, enter the blood from the intestine or the liver. Because glucose molecules are polar, they dissolve in body fluids and are circulated to tissues throughout the body. As noted above, nonpolar nutrients are transported with the help of carrier proteins. Similarly, nitrogen-containing waste products, such as urea, are passed from cells to the blood. They are continuously and efficiently removed from the blood by the kidneys.

In cases of loss of kidney function, mechanical devices—dialysis machines—mimic the action of the kidney. The process of blood dialysis—hemodialysis—is discussed in A Medical Perspective: Hemodialysis on page 201.

A Medical Perspective

Hemodialysis

As we have seen in Section 6.6, blood is the medium for exchange of both nutrients and waste products. The membranes of the kidneys remove waste materials such as urea and uric acid (Chapter 22) and excess salts and large quantities of water. This process of waste removal is termed **dialysis**, a process similar in function to osmosis (Section 6.4). Semipermeable membranes in the kidneys, dialyzing membranes, allow small molecules (principally water and urea) and ions in solution to pass through and ultimately collect in the bladder. From there they can be eliminated from the body.

Unfortunately, a variety of diseases can cause partial or complete kidney failure. Should the kidneys fail to perform their primary function, dialysis of waste products, urea and other waste products rapidly increase in concentration in the blood. This can become a life-threatening situation in a very short time.

The most effective treatment of kidney failure is the use of a machine, an artificial kidney, that mimics the function of the kidney. The artificial kidney removes waste from the blood using the process of hemodialysis (blood dialysis). The blood is pumped through a long semipermeable membrane, the dialysis membrane. The dialysis process is similar to osmosis. However, in addition to water molecules, larger molecules (including the waste products in the blood) and ions can pass across the membrane from the blood into a dialyzing fluid. The dialyzing fluid is isotonic with normal blood; it also is similar in its concentration of all other essential blood components.

The waste materials move across the dialysis membrane (from a higher to a lower concentration, as in diffusion). A successful dialysis procedure selectively removes the waste from the body without upsetting the critical electrolyte balance in the blood.

Hemodialysis, although lifesaving, is not by any means a pleasant experience. The patient's water intake must be severely limited to minimize the number of times each week that treatment must be used. Many dialysis patients require two or three treatments per week and each session may require one-half (or more) day of hospitalization, especially when the patient suffers from complicating conditions such as diabetes.

Improvements in technology, as well as the growth and sophistication of our health care delivery systems over the past several years, have made dialysis treatment much more patient friendly. Dialysis centers, specializing in the treatment of kidney patients, are now found in most major population centers. Smaller, more automated dialysis units are available for home use, under the supervision of a nursing practitioner. With the remarkable progress in kidney transplant success, dialysis is becoming, more and more, a temporary solution, sustaining life until a suitable kidney donor match can be found.

For Further Understanding

In what way is dialysis similar to osmosis?

In what way does dialysis differ from osmosis?

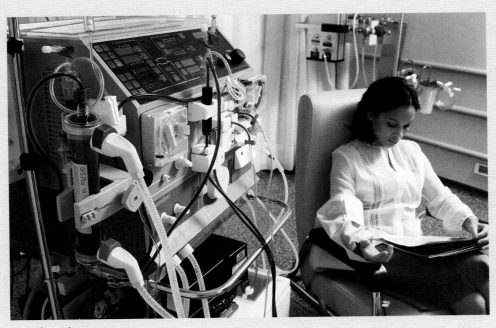

Dialysis patient.

EXAMPLE 6.14 Calculating Electrolyte Concentrations

A typical concentration of calcium ion in blood plasma is 4 meq/L. Represent this concentration in moles/L.

Solution

The calcium ion has a 2+ charge (recall that calcium is in Group IIA of the periodic table; hence, a 2+ charge on the calcium ion).

We will need three conversion factors:

$$\text{meq (milliequivalents)} \longrightarrow \text{eq (equivalents)}$$
$$\text{eq (equivalents)} \longrightarrow \text{moles of charge}$$
$$\text{moles of charge} \longrightarrow \text{moles of calcium ion}$$

Using dimensional analysis as in Example 6.10,

$$\frac{4 \text{ meq Ca}^{2+}}{1 \text{ L}} \times \frac{1 \text{ eq Ca}^{2+}}{10^3 \text{ meq Ca}^{2+}} \times \frac{1 \text{ mol charge}}{1 \text{ eq Ca}^{2+}} \times \frac{1 \text{ mol Ca}^{2+}}{2 \text{ mol charge}} = \frac{2 \times 10^{-3} \text{ mol Ca}^{2+}}{\text{L}}$$

Question 6.27

Sodium chloride [0.9% (W/V)] is a solution administered intravenously to replace fluid loss. It is frequently used to avoid dehydration. The sodium ion concentration is 15.4 meq/L. Calculate the sodium ion concentration in moles/L.

Question 6.28

A potassium chloride solution that also contains 5% (W/V) dextrose is administered intravenously to treat some forms of malnutrition. The potassium ion concentration in this solution is 40 meq/L. Calculate the potassium ion concentration in moles/L.

SUMMARY

6.1 Properties of Solutions

A majority of chemical reactions, and virtually all important organic and biochemical reactions, take place not as a combination of two or more pure substances, but rather as reactants dissolved in solution, *solution reactions*.

A *solution* is a homogeneous (or uniform) mixture of two or more substances. A solution is composed of one or more *solutes*, dissolved in a *solvent*. When the solvent is water, the solution is called an *aqueous solution*.

Liquid solutions are clear and transparent with no visible particles of solute. They may be colored or colorless, depending on the properties of the solute and solvent.

In solutions of *electrolytes* the solutes are ionic compounds that dissociate in solution to produce ions. They are good conductors of electricity. Solutions of *nonelectrolytes* are formed from nondissociating molecular solutes (nonelectrolytes), and their solutions are nonconducting.

The rule "like dissolves like" is the fundamental condition for solubility. Polar solutes are soluble in polar solvents, and nonpolar solutes are soluble in nonpolar solvents.

The degree of solubility depends on the difference between the polarity of solute and solvent, the temperature, and the pressure. Pressure considerations are significant only for solutions of gases.

When a solution contains all the solute that can be dissolved at a particular temperature, it is *saturated*. Excess solute falls to the bottom of the container as a *precipitate*. Occasionally, on cooling, the excess solute may remain in solution for a time before precipitation. Such a solution is a *supersaturated solution*. When excess solute, the precipitate, contacts solvent, the dissolution process reaches a state of dynamic equilibrium. *Colloidal suspensions* have particle sizes between those of true solutions and precipitates. A *suspension* is a heterogeneous mixture that contains particles much larger than a colloidal suspension. Over time, these particles may settle, forming a second phase.

Henry's law describes the solubility of gases in liquids. At a given temperature the solubility of a gas is proportional to the partial pressure of the gas.

6.2 Concentration Based on Mass

The amount of solute dissolved in a given amount of solution is the solution *concentration*. The more widely used percentage-based concentration units are *weight/volume percent* and *weight/weight percent*. Parts per thousand (ppt) and parts per million (ppm) are used with very dilute solutions.

6.3 Concentration of Solutions: Moles and Equivalents

Molarity, symbolized M, is defined as the number of moles of solute per liter of solution.

Dilution is often used to prepare less concentrated solutions. The expression for this calculation is $(M_1)(V_1) = (M_2)(V_2)$. Knowing any three of these terms enables one to calculate the fourth. The concentration of solute may be represented as moles per liter (molarity) or any other suitable concentration units. However, both concentrations must be in the same units when using the dilution equation.

When discussing solutions of ionic compounds, molarity emphasizes the number of individual ions. A 1 M solution of Na^+ contains Avogadro's number of sodium ions. In contrast, equivalents per liter emphasizes charge; a solution containing one equivalent of Na^+ per liter contains Avogadro's number of positive charge.

One *equivalent* of an ion is the number of grams of the ion corresponding to Avogadro's number of electrical charges. Changing from moles per liter to equivalents per liter (or the reverse) is done using conversion factors.

6.4 Concentration-Dependent Solution Properties

Solution properties that depend on the concentration of solute particles, rather than the identity of the solute, are *colligative properties*.

There are four colligative properties of solutions, all of which depend on the concentration of *particles* in solution.

1. *Vapor pressure lowering.* Raoult's law states that when a solute is added to a solvent, the vapor pressure of the solvent decreases in proportion to the concentration of the solute.
2. and 3. *Freezing point depression and boiling point elevation.* When a nonvolatile solid is added to a solvent, the freezing point of the resulting solution decreases, and the boiling point increases. The magnitudes of both the freezing point depression (ΔT_f) and the boiling point elevation (ΔT_b) are proportional to the solute concentration over a limited range of concentrations. The mole-based concentration unit, molality, is more commonly used in calculations involving colligative properties. This is due to the fact that molality is temperature independent. *Molality* (symbolized m) is defined as the number of moles of solute per kilogram of solvent in a solution.
4. *Osmosis and osmotic pressure.* Osmosis is the movement of solvent from a dilute solution to a more concentrated solution through a *semipermeable membrane*. The pressure that must be applied to the more concentrated solution to stop this flow is the *osmotic pressure*. The osmotic pressure, like the pressure exerted by a gas, may be treated quantitatively by using an equation similar in form to the ideal gas equation: $\pi = MRT$. By convention the molarity of particles that is used for osmotic pressure calculations is termed *osmolarity (osmol)*.

In biological systems, if the concentration of the fluid surrounding red blood cells is higher than that inside the cell (a *hypertonic* solution), water flows from the cell, causing it to collapse (*crenation*). Too low a concentration of this fluid relative to the solution within the cell (a *hypotonic* solution) will cause cell rupture (*hemolysis*).

Two solutions are *isotonic* if they have identical osmotic pressures. In that way the osmotic pressure differential across the cell is zero, and no cell disruption occurs.

6.5 Water as a Solvent

The role of water in the solution process deserves special attention. It is often referred to as the "universal solvent" because of the large number of ionic and polar covalent compounds that are at least partially soluble in water. It is the principal biological solvent. These characteristics are a direct consequence of the molecular geometry and structure of water and its ability to undergo hydrogen bonding.

6.6 Electrolytes in Body Fluids

The concentrations of cations, anions, and other substances in biological fluids are critical to health. As a result, the osmolarity of body fluids is carefully regulated by the kidney using the process of *dialysis*.

KEY TERMS

aqueous solution (6.1)
colligative property (6.4)
colloidal suspension (6.1)
concentration (6.2)
crenation (6.4)
dialysis (6.6)
electrolyte (6.1)
equivalent (6.3)
hemolysis (6.4)
Henry's law (6.1)

hypertonic solution (6.4)
hypotonic solution (6.4)
isotonic solution (6.4)
molality (6.4)
molarity (6.3)
nonelectrolyte (6.1)
osmolarity (6.4)
osmosis (6.4)
osmotic pressure (6.4)
precipitate (6.1)

Raoult's law (6.4)
saturated solution (6.1)
semipermeable membrane (6.4)
solubility (6.1)
solute (6.1)
solution (6.1)
solvent (6.1)
supersaturated solution (6.1)
suspension (6.1)
weight/volume percent (% [W/V]) (6.2)
weight/weight percent (% [W/W]) (6.2)

QUESTIONS AND PROBLEMS

Concentration Based on Mass

Fundamentals

6.29 Calculate the composition of each of the following solutions in weight/volume %:
  a. 20.0 g NaCl in 1.00 L solution
  b. 33.0 g sugar, $C_6H_{12}O_6$, in 5.00×10^2 mL solution

6.30 Calculate the composition of each of the following solutions in weight/volume %:
  a. 0.700 g KCl per 1.00 mL
  b. 1.00 mol $MgCl_2$ in 2.50×10^2 mL solution

6.31 Calculate the composition of each of the following solutions in weight/volume %:
  a. 50.0 g ethanol dissolved in 1.00 L solution
  b. 50.0 g ethanol dissolved in 5.00×10^2 mL solution

6.32 Calculate the composition of each of the following solutions in weight/volume %:
  a. 20.0 g acetic acid dissolved in 2.50 L solution
  b. 20.0 g benzene dissolved in 1.00×10^2 mL solution

6.33 Calculate the composition of each of the following solutions in weight/weight %:
  a. 21.0 g NaCl in 1.00×10^2 g solution
  b. 21.0 g NaCl in 5.00×10^2 mL solution (d = 1.12 g/mL)

6.34 Calculate the composition of each of the following solutions in weight/weight %:
  a. 1.00 g KCl in 1.00×10^2 g solution
  b. 50.0 g KCl in 5.00×10^2 mL solution (d = 1.14 g/mL)

Applications

6.35 How many grams of solute are needed to prepare each of the following solutions?
  a. 2.50×10^2 g of 0.900% (W/W) NaCl
  b. 2.50×10^2 g of 1.25% (W/W) $NaC_2H_3O_2$ (sodium acetate)

6.36 How many grams of solute are needed to prepare each of the following solutions?
  a. 2.50×10^2 g of 5.00% (W/W) NH_4Cl (ammonium chloride)
  b. 2.50×10^2 g of 3.50% (W/W) Na_2CO_3

6.37 A solution was prepared by dissolving 14.6 g of KNO_3 in sufficient water to produce 75.0 mL of solution. What is the weight/volume % of this solution?

6.38 A solution was prepared by dissolving 12.4 g of $NaNO_3$ in sufficient water to produce 95.0 mL of solution. What is the weight/volume % of this solution?

6.39 How many grams of sugar would you use to prepare 100 mL of a 1.00 weight/volume % solution?

6.40 How many mL of 4.0 weight/volume % $Mg(NO_3)_2$ solution would contain 1.2 g of magnesium nitrate?

6.41 Which solution is more concentrated: a 0.04% (W/W) solution or a 50 ppm solution?

6.42 Which solution is more concentrated: a 20 ppt solution or a 200 ppm solution?

6.43 A solution contains 1.0 mg of Cu^{2+} per 0.50 kg solution. Calculate the concentration in ppt.

6.44 A solution contains 1.0 mg of Cu^{2+} per 0.50 kg solution. Calculate the concentration in ppm.

Concentration of Solutions: Moles and Equivalents

6.45 Calculate the molarity of each solution in Problem 6.29.

6.46 Calculate the molarity of each solution in Problem 6.30.

6.47 Calculate the number of grams of solute that would be needed to make each of the following solutions:
  a. 2.50×10^2 mL of 0.100 M NaCl
  b. 2.50×10^2 mL of 0.200 M $C_6H_{12}O_6$ (glucose)

6.48 Calculate the number of grams of solute that would be needed to make each of the following solutions:
  a. 2.50×10^2 mL of 0.100 M NaBr
  b. 2.50×10^2 mL of 0.200 M KOH

6.49 Calculate the molarity of a sucrose (table sugar, $C_{12}H_{22}O_{11}$) solution that contains 50.0 g of sucrose per liter.

6.50 A saturated silver chloride solution is 1.58×10^{-4} g of silver chloride per 1.00×10^2 mL of solution. What is the molarity of this solution?

6.51 It is desired to prepare 0.500 L of a 0.100 M solution of NaCl from a 1.00 M stock solution. How many milliliters of the stock solution must be taken for the dilution?

6.52 50.0 mL of a 0.250 M sucrose solution was diluted to 5.00×10^2 mL. What is the molar concentration of the resulting solution?

6.53 A 50.0-mL portion of a stock solution was diluted to 500.0 mL. If the resulting solution was 2.00 M, what was the molarity of the original stock solution?

6.54 A 6.00-mL portion of an 8.00 M stock solution is to be diluted to 0.400 M. What will be the final volume after dilution?

6.55 Calculate the molarity of a solution that contains 2.25 mol of $NaNO_3$ dissolved in 2.50 L.

6.56 Calculate the molarity of a solution that contains 1.75 mol of KNO_3 dissolved in 3.00 L.

6.57 How many grams of glucose ($C_6H_{12}O_6$) are present in 1.75 L of a 0.500 M solution?

6.58 How many grams of sodium hydroxide are present in 675 mL of a 0.500 M solution?

6.59 50.0 mL of 0.500 M NaOH were diluted to 500.0 mL. What is the new molarity?

6.60 300.0 mL of H_2O are added to 300.0 mL of 0.250 M H_2SO_4. What is the new molarity?

Concentration-Dependent Solution Properties

Foundations

6.61 What is meant by the term *colligative property*?

6.62 Name and describe four colligative solution properties.

6.63 Explain, in terms of solution properties, why salt is used to melt ice in the winter.

6.64 Explain, in terms of solution properties, why a wilted plant regains its "health" when watered.

Applications

6.65 In what way do colligative properties and chemical properties differ?

6.66 Look up the meaning of the term "colligative." Why is it an appropriate title for these properties?

6.67 State Raoult's law.

6.68 What is the major importance of Raoult's law?

6.69 Why does one mole of $CaCl_2$ lower the freezing point of water more than one mole of NaCl?

6.70 Using salt to try to melt ice on a day when the temperature is −20°C will be unsuccessful. Why?

Answer questions 6.71–6.76 by comparing two solutions: 0.50 M sodium chloride (an ionic compound) and 0.50 M sucrose (a covalent compound).

6.71 Which solution has the higher melting point?
6.72 Which solution has the higher boiling point?
6.73 Which solution has the higher vapor pressure?
6.74 Each solution is separated from water by a semipermeable membrane. Which solution has the higher osmotic pressure?
6.75 Calculate the osmotic pressure of $0.50\ M$ sodium chloride.
6.76 Calculate the osmotic pressure of $0.50\ M$ sucrose.

Answer questions 6.77–6.80 based on the following scenario: Two solutions, A and B, are separated by a semipermeable membrane. For each case, predict whether there will be a net flow of water in one direction and, if so, which direction.

6.77 A is pure water and B is 5% glucose.
6.78 A is $0.10\ M$ glucose and B is $0.10\ M$ KCl.
6.79 A is $0.10\ M$ NaCl and B is $0.10\ M$ KCl.
6.80 A is $0.10\ M$ NaCl and B is $0.20\ M$ glucose.

In questions 6.81–6.84, label each solution as isotonic, hypotonic, or hypertonic in comparison to 0.9% NaCl ($0.15\ M$ NaCl).

6.81 $0.15\ M\ CaCl_2$
6.82 $0.35\ M$ glucose
6.83 $0.15\ M$ glucose
6.84 3% NaCl

Water as a Solvent

6.85 What properties make water such a useful solvent?
6.86 Sketch the "interactive network" of water molecules in the liquid state.
6.87 Solutions of ammonia in water are sold as window cleaner. Why do these solutions have a long "shelf life"?
6.88 Why does water's abnormally high boiling point help to make it a desirable solvent?
6.89 Sketch the interaction of a water molecule with a sodium ion.
6.90 Sketch the interaction of a water molecule with a chloride ion.
6.91 What type of solute dissolves readily in water?
6.92 What type of solute dissolves readily in gasoline?
6.93 Sketch the interaction of water with an ammonia molecule.
6.94 Sketch the interaction of water with an ethanol molecule.

Electrolytes in Body Fluids

6.95 Explain why a dialysis solution must have a low sodium ion concentration if it is designed to remove excess sodium ion from the blood.
6.96 Explain why a dialysis solution must have an elevated potassium ion concentration when loss of potassium ion from the blood is a concern.
6.97 Describe the clinical effects of elevated concentrations of sodium ion in the blood.
6.98 Describe the clinical effects of depressed concentrations of potassium ion in the blood.
6.99 Describe conditions that can lead to elevated concentrations of sodium in the blood.
6.100 Describe conditions that can lead to dangerously low concentrations of potassium in the blood.

CRITICAL THINKING PROBLEMS

1. Which of the following compounds would cause the greater freezing point depression, per mole, in H_2O: $C_6H_{12}O_6$ (glucose) or NaCl?
2. Which of the following compounds would cause the greater boiling point elevation, per mole, in H_2O: $MgCl_2$ or $HOCH_2CH_2OH$ (ethylene glycol, antifreeze)? (Hint: $HOCH_2CH_2OH$ is covalent.)
3. Analytical chemists often take advantage of differences in solubility to separate ions. For example, adding Cl^- to a solution of Cu^{2+} and Ag^+ causes AgCl to precipitate; Cu^{2+} remains in solution. Filtering the solution results in a separation. Design a scheme to separate the cations Ca^{2+} and Pb^{2+}.
4. Using the strategy outlined in the above problem, design a scheme to separate the anions S^{2-} and CO_3^{2-}.
5. Design an experiment that would enable you to measure the degree of solubility of a salt such as KI in water.
6. How could you experimentally distinguish between a saturated solution and a supersaturated solution?
7. Blood is essentially an aqueous solution, but it must transport a variety of nonpolar substances (hormones, for example). Colloidal proteins, termed *albumins*, facilitate this transport. Must these albumins be polar or nonpolar? Why?

GENERAL CHEMISTRY 7

Energy, Rate, and Equilibrium

Learning Goals

1. Correlate the terms *endothermic* and *exothermic* with heat flow between a *system* and its *surroundings*.

2. State the meaning of the terms *enthalpy*, *entropy*, and *free energy* and know their implications.

3. Describe experiments that yield thermochemical information and calculate fuel values based on experimental data.

4. Describe the concept of reaction rate and the role of kinetics in chemical and physical change.

5. Describe the importance of *activation energy* and the *activated complex* in determining reaction rate.

6. Predict the way reactant structure, concentration, temperature, and catalysis affect the rate of a chemical reaction.

7. Write rate equations for elementary processes.

8. Recognize and describe equilibrium situations.

9. Write equilibrium-constant expressions and use these expressions to calculate equilibrium constants.

10. Use LeChatelier's principle to predict changes in equilibrium position.

The reaction of magnesium metal and oxygen (in air) is a graphic example of a highly exothermic reaction.

Outline

Chemistry Connection:
The Cost of Energy? More Than You Imagine

7.1 Thermodynamics

A Human Perspective:
Triboluminescence: Sparks in the Dark with Candy

7.2 Experimental Determination of Energy Change in Reactions

7.3 Kinetics

A Medical Perspective:
Hot and Cold Packs

7.4 Equilibrium

Chemistry Connection

The Cost of Energy? More Than You Imagine

When we purchase gasoline for our automobiles or oil for the furnace, we are certainly buying matter. That matter is only a storage device; we are really purchasing the energy stored in the chemical bonds. Combustion, burning in oxygen, releases the stored potential energy in a form suited to its function: mechanical energy to power a vehicle or heat energy to warm a home.

Energy release is a consequence of change. In fuel combustion, this change results in the production of waste products that may be detrimental to our environment. This necessitates the expenditure of time, money, and *more* energy to clean up our surroundings.

If we are paying a considerable price for our energy supply, it would be nice to believe that we are at least getting full value for our expenditure. Even that is not the case. Removal of energy from molecules also extracts a price. For example, a properly tuned automobile engine is perhaps 30% efficient. That means that less than one-third of the available energy actually moves the car. The other two-thirds is released into the atmosphere as wasted energy, mostly heat energy. The law of conservation of energy tells us that the energy is not destroyed, but it is certainly not available to us in a useful form.

Can we build a 100% efficient energy transfer system? Is there such a thing as cost-free energy? No, on both counts. It is theoretically impossible, and the laws of thermodynamics, which we discuss in this chapter, tell us why this is so.

Introduction

In Chapter 4 we calculated quantities of matter involved in chemical change, assuming that all of the reacting material was consumed and that only products of the reaction remain at the end of the reaction. Often this is not true. Furthermore, not all chemical reactions take place at the same speed; some occur almost instantaneously (explosions), whereas others may proceed for many years (corrosion).

Two concepts play important roles in determining the extent and speed of a chemical reaction: thermodynamics, which deals with energy changes in chemical reactions, and kinetics, which describes the rate or speed of a chemical reaction.

Although both thermodynamics and kinetics involve energy, they are two separate considerations. A reaction may be thermodynamically favored but very slow; conversely, a reaction may be very fast because it is kinetically favorable yet produce very little (or no) product because it is thermodynamically unfavorable.

In this chapter we investigate the fundamentals of thermodynamics and kinetics, with an emphasis on the critical role that energy changes play in chemical reactions. We consider physical change and chemical change, including the conversions that take place among the states of matter (solid, liquid, and gas). We use these concepts to explain the behavior of reactions that do not go to completion, equilibrium reactions. We develop the equilibrium-constant expression and demonstrate how equilibrium composition can be altered using LeChatelier's principle.

7.1 Thermodynamics

Thermodynamics is the study of energy, work, and heat. It may be applied to chemical change, such as the calculation of the quantity of heat obtainable from the combustion of one gallon of fuel oil. Similarly, energy released or consumed in physical change, such as the boiling or freezing of water, may be determined.

There are three basic laws of thermodynamics, but only the first two will be of concern here. They help us to understand why some chemical reactions may occur spontaneously while others do not.

The Chemical Reaction and Energy

John Dalton believed that chemical change involved joining, separating, or rearranging atoms. Two centuries later, this statement stands as an accurate description of chemical reactions. However, we now know much more about the nonmaterial energy changes that are an essential part of every reaction.

Throughout the discussion of thermodynamics and kinetics it will be useful to remember the basic ideas of the kinetic molecular theory (Section 5.1):

- molecules and atoms in a reaction mixture are in constant, random motion;
- these molecules and atoms frequently collide with each other;
- only some collisions, those with sufficient energy, will break bonds in molecules; and
- when reactant bonds are broken, new bonds may be formed and products result.

It is worth noting that we cannot measure an absolute value for energy stored in a chemical system. We can only measure the *change* in energy (energy absorbed or released) as a chemical reaction occurs. Also, it is often both convenient and necessary to establish a boundary between the *system* and its *surroundings*.

The **system** contains the process under study. The **surroundings** encompass the rest of the universe. Energy is lost from the system to the surroundings or energy may be gained by the system at the expense of the surroundings. This energy change, most often in the form of heat, may be determined because the temperature of the system or surroundings will change, and this change can be measured. This process is illustrated in Figure 7.1.

Consider the combustion of methane in a Bunsen burner, the system. The temperature of the air surrounding the burner increases, indicating that some of the potential energy of the system has been converted to heat energy. The heat energy of the system (methane, oxygen, and the Bunsen burner) is being lost to the surroundings.

Now, an exact temperature measurement of the air before and after the reaction is difficult. However, if we could insulate a portion of the surroundings, to isolate and trap the heat, we could calculate a useful quantity, the heat of the reaction. Experimental strategies for measuring temperature change and calculating heats of reactions, termed *calorimetry*, are discussed in Section 7.2.

Exothermic and Endothermic Reactions

The first law of thermodynamics states that the energy of the universe is constant. It is the law of conservation of energy. The study of energy changes that occur in

LEARNING GOAL

Figure 7.1
Illustration of heat flow in (a) exothermic and (b) endothermic reactions.

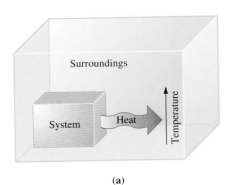
(a)

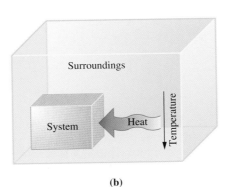
(b)

chemical reactions is a very practical application of the first law. Consider, for example, the generalized reaction:

$$A\text{---}B + C\text{---}D \longrightarrow A\text{---}D + C\text{---}B$$

An **exothermic reaction** releases energy to the surroundings. The surroundings become warmer.

Each chemical bond is stored chemical energy (potential energy). For the reaction to take place, bond A—B and bond C—D must break; this process *always* requires energy. At the same time, bonds A—D and C—B must form; this process always releases energy.

If the energy required to break the A—B and C—D bonds is *less* than the energy given off when the A—D and C—B bonds form, the reaction will release the excess energy. The energy is a *product*, and the reaction is called an exothermic (Gr. *exo*, out, and Gr. *therm*, heat) reaction. This conversion of chemical energy to heat is represented in Figure 7.2a.

An example of an exothermic reaction is the combustion of methane, represented by a *thermochemical equation*:

$$CH_4(g) + 2O_2(g) \longrightarrow CO_2(g) + 2H_2O(g) + \boxed{211 \text{ kcal}}$$

Exothermic reaction

This thermochemical equation reads: the combustion of one mole of methane releases 211 kcal of heat.

An **endothermic reaction** absorbs energy from the surroundings. The surroundings become colder.

If the energy required to break the A—B and C—D bonds is *greater* than the energy released when the A—D and C—B bonds form, the reaction will need an external supply of energy (perhaps from a Bunsen burner). Insufficient energy is available in the system to initiate the bond-breaking process. Such a reaction is called an endothermic (Gr. *endo*, to take on, and Gr. *therm*, heat) reaction, and energy is a *reactant*. The conversion of heat energy into chemical energy is represented in Figure 7.2b.

The decomposition of ammonia into nitrogen and hydrogen is one example of an endothermic reaction:

$$\boxed{22 \text{ kcal}} + 2NH_3(g) \longrightarrow N_2(g) + 3H_2(g)$$

Endothermic reaction

This thermochemical equation reads: the decomposition of two moles of ammonia requires 22 kcal of heat.

The examples used here show the energy absorbed or released as heat energy. Depending on the reaction and the conditions under which the reaction is run, the energy may take the form of light energy or electrical energy. A firefly releases energy as a soft glow of light on a summer evening. An electrical current results from a chemical reaction in a battery, enabling your car to start.

> In an exothermic reaction, heat is released *from the system* to the surroundings. In an endothermic reaction, heat is absorbed *by the system* from the surroundings.

Figure 7.2
(a) An exothermic reaction. ΔE represents the energy released during the progress of the exothermic reaction: $A + B \longrightarrow C + D + \Delta E$. (b) An endothermic reaction. ΔE represents the energy absorbed during the progress of the endothermic reaction: $\Delta E + A + B \longrightarrow C + D$.

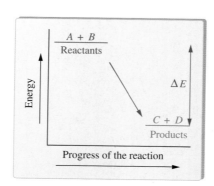

(a)

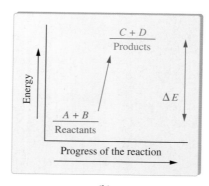

(b)

Enthalpy

Enthalpy is the term used to represent heat. The *change in enthalpy* is the energy difference between the products and reactants of a chemical reaction and is symbolized as ΔH. By convention, energy released is represented with a negative sign (indicating an exothermic reaction), and energy absorbed is shown with a positive sign (indicating an endothermic reaction).

For the combustion of methane, an exothermic process, energy is a *product* in the thermochemical equation, and

$$\Delta H = -211 \text{ kcal}$$

For the decomposition of ammonia, an endothermic process, energy is a *reactant* in the thermochemical equation, and

$$\Delta H = +22 \text{ kcal}$$

In these discussions, we consider the enthalpy change and energy change to be identical. This is true for most common reactions carried out in lab, with minimal volume change.

Spontaneous and Nonspontaneous Reactions

It seems that all exothermic reactions should be spontaneous. After all, an external supply of energy does not appear to be necessary; in fact, energy is a product of the reaction. It also seems that all endothermic reactions should be nonspontaneous: energy is a reactant that we must provide. However, these hypotheses are not supported by experimentation.

Experimental measurement has shown that most *but not all* exothermic reactions are spontaneous. Likewise, most *but not all*, endothermic reactions are not spontaneous. There must be some factor in addition to enthalpy that will help us to explain the less obvious cases of nonspontaneous exothermic reactions and spontaneous endothermic reactions. This other factor is entropy.

Entropy

The first law of thermodynamics considers the enthalpy of chemical reactions. The second law states that the universe spontaneously tends toward increasing disorder or randomness.

A measure of the randomness of a chemical system is its **entropy**. The entropy of a substance is represented by the symbol S. A random, or disordered, system is characterized by *high entropy*; a well-organized system has *low entropy*.

What do we mean by disorder in chemical systems? Disorder is simply the absence of a regular repeating pattern. Disorder or randomness increases as we convert from the solid to the liquid to the gaseous state. As we have seen, solids often have an ordered crystalline structure, liquids have, at best, a loose arrangement, and gas particles are virtually random in their distribution. Therefore gases have high entropy, and crystalline solids have very low entropy. Figures 7.3 and 7.4 illustrate properties of entropy in systems.

A *system* is a part of the universe upon which we wish to focus our attention. For example, it may be a beaker containing reactants and products.

Chapter 5 compares the physical properties of solids, liquids, and gases.

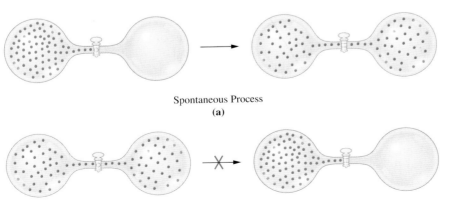

Spontaneous Process
(a)

Nonspontaneous Process
(b)

Figure 7.3

(a) Gas particles, trapped in the left chamber, spontaneously diffuse into the right chamber, initially under vacuum, when the valve is opened. (b) It is unimaginable that the gas particles will rearrange themselves and reverse the process to create a vacuum. This can only be accomplished using a pump, that is, by doing work on the system.

Figure 7.4
Processes such as (a) melting, (b) vaporization, and (c) dissolution increase entropy, or randomness, of the particles.

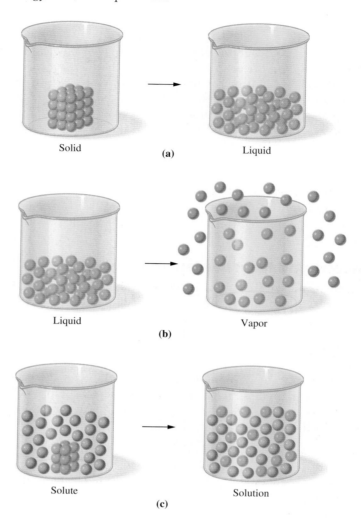

Question 7.1

Are the following processes exothermic or endothermic?

a. Fuel oil is burned in a furnace.
b. $C_6H_{12}O_6(s) \longrightarrow 2C_2H_5OH(l) + 2CO_2(g)$, $\Delta H = -16$ kcal
c. $N_2O_5(g) + H_2O(l) \longrightarrow 2HNO_3(l) + 18.3$ kcal

Question 7.2

Are the following processes exothermic or endothermic?

a. When solid NaOH is dissolved in water, the solution gets hotter.
b. $S(s) + O_2(g) \longrightarrow SO_2(g)$, $\Delta H = -71$ kcal
c. $N_2(g) + 2O_2(g) + 16.2$ kcal $\longrightarrow 2NO_2(g)$

The second law describes the entire universe or any isolated system within the universe. On a more personal level, we all fall victim to the law of increasing disorder. Chaos in our room or workplace is certainly not our intent! It happens almost effortlessly. However, reversal of this process requires work and energy. The same is true at the molecular level. The gradual deterioration of our cities' infrastructure (roads, bridges, water mains, and so forth) is an all-too-familiar example.

A Human Perspective

Triboluminescence: Sparks in the Dark with Candy

Generations of children have inadvertently discovered the phenomenon of triboluminescence. Crushing a wintergreen candy (Lifesavers) with the teeth in a dark room (in front of a few friends or a mirror) or simply rubbing two pieces of candy together may produce the effect—transient sparks of light!

Triboluminescence is simply the production of light upon fracturing a solid. It is easily observed and straightforward to describe but difficult to explain. It is believed to result from charge separation produced by the disruption of a crystal lattice. The charge separation has a very short lifetime. When the charge distribution returns to equilibrium, energy is released, and that energy is the light that is observed.

Dr. Linda M. Sweeting and several other groups of scientists have tried to reproduce these events under controlled circumstances. Crystals similar to the sugars in wintergreen candy are prepared with a very high level of purity. Some theories attribute the light emission to impurities in a crystal rather than to the crystal itself. Devices have been constructed that will crush the crystal with a uniform and reproducible force. Light-measuring devices, spectrophotometers, accurately measure the various wavelengths of light and the intensity of the light at each wavelength.

Through the application of careful experimentation and measurement of light-emitting properties of a variety of related compounds, these scientists hope to develop a theory of light emission from fractured solids.

This is one more example of the scientific method improving our understanding of everyday occurrences.

Charles Schulz's "Peanuts" vision of triboluminescence.
Reprinted by permission of UFS, Inc.

For Further Understanding

Can you suggest an experiment that would support or refute the hypothesis that impurities in crystals are responsible for triboluminescence?

Would you expect that amorphous substances would exhibit triboluminescence? Why or why not?

Millions of dollars (translated into energy and work) are needed annually just to try to maintain the status quo.

The entropy of a reaction is measured as a difference, ΔS, between the entropies, S, of products and reactants.

The drive toward increased entropy, along with a tendency to achieve a lower potential energy, is responsible for spontaneous chemical reactions. Reactions that are exothermic and whose products are more disordered (higher in entropy) will occur spontaneously, whereas endothermic reactions producing products of lower entropy will not be spontaneous. If they are to take place at all, they will need some energy input.

Which substance has the greatest entropy, He(g) or Na(s)? Explain your reasoning.

Which substance has the greatest entropy, $H_2O(l)$ or $H_2O(g)$? Explain your reasoning.

Free Energy

LEARNING GOAL

The two situations described above are clear-cut and unambiguous. In any other situation the reaction may or may not be spontaneous. It depends on the relative size of the enthalpy and entropy values.

Free energy, symbolized by ΔG, represents the combined contribution of the enthalpy *and* entropy values for a chemical reaction. Thus free energy is the ultimate predictor of reaction spontaneity and is expressed as

$$\Delta G = \Delta H - T\Delta S$$

ΔH represents the change in enthalpy between products and reactants, ΔS represents the change in entropy between products and reactants, and T is the Kelvin temperature of the reaction. A reaction with a negative value of ΔG will *always* be spontaneous. Reactions with a positive ΔG will *always* be nonspontaneous.

We need to know both ΔH and ΔS in order to predict the sign of ΔG and make a definitive statement regarding the spontaneity of the reaction. Additionally, the temperature may determine the direction of spontaneity. Consider the four possible situations:

- ΔH positive and ΔS negative: ΔG is always positive, regardless of the temperature. The reaction is always nonspontaneous.
- ΔH negative and ΔS positive: ΔG is always negative, regardless of the temperature. The reaction is always spontaneous.
- Both ΔH and ΔS positive: The sign of ΔG depends on the temperature.
- Both ΔH and ΔS negative: The sign of ΔG depends on the temperature.

EXAMPLE 7.1 Determining Whether a Process Is Exothermic or Endothermic

An ice cube is dropped into a glass of water at room temperature. The ice cube melts. Is the melting of the ice exothermic or endothermic?

Solution

Consider the ice cube to be the system and the water, the surroundings. For the cube to melt, it must gain energy and its energy source must be the water. The heat flow is from surroundings to system. The system gains energy (+energy); hence, the melting process (physical change) is endothermic.

Question 7.5

Predict whether a reaction with positive ΔH and negative ΔS will be spontaneous, nonspontaneous, or temperature dependent. Explain your reasoning.

Question 7.6

Predict whether a reaction with positive ΔH and positive ΔS will be spontaneous, nonspontaneous, or temperature dependent. Explain your reasoning.

7.2 Experimental Determination of Energy Change in Reactions

The measurement of heat energy changes in a chemical reaction is **calorimetry**. This technique involves the measurement of the change in the temperature of a quantity of water or solution that is in contact with the reaction of interest and isolated from the surroundings. A device used for these measurements is a *calorimeter*, which measures heat changes in calories.

A Styrofoam coffee cup is a simple design for a calorimeter, and it produces surprisingly accurate results. It is a good insulator, and, when filled with solution, it can be used to measure temperature changes taking place as the result of a chemical reaction occurring in that solution (Figure 7.5). The change in the temperature of the solution, caused by the reaction, can be used to calculate the gain or loss of heat energy for the reaction.

For an exothermic reaction, heat released by the reaction is absorbed by the surrounding solution. For an endothermic reaction, the reactants absorb heat from the solution.

The **specific heat** of a substance is defined as the number of calories of heat needed to raise the temperature of 1 g of the substance 1 degree Celsius. Knowing the specific heat of the water or the aqueous solution along with the total number of grams of solution and the temperature increase (measured as the difference between the final and initial temperatures of the solution), enables the experimenter to calculate the heat released during the reaction.

The solution behaves as a "trap" or "sink" for energy released in the exothermic process. The temperature increase indicates a gain in heat energy. Endothermic reactions, on the other hand, take heat energy away from the solution, lowering its temperature.

The quantity of heat absorbed or released by the reaction (Q) is the product of the mass of solution in the calorimeter (m_s), the specific heat of the solution (SH_s), and the change in temperature (ΔT_s) of the solution as the reaction proceeds from the initial to final state.

The heat is calculated by using the following equation:

$$Q = m_s \times \Delta T_s \times SH_s$$

with units

$$\text{calories} = \cancel{\text{gram}} \times \cancel{°C} \times \frac{\text{calories}}{\cancel{\text{gram-°C}}}$$

The details of the experimental approach are illustrated in Example 7.2.

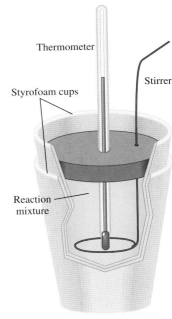

Figure 7.5
A "coffee cup" calorimeter used for the measurement of heat change in chemical reactions. The concentric Styrofoam cups insulate the system from its surroundings. Heat released by the chemical reaction enters the solution, raising its temperature, which is measured by using a thermometer.

EXAMPLE 7.2 Calculating Energy Involved in Calorimeter Reactions

If 0.050 mol of hydrochloric acid (HCl) is mixed with 0.050 mol of sodium hydroxide (NaOH) in a "coffee cup" calorimeter, the temperature of 1.00×10^2 g of the resulting solution increases from 25.0°C to 31.5°C. If the specific heat of the solution is 1.00 cal/g solution °C, calculate the quantity of energy involved in the reaction. Also, is the reaction endothermic or exothermic?

Continued—

EXAMPLE 7.2 —Continued

Solution

The change in temperature is

$$\Delta T_s = T_{s\,final} - T_{s\,initial}$$
$$= 31.5°C - 25.0°C = 6.5°C$$

$$Q = m_s \times \Delta T_s \times SH_s$$

$$Q = 1.00 \times 10^2 \text{ g solution} \times 6.5°C \times \frac{1.00 \text{ cal}}{\text{g solution } °C}$$

$$= 6.5 \times 10^2 \text{ cal}$$

6.5×10^2 cal (or 0.65 kcal) of heat energy were released by this acid-base reaction to the surroundings, the solution; the reaction is exothermic.

EXAMPLE 7.3 Calculating Energy Involved in Calorimeter Reactions

If 0.10 mol of ammonium chloride (NH_4Cl) is dissolved in water producing 1.00×10^2 g solution, the water temperature decreases from 25.0°C to 18.0°C. If the specific heat of the resulting solution is 1.00 cal/g-°C, calculate the quantity of energy involved in the process. Also, is the dissolution of ammonium chloride endothermic or exothermic?

Solution

The change in temperature is

$$\Delta T = T_{s\,final} - T_{s\,initial}$$
$$= 18.0°C - 25.0°C = -7.0°C$$

$$Q = m_s \times \Delta T_s \times SH_s$$

$$Q = 1.00 \times 10^2 \text{ g solution} \times (-7.0°C) \times \frac{1.00 \text{ cal}}{\text{g solution } °C}$$

$$= -7.0 \times 10^2 \text{ cal}$$

7.0×10^2 cal (or 0.70 kcal) of heat energy were absorbed by the dissolution process because the solution lost (– sign) 7.0×10^2 cal of heat energy to the system. The reaction is endothermic.

Question 7.7

Refer to Example 7.2 and calculate the temperature change that would have been observed if 50.0 g solution were in the calorimeter instead of 1.00×10^2 g solution.

Question 7.8

Refer to Example 7.2 and calculate the temperature change that would have been observed if 1.00×10^2 g of another liquid, producing a solution with a specific heat capacity of 0.800 cal/g-°C, was substituted for the water in the calorimeter.

7.2 Experimental Determination of Energy Change in Reactions

Question 7.9

Convert the energy released in Example 7.2 to joules (recall the conversion factor for calories and joules, Chapter 1).

Question 7.10

Convert the energy absorbed in Example 7.3 to joules (recall the conversion factor for calories and joules, Chapter 1).

Many chemical reactions that produce heat are combustion reactions. In our bodies many food substances (carbohydrates, proteins, and fats, Chapters 21 and 22) are oxidized to release energy. **Fuel value** is the amount of energy per gram of food.

The fuel value of food is an important concept in nutrition science. The fuel value is generally reported in units of *nutritional Calories*. One **nutritional Calorie** is equivalent to one kilocalorie (1000 calories). It is also known as the *large Calorie* (uppercase C).

Energy necessary for our daily activity and bodily function comes largely from the "combustion" of carbohydrates. Chemical energy from foods that is not used to maintain normal body temperature or in muscular activity is stored in the bonds of chemical compounds known collectively as fat. Thus "high-calorie" foods are implicated in obesity.

A special type of calorimeter, a *bomb calorimeter*, is useful for the measurement of the fuel value (Calories) of foods. Such a device is illustrated in Figure 7.6. Its design is similar, in principle, to that of the "coffee cup" calorimeter discussed earlier. It incorporates the insulation from the surroundings, solution pool, reaction chamber, and thermometer. Oxygen gas is added as one of the reactants, and an electrical igniter is inserted to initiate the reaction. However, it is not open to the atmosphere. In the sealed container the reaction continues until the sample is completely oxidized. All of the heat energy released during the reaction is captured in the water.

Note: Refer to A Human Perspective: Food Calories, Section 1.5.

  LEARNING GOAL

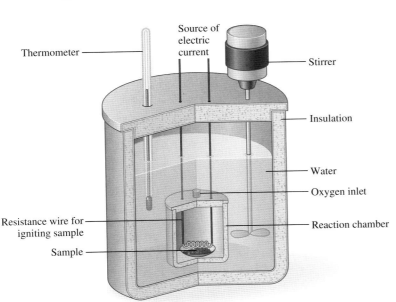

Figure 7.6
A bomb calorimeter that may be used to measure heat released upon combustion of a sample. This device is commonly used to determine the fuel value of foods. The bomb calorimeter is similar to the "coffee cup" calorimeter. However, note the electrical component necessary to initiate the combustion reaction.

EXAMPLE 7.4 Calculating the Fuel Value of Foods

One gram of glucose (a common sugar or carbohydrate) was burned in a bomb calorimeter. The temperature of 1.00×10^3 g H_2O was raised from 25.0°C to 28.8°C ($\Delta T_w = 3.8°C$). Calculate the fuel value of glucose.

Solution

Recall that the fuel value is the number of nutritional Calories liberated by the combustion of 1 g of material and 1 g of material was burned in the calorimeter. Then

$$\text{Fuel value} = Q = m_w \times \Delta T_w \times SH_w$$

Water is the surroundings in the calorimeter; it has a specific heat capacity equal to 1.00 cal/g H_2O °C.

$$\text{Fuel value} = Q = \text{g } H_2O \times °C \times \frac{1.00 \text{ cal}}{\text{g } H_2O \text{ °C}}$$

$$= 1.00 \times 10^3 \text{ g } H_2O \times 3.8°C \times \frac{1.00 \text{ cal}}{\text{g } H_2O \text{ °C}}$$

$$= 3.8 \times 10^3 \text{ cal}$$

and

$$3.8 \times 10^3 \text{ cal} \times \frac{1 \text{ nutritional Calorie}}{10^3 \text{ cal}} = 3.8 \text{ C (nutritional Calories, or kcal)}$$

The fuel value of glucose is 3.8 kcal/g.

Question 7.11

A 1.0-g sample of a candy bar (which contains lots of sugar!) was burned in a bomb calorimeter. A 3.0°C temperature increase was observed for 1.00×10^3 g of water. The entire candy bar weighed 2.5 ounces. Calculate the fuel value (in nutritional Calories) of the sample and the total caloric content of the candy bar.

Question 7.12

If the fuel value of 1.00 g of a certain carbohydrate (sugar) is 3.00 nutritional Calories, how many grams of water must be present in the calorimeter to record a 5.00°C change in temperature?

7.3 Kinetics

LEARNING GOAL

Thermodynamics help us to decide whether a chemical reaction is spontaneous. Knowing that a reaction can occur spontaneously tells us nothing about the time that it may take.

Chemical **kinetics** is the study of the **rate** (or speed) **of chemical reactions.** Kinetics also gives an indication of the *mechanism* of a reaction, a step-by-step description of how reactants become products. Kinetic information may be represented as the *disappearance* of reactants or *appearance* of product over time. A typical graph of concentration versus time is shown in Figure 7.7.

Information about the rate at which various chemical processes occur is useful. For example, what is the "shelf life" of processed foods? When will slow changes in composition make food unappealing or even unsafe? Many drugs lose their potency with time because the active ingredient decomposes into other substances. The rate of hardening of dental filling material (via a chemical reaction) influences the dentist's technique. Our very lives depend on the efficient transport of oxygen to each of our cells and the rapid use of the oxygen for energy-harvesting reactions.

The diagram in Figure 7.8 is a useful way of depicting the kinetics of a reaction at the molecular level.

Often a color change, over time, can be measured. Such changes are useful in assessing the rate of a chemical reaction (Figure 7.9).

Let's see what actually happens when two chemical compounds react and what experimental conditions affect the rate of a reaction.

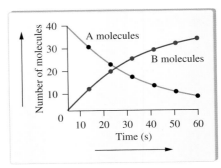

Figure 7.7
For a hypothetical reaction $A \longrightarrow B$ the concentration of A molecules (reactant molecules) decreases over time and B molecules (product molecules) increase in concentration over time.

The Chemical Reaction

Consider the exothermic reaction that we discussed in Section 7.1:

$$CH_4(g) + 2O_2(g) \longrightarrow CO_2(g) + 2H_2O(l) + \boxed{211 \text{ kcal}}$$

For the reaction to proceed, C—H and O—O bonds must be broken, and C—O and H—O bonds must be formed. Sufficient energy must be available to cause the bonds to break if the reaction is to take place. This energy is provided by the collision of molecules. If sufficient energy is available at the temperature of the reaction, one or more bonds will break, and the atoms will recombine in a lower energy arrangement, in this case as carbon dioxide and water. A collision producing product molecules is termed an *effective collision*. Only effective collisions lead to chemical reaction.

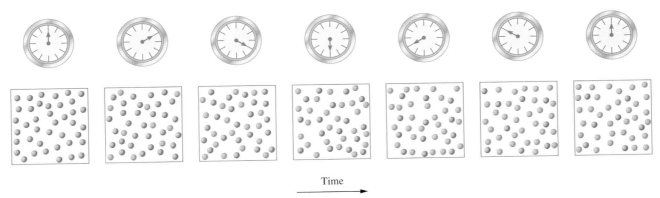

Figure 7.8
An alternate way of representing the information contained in Figure 7.7.

Figure 7.9
The conversion of reddish brown Br_2 in solution to colorless Br^- over time.

Activation Energy and the Activated Complex

LEARNING GOAL 5

The minimum amount of energy required to initiate a chemical reaction is called the **activation energy** for the reaction.

We can picture the chemical reaction in terms of the changes in potential energy that occur during the reaction. Figure 7.10a graphically shows these changes for an exothermic reaction. Important characteristics of this graph include the following:

- The reaction proceeds from reactants to products through an extremely unstable state that we call the **activated complex.** The activated complex cannot be isolated from the reaction mixture but may be thought of as a short-lived group of atoms structured in such a way that it quickly and easily breaks apart into the products of the reaction.
- Formation of the activated complex requires energy. The difference between the energy of reactants and that of the activated complex is the activation energy. This energy must be provided by the collision of the reacting molecules or atoms at the temperature of the reaction.
- Because this is an exothermic reaction, the overall energy change must be a *net* release of energy. The *net* release of energy is the difference in energy between products and reactants.

For an endothermic reaction, such as the decomposition of water,

$$\boxed{\text{energy}} + 2H_2O(l) \longrightarrow 2H_2(g) + O_2(g)$$

the change of potential energy with reaction time is shown in Figure 7.10b.

The reaction takes place slowly because of the large activation energy required for the conversion of water into the elements hydrogen and oxygen.

This reaction will take place when an electrical current is passed through water. The process is called *electrolysis*.

Question 7.13

The act of striking a match illustrates the role of activation energy in a chemical reaction. Explain.

Question 7.14

Distinguish between the terms *net energy* and *activation energy*.

Figure 7.10
(a) The change in potential energy as a function of reaction time for an exothermic chemical reaction. Note particularly the energy barrier associated with the formation of the activated complex. This energy barrier (E_a) *is* the activation energy. (b) The change in potential energy as a function of reaction time for an endothermic chemical reaction. In contrast to the exothermic reaction in (a), the energy of the products is greater than the energy of the reactants.

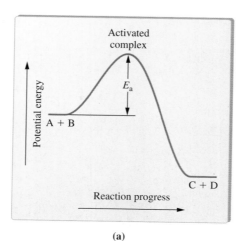
(a)

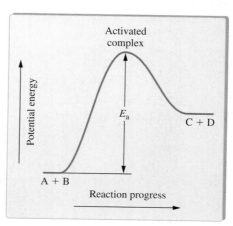
(b)

A Medical Perspective

Hot and Cold Packs

Hot packs provide "instant warmth" for hikers and skiers and are used in treatment of injuries such as pulled muscles. Cold packs are in common use today for the treatment of injuries and the reduction of swelling.

These useful items are an excellent example of basic science producing a technologically useful product. (Recall our discussion in Chapter 1 of the relationship of science and technology.)

Both hot and cold packs depend on large energy changes taking place during a chemical reaction. Cold packs rely on an endothermic reaction, and hot packs generate heat energy from an exothermic reaction.

A cold pack is fabricated as two separate compartments within a single package. One compartment contains NH_4NO_3, and the other contains water. When the package is squeezed, the inner boundary between the two compartments ruptures, allowing the components to mix, and the following reaction occurs:

$$6.7 \text{ kcal/mol} + NH_4NO_3(s) \longrightarrow NH_4^+(aq) + NO_3^-(aq)$$

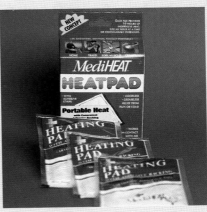

Heating pads.

This reaction is endothermic; heat taken from the surroundings produces the cooling effect.

The design of a hot pack is similar. Here, finely divided iron powder is mixed with oxygen. Production of iron oxide results in the evolution of heat:

$$4Fe + 3O_2 \longrightarrow 2Fe_2O_3 + 198 \text{ kcal/mol}$$

This reaction occurs via an oxidation-reduction mechanism (see Chapter 8). The iron atoms are oxidized, O_2 is reduced. Electrons are transferred from the iron atoms to O_2 and Fe_2O_3 forms exothermically. The rate of the reaction is slow; therefore the heat is liberated gradually over a period of several hours.

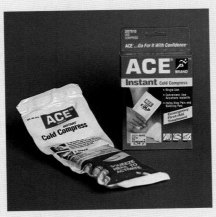

A cold pack.

For Further Understanding

What is the sign of ΔH for each equation in this story?

Would reactions with small rate constants be preferred for applications such as those described here? Why or why not?

Factors That Affect Reaction Rate

Five major factors influence reaction rate:

- structure of the reacting species,
- concentration of reactants,
- temperature of reactants,
- physical state of reactants, and
- presence of a catalyst.

Structure of the Reacting Species

Reactions among ions in solution are usually very rapid. Ionic compounds in solution are dissociated; consequently, their bonds are already broken, and the activation energy for their reaction should be very low. On the other hand, reactions involving covalently bonded compounds may proceed more slowly. Covalent bonds must be broken and new bonds formed. The activation energy for this process would be significantly higher than that for the reaction of free ions.

Bond strengths certainly play a role in determining reaction rates for the magnitude of the activation energy, or energy barrier, is related to bond strength.

The size and shape of reactant molecules influence the rate of the reaction. Large molecules, containing bulky groups of atoms, may block the reactive part of the molecule from interacting with another reactive substance, causing the reaction to proceed slowly. Only molecular collisions that have the correct collision orientation lead to product formation. These collisions are termed *effective collisions*.

The Concentration of Reactants

The rate of a chemical reaction is often a complex function of the concentration of one or more of the reacting substances. The rate will generally *increase* as concentration *increases* simply because a higher concentration means more reactant molecules in a given volume and therefore a greater number of collisions per unit time. If we assume that other variables are held constant, a larger number of collisions leads to a larger number of effective collisions. For example, the rate at which a fire burns depends on the concentration of oxygen in the atmosphere surrounding the fire, as well as the concentration of the fuel (perhaps methane or propane). A common fire-fighting strategy is the use of fire extinguishers filled with carbon dioxide. The carbon dioxide dilutes the oxygen to a level where the combustion process can no longer be sustained.

The Temperature of Reactants

The rate of a reaction *increases* as the temperature increases, because the average kinetic energy of the reacting particles is directly proportional to the Kelvin temperature. Increasing the speed of particles increases the likelihood of collision, and the higher kinetic energy means that a higher percentage of these collisions will result in product formation (effective collisions). A 10°C rise in temperature has often been found to double the reaction rate.

The Physical State of Reactants

The rate of a reaction depends on the physical state of the reactants: solid, liquid, or gas. For a reaction to occur the reactants must collide frequently and have sufficient energy to react. In the solid state, the atoms, ions, or molecules are restricted in their motion. In the gaseous and liquid states the particles have both free motion and proximity to each other. Hence reactions tend to be fastest in the liquid state and slowest in the solid state.

The Presence of a Catalyst

A **catalyst** is a substance that *increases* the reaction rate. If added to a reaction mixture, the catalytic substance undergoes no net change, nor does it alter the outcome of the reaction. However, the catalyst interacts with the reactants to create an alternative pathway for production of products. This alternative path has a lower activation energy. This makes it easier for the reaction to take place and thus increases the rate. This effect is illustrated in Figure 7.11.

Catalysis is important industrially; it may often make the difference between profit and loss in the sale of a product. For example, catalysis is useful in converting double bonds to single bonds. An important application of this principle involves the process of hydrogenation. Hydrogenation converts one or more of the carbon-carbon double bonds of unsaturated fats (e.g., corn oil, olive oil) to single bonds characteristic of saturated fats (such as margarine). The use of a metal catalyst, such as nickel, in contact with the reaction mixture dramatically increases the rate of the reaction.

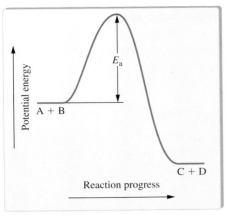

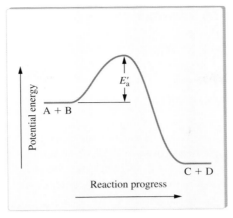

(a) Uncatalyzed reaction

(b) Catalyzed reaction

Figure 7.11
The effect of a catalyst on the magnitude of the activation energy of a chemical reaction: (a) uncatalyzed reaction, (b) catalyzed reaction. Note that the presence of a catalyst decreases the activation energy ($E'_a < E_a$), thus increasing the rate of the reaction.

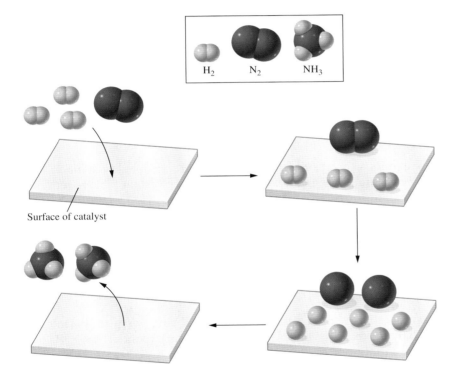

Figure 7.12
The synthesis of ammonia, an important industrial product, is facilitated by a solid phase catalyst (the Haber process). H_2 and N_2 bind to the surface, their bonds are weakened, dissociation and reformation as ammonia occur, and the newly formed ammonia molecules leave the surface. This process is repeated over and over, with no change in the catalyst.

Thousands of essential biochemical reactions in our bodies are controlled and speeded up by biological catalysts called *enzymes*.

A molecular level view of the action of a solid catalyst widely used in industrial synthesis of ammonia is presented in Figure 7.12.

Question 7.15
Would you imagine that a substance might act as a poison if it interfered with the function of an enzyme? Why?

Question 7.16
Bacterial growth decreases markedly in a refrigerator. Why?

Mathematical Representation of Reaction Rate

LEARNING GOAL

Consider the decomposition reaction of N_2O_5 (dinitrogen pentoxide) in the gas phase. When heated, N_2O_5 decomposes and forms two products: NO_2 (nitrogen dioxide) and O_2 (diatomic oxygen). The balanced chemical equation for the reaction is

$$2N_2O_5(g) \xrightarrow{\Delta} 4NO_2(g) + O_2(g)$$

When all of the factors that affect the rate of the reaction (except concentration) are held constant (i.e., the nature of the reactant, temperature and physical state of the reactant, and the presence or absence of a catalyst) the rate of the reaction is proportional to the concentration of N_2O_5.

$$\text{rate} \propto \text{concentration } N_2O_5$$

We will represent the concentration of N_2O_5 in units of molarity and represent molar concentration using brackets.

$$\text{concentration } N_2O_5 = [N_2O_5]$$

Then,

$$\text{rate} \propto [N_2O_5]$$

Laboratory measurement shows that the rate of the reaction depends on the molar concentration raised to an experimentally determined exponent that we will symbolize as n

$$\text{rate} \propto [N_2O_5]^n$$

In expressions such as the one shown, the proportionality symbol, $\propto$, may be replaced by an equality sign and a proportionality constant that we represent as k, the **rate constant.**

$$\text{rate} = k[N_2O_5]^n$$

The exponent, n, is the **order of the reaction.** For the reaction described here, which has been studied in great detail, n is numerically equal to 1, hence the reaction is first order in N_2O_5 and the rate equation for the reaction is:

$$\text{rate} = k[N_2O_5]$$

Equations that follow this format, the rate being equal to the rate constant multiplied by the reactant concentration raised to an exponent that is the order, are termed **rate equations.**

Note that the exponent, n, in the rate equation is not the same as the coefficient of N_2O_5 in the balanced equation. However, in reactions that occur in a single step, the coefficient in the balanced equation and the exponent n (the order of the reaction) are numerically the same.

In general, the rate of reaction for an equation of the general form:

$$A \longrightarrow \text{product}$$

is

$$\text{rate} = k[A]^n$$

in which

$$n = \text{order of the reaction}$$
$$k = \text{the rate constant of the reaction}$$

An equation of the form

$$A + B \longrightarrow \text{products}$$

has a rate expression

$$\text{rate} = k[A]^n[B]^{n'}$$

Both the value of the rate constant and the order of the reaction are deduced from a series of experiments. We cannot predict them by simply looking at the chemical equation. Only the *form* of the rate expression can be found by inspection of the chemical equation, and, even then, only for reactions that occur in a single step.

EXAMPLE 7.5

Writing Rate Equations

Write the form of the rate equation for the oxidation of ethanol (C_2H_5OH). The reaction has been experimentally determined to be first order in ethanol and third order in oxygen (O_2).

Solution

The rate expression involves only the reactants, C_2H_5OH and O_2. Depict their concentration as

$$[C_2H_5OH][O_2]$$

and raise each to an exponent corresponding to its experimentally determined order

$$[C_2H_5OH][O_2]^3$$

and this is proportional to the rate:

$$\text{rate} \propto [C_2H_5OH][O_2]^3$$

or

$$\text{rate} = k[C_2H_5OH][O_2]^3$$

is the rate expression. (Remember that 1 is understood as an exponent; $[C_2H_5OH]$ is correct and $[C_2H_5OH]^1$ is not.)

Question 7.17

Write the general form of the rate equation for each of the following processes.

a. $N_2(g) + O_2(g) \longrightarrow 2NO(g)$
b. $2C_4H_6(g) \longrightarrow C_8H_{12}(g)$

Question 7.18

Write the general form of the rate equation for each of the following processes.

a. $CH_4(g) + 2O_2(g) \longrightarrow CO_2(g) + 2H_2O(g)$
b. $2NO_2(g) \longrightarrow 2NO(g) + O_2(g)$

Knowledge of the form of the rate equation, coupled with the experimental determination of the value of the rate constant, k, and the order, n, are valuable in a number of ways. Industrial chemists use this information to establish optimum conditions for preparing a product in the shortest practical time. The design of an entire manufacturing facility may, in part, depend on the rates of the critical reactions.

In Section 7.4 we will see how the rate equation forms the basis for describing equilibrium reactions.

7.4 Equilibrium

Rate and Reversibility of Reactions

LEARNING GOAL

We have assumed that most chemical and physical changes considered thus far proceed to completion. A complete reaction is one in which all reactants have been converted to products. However, many important chemical reactions do not go to completion. As a result, after no further obvious change is taking place, measurable quantities of reactants and products remain. Reactions of this type (incomplete reactions) are called **equilibrium reactions.**

Examples of physical and chemical equilibria abound in nature. Many environmental systems depend on fragile equilibria. The amount of oxygen dissolved in a certain volume of lake water (the oxygen concentration) is governed by the principles of equilibrium. The lives of plants and animals within this system are critically related to the levels of dissolved oxygen in the water.

The very form and function of the earth is a consequence of a variety of complex equilibria. Stalactite and stalagmite formations in caves are made up of solid calcium carbonate ($CaCO_3$). They owe their existence to an equilibrium process described by the following equation:

$$Ca^{2+}(aq) + 2HCO_3^-(aq) \rightleftharpoons CaCO_3(s) + CO_2(aq) + H_2O(l)$$

Physical Equilibrium

A physical equilibrium, such as sugar dissolving in water, is a reversible reaction. A **reversible reaction** is a process that can occur in both directions. It is indicated by using a double arrow ($\rightleftharpoons$) symbol.

Dissolution of sugar in water, producing a saturated solution, is a convenient illustration of a state of *dynamic equilibrium.*

A **dynamic equilibrium** is a situation in which the rate of the forward process in a reversible reaction is exactly balanced by the rate of the reverse process.

Let's now look at the sugar and water equilibrium in more detail.

Sugar in Water

Imagine that you mix a small amount of sugar (2 or 3g) in 100 mL of water. After you have stirred it for a short time, all of the sugar dissolves; there is no residual solid sugar because the sugar has dissolved *completely*. The reaction clearly has converted all solid sugar to its dissolved state, an aqueous solution of sugar, or

$$sugar(s) \longrightarrow sugar(aq)$$

Now, suppose that you add a very large amount of sugar (100 g), more than can possibly dissolve, to the same volume of water. As you stir the mixture you observe more and more sugar dissolving. After some time the amount of solid sugar remaining in contact with the solution appears constant. Over time, you observe no further change in the amount of dissolved sugar. At this point, although nothing further appears to be happening, in reality a great deal of activity is taking place!

An equilibrium situation has been established. Over time the amount of sugar dissolved in the measured volume of water (the concentration of sugar in water) does not change. Hence the amount of undissolved sugar remains the same. However, if you could look at the individual sugar molecules, you would see something quite amazing. Rather than sugar molecules in the solid simply staying in place, you would see them continuing to leave the solid state and go into solution. At the same time, a like number of dissolved sugar molecules would leave the water and form more solid. This active process is described as a *dynamic equilibrium.*

The reaction is proceeding in a forward (left to right) and a reverse (right to left) direction at the same time and is a reversible reaction:

$$\text{sugar}(s) \rightleftarrows \text{sugar}(aq)$$

The double arrow serves as

- an indicator of a reversible process,
- an indicator of an equilibrium process, and
- a reminder of the dynamic nature of the process.

How can we rationalize the apparent contradiction: continuous change is taking place yet no observable change in the amount of sugar in either the solid or dissolved form is observed.

The only possible explanation is that the rate of the forward process

$$\text{sugar}(s) \longrightarrow \text{sugar}(aq)$$

must be equal to the rate of the reverse process

$$\text{sugar}(s) \longleftarrow \text{sugar}(aq)$$

Under this condition, the number of sugar molecules leaving the solid in a given time interval is identical to the number of sugar molecules returning to the solid state.

> Dynamic equilibrium can be particularly dangerous for living cells because it represents a situation in which nothing is getting done. There is no gain. Let's consider an exothermic reaction designed to produce a net gain of energy for the cell. In a dynamic equilibrium the rate of the forward (energy-releasing) reaction is equal to the rate of the backward (energy-requiring) reaction. Thus there is no net gain of energy to fuel cellular activity, and the cell will die.

Question 7.19

Construct an example of a dynamic equilibrium using a subway car at rush hour.

Question 7.20

A certain change in reaction conditions for a process was found to increase the rate of the forward reaction much more than that of the reverse reaction. Did the amount of product increase, decrease, or remain the same? Why?

Chemical Equilibrium

The Reaction of N_2 and H_2

When we mix nitrogen gas (N_2) and hydrogen gas (H_2) at an elevated temperature (perhaps 500°C), some of the molecules will collide with sufficient energy to break N—N and H—H bonds. Rearrangement of the atoms will produce the product (NH_3):

$$N_2(g) + 3H_2(g) \rightleftarrows 2NH_3(g)$$

Beginning with a mixture of hydrogen and nitrogen, the rate of the reaction is initially rapid, because the reactant concentration is high; as the reaction proceeds, the concentration of reactants decreases. At the same time the concentration of the product, ammonia, is increasing. At equilibrium the *rate of depletion* of hydrogen and nitrogen *is equal to* the *rate of depletion* of ammonia. In other words, *the rates of the forward and reverse reactions are equal.*

The concentration of the various species is fixed at equilibrium because product is being *consumed and formed at the same rate*. In other words, the reaction continues indefinitely (dynamic), but the concentration of products and reactants is fixed (equilibrium). This is a *dynamic equilibrium*. The composition of this reaction mixture as a function of time is depicted in Figure 7.13.

For systems such as the ammonia/hydrogen/nitrogen equilibrium, an **equilibrium constant** expression can be written; it summarizes the relationship between the concentration of reactants and products in an equilibrium reaction.

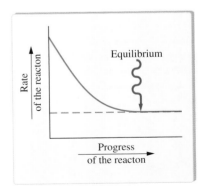

Figure 7.13
The change of the rate of reaction as a function of time. The rate of reaction, initially rapid, decreases as the concentration of reactant decreases and approaches a limiting value at equilibrium.

Products of the overall equilibrium reaction are in the numerator, and *reactants* are in the denominator.

[ ] represents molar concentration, *M*.

The Generalized Equilibrium-Constant Expression for a Chemical Reaction

We write the general form of an equilibrium chemical reaction as

$$aA + bB \rightleftarrows cC + dD$$

in which A and B represent reactants, C and D represent products, and a, b, c, and d are the coefficients of the balanced equation. The equilibrium constant expression for this general case is

$$K_{eq} = \frac{[C]^c[D]^d}{[A]^a[B]^b}$$

For the ammonia system, it follows that the appropriate equilibrium expression is:

$$K_{eq} = \frac{[NH_3]^2}{[N_2][H_2]^3}$$

It does not matter what initial amounts (concentrations) of reactants or products we choose. When the system reaches equilibrium, the calculated value of K_{eq} will not change. The magnitude of K_{eq} can be altered only by changing the temperature. Thus K_{eq} is temperature dependent. The chemical industry uses this fact to advantage by choosing a reaction temperature that will maximize the yield of a desired product.

Question 7.21

How could one determine when a reaction has reached equilibrium?

Question 7.22

Does the attainment of equilibrium imply that no further change is taking place in the system?

LEARNING GOAL

The exponents corresponds to the *coefficients* of the balanced equation.

Writing Equilibrium-Constant Expressions

An equilibrium-constant expression can be written only after a correct, balanced chemical equation that describes the equilibrium system has been developed. A balanced equation is essential because the *coefficients* in the equation become the *exponents* in the equilibrium-constant expression.

Each chemical reaction has a unique equilibrium constant value at a specified temperature. Equilibrium constants listed in the chemical literature are often reported at 25°C, to allow comparison of one system with any other. For any equilibrium reaction, the value of the equilibrium constant changes with temperature.

The brackets represent molar concentration or molarity; recall that molarity has units of mol/L. Although the equilibrium constant may have units (owing to the units on each concentration term), by convention units are usually not used. In our discussion of equilibrium, all equilibrium constants are shown as *unitless*.

A properly written equilibrium-constant expression may not include all of the terms in the chemical equation upon which it is based. Only the concentration of gases and substances in solution are shown, because their concentrations can change. Concentration terms for liquids and solids are *not* shown. The concentration of a liquid is constant. Most often, the liquid is the solvent for the reaction under consideration. A solid also has a fixed concentration and, for solution reactions, is not really a part of the solution. When a solid is formed it exists as a solid phase in contact with a liquid phase (the solution).

EXAMPLE 7.6 Writing an Equilibrium-Constant Expression

Write an equilibrium-constant expression for the reversible reaction:

$$H_2(g) + F_2(g) \rightleftharpoons 2HF(g)$$

Solution

Inspection of the chemical equation reveals that no solids or liquids are present. Hence all reactants and products appear in the equilibrium-constant expression:

The numerator term is the product term $[HF]^2$.

The denominator terms are the reactants $[H_2]$ and $[F_2]$.

Note that each term contains an exponent identical to the corresponding coefficient in the balanced equation. Arranging the numerator and denominator terms as a fraction and setting the fraction equal to K_{eq} yields

$$K_{eq} = \frac{[HF]^2}{[H_2][F_2]}$$

EXAMPLE 7.7 Writing an Equilibrium-Constant Expression

Write an equilibrium-constant expression for the reversible reaction:

$$MnO_2(s) + 4H^+(aq) + 2Cl^-(aq) \rightleftharpoons Mn^{2+}(aq) + Cl_2(g) + 2H_2O(l)$$

Solution

MnO_2 is a solid and H_2O, although a product, is negligible compared with the water solvent. Thus they are not written in the equilibrium-constant expression.

$$\boxed{MnO_2(s)} + 4H^+(aq) + 2Cl^-(aq) \rightleftharpoons Mn^{2+}(aq) + Cl_2(g) + \boxed{2H_2O(l)}$$

— Not a part of the K_{eq} expression —

The numerator term includes the remaining products:

$$[Mn^{2+}] \quad \text{and} \quad [Cl_2]$$

The denominator term includes the remaining reactants:

$$[H^+]^4 \quad \text{and} \quad [Cl^-]^2$$

Note that each exponent is identical to the corresponding coefficient in the chemical equation.
  Arranging the numerator and denominator terms as a fraction and setting the fraction equal to K_{eq} yields

$$K_{eq} = \frac{[Mn^{2+}][Cl_2]}{[H^+]^4[Cl^-]^2}$$

Question 7.23

Write an equilibrium-constant expression for each of the following reversible reactions.

a. $2NO_2(g) \rightleftharpoons N_2(g) + 2O_2(g)$

b. $2H_2O(l) \rightleftharpoons 2H_2(g) + O_2(g)$

Question 7.24

Write an equilibrium-constant expression for each of the following reversible reactions.

a. $2HI(g) \rightleftharpoons H_2(g) + I_2(g)$

b. $PCl_5(s) \rightleftharpoons PCl_3(g) + Cl_2(g)$

Interpreting Equilibrium Constants

What utility does the equilibrium constant have? The reversible arrow in the chemical equation alerts us to the fact that an equilibrium exists. Some measurable quantity of the product and reactant remain. However, there is no indication whether products predominate, reactants predominate, or significant concentrations of both products and reactants are present at equilibrium.

The numerical value of the equilibrium constant provides this additional information. It tells us the extent to which reactants have converted to products. This is important information for anyone who wants to manufacture and sell the product. It also is important to anyone who studies the effect of equilibrium reactions on environmental systems and living organisms.

Although an absolute interpretation of the numerical value of the equilibrium constant depends on the form of the equilibrium-constant expression, the following generalizations are useful:

- K_{eq} greater than 1×10^2. A large numerical value of K_{eq} indicates that the numerator (product term) is much larger than the denominator (reactant term) and that at equilibrium mostly product is present.
- K_{eq} less than 1×10^{-2}. A small numerical value of K_{eq} indicates that the numerator (product term) is much smaller than the denominator (reactant term) and that at equilibrium mostly reactant is present.
- K_{eq} between 1×10^{-2} and 1×10^2. In this case the equilibrium mixture contains significant concentrations of both reactants and products.

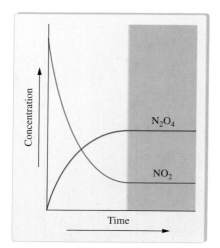

Figure 7.14
The combination reaction of NO_2 molecules produces N_2O_4. Initially, the concentration of reactant (NO_2) diminishes rapidly while the N_2O_4 concentration builds. Eventually, the concentrations of both reactant and product become constant over time (blue area). The equilibrium condition has been attained.

Question 7.25

At a given temperature, the equilibrium constant for a certain reaction is 1×10^{20}. Does this equilibrium favor products or reactants? Why?

Question 7.26

At a given temperature, the equilibrium constant for a certain reaction is 1×10^{-18}. Does this equilibrium favor products or reactants? Why?

Calculating Equilibrium Constants

LEARNING GOAL

The magnitude of the equilibrium constant for a chemical reaction is determined experimentally. The reaction under study is allowed to proceed until the composition of products and reactants no longer changes (Figure 7.14). This may be a matter of seconds, minutes, hours, or even months or years, depending on the rate of the reaction. The reaction mixture is then analyzed to determine the molar concentration of each of the products and reactants. These concentrations are substituted in the equilibrium-constant expression and the equilibrium constant is calculated. The following example illustrates this process.

Section 6.3 describes molar concentration.

7.4 Equilibrium

EXAMPLE 7.8

Calculating an Equilibrium Constant

Hydrogen iodide is placed in a sealed container and allowed to come to equilibrium. The equilibrium reaction is:

$$2HI(g) \rightleftharpoons H_2(g) + I_2(g)$$

and the equilibrium concentrations are:

$$[HI] = 0.54\ M$$
$$[H_2] = 1.72\ M$$
$$[I_2] = 1.72\ M$$

Calculate the equilibrium constant.

Solution

First, write the equilibrium-constant expression:

$$K_{eq} = \frac{[H_2][I_2]}{[HI]^2}$$

Then substitute the equilibrium concentrations of products and reactants to obtain

$$K_{eq} = \frac{[1.72][1.72]}{[0.54]^2} = \frac{2.96}{0.29}$$

$$= 10.1\ \text{or}\ 1.0 \times 10^1\ \text{(two significant figures)}$$

Question 7.27

A reaction chamber contains the following mixture at equilibrium:

$$[NH_3] = 0.25\ M$$
$$[N_2] = 0.11\ M$$
$$[H_2] = 1.91\ M$$

If the reaction is:

$$N_2(g) + 3H_2(g) \rightleftharpoons 2NH_3(g)$$

Calculate the equilibrium constant.

Question 7.28

A reaction chamber contains the following mixture at equilibrium:

$$[H_2] = 0.22\ M$$
$$[S_2] = 1.0 \times 10^{-6}\ M$$
$$[H_2S] = 0.80\ M$$

If the reaction is:

$$2H_2(g) + S_2(g) \rightleftharpoons 2H_2S(g)$$

Calculate the equilibrium constant.

LeChatelier's Principle

LEARNING GOAL 10

In the nineteenth century the French chemist LeChatelier discovered that changes in equilibrium depend on the amount of "stress" applied to the system. The stress may take the form of an increase or decrease of the temperature of the system at equilibrium or perhaps a change in the amount of reactant or product present in a fixed volume (the concentration of reactant or product).

LeChatelier's principle states that if a stress is placed on a system at equilibrium, the system will respond by altering the equilibrium composition in such a way as to minimize the stress.

Consider the equilibrium situation discussed earlier:

$$N_2(g) + 3H_2(g) \rightleftharpoons 2NH_3(g)$$

If the reactants and products are present in a fixed volume (such as 1L) and more NH_3 (the *product*) is introduced into the container, the system will be stressed—the equilibrium will be disturbed. The system will try to alleviate the stress (as we all do) by *removing* as much of the added material as possible. How can it accomplish this? By converting some NH_3 to H_2 and N_2. The equilibrium shifts to the left, and the dynamic equilibrium is soon reestablished.

Adding extra H_2 or N_2 would apply the stress to the other side of the equilibrium. To minimize the stress, the system would "use up" some of the excess H_2 or N_2 to make product, NH_3. The equilibrium would shift to the right.

In summary,

$$N_2(g) + 3H_2(g) \rightleftharpoons 2NH_3(g)$$

Product introduced: Equilibrium shifted ←

Reactant introduced: Equilibrium shifted →

What would happen if some of the ammonia molecules were *removed* from the system? The loss of ammonia represents a stress on the system; to relieve that stress, the ammonia would be replenished by the reaction of hydrogen and nitrogen. The equilibrium would shift to the right.

Effect of Concentration

Addition of extra product or reactant to a fixed reaction volume is just another way of saying that we have increased the concentration of product or reactant. Removal of material from a fixed volume decreases the concentration. Therefore changing the concentration of one or more components of a reaction mixture is a way to alter the equilibrium composition of an equilibrium mixture (Figure 7.15). Let's look at some additional experimental variables that may change equilibrium composition.

Addition of products or reactants may have a profound effect on the composition of a reaction mixture but does not affect the value of the equilibrium constant.

Figure 7.15

The effect of concentration on equilibrium position of the reaction:

$FeSCN^{2+}(aq) \rightleftharpoons Fe^{3+}(aq) + SCN^-(aq)$
(red)   (yellow)   (colorless)

Solution (a) represents this reaction at equilibrium; addition of SCN^- shifts the equilibrium to the left (b) intensifying the red color. Removal of SCN^- shifts the equilibrium to the right (c) shown by the disappearance of the red color.

(a)   (b)   (c)

Effect of Heat

The change in equilibrium composition caused by the addition or removal of heat from an equilibrium mixture can be explained by treating heat as a product or reactant. The reaction of nitrogen and hydrogen is an exothermic reaction:

$$N_2(g) + 3H_2(g) \rightleftharpoons 2NH_3(g) + 22 \text{ kcal}$$

Adding heat to the reaction is similar to increasing the amount of product. The equilibrium will shift to the left, increasing the amounts of N_2 and H_2 and decreasing the amount of NH_3. If the reaction takes place in a fixed volume, the concentrations of N_2 and H_2 increase and the NH_3 concentration decreases.

Removal of heat produces the reverse effect. More ammonia is produced from N_2 and H_2, and the concentrations of these reactants must decrease.

In the case of an endothermic reaction such as

$$39 \text{ kcal} + 2N_2(g) + O_2(g) \rightleftharpoons 2N_2O(g)$$

addition of heat is analogous to the addition of reactant, and the equilibrium shifts to the right. Removal of heat would shift the reaction to the left, favoring the formation of reactants.

The dramatic effect of heat on the position of equilibrium is shown in Figure 7.16.

Effect of Pressure

Only gases are affected significantly by changes in pressure because gases are free to expand and compress in accordance with Boyle's law. However, liquids and solids are not compressible, so their volumes are unaffected by pressure.

Therefore pressure changes will alter equilibrium composition only in reactions that involve a gas or variety of gases as products and/or reactants. Again, consider the ammonia example,

$$N_2(g) + 3H_2(g) \rightleftharpoons 2NH_3(g)$$

One mole of N_2 and three moles of H_2 (total of four moles of reactants) convert to two moles of NH_3 (two moles of product). An increase in pressure favors a decrease in volume and formation of product. This decrease in volume is made possible by a shift to the right in equilibrium composition. Two moles of ammonia require less volume than four moles of reactant.

A decrease in pressure allows the volume to expand. The equilibrium composition shifts to the left and ammonia decomposes to form more nitrogen and hydrogen.

Expansion and compression of gases and Boyle's law are discussed in Section 5.1.

The industrial process for preparing ammonia, the Haber process, uses pressures of several hundred atmospheres to increase the yield.

Figure 7.16
The effect of heat on equilibrium position. For the reaction:

$CoCl_4^{2-}(aq) + 6H_2O(l) \rightleftharpoons$
(blue)
$\quad\quad Co(H_2O)_6^{2+}(aq) + 4Cl^-(aq)$
$\quad\quad$ (pink)

Heating the solution favors the blue $CoCl_4^{2-}$ species; cooling favors the pink $Co(H_2O)_6^{2+}$ species.

In contrast, the decomposition of hydrogen iodide,

$$2HI(g) \rightleftharpoons H_2(g) + I_2(g)$$

is unaffected by pressure. The number of moles of gaseous product and reactant are identical. No volume advantage is gained by a shift in equilibrium composition.

In summary:

- Pressure affects the equilibrium composition only of reactions that involve at least one gaseous substance.
- Additionally, the relative number of moles of gaseous products and reactants must differ.
- The equilibrium composition will shift to increase the number of moles of gas when the pressure decreases; it will shift to decrease the number of moles of gas when the pressure increases.

Effect of a Catalyst

A catalyst has no effect on the equilibrium composition. A catalyst increases the rates of both forward and reverse reactions to the same extent. The equilibrium composition *and* equilibrium concentration do not change when a catalyst is used, but the equilibrium composition is achieved in a shorter time. The role of a solid-phase catalyst in the synthesis of ammonia is shown in Figure 7.12.

EXAMPLE 7.9 Predicting Changes in Equilibrium Composition

Earlier in this section we considered the geologically important reaction that occurs in rock and soil.

$$Ca^{2+}(aq) + 2HCO_3^-(aq) \rightleftharpoons CaCO_3(s) + CO_2(aq) + H_2O(l)$$

Predict the effect on the equilibrium composition for each of the following changes.

a. The $[Ca^{2+}]$ is increased.
b. The amount of $CaCO_3$ is increased.
c. The $[HCO_3^-]$ is decreased.
d. A catalyst is added.

Solution

a. The concentration of reactant increases; the equilibrium shifts to the right, and more products are formed.
b. $CaCO_3$ is a solid; solids are not written in the equilibrium-constant expression, so there is no effect on the equilibrium composition.
c. The concentration of reactant decreases; the equilibrium shifts to the left, and more reactants are formed.
d. A catalyst has no effect on the equilibrium composition.

Question 7.29

For the hypothetical equilibrium reaction

$$A(g) + B(g) \rightleftharpoons C(g) + D(g)$$

predict whether the amount of A in a 5.0-L container would increase, decrease, or remain the same for each of the following changes.

a. Addition of excess B
b. Addition of excess C
c. Removal of some D
d. Addition of a catalyst

Question 7.30

For the hypothetical equilibrium reaction

$$A(g) + B(g) \rightleftharpoons C(g) + D(g)$$

predict whether the amount of A in a 5.0-L container would increase, decrease, or remain the same for each of the following changes.

a. Removal of some B
b. Removal of some C
c. Addition of excess D
d. Removal of a catalyst

SUMMARY

7.1 Thermodynamics

Thermodynamics is the study of energy, work, and heat. Thermodynamics can be applied to the study of chemical reactions because we can determine the quantity of heat flow (by measuring the temperature change) between the *system* and the *surroundings*. *Exothermic reactions* release energy and products that are lower in energy than the reactants. *Endothermic reactions* require energy input. Heat energy is represented as *enthalpy, H*. The energy gain or loss is the change in enthalpy, ΔH, and is one factor that is useful in predicting whether a reaction is spontaneous or nonspontaneous.

Entropy, S, is a measure of the randomness of a system. A random, or disordered system has high entropy; a well-ordered system has low entropy. The change in entropy in a chemical reaction, ΔS, is also a factor in predicting reaction spontaneity.

Free energy, ΔG, incorporates both factors, enthalpy and entropy; as such, it is an absolute predictor of the spontaneity of a chemical reaction.

7.2 Experimental Determination of Energy Change in Reactions

A *calorimeter* measures heat changes (in calories or joules) that occur in chemical reactions.

The *specific heat* of a substance is the number of calories of heat needed to raise the temperature of 1 g of the substance 1 degree Celsius.

The amount of energy per gram of food is referred to as its *fuel value*. Fuel values are commonly reported in units of *nutritional Calories* (1 nutritional Calorie = 1 kcal). A bomb calorimeter is useful for measurement of the fuel value of foods.

7.3 Kinetics

Chemical *kinetics* is the study of the *rate* or speed of a chemical reaction. Energy for reactions is provided by molecular collisions. If this energy is sufficient, bonds may break, and atoms may recombine in a different arrangement, producing product. A collision producing one or more product molecules is termed an effective collision.

The minimum amount of energy needed for a reaction is the *activation energy*. The reaction proceeds from reactants to products through an intermediate state, the *activated complex*.

Experimental conditions influencing the reaction rate include the structure of the reacting species, the concentration of reactants, the temperature of reactants, the physical state of reactants, and the presence or absence of a catalyst.

A *catalyst* increases the rate of a reaction. The catalytic substance undergoes no net change in the reaction, nor does it alter the outcome of the reaction.

7.4 Equilibrium

Many chemical reactions do not completely convert reactants to products. A mixture of products and reactants exists, and its composition will remain constant until the experimental conditions are changed. This mixture is in a state of *chemical equilibrium*. The reaction continues indefinitely (dynamic), but the concentrations of products and reactants are fixed (equilibrium) because the rates of the forward and reverse reactions are equal. This is a *dynamic equilibrium*.

LeChatelier's principle states that if a stress is placed on an equilibrium system, the system will respond by altering the equilibrium in such a way as to minimize the stress.

KEY TERMS

activated complex (7.3)
activation energy (7.3)
calorimetry (7.2)
catalyst (7.3)
dynamic equilibrium (7.4)
endothermic reaction (7.1)
enthalpy (7.1)
entropy (7.1)
equilibrium constant (7.4)
equilibrium reaction (7.4)
exothermic reaction (7.1)
free energy (7.1)
fuel value (7.2)
kinetics (7.3)
LeChatelier's principle (7.4)
nutritional Calorie (7.2)
order of the reaction (7.3)
rate constant (7.3)
rate equation (7.3)
rate of chemical reaction (7.3)
reversible reaction (7.4)
specific heat (7.2)
surroundings (7.1)
system (7.1)
thermodynamics (7.1)

QUESTIONS AND PROBLEMS

Energy and Thermodynamics

Fundamentals

7.31 What is the energy unit most commonly employed in chemistry?
7.32 What energy unit is commonly employed in nutrition science?
7.33 Describe what is meant by an exothermic reaction.
7.34 Describe what is meant by an endothermic reaction.
7.35 The oxidation of fuels (coal, oil, gasoline) are exothermic reactions. Why?
7.36 Provide an explanation for the fact that most decomposition reactions are endothermic but most combination reactions are exothermic.
7.37 Describe how a calorimeter is used to distinguish between exothermic and endothermic reactions.
7.38 Construct a diagram of a coffee-cup calorimeter.
7.39 Why does a calorimeter have a "double-walled" container?
7.40 Explain why the fuel value of foods is an important factor in nutrition science.
7.41 Explain what is meant by the term *free energy*.
7.42 Explain what is meant by the term *specific heat*.
7.43 State the first law of thermodynamics.
7.44 State the second law of thermodynamics.
7.45 Explain what is meant by the term *enthalpy*.
7.46 Explain what is meant by the term *entropy*.

Applications

7.47 5.00 g of octane are burned in a bomb calorimeter containing 2.00×10^2 g H_2O. How much energy, in calories, is released if the water temperature increases 6.00°C?
7.48 0.0500 mol of a nutrient substance is burned in a bomb calorimeter containing 2.00×10^2 g H_2O. If the formula weight of this nutrient substance is 114 g/mol, what is the fuel value (in nutritional Calories) if the temperature of the water increased 5.70°C.
7.49 Calculate the energy released, in joules, in Question 7.47 (recall conversion factors, Chapter 1).
7.50 Calculate the fuel value, in kilojoules, in Question 7.48 (recall conversion factors, Chapter 1).
7.51 Predict whether each of the following processes increases or decreases entropy, and explain your reasoning.
   a. melting of a solid metal
   b. boiling of water
7.52 Predict whether each of the following processes increases or decreases entropy, and explain your reasoning.
   a. burning a log in a fireplace
   b. condensation of water vapor on a cold surface
7.53 Predict whether a reaction with a negative ΔH and a positive ΔS will be spontaneous, nonspontaneous, or temperature dependent. Explain your reasoning.
7.54 Predict whether a reaction with a negative ΔH and a negative ΔS will be spontaneous, nonspontaneous, or temperature dependent. Explain your reasoning.
7.55 Isopropyl alcohol, commonly known as rubbing alcohol, feels cool when applied to the skin. Explain why.
7.56 Energy is required to break chemical bonds during the course of a reaction. When is energy released?

Kinetics

Fundamentals

7.57 Provide an example of a reaction that is extremely slow, taking days, weeks, or years to complete.
7.58 Provide an example of a reaction that is extremely fast, perhaps quicker than the eye can perceive.
7.59 Define the term *activated complex* and explain its significance in a chemical reaction.
7.60 Define and explain the term *activation energy* as it applies to chemical reactions.

Applications

7.61 Distinguish among the terms *rate*, *rate constant*, and *order*.
7.62 Write the rate equation for:

$$CH_4(g) + 2O_2(g) \longrightarrow 2H_2O(l) + CO_2(g)$$

if the order of all reactants is one.
7.63 Will the rate of the reaction in Question 7.62 increase, decrease, or remain the same if the rate constant doubles?
7.64 Will the rate of the reaction in Question 7.62 increase, decrease, or remain the same if the concentration of methane increases?
7.65 Describe the general characteristics of a catalyst.
7.66 Select one enzyme from a later chapter in this book and describe its biochemical importance.
7.67 Sketch a potential energy diagram for a reaction that shows the effect of a catalyst on an exothermic reaction.
7.68 Sketch a potential energy diagram for a reaction that shows the effect of a catalyst on an endothermic reaction.
7.69 Give at least two examples from the life sciences in which the rate of a reaction is critically important.
7.70 Give at least two examples from everyday life in which the rate of a reaction is an important consideration.
7.71 Describe how an increase in the concentration of reactants increases the rate of a reaction.

7.72 Describe how an increase in the temperature of reactants increases the rate of a reaction.

7.73 Write the rate expression for the single-step reaction:
$$N_2O_4(g) \rightleftharpoons 2NO_2(g)$$

7.74 Write the rate expression for the single-step reaction:
$$H_2S(aq) + Cl_2(aq) \rightleftharpoons S(s) + 2HCl(aq)$$

7.75 Describe how a catalyst speeds up a chemical reaction.

7.76 Explain how a catalyst can be involved in a chemical reaction without being consumed in the process.

Equilibrium

Fundamentals

7.77 Does a large equilibrium constant mean that products or reactants are favored?

7.78 Does a large equilibrium constant mean that the reaction must be rapid?

7.79 Provide an example of a physical equilibrium.

7.80 Provide an example of a chemical equilibrium.

7.81 Explain LeChatelier's principle.

7.82 How can LeChatelier's principle help us to increase yields of chemical reactions?

7.83 Describe the meaning of the term *dynamic equilibrium*.

7.84 What is the relationship between the forward and reverse rates for a reaction at equilibrium?

Applications

7.85 Write a valid equilibrium constant for the reaction shown in Question 7.73.

7.86 Write a valid equilibrium constant for the reaction shown in Question 7.74.

7.87 Distinguish between a physical equilibrium and a chemical equilibrium.

7.88 Distinguish between the rate constant and the equilibrium constant for a reaction.

7.89 For the reaction
$$CH_4(g) + Cl_2(g) \rightleftharpoons CH_3Cl(g) + HCl(g) + 26.4 \text{ kcal}$$
predict the effect on the equilibrium (will it shift to the left or to the right, or will there be no change?) for each of the following changes.
a. The temperature is increased.
b. The pressure is increased by decreasing the volume of the container.
c. A catalyst is added.

7.90 For the reaction
$$47 \text{ kcal} + 2SO_3(g) \rightleftharpoons 2SO_2(g) + O_2(g)$$
predict the effect on the equilibrium (will it shift to the left or to the right, or will there be no change?) for each of the following changes.
a. The temperature is increased.
b. The pressure is increased by decreasing the volume of the container.
c. A catalyst is added.

7.91 Label each of the following statements as true or false and explain why.
a. A slow reaction is an incomplete reaction.
b. The rates of forward and reverse reactions are never the same.

7.92 Label each of the following statements as true or false and explain why.
a. A reaction is at equilibrium when no reactants remain.
b. A reaction at equilibrium is undergoing continual change.

7.93 Use LeChatelier's principle to predict whether the amount of PCl_3 in a 1.00-L container is increased, is decreased, or remains the same for the equilibrium
$$PCl_3(g) + Cl_2(g) \rightleftharpoons PCl_5(g) + \text{heat}$$
when each of the following changes is made.
a. PCl_5 is added.
b. Cl_2 is added.
c. PCl_5 is removed.
d. The temperature is decreased.
e. A catalyst is added.

7.94 Use LeChatelier's principle to predict the effects, if any, of each of the following changes on the equilibrium system, described below, in a closed container.
$$C(s) + 2H_2(g) \rightleftharpoons CH_4(g) + 18 \text{ kcal}$$
a. adding more C.
b. adding more H_2.
c. removing CH_4.
d. increasing the temperature.
e. adding a catalyst.

7.95 Will an increase in pressure increase, decrease, or have no effect on the concentration of $H_2(g)$ in the reaction:
$$C(s) + H_2O(g) \rightleftharpoons CO(g) + H_2(g)$$

7.96 Will an increase in pressure increase, decrease, or have no effect on the concentration of $NO(g)$ in the reaction:
$$N_2(g) + O_2(g) \rightleftharpoons 2NO(g)$$

7.97 Write the equilibrium-constant expression for the reaction described in Question 7.95.

7.98 Write the equilibrium-constant expression for the reaction described in Question 7.96.

7.99 True or false: The equilibrium will shift to the right when a catalyst is added to the mixture described in Question 7.95. Explain your reasoning.

7.100 True or false: The equilibrium for an endothermic reaction will shift to the right when the reaction mixture is heated. Explain your reasoning.

7.101 A bottle of carbonated beverage slowly goes "flat" (loses CO_2) after it is opened. Explain, using LeChatelier's principle.

7.102 Carbonated beverages quickly go flat (lose CO_2) when heated. Explain, using LeChatelier's principle.

7.103 Write the equilibrium constant expression for the reaction:
$$N_2(g) + 3H_2(g) \rightleftharpoons 2NH_3(g)$$

7.104 Using the equilibrium constant expression in Question 7.103, calculate the equilibrium constant if:
$$[N_2] = 0.071 \text{ } M$$
$$[H_2] = 9.2 \times 10^{-3} \text{ } M$$
$$[NH_3] = 1.8 \times 10^{-4} \text{ } M$$

CRITICAL THINKING PROBLEMS

1. Predict the sign of ΔG for perspiration evaporating. Would you expect the ΔH term or the ΔS term to be more dominant? Explain your reasoning.

2. Can the following statement ever be true? "Heating a reaction mixture increases the rate of a certain reaction but decreases the yield of product from the reaction." Explain why or why not.

3. Molecules must collide for a reaction to take place. Sketch a model of the orientation and interaction of HI and Cl that is most favorable for the reaction:
$$HI(g) + Cl(g) \longrightarrow HCl(g) + I(g)$$

4. Silver ion reacts with chloride ion to form the precipitate, silver chloride:

$$Ag^+(aq) + Cl^-(aq) \rightleftharpoons AgCl(s)$$

After the reaction reached equilibrium, the chemist filtered 99% of the solid silver chloride from the solution, hoping to shift the equilibrium to the right, to form more product. Critique the chemist's experiment.

5. Human behavior often follows LeChatelier's principle. Provide one example and explain in terms of LeChatelier's principle.

6. A clever device found in some homes is a figurine that is blue on dry, sunny days and pink on damp, rainy days. These figurines are coated with substances containing chemical species that undergo the following equilibrium reaction:

$$Co(H_2O)_6^{2+}(aq) + 4Cl^-(aq) \rightleftharpoons CoCl_4^{2-}(aq) + 6H_2O(l)$$

   a. Which substance is blue?
   b. Which substance is pink?
   c. How is LeChatelier's principle applied here?

7. You have spent the entire morning in a 20°C classroom. As you ride the elevator to the cafeteria, six persons enter the elevator after being outside on a subfreezing day. You suddenly feel chilled. Explain the heat flow situation in the elevator in thermodynamic terms.

GENERAL CHEMISTRY

8

Acids and Bases and Oxidation-Reduction

Solution properties, including color, are often pH dependent.

Learning Goals

1. Identify acids and bases and acid-base reactions.
2. Describe the role of the solvent in acid-base reactions.
3. Write equations describing acid-base dissociation and label the conjugate acid-base pairs.
4. Calculate pH from concentration data.
5. Calculate hydronium and/or hydroxide ion concentration from pH data.
6. Provide examples of the importance of pH in chemical and biochemical systems.
7. Describe the meaning and utility of neutralization reactions.
8. State the meaning of the term *buffer* and describe the applications of buffers to chemical and biochemical systems, particularly blood chemistry.
9. Describe *oxidation* and *reduction*, and describe some practical examples of redox processes.
10. Diagram a voltaic cell and describe its function.
11. Compare and contrast voltaic and electrolytic cells.

Outline

Chemistry Connection:
Drug Delivery

8.1 Acids and Bases
8.2 pH: A Measurement Scale for Acids and Bases
8.3 Reactions Between Acids and Bases

An Environmental Perspective:
Acid Rain

8.4 Acid-Base Buffers

A Medical Perspective:
Control of Blood pH

8.5 Oxidation-Reduction Processes

A Medical Perspective:
Oxidizing Agents for Chemical Control of Microbes

A Medical Perspective:
Electrochemical Reactions in the Statue of Liberty and in Dental Fillings

A Medical Perspective:
Turning the Human Body into a Battery

Chemistry Connection

Drug Delivery

When a doctor prescribes medicine to treat a disease or relieve its symptoms, the medication may be administered in a variety of ways. Drugs may be taken orally, injected into a muscle or a vein, or absorbed through the skin. Specific instructions are often provided to regulate the particular combination of drugs that can or cannot be taken. The diet, both before and during the drug therapy, may be of special importance.

To appreciate why drugs are administered in a specific way, it is necessary to understand a few basic facts about medications and how they interact with the body.

Drugs function by undergoing one or more chemical reactions in the body. Few compounds react in only one way, to produce a limited set of products, even in the simple environment of a beaker or flask. Imagine the number of possible reactions that a drug can undergo in a complex chemical factory like the human body. In many cases a drug can react in a variety of ways other than its intended path. These alternative paths are side reactions, sometimes producing *side effects* such as nausea, vomiting, insomnia, or drowsiness. Side effects may be unpleasant and may actually interfere with the primary function of the drug.

The development of safe, effective medication, with minimal side effects, is a slow and painstaking process and determining the best drug delivery system is a critical step. For example, a drug that undergoes an unwanted side reaction in an acidic solution would not be very effective if administered orally. The acidic digestive fluids in the stomach could prevent the drug from even reaching the intended organ, let alone retaining its potency. The drug could be administered through a vein into the blood; blood is not acidic, in contrast to digestive fluids. In this way the drug may be delivered intact to the intended site in the body, where it is free to undergo its primary reaction.

Drug delivery has become a science in its own right. Pharmacology, the study of drugs and their uses in the treatment of disease, has a goal of creating drugs that are highly selective. In other words, they will undergo only one reaction, the intended reaction. Encapsulation of drugs, enclosing them within larger molecules or collections of molecules, may protect them from unwanted reactions as they are transported to their intended site.

In this chapter we will explore the fundamentals of solutions and solution reactions, including acid-base and oxidation-reduction reactions. Knowing a few basic concepts that govern reactions in beakers will help us to understand the conditions that affect the reactivity of a host of biochemically interesting molecules that we will encounter in later chapters.

Introduction

In this chapter we will learn about two general classes of chemical change: acid-base reactions and oxidation-reduction reactions. Although superficially quite different, their underlying similarity is that both are essentially charge-transfer processes. An acid-base reaction involves the transfer of one or more positively charged units, protons or hydrogen ions; an oxidation-reduction reaction involves the transfer of one or more negatively charged particles, electrons.

The effects of pH on enzyme activity are discussed in Chapter 19.

Acids and bases include some of the most important compounds in nature. Historically, it was recognized that certain compounds, acids, had a sour taste, were able to dissolve some metals, and caused vegetable dyes to change color. Bases have long been recognized by their bitter taste, slippery feel, and corrosive nature. Bases react strongly with acids and cause many metal ions in solution to form a solid precipitate.

Buffers are discussed in Section 8.4.

Digestion of proteins is aided by stomach acid (hydrochloric acid) and many biochemical processes such as enzyme catalysis depend on the proper level of acidity. Indeed, a wide variety of chemical reactions critically depend on the acid-base composition of the solution (Figure 8.1). This is especially true of the biochemical reactions occurring in the cells of our bodies. For this reason the level of acidity must be very carefully regulated. This is done with substances called buffers.

Oxidation-reduction processes are also common in living systems. Respiration is driven by oxidation-reduction reactions. Additionally, oxidation-reduction reactions generate heat that warms our homes and workplaces and fuels our industrial civilization. Moreover, oxidation-reduction is the basis for battery design. Batteries are found in automobiles and electronic devices such as cameras and radios, and are even implanted in the human body to regulate heart rhythm.

8.1 Acids and Bases

LEARNING GOAL

The properties of acids and bases are related to their chemical structure. All acids have common characteristics that enable them to increase the hydrogen ion concentration in water. All bases lower the hydrogen ion concentration in water.

Two theories, one developed from the other, help us to understand the unique chemistry of acids and bases.

Arrhenius Theory of Acids and Bases

One of the earliest definitions of acids and bases is the **Arrhenius theory.** According to this theory, an **acid,** dissolved in water, dissociates to form *hydrogen ions or protons* (H^+), and a **base,** dissolved in water, dissociates to form *hydroxide ions* (OH^-). For example, hydrochloric acid dissociates in solution according to the reaction

$$HCl(aq) \longrightarrow H^+(aq) + Cl^-(aq)$$

Sodium hydroxide, a base, produces hydroxide ions in solution:

$$NaOH(aq) \longrightarrow Na^+(aq) + OH^-(aq)$$

The Arrhenius theory satisfactorily explains the behavior of many acids and bases. However, a substance such as ammonia, NH_3, has basic properties but cannot be an Arrhenius base, because it contains no OH^-. The *Brønsted-Lowry theory* explains this mystery and gives us a broader view of acid-base theory by considering the central role of the solvent in the dissociation process.

Brønsted-Lowry Theory of Acids and Bases

The **Brønsted-Lowry theory** defines an **acid** as a proton (H^+) donor and a **base** as a proton acceptor.

Hydrochloric acid in solution *donates* a proton to the solvent water thus behaving as a Brønsted-Lowry acid:

$$HCl(aq) + H_2O(l) \longrightarrow H_3O^+(aq) + Cl^-(aq)$$

H_3O^+ is referred to as the hydrated proton or **hydronium ion.**

The basic properties of ammonia are clearly accounted for by the Brønsted-Lowry theory. Ammonia *accepts* a proton from the solvent water, producing OH^-. An equilibrium mixture of NH_3, H_2O, NH_4^+, and OH^- results.

$$NH_3(aq) + H-OH(l) \rightleftharpoons NH_4^+(aq) + OH^-(aq)$$

For aqueous solutions, the Brønsted-Lowry theory adequately describes the behavior of acids and bases. We shall limit our discussion of acid-base chemistry to aqueous solutions and use the Brønsted-Lowry definition described here.

Figure 8.1
The yellow solution on the left, containing CrO_4^{2-} (chromate ion), was made acidic producing the reddish brown solution on the right. The principal component in solution is now $Cr_2O_7^{2-}$. Addition of base to this solution removes H^+ ions and regenerates the yellow CrO_4^{2-}. This is an example of an acid-base dependent chemical equilibrium.

Acid-Base Properties of Water

LEARNING GOAL 2

The role that the solvent, water, plays in acid-base reactions is noteworthy. In the example above, the water molecule accepts a proton from the HCl molecule. The water is behaving as a proton acceptor, a base.

However, when water is a solvent for ammonia (NH_3), a base, the water molecule donates a proton to the ammonia molecule. The water, in this situation, is acting as a proton donor, an acid.

Water, owing to the fact that it possesses *both* acid and base properties, is termed **amphiprotic**. Water is the most commonly used solvent for acids and bases. Solute-solvent interactions between water and acids or bases promote both the solubility and the dissociation of acids and bases.

Acid and Base Strength

Concentration of solutions is discussed in Section 6.3.

The terms *acid or base strength* and *acid or base concentration* are easily confused. *Strength* is a measure of the *degree of dissociation* of an acid or base in solution, independent of its concentration. The degree of dissociation is the fraction of acid or base molecules that produces ions in solution. Concentration, as we have learned, refers to the amount of solute (in this case, the amount of acid or base) per quantity of solution.

The concentration of an acid or base does affect the degree of dissociation. However, the major factor in determining the degree of dissociation is the strength of the acid or base.

The strength of acids and bases in water depends on the extent to which they react with the solvent, water. Acids and bases are classified as *strong* when the reaction with water is virtually 100% complete and as *weak* when the reaction with water is much less than 100% complete.

Important strong acids include:

Hydrochloric acid	$HCl(aq) + H_2O(l) \longrightarrow H_3O^+(aq) + Cl^-(aq)$
Nitric acid	$HNO_3(aq) + H_2O(l) \longrightarrow H_3O^+(aq) + NO_3^-(aq)$
Sulfuric acid	$H_2SO_4(aq) + H_2O(l) \longrightarrow H_3O^+(aq) + HSO_4^-(aq)$

Reversibility of reactions is discussed in Section 7.4.

Note that the equation for the dissociation of each of these acids is written with a single arrow. This indicates that the reaction has little or no tendency to proceed in the reverse direction to establish equilibrium. Virtually all of the acid molecules are dissociated to form ions.

All common strong bases are *metal hydroxides*. Strong bases completely dissociate in aqueous solution to produce hydroxide ions and metal cations. Of the common metal hydroxides, only NaOH and KOH are soluble in water and are readily usable strong bases:

Sodium hydroxide	$NaOH(aq) \longrightarrow Na^+(aq) + OH^-(aq)$
Potassium hydroxide	$KOH(aq) \longrightarrow K^+(aq) + OH^-(aq)$

Weak acids and weak bases dissolve in water principally in the molecular form. Only a small percentage of the molecules dissociate to form the hydronium or hydroxide ion.

Two important weak acids are:

The double arrow implies an equilibrium between dissociated and undissociated species.

Acetic acid	$CH_3COOH(aq) + H_2O(l) \rightleftharpoons H_3O^+(aq) + CH_3COO^-(aq)$
Carbonic acid	$H_2CO_3(aq) + H_2O(l) \rightleftharpoons H_3O^+(aq) + HCO_3^-(aq)$

We have already mentioned the most common weak base, ammonia. Many organic compounds function as weak bases. Several examples of weak bases follow:

Many organic compounds are acid or base properties. The chemistry of organic acids and bases will be discussed in Chapter 14 (Carboxylic Acids and Carboxylic Acid Derivatives) and 15 (Amines and Amides).

Pyridine	$C_5H_5N(aq) + H_2O(l) \rightleftharpoons C_5H_5NH^+(aq) + OH^-(aq)$
Aniline	$C_6H_5NH_2(aq) + H_2O(l) \rightleftharpoons C_6H_5NH_3^+(aq) + OH^-(aq)$
Methylamine	$CH_3NH_2(aq) + H_2O(l) \rightleftharpoons CH_3NH_3^+(aq) + OH^-(aq)$

The fundamental chemical difference between strong and weak acids or bases is their equilibrium ion concentration. A strong acid, such as HCl, does not, in aqueous solution, exist to any measurable degree in equilibrium with its ions, H_3O^+ and Cl^-. On the other hand, a weak acid, such as acetic acid, establishes a dynamic equilibrium with its ions, H_3O^+ and CH_3COO^-.

Conjugate Acids and Bases

3 LEARNING GOAL

The Brønsted-Lowry theory contributed several fundamental ideas that broadened our understanding of solution chemistry. First of all, an acid-base reaction is a charge-transfer process. Second, the transfer process usually involves the solvent. Water may, in fact, accept or donate a proton. Last, and perhaps most important, the acid-base reaction is seen as a reversible process. This leads to the possibility of a reversible, dynamic equilibrium (see Section 7.4).

Consequently, any acid-base reaction can be represented by the general equation

$$HA + B \rightleftharpoons BH^+ + A^-$$
$$\text{(acid)} \quad \text{(base)}$$

In the forward reaction, the acid (HA) donates a proton (H^+) to the base (B) leading to the formation of BH^+ and A^-. However, in the reverse reaction, it is the BH^+ that behaves as an acid; it donates its "extra" proton to A^-. A^- is therefore a base in its own right because it accepts the proton.

These *product* acids and bases are termed *conjugate acids and bases*.

A **conjugate acid** is the species formed when a base accepts a proton.

A **conjugate base** is the species formed when an acid donates a proton.

The acid and base on the opposite sides of the equation are collectively termed a **conjugate acid-base pair**. In the above equation:

BH^+ is the conjugate acid of the base B.

A^- is the conjugate base of the acid HA.

B and BH^+ constitute a conjugate acid-base pair.

HA and A^- constitute a conjugate acid-base pair.

Rewriting our model equation:

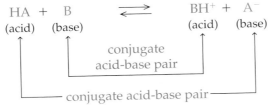

Although we show the forward and reverse arrows to indicate the reversibility of the reaction, seldom are the two processes "equal but opposite." One reaction, either forward or reverse, is usually favored. Consider the reaction of hydrochloric acid in water:

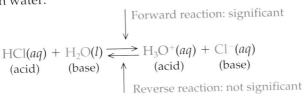

Chapter 8 Acids and Bases and Oxidation-Reduction

HCl is a much better proton donor than H_3O^+. Consequently the forward reaction predominates, the reverse reaction is inconsequential, and hydrochloric acid is termed a *strong acid*. As we learned in Chapter 7, reactions in which the forward reaction is strongly favored have large equilibrium constants. The dissociation of hydrochloric acid is so favorable that we describe it as 100% dissociated and use only a single forward arrow to represent its behavior in water:

$$HCl(aq) + H_2O(l) \longrightarrow H_3O^+(aq) + Cl^-(aq)$$

The degree of dissociation, or strength, of acids and bases has a profound influence on their aqueous chemistry. For example, vinegar (a 5% [W/V] solution of acetic acid in water) is a consumable product; aqueous hydrochloric acid in water is not. Why? Acetic acid is a *weak* acid and, as a result, a dilute solution does no damage to the mouth and esophagus. The following section looks at the strength of acids and bases in solution in more detail.

Question 8.1

Write an equation for the reaction of each of the following with water:

a. HF (a weak acid)
b. NH_3 (a weak base)

Question 8.2

Write an equation for the reaction of each of the following with water:

a. H_2S (a weak acid)
b. CH_3NH_2 (a weak base)

Question 8.3

Select the conjugate acid-base pairs for each reaction in Question 8.1.

Question 8.4

Select the conjugate acid-base pairs for each reaction in Question 8.2.

The relative strength of an acid or base is determined by the ease with which it donates or accepts a proton. Acids with the greatest proton-donating capability (strongest acids) have the weakest conjugate bases. Good proton acceptors (strong bases) have weak conjugate acids. This relationship is clearly indicated in Figure 8.2. This figure can be used to help us compare and predict relative acid-base strength.

EXAMPLE 8.1 Predicting Relative Acid-Base Strengths

a. Write the conjugate acid of HS^-.

Solution

The conjugate acid may be constructed by adding a proton (H^+) to the base structure, consequently, H_2S.

b. Using Figure 8.2, identify the stronger base, HS^- or F^-.

Continued—

EXAMPLE 8.1 —Continued

Solution

HS⁻ is the stronger base because it is located farther down the right-hand column.

c. Using Figure 8.2 identify the stronger acid, H_2S or HF.

Solution

HF is the stronger acid because its conjugate base is weaker *and* because it is located farther up the left-hand column.

Question 8.5

In each pair, identify the stronger acid.

a. H_2O or NH_4^+
b. H_2SO_4 or H_2SO_3

Question 8.6

In each pair, identify the stronger base.

a. CO_3^{2-} or PO_4^{3-}
b. HCO_3^- or HPO_4^{2-}

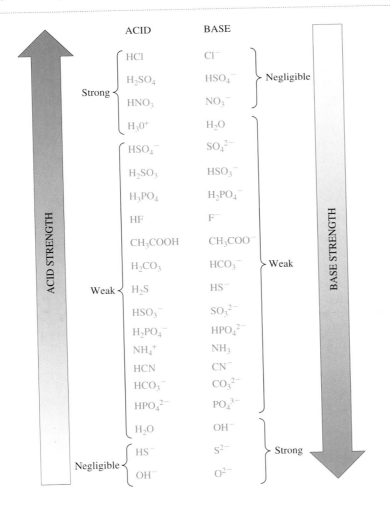

Figure 8.2
Conjugate acid-base pairs. Strong acids have weak conjugate bases; strong bases have weak conjugate acids. Note that, in every case, the conjugate base has one fewer H⁺ than the corresponding conjugate acid.

Solutions of acids and bases used in the laboratory must be handled with care. Acids burn because of their exothermic reaction with water present on and in the skin. Bases react with proteins, which are principal components of the skin and eyes.

Such solutions are more hazardous if they are strong or concentrated. A strong acid or base produces more H_3O^+ or OH^- than does the corresponding weak acid or base. More-concentrated acids or bases contain more H_3O^+ or OH^- than do less-concentrated solutions of the same strength.

The Dissociation of Water

Solutions of electrolytes are discussed in Section 6.3.

Aqueous solutions of acids and bases are electrolytes. The dissociation of the acid or base produces ions that can conduct an electrical current. As a result of the differences in the degree of dissociation, *strong acids and bases are strong electrolytes; weak acids and bases are weak electrolytes.* The conductivity of these solutions is principally dependent on the solute and not the solvent (water).

Although pure water is virtually 100% molecular, a small number of water molecules do ionize. This process occurs by the transfer of a proton from one water molecule to another, producing a hydronium ion and a hydroxide ion:

$$H_2O(l) + H_2O(l) \rightleftharpoons H_3O^+(aq) + OH^-(aq)$$

This process is the **autoionization,** or self-ionization, of water. Water is therefore a *very* weak electrolyte and a very poor conductor of electricity. Water has *both* acid and base properties; dissociation produces both the hydronium and hydroxide ion.

Pure water at room temperature has a hydronium ion concentration of 1.0×10^{-7} M. One hydroxide ion is produced for each hydronium ion. Therefore, the hydroxide ion concentration is also 1.0×10^{-7} M. Molar equilibrium concentration is conveniently indicated by brackets around the species whose concentration is represented:

$$[H_3O^+] = 1.0 \times 10^{-7} M$$
$$[OH^-] = 1.0 \times 10^{-7} M$$

The product of hydronium and hydroxide ion concentration in pure water is referred to as the **ion product for water,** symbolized by K_w.

$$K_w = \text{ion product} = [H_3O^+][OH^-]$$
$$= [1.0 \times 10^{-7}][1.0 \times 10^{-7}]$$
$$= 1.0 \times 10^{-14}$$

The ion product is constant because its value does not depend on the nature or concentration of the solute, as long as the temperature does not change. The ion product is a temperature-dependent quantity.

The nature and concentration of the solutes added to water do alter the relative concentrations of H_3O^+ and OH^- present, but the product, $[H_3O^+][OH^-]$, always equals 1.0×10^{-14} at 25°C. This relationship is the basis for a scale that is useful in the measurement of the level of acidity or basicity of solutions. This scale, the pH scale, is discussed next.

8.2 pH: A Measurement Scale for Acids and Bases

A Definition of pH

LEARNING GOAL 3

The **pH scale** gauges the hydronium ion concentration and reflects the degree of acidity or basicity of a solution. The pH scale is somewhat analogous to the temperature scale used for assignment of relative levels of hot or cold. The tempera-

8.2 pH: A Measurement Scale for Acids and Bases

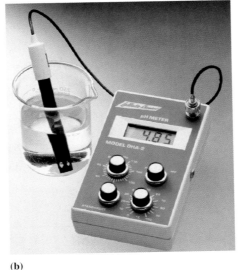

(a) (b)

Figure 8.3
The measurement of pH. (a) A strip of test paper impregnated with indicator (a material that changes color as the acidity of the surroundings changes) is put in contact with the solution of interest. The resulting color is matched with a standard color chart (colors shown as a function of pH) to obtain the approximate pH. (b) A pH meter uses a sensor (a pH electrode) that develops an electrical potential that is proportional to the pH of the solution.

ture scale was developed to allow us to indicate how cold or how hot an object is. The pH scale specifies how acidic or how basic a solution is. The pH scale has values that range from 0 (very acidic) to 14 (very basic). A pH of 7, the middle of the scale, is neutral, neither acidic nor basic.

To help us to develop a concept of pH, let's consider the following:

- Addition of an acid (proton donor) to water *increases* the [H_3O^+] and decreases the [OH^-].
- Addition of a base (proton acceptor) to water *decreases* the [H_3O^+] by increasing the [OH^-].
- [H_3O^+] = [OH^-] when *equal* amounts of acid and base are present.
- In all three cases, [H_3O^+][OH^-] = 1.0×10^{-14} = the ion product for water at 25°C.

pH values greater than 14 and less than zero are possible, but largely meaningless, due to ion association characteristics of very concentrated solutions.

Measuring pH

The pH of a solution can be calculated if the concentration of either H_3O^+ or OH^- is known. Alternatively, measurement of pH allows the calculation of H_3O^+ or OH^- concentration. The pH of aqueous solutions may be approximated by using indicating paper (pH paper) that develops a color related to the solution pH. Alternatively, a pH meter can give us a much more exact pH measurement. A sensor measures an electrical property of a solution that is proportional to pH (Figure 8.3).

Calculating pH

One of our objectives in this chapter is to calculate the pH of a solution when the hydronium or hydroxide ion concentration is known, and to calculate [H_3O^+] or [OH^-] from the pH.

The pH of a solution is defined as the negative logarithm of the molar concentration of the hydronium ion:

$$pH = -\log [H_3O^+]$$

4 LEARNING GOAL

5 LEARNING GOAL

EXAMPLE 8.2

Calculating pH from Acid Molarity

Calculate the pH of a 1.0×10^{-3} M solution of HCl.

Continued—

EXAMPLE 8.2 —Continued

Solution

HCl is a strong acid. If 1 mol HCl dissolves and dissociates in 1L of aqueous solution, it produces 1 mol H_3O^+ (a 1 M solution of H_3O^+). Therefore a 1.0×10^{-3} M HCl solution has $[H_3O^+] = 1.0 \times 10^{-3}$ M, and

$$pH = -\log [H_3O^+]$$
$$= -\log [1.0 \times 10^{-3}]$$
$$= -[-3.00] = 3.00$$

EXAMPLE 8.3 Calculating [H₃O⁺] from pH

Calculate the $[H_3O^+]$ of a solution of hydrochloric acid with pH = 4.00.

Solution

We use the pH expression:

$$pH = -\log [H_3O^+]$$
$$4.00 = -\log [H_3O^+]$$

Multiplying both sides of the equation by −1, we get

$$-4.00 = \log [H_3O^+]$$

Taking the antilogarithm of both sides (the reverse of a logarithm), we have

$$\text{antilog} -4.00 = [H_3O^+]$$

The antilog is the exponent of 10; therefore

$$1.0 \times 10^{-4} M = [H_3O^+]$$

EXAMPLE 8.4 Calculating the pH of a Base

Calculate the pH of a 1.0×10^{-5} M solution of NaOH.

Solution

NaOH is a strong base. If 1 mol NaOH dissolves and dissociates in 1L of aqueous solution, it produces 1 mol OH^- (a 1 M solution of OH^-). Therefore a 1.0×10^{-5} M NaOH solution has $[OH^-] = 1.0 \times 10^{-5}$ M. To calculate pH, we need $[H_3O^+]$. Recall that

$$[H_3O^+][OH^-] = 1.0 \times 10^{-14}$$

Solving this equation for $[H_3O^+]$,

$$[H_3O^+] = \frac{1.0 \times 10^{-14}}{[OH^-]}$$

Continued—

EXAMPLE 8.4 —Continued

substituting the information provided in the problem,

$$= \frac{1.0 \times 10^{-14}}{1.0 \times 10^{-5}}$$

$$= 1.0 \times 10^{-9} \, M$$

The solution is now similar to that in Example 8.2:

$$pH = -\log [H_3O^+]$$
$$= -\log [1.0 \times 10^{-9}]$$
$$= 9.00$$

EXAMPLE 8.5 Calculating Both Hydronium and Hydroxide Ion Concentrations from pH

Calculate the $[H_3O^+]$ and $[OH^-]$ of a sodium hydroxide solution with a pH = 10.00.

Solution

First, calculate $[H_3O^+]$:

$$pH = -\log [H_3O^+]$$
$$10.00 = -\log [H_3O^+]$$
$$-10.00 = \log [H_3O^+]$$
$$\text{antilog} -10 = [H_3O^+]$$
$$1.0 \times 10^{-10} \, M = [H_3O^+]$$

To calculate the $[OH^-]$, we need to solve for $[OH^-]$ by using the following expression:

$$K_w = [H_3O^+][OH^-] = 1.0 \times 10^{-14}$$

$$[OH^-] = \frac{1.0 \times 10^{-14}}{[H_3O^+]}$$

Substituting the $[H_3O^+]$ from the first part, we have

$$[OH^-] = \frac{1.0 \times 10^{-14}}{[1.0 \times 10^{-10}]}$$

$$= 1.0 \times 10^{-4} \, M$$

Often, the pH or $[H_3O^+]$ will not be a whole number (pH = 1.5, pH = 5.3, $[H_3O^+] = 1.5 \times 10^{-3}$ and so forth). With the advent of inexpensive and versatile calculators, calculations with noninteger numbers pose no great problems. Consider Examples 8.6 and 8.7.

EXAMPLE 8.6 Calculating pH with Noninteger Numbers

Calculate the pH of a sample of lake water that has a $[H_3O^+] = 6.5 \times 10^{-5}$ M.

Solution

$$pH = -\log[H_3O^+]$$
$$= -\log[6.5 \times 10^{-5}]$$
$$= 4.19$$

The pH, 4.19, is low enough to suspect acid rain. (See An Environmental Perspective: Acid Rain in this chapter.)

EXAMPLE 8.7 Calculating [H₃O⁺] from pH

The measured pH of a sample of lake water is 6.40. Calculate $[H_3O^+]$.

Solution

An alternative mathematical form of

$$pH = -\log[H_3O^+]$$

is the expression

$$[H_3O^+] = 10^{-pH}$$

which we will use when we must solve for $[H_3O^+]$.

$$[H_3O^+] = 10^{-6.40}$$

Performing the calculation on your calculator results in 3.98×10^{-7} or 4.0×10^{-7} M = $[H_3O^+]$.

Examples 8.2–8.7 illustrate the most frequently used pH calculations. It is important to remember that in the case of a base you must convert the $[OH^-]$ to $[H_3O^+]$, using the expression for the ion product for the solvent, water. It is also useful to remember the following points:

- The pH of a 1 M solution of any strong acid is 0.
- The pH of a 1 M solution of any strong base is 14.
- Each tenfold change in concentration changes the pH by one unit. A tenfold change in concentration is equivalent to moving the decimal point one place.
- A *decrease* in acid concentration *increases* the pH.
- A *decrease* in base concentration *decreases* the pH.

Figure 8.4 provides a convenient overview of solution pH.

Question 8.7

Calculate the $[OH^-]$ of the solution in Example 8.2.

Question 8.8

Calculate the $[OH^-]$ of the solution in Example 8.3.

8.2 pH: A Measurement Scale for Acids and Bases

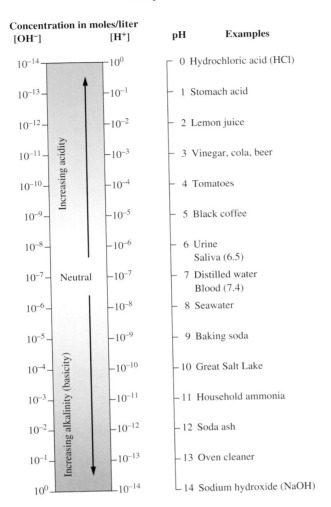

Figure 8.4
The pH scale. A pH of 7 is neutral ($[H_3O^+] = [OH^-]$). Values less than 7 are acidic (H_3O^+ predominates) and values greater than 7 are basic (OH^- predominates).

Question 8.9
Calculate the pH corresponding to a solution of sodium hydroxide with a $[OH^-]$ of 1.0×10^{-2} M.

Question 8.10
Calculate the pH corresponding to a solution of sodium hydroxide with a $[OH^-]$ of 1.0×10^{-6} M.

Question 8.11
Calculate the $[H_3O^+]$ corresponding to pH = 8.50.

Question 8.12
Calculate the $[H_3O^+]$ corresponding to pH = 4.50.

The Importance of pH and pH Control

LEARNING GOAL

Solution pH and pH control play a major role in many facets of our lives. Consider a few examples:

- *Agriculture:* Crops grow best in a soil of proper pH. Proper fertilization involves the maintenance of a suitable pH.
- *Physiology:* If the pH of our blood were to shift by one unit, we would die. Many biochemical reactions in living organisms are extremely pH dependent.
- *Industry:* From manufacture of processed foods to the manufacture of automobiles, industrial processes often require rigorous pH control.
- *Municipal services:* Purification of drinking water and treatment of sewage must be carried out at their optimum pH.
- *Acid rain:* Nitric acid and sulfuric acid, resulting largely from the reaction of components of vehicle emissions and electric power generation (nitrogen and sulfur oxides) with water, are carried down by precipitation and enter aquatic systems (lakes and streams), lowering the pH of the water. A less than optimum pH poses serious problems for native fish populations.

See An Environmental Perspective: Acid Rain in this chapter.

The list could continue on for many pages. However, in summary, any change that takes place in aqueous solution generally has at least some pH dependence.

8.3 Reactions Between Acids and Bases

Neutralization

LEARNING GOAL

The reaction of an acid with a base to produce a salt and water is referred to as **neutralization.** In the strictest sense, neutralization requires equal numbers of moles of H_3O^+ and OH^- to produce a neutral solution (no excess acid or base).

Consider the reaction of a solution of hydrochloric acid and sodium hydroxide:

$$HCl(aq) + NaOH(aq) \longrightarrow NaCl(aq) + H_2O(l)$$
$$\text{Acid} \qquad \text{Base} \qquad\qquad \text{Salt} \qquad \text{Water}$$

Equation balancing is discussed in Chapter 4.

Our objective is to make the balanced equation represent the process actually occurring. We recognize that HCl, NaOH, and NaCl are dissociated in solution:

$$H^+(aq) + Cl^-(aq) + Na^+(aq) + OH^-(aq) \longrightarrow Na^+(aq) + Cl^-(aq) + H_2O(l)$$

We further know that Na^+ and Cl^- are unchanged in the reaction; they are termed *spectator ions.* If we write only those components that actually change, ignoring the spectator ions, we produce a *net, balanced ionic equation:*

$$H^+(aq) + OH^-(aq) \longrightarrow H_2O(l)$$

You can find further information online at www.mhhe.com/denniston5e in "Writing Net Ionic Equations."

If we realize that the H^+ occurs in aqueous solution as the hydronium ion, H_3O^+, the most correct form of the net, balanced ionic equation is

$$H_3O^+(aq) + OH^-(aq) \longrightarrow 2H_2O(l)$$

The equation for any strong acid/strong base neutralization reaction is the same as this equation.

A neutralization reaction may be used to determine the concentration of an unknown acid or base solution. The technique of **titration** involves the addition of measured amounts of a **standard solution** (one whose concentration is known with certainty) to neutralize the second, unknown solution. From the volumes of the two solutions and the concentration of the standard solution the concentration of the unknown solution may be determined. Consider the following application.

EXAMPLE 8.8

Determining the Concentration of a Solution of Hydrochloric Acid

Step 1. A known volume (perhaps 25.00 mL) of the unknown acid is measured into a flask using a pipet.

Step 2. An **indicator**, a substance that changes color as the solution reaches a certain pH (Figures 8.5 and 8.6), is added to the unknown solution. We must know, from prior experience, the expected pH at the equivalence point (see step 4, below). For this titration, phenolphthalein or phenol red would be a logical choice, since the equivalence-point pH is known to be seven.

Step 3. A solution of sodium hydroxide (perhaps 0.1000 M) is carefully added to the unknown solution using a **buret** (Figure 8.7), which is a long glass tube calibrated in milliliters. A stopcock at the bottom of the buret regulates the amount of liquid dispensed. The standard solution is added until the indicator changes color.

Continued—

(a)

(b)

Figure 8.5
The color of the petals of the hydrangea is formed by molecules that behave as acid-base indicators. The color is influenced by the pH of the soil in which the hydrangea is grown. The plant (a) was grown in soil with lower pH (more acidic) than plant (b).

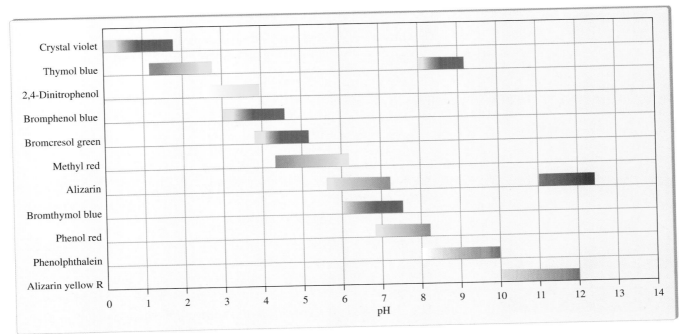

Figure 8.6
The relationship between pH and color of a variety of compounds, some of which are commonly used as acid-base indicators. Many indicators are naturally occurring substances.

EXAMPLE 8.8 —Continued

Step 4. At this point, the **equivalence point,** the number of moles of hydroxide ion added is equal to the number of moles of hydronium ion present in the unknown acid.

Step 5. The volume dispensed by the buret (perhaps 35.00 mL) is measured and used in the calculation of the unknown acid concentration.

Step 6. The calculation is as follows:

Pertinent information for this titration includes:

Volume of the unknown acid solution, 25.00 mL

Volume of sodium hydroxide solution added, 35.00 mL

Concentration of the sodium hydroxide solution, 0.1000 M

Furthermore, from the balanced equation, we know that HCl and NaOH react in a 1:1 combining ratio.

Using a strategy involving conversion factors

$$35.00 \text{ mL NaOH} \times \frac{1 \text{ L NaOH}}{10^3 \text{ mL NaOH}} \times \frac{0.1000 \text{ mol NaOH}}{\text{L NaOH}} = 3.500 \times 10^{-3} \text{ mol NaOH}$$

Knowing that HCl and NaOH undergo a 1:1 reaction,

$$3.500 \times 10^{-3} \text{ mol NaOH} \times \frac{1 \text{ mol HCl}}{1 \text{ mol NaOH}} = 3.500 \times 10^{-3} \text{ mol HCl}$$

3.500×10^{-3} mol HCl are contained in 25.00 mL of HCl solution. Thus,

$$\frac{3.500 \times 10^{-3} \text{ mol HCl}}{25.00 \text{ mL HCl soln}} \times \frac{10^3 \text{ mL HCl soln}}{1 \text{ L HCl soln}} = 1.400 \times 10^{-1} \text{ mol HCl/L HCl soln}$$

$$= 0.1400 M$$

The titration of an acid with a base is depicted in Figure 8.7.

An alternate problem-solving strategy produces the same result:

$$(M_{acid})(V_{acid}) = (M_{base})(V_{base})$$

and

$$M_{acid} = M_{base}\left(\frac{V_{base}}{V_{acid}}\right)$$

$$M_{acid} = (0.1000 M)\left(\frac{35.00 \text{ mL}}{25.00 \text{ mL}}\right)$$

$$M_{acid} = 0.1400 M$$

Question 8.13

Calculate the molar concentration of a sodium hydroxide solution if 40.00 mL of this solution were required to neutralize 20.00 mL of a 0.2000 M solution of hydrochloric acid.

8.3 Reactions Between Acids and Bases

Question 8.14

Calculate the molar concentration of a sodium hydroxide solution if 36.00 mL of this solution were required to neutralize 25.00 mL of a 0.2000 M solution of hydrochloric acid.

Polyprotic Substances

Not all acid-base reactions occur in a 1:1 combining ratio (as hydrochloric acid and sodium hydroxide in the previous example). Acid-base reactions with other than 1:1 combining ratios occur between what are termed *polyprotic substances*. **Polyprotic substances** donate (as acids) or accept (as bases) more than one proton per formula unit.

Reactions of Polyprotic Substances

HCl dissociates to produce one H^+ ion for each HCl. For this reason, it is termed a *monoprotic acid*. Its reaction with sodium hydroxide is:

$$HCl(aq) + NaOH(aq) \longrightarrow H_2O(l) + Na^+(aq) + Cl^-(aq)$$

Sulfuric acid, in contrast, is a *diprotic acid*. Each unit of H_2SO_4 produces two H^+ ions (the prefix *di-* indicating two). Its reaction with sodium hydroxide is:

$$H_2SO_4(aq) + 2NaOH(aq) \longrightarrow 2H_2O(l) + 2Na^+(aq) + SO_4^{2-}(aq)$$

Phosphoric acid is a *triprotic acid*. Each unit of H_3PO_4 produces three H^+ ions. Its reaction with sodium hydroxide is:

$$H_3PO_4(aq) + 3NaOH(aq) \longrightarrow 3H_2O(l) + 3Na^+(aq) + PO_4^{3-}(aq)$$

Dissociation of Polyprotic Substances

Sulfuric acid, and other diprotic acids, dissociate in two steps:

Step 1. $H_2SO_4(aq) + H_2O(l) \longrightarrow H_3O^+(aq) + HSO_4^-(aq)$

Step 2. $HSO_4^-(aq) + H_2O(l) \rightleftharpoons H_3O^+(aq) + SO_4^{2-}(aq)$

Notice that H_2SO_4 behaves as a strong acid (Step 1) and HSO_4^- behaves as a weak acid, indicated by a double arrow (Step 2).

Phosphoric acid dissociates in three steps, all forms behaving as weak acids.

Step 1. $H_3PO_4(aq) + H_2O(l) \rightleftharpoons H_3O^+(aq) + H_2PO_4^-(aq)$

Step 2. $H_2PO_4^-(aq) + H_2O(l) \rightleftharpoons H_3O^+(aq) + HPO_4^{2-}(aq)$

Step 3. $HPO_4^{2-}(aq) + H_2O(l) \rightleftharpoons H_3O^+(aq) + PO_4^{3-}(aq)$

Bases exhibit this property as well.
NaOH produces one OH^- ion per formula unit:

$$NaOH(aq) \longrightarrow Na^+(aq) + OH^-(aq)$$

$Ba(OH)_2$, barium hydroxide, produces two OH^- ions per formula unit:

$$Ba(OH)_2(aq) \longrightarrow Ba^{2+}(aq) + 2OH^-(aq)$$

(a)

(b)

Figure 8.7
An acid-base titration. (a) An exact volume of a standard solution (in this example, a base) is added to a solution of unknown concentration (in this example, an acid). (b) From the volume (read from the buret) and concentration of the standard solution, coupled with the mass or volume of the unknown, the concentration of the unknown may be calculated.

An Environmental Perspective

Acid Rain

Acid rain is a global environmental problem that has raised public awareness of the chemicals polluting the air through the activities of our industrial society. Normal rain has a pH of about 5.6 as a result of the chemical reaction between carbon dioxide gas and water in the atmosphere. The following equation shows this reaction:

$$CO_2(g) + H_2O(l) \rightleftharpoons H_2CO_3(aq)$$
Carbon dioxide   Water   Carbonic acid

Acid rain refers to conditions that are much more acidic than this. In upstate New York the rain has as much as 25 times the acidity of normal rainfall. One rainstorm, recorded in West Virginia, produced rainfall that measured 1.5 on the pH scale. This is approximately the pH of stomach acid or about ten thousand times more acidic than "normal rain" (remember that the pH scale is logarithmic; a 1 pH unit decrease represents a tenfold increase in hydronium ion concentration).

Acid rain is destroying life in streams and lakes. More than half the highland lakes in the western Adirondack Mountains have no native game fish. In addition to these 300 lakes, 140 lakes in Ontario have suffered a similar fate. It is estimated that 48,000 other lakes in Ontario and countless others in the northeastern and central United States are threatened. Our forests are endangered as well. The acid rain decreases soil pH, which in turn alters the solubility of minerals needed by plants. Studies have shown that about 40% of the red spruce and maple trees in New England have died. Increased acidity of rainfall appears to be the major culprit.

What is the cause of this acid rain? The combustion of fossil fuels (gas, oil, and coal) by power plants produces oxides of sulfur and nitrogen. Nitrogen oxides, in excess of normal levels, arise mainly from conversion of atmospheric nitrogen to nitrogen oxides in the engines of gasoline and diesel powered vehicles. Sulfur oxides result from the oxidation of sulfur in fossil fuels. The sulfur atoms were originally a part of the amino acids and proteins of plants and animals that became, over the millenia, our fuel. These react with water, as does the CO_2 in normal rain, but the products are strong acids: sulfuric and nitric acids. Let's look at the equations for these processes.

pH Values for a Variety of Substances Compared with the pH of Acid Rain

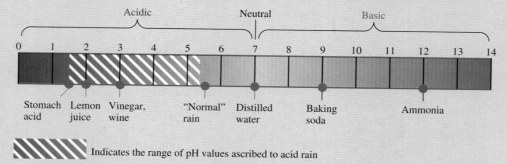

8.4 Acid-Base Buffers

LEARNING GOAL

A **buffer solution** contains components that enable the solution to resist large changes in pH when either acids or bases are added. Buffer solutions may be prepared in the laboratory to maintain optimum conditions for a chemical reaction. Buffers are routinely used in commercial products to maintain optimum conditions for product behavior (Figure 8.8).

Buffer solutions also occur naturally. Blood, for example, is a complex natural buffer solution maintaining a pH of approximately 7.4, optimum for oxygen transport. The major buffering agent in blood is the mixture of carbonic acid (H_2CO_3) and bicarbonate ions (HCO_3^-).

A similar chemistry is seen with the sulfur oxides. Coal may contain as much as 3% sulfur. When the coal is burned, the sulfur also burns. This produces choking, acrid sulfur dioxide gas:

$$S(s) + O_2(g) \longrightarrow SO_2(g)$$

By itself, sulfur dioxide can cause serious respiratory problems for people with asthma or other lung diseases, but matters are worsened by the reaction of SO_2 with atmospheric oxygen:

$$2SO_2(g) + O_2(g) \longrightarrow 2SO_3(g)$$

Sulfur trioxide will react with water in the atmosphere:

$$SO_3(g) + H_2O(l) \longrightarrow H_2SO_4(aq)$$

The product, sulfuric acid, is even more irritating to the respiratory tract. When the acid rain created by the reactions shown above falls to earth, the impact is significant.

It is easy to balance these chemical equations, but decades could be required to balance the ecological systems that we have disrupted by our massive consumption of fossil fuels. A sudden decrease of even 25% in the use of fossil fuels would lead to worldwide financial chaos. Development of alternative fuel sources, such as solar energy and safe nuclear power, will help to reduce our dependence on fossil fuels and help us to balance the global equation.

Damage caused by acid rain.

In the atmosphere, nitric oxide (NO) can react with oxygen to produce nitrogen dioxide as shown:

$$2NO(g) + O_2(g) \longrightarrow 2NO_2(g)$$
$$\text{Nitric oxide} \quad \text{Oxygen} \quad \text{Nitrogen dioxide}$$

Nitrogen dioxide (which causes the brown color of smog) then reacts with water to form nitric acid:

$$3NO_2(g) + H_2O(l) \longrightarrow 2HNO_3(aq) + NO(g)$$

For Further Understanding

Criticize this statement: "Passing and enforcing strong legislation against sulfur and nitrogen oxide emission will solve the problem of acid rain in the United States".

Research the literature to determine the percentage of electricity that is produced from coal in your state of residence.

The Buffer Process

The basis of buffer action is the establishment of an equilibrium between either a weak acid and its conjugate base or a weak base and its conjugate acid. Let's consider the case of a weak acid and its salt.

A common buffer solution may be prepared from acetic acid (CH_3COOH) and sodium acetate (CH_3COONa). Sodium acetate is a salt that is the source of the conjugate base CH_3COO^-. An *equilibrium* is established in solution between the weak acid and the conjugate base.

$$\underset{\substack{\text{Acetic acid}\\\text{(weak acid)}}}{CH_3COOH(aq)} + \underset{\text{Water}}{H_2O(l)} \rightleftarrows \underset{\text{Hydronium ion}}{H_3O^+(aq)} + \underset{\substack{\text{Acetate ion}\\\text{(conjugate base)}}}{CH_3COO^-(aq)}$$

We ignore Na^+ in the description of the buffer. Na^+ does not actively participate in the reaction.

The acetate ion is the conjugate base of acetic acid.

Figure 8.8
Commercial products that claim improved function owing to their ability to control pH.

A buffer solution functions in accordance with LeChatelier's principle, which states that an equilibrium system, when stressed, will shift its equilibrium to relieve that stress. This principle is illustrated by the following examples.

Addition of Base (OH⁻) to a Buffer Solution

Addition of a basic substance to a buffer solution causes the following changes.

- OH^- from the base reacts with H_3O^+ producing water.
- Molecular acetic acid *dissociates* to replace the H_3O^+ consumed by the base, maintaining the pH close to the initial level.

This is an example of LeChatelier's principle, because the loss of H_3O^+ (the *stress*) is compensated by the dissociation of acetic acid to produce more H_3O^+.

Addition of Acid (H₃O⁺) to a Buffer Solution

Addition of an acidic solution to a buffer results in the following changes.

- H_3O^+ from the acid increases the overall $[H_3O^+]$.
- The system reacts to this stress, in accordance with LeChatelier's principle, to form more molecular acetic acid; the acetate ion combines with H_3O^+. Thus, the H_3O^+ concentration and therefore, the pH, remain close to the initial level.

These effects may be summarized as follows:

$$CH_3COOH(aq) + H_2O(l) \rightleftharpoons H_3O^+(aq) + CH_3COO^-(aq)$$

OH^- added, equilibrium shifts to the right $\longrightarrow$

H_3O^+ added, equilibrium shifts to the left $\longleftarrow$

Buffer Capacity

Buffer capacity is a measure of the ability of a solution to resist large changes in pH when a strong acid or strong base is added. More specifically, buffer capacity is described as the amount of strong acid or strong base that a buffer can neutralize without significantly changing its pH. Buffering capacity against base is a function of the concentration of the weak acid (in this case CH_3COOH). Buffering capacity against acid is dependent on the concentration of the anion of the salt, the conjugate base (CH_3COO^- in this example). Buffer solutions are often designed to have identical buffer capacity for both acids and bases. This is achieved when, in the above example, $[CH_3COO^-]/[CH_3COOH] = 1$. As an added bonus, making the $[CH_3COO^-]$ and $[CH_3COOH]$ as large as is practical ensures a high buffer capacity for both added acid and added base.

Question 8.15

Explain how the molar concentration of H_2CO_3 in the blood would change if the partial pressure of CO_2 in the lungs were to increase. (Refer to A Medical Perspective: Control of Blood pH on page 262.)

Question 8.16

Explain how the molar concentration of H_2CO_3 in the blood would change if the partial pressure of CO_2 in the lungs were to decrease. (Refer to A Medical Perspective: Control of Blood pH on page 262.)

Question 8.17

Explain how the molar concentration of hydronium ion in the blood would change under each of the conditions described in Questions 8.15 and 8.16.

Question 8.18

Explain how the pH of blood would change under each of the conditions described in Questions 8.15 and 8.16.

Preparation of a Buffer Solution

It is useful to understand how to prepare a buffer solution and how to determine the pH of the resulting solution. Many chemical reactions produce the largest amount of product only when they are run at an optimal, constant pH. The study of biologically important processes in the laboratory often requires conditions that approximate the composition of biological fluids. A constant pH would certainly be essential.

The buffer process is an equilibrium reaction and is described by an equilibrium-constant expression. For acids, the equilibrium constant is represented as K_a, the subscript a implying an acid equilibrium. For example, the acetic acid/sodium acetate system is described by

$$CH_3COOH(aq) + H_2O(l) \rightleftharpoons H_3O^+(aq) + CH_3COO^-(aq)$$

and

$$K_a = \frac{[H_3O^+][CH_3COO^-]}{[CH_3COOH]}$$

Using a few mathematical maneuvers we can turn this equilibrium-constant expression into one that will allow us to calculate the pH of the buffer if we know how much acid (acetic acid) and salt (sodium acetate) are present in a known volume of the solution.

First, multiply both sides of the equation by the concentration of acetic acid, [CH$_3$COOH]. This will eliminate the denominator on the right side of the equation.

$$[CH_3COOH]K_a = \frac{[H_3O^+][CH_3COO^-][CH_3COOH]}{[CH_3COOH]}$$

or

$$[CH_3COOH]K_a = [H_3O^+][CH_3COO^-]$$

Now, dividing both sides of the equation by the acetate ion concentration [CH$_3$COO$^-$] will give us an expression for the hydronium ion concentration [H$_3$O$^+$].

The calculation of pH from $[H_3O^+]$ is discussed in Section 8.2.

$$\frac{[CH_3COOH]K_a}{[CH_3COO^-]} = [H_3O^+]$$

Once we know the value for [H$_3$O$^+$], we can easily find the pH.

To use this equation:

- assume that [CH$_3$COOH] represents the concentration of the acid component of the buffer.
- assume that [CH$_3$COO$^-$] represents the concentration of the conjugate base (principally from the dissociation of the salt, sodium acetate) component of the buffer.

$$\frac{[CH_3COOH]K_a}{[CH_3COO^-]} = [H_3O^+]$$

$$\frac{[acid]K_a}{[conjugate\ base]} = [H_3O^+]$$

Let's look at examples of practical applications of this equation.

EXAMPLE 8.9 **Calculating the pH of a Buffer Solution**

Calculate the pH of a buffer solution in which both the acetic acid and sodium acetate concentrations are 1.0×10^{-1} M. The equilibrium constant, K_a, for acetic acid is 1.8×10^{-5}.

Solution

Acetic acid is the acid; [acid] = 1.0×10^{-1} M
   Sodium acetate is the salt, furnishing the conjugate base; [conjugate base] = 1.0×10^{-1} M
      The equilibrium is

$$CH_3COOH(aq) + H_2O(l) \rightleftharpoons H_3O^+(aq) + CH_3COO^-(aq)$$
$$\text{acid} \hspace{4cm} \text{conjugate base}$$

and the hydronium ion concentration,

$$[H_3O^+] = \frac{[\text{acid}]K_a}{[\text{conjugate base}]}$$

Substituting the values given in the problem

$$[H_3O^+] = \frac{[1.0 \times 10^{-1}]1.8 \times 10^{-5}}{[1.0 \times 10^{-1}]}$$

$$[H_3O^+] = 1.8 \times 10^{-5}$$

and because

$$pH = -\log[H_3O^+]$$
$$pH = -\log 1.8 \times 10^{-5}$$
$$= 4.74$$

The pH of the buffer solution is 4.74.

EXAMPLE 8.10 **Calculating the pH of a Buffer Solution**

Calculate the pH of a buffer solution similar to that described in Example 8.9 except that the acid concentration is doubled, while the salt concentration remains the same.

Solution

Acetic acid is the acid; [acid] = 2.0×10^{-1} M (remember, the acid concentration is twice that of Example 8.9; $2 \times [1.0 \times 10^{-1}] = 2.0 \times 10^{-1}$ M
   Sodium acetate is the salt, furnishing the conjugate base; [conjugate base] = 1.0×10^{-1} M
      The equilibrium is

$$CH_3COOH(aq) + H_2O(l) \rightleftharpoons H_3O^+(aq) + CH_3COO^-(aq)$$
$$\text{acid} \hspace{4cm} \text{conjugate base}$$

Continued—

EXAMPLE 8.10 —Continued

and the hydronium ion concentration,

$$[H_3O^+] = \frac{[\text{acid}]K_a}{[\text{conjugate base}]}$$

Substituting the values given in the problem

$$[H_3O^+] = \frac{[2.0 \times 10^{-1}]1.8 \times 10^{-5}}{[1.00 \times 10^{-1}]}$$

$$[H_3O^+] = 3.60 \times 10^{-5}$$

and because

$$pH = -\log[H_3O^+]$$
$$pH = -\log 3.60 \times 10^{-5}$$
$$= 4.44$$

The pH of the buffer solution is 4.44.

A comparison of the two solutions described in Examples 8.9 and 8.10 demonstrates a buffer solution's most significant attribute: the ability to stabilize pH. Although the acid concentration of these solutions differs by a factor of two, the difference in their pH is only 0.30 units.

Question 8.19

A buffer solution is prepared in such a way that the concentrations of propanoic acid and sodium propanoate are each 2.00×10^{-1} M. If the buffer equilibrium is described by

$$C_2H_5COOH(aq) + H_2O(l) \rightleftharpoons H_3O^+(aq) + C_2H_5COO^-(aq)$$

Propanoic acid   Propanoate anion

with $K_a = 1.34 \times 10^{-5}$, calculate the pH of the solution.

Question 8.20

Calculate the pH of the buffer solution in Question 8.19 if the concentration of the sodium propanoate were doubled while the acid concentration remained the same.

The Henderson-Hasselbalch Equation

The solution of the equilibrium-constant expression and the pH are sometimes combined into one operation. The combined expression is termed the **Henderson-Hasselbalch equation**.

For the acetic acid/sodium acetate buffer system,

$$CH_3COOH(aq) + H_2O(l) \rightleftharpoons H_3O^+(aq) + CH_3COO^-(aq)$$

$$K_a = \frac{[H_3O^+][CH_3COO^-]}{[CH_3COOH]}$$

A Medical Perspective

Control of Blood pH

A pH of 7.4 is maintained in blood partly by a carbonic acid–bicarbonate buffer system based on the following equilibrium:

$$H_2CO_3(aq) + H_2O(l) \rightleftharpoons H_3O^+(aq) + HCO_3^-(aq)$$

Carbonic acid (weak acid)   Bicarbonate ion (salt)

The regulation process based on LeChatelier's principle is similar to the acetic acid–sodium acetate buffer, which we have already discussed.

Red blood cells transport O_2, bound to hemoglobin, to the cells of body tissue. The metabolic waste product, CO_2, is picked up by the blood and delivered to the lungs.

The CO_2 in the blood also participates in the carbonic acid–bicarbonate buffer equilibrium. Carbon dioxide reacts with water in the blood to form carbonic acid:

$$CO_2(aq) + H_2O(l) \rightleftharpoons H_2CO_3(aq)$$

As a result the buffer equilibrium becomes more complex:

$$CO_2(aq) + 2H_2O(l) \rightleftharpoons H_2CO_3(aq) + H_2O(l) \rightleftharpoons H_3O^+(aq) + HCO_3^-(aq)$$

Through this sequence of relationships the concentration of CO_2 in the blood affects the blood pH.

Higher than normal CO_2 concentrations shift the above equilibrium to the right (LeChatelier's principle), increasing $[H_3O^+]$ and lowering the pH. The blood becomes too acidic, leading to numerous medical problems. A situation of high blood CO_2 levels and low pH is termed *acidosis*. Respiratory acidosis results from various diseases (emphysema, pneumonia) that restrict the breathing process, causing the buildup of waste CO_2 in the blood.

Lower than normal CO_2 levels, on the other hand, shift the equilibrium to the left, decreasing $[H_3O^+]$ and making the pH more basic. This condition is termed *alkalosis* (from "alkali," implying basic). Hyperventilation, or rapid breathing, is a common cause of respiratory alkalosis.

For Further Understanding

Write the Henderson-Hasselbalch expression for the equilibrium between carbonic acid and the bicarbonate ion.

Calculate the $[HCO_3^-]/[H_2CO_3]$ that corresponds to a pH of 7.4. The K_a for carbonic acid is 4.2×10^{-7}.

Taking the −log of both sides of the equation:

$$-\log K_a = -\log [H_3O^+] - \log \frac{[CH_3COO^-]}{[CH_3COOH]}$$

pKa = −log Ka, analogous to pH = −log [H₃O⁺].

$$pK_a = pH - \log \frac{[CH_3COO^-]}{[CH_3COOH]}$$

the Henderson-Hasselbalch expression is:

$$pH = pK_a + \log \frac{[CH_3COO^-]}{[CH_3COOH]}$$

The form of this equation is especially amenable to buffer problem calculations. In this expression, $[CH_3COOH]$ represents the molar concentration of the weak acid and $[CH_3COO^-]$ is the molar concentration of the conjugate base of the weak acid. The generalized expression is:

$$pH = pK_a + \log \frac{[\text{conjugate base}]}{[\text{weak acid}]}$$

Substituting concentrations along with the value for the pK_a of the acid allows the calculation of the pH of the buffer solution in problems such as those shown in Examples 8.9 and 8.10 as well as Questions 8.19 and 8.20.

Question 8.21 Solve the problem in Example 8.9 using the Henderson-Hasselbalch equation.

Question 8.22 Solve the problem in Example 8.10 using the Henderson-Hasselbalch equation.

Question 8.23 Solve Question 8.19 using the Henderson-Hasselbalch equation.

Question 8.24 Solve Question 8.20 using the Henderson-Hasselbalch equation.

8.5 Oxidation-Reduction Processes

Oxidation-reduction processes are responsible for many types of chemical change. Corrosion, the operation of a battery, and biochemical energy-harvesting reactions are a few examples. In this section we explore the basic concepts underlying this class of chemical reactions.

Oxidation and Reduction

Oxidation is defined as a loss of electrons, loss of hydrogen atoms, or gain of oxygen atoms. *Sodium metal*, is, for example, oxidized to a *sodium ion*, losing one electron when it reacts with a nonmetal such as chlorine:

$$Na \longrightarrow Na^+ + e^-$$

Reduction is defined as a gain of electrons, gain of hydrogen atoms, or loss of oxygen atoms. A *chlorine atom* is reduced to a *chloride ion* by gaining one electron when it reacts with a metal such as sodium:

$$Cl + e^- \longrightarrow Cl^-$$

Oxidation and reduction are complementary processes. The *oxidation half-reaction* produces an electron that is the reactant for the *reduction half-reaction*. The combination of two half-reactions, one oxidation and one reduction, produces the complete reaction:

Oxidation half-reaction: $\quad Na \longrightarrow Na^+ + e^-$

Reduction half-reaction: $\quad Cl + e^- \longrightarrow Cl^-$

Complete reaction: $\quad Na + Cl \longrightarrow Na^+ + Cl^-$

Half-reactions, one oxidation and one reduction, are exactly that: one-half of a complete reaction. The two half-reactions combine to produce the complete reaction. Note that the electrons cancel: in the electron transfer process, no free electrons remain.

In the preceding reaction, sodium metal is the **reducing agent.** It releases electrons for the reduction of chlorine. Chlorine is the **oxidizing agent.** It accepts electrons from the sodium, which is oxidized.

Oxidation-reduction reactions are often termed *redox reactions*.

The reducing agent becomes oxidized and the oxidizing agent becomes reduced.

A Medical Perspective

Oxidizing Agents for Chemical Control of Microbes

Before the twentieth century, hospitals were not particularly sanitary establishments. Refuse, including human waste, was disposed of on hospital grounds. Because many hospitals had no running water, physicians often cleaned their hands and instruments by wiping them on their lab coats and then proceeded to treat the next patient! As you can imagine, many patients died of infections in hospitals.

By the late nineteenth century a few physicians and microbiologists had begun to realize that infectious diseases are transmitted by microbes, including bacteria and viruses. To decrease the number of hospital-acquired infections, physicians like Joseph Lister and Ignatz Semmelweis experimented with chemicals and procedures that were designed to eliminate pathogens from environmental surfaces and from wounds.

Many of the common disinfectants and antiseptics are oxidizing agents. A disinfectant is a chemical that is used to kill or inhibit the growth of pathogens, disease-causing microorganisms, on environmental surfaces. An antiseptic is a milder chemical that is used to destroy pathogens associated with living tissue.

Hydrogen peroxide is an effective antiseptic that is commonly used to cleanse cuts and abrasions. We are all familiar with the furious bubbling that occurs as the enzyme catalase from our body cells catalyzes the breakdown of H_2O_2:

$$2H_2O_2(aq) \longrightarrow 2H_2O(l) + O_2(g)$$

A highly reactive and deadly form of oxygen, the superoxide radical (O_2^-), is produced during this reaction. This superoxide inactivates proteins, especially critical enzyme systems.

At higher concentrations (3–6%), H_2O_2 is used as a disinfectant. It is particularly useful for disinfection of soft contact lenses, utensils, and surgical implants because there is no residual toxicity. Concentrations of 6–25% are even used for complete sterilization of environmental surfaces.

Benzoyl peroxide is another powerful oxidizing agent. Ointments containing 5–10% benzoyl peroxide have been used as antibacterial agents to treat acne. The compound is currently found in over-the-counter facial scrubs because it is also an exfoliant, causing sloughing of old skin and replacement with smoother-looking skin. A word of caution is in order: in sensitive individuals, benzoyl peroxide can cause swelling and blistering of tender facial skin.

Chlorine is a very widely used disinfectant and antiseptic. Calcium hypochlorite [$Ca(OCl)_2$] was first used in hospital maternity wards in 1847 by the pioneering Hungarian physician Ignatz Semmelweis. Semmelweis insisted that hospital workers cleanse their hands in a $Ca(OCl)_2$ solution and dramatically reduced the incidence of infection. Today, calcium hypochlorite is more commonly used to disinfect bedding, clothing, restaurant eating utensils, slaughterhouses, barns, and dairies.

Sodium hypochlorite (NaOCl), sold as Clorox, is used as a household disinfectant and deodorant but is also used to disinfect swimming pools, dairies, food-processing equipment, and kidney dialysis units. It can be used to treat drinking water of questionable quality. Addition of 1/2 teaspoon of household bleach (5.25% NaOCl) to 2 gallons of clear water renders it drinkable after 1/2 hour. The Centers for Disease Control even recommend a 1:10 dilution of bleach as an effective disinfectant against human immunodeficiency virus, the virus that causes acquired immune deficiency syndrome (AIDS).

Chlorine gas (Cl_2) is used to disinfect swimming pool water, sewage, and municipal water supplies. This treatment has successfully eliminated epidemics of waterborne diseases. However, chlorine is inactivated in the presence of some organic materials and, in some cases, may form toxic chlorinated organic compounds. For these reasons, many cities are considering the use of ozone (O_3) rather than chlorine.

Ozone is produced from O_2 by high-voltage electrical discharges. (That fresh smell in the air after an electrical storm is ozone.) Several European cities use ozone to disinfect drinking water. It is a more effective killing agent than chlorine, especially with some viruses: less ozone is required for disinfection; there is no unpleasant residual odor or flavor; and there appear to be fewer toxic by-products. However, ozone is more expensive than chlorine, and maintaining the required concentration in the water is more difficult. Nonetheless, the benefits seem to outweigh the drawbacks, and many U.S. cities may soon follow the example of European cities and convert to the use of ozone for water treatment.

For Further Understanding

Describe the difference between the terms *disinfectant* and *antiseptic*.

Explain why hydrogen peroxide, at higher concentration, is used as a disinfectant, whereas lower concentrations are used as antiseptics.

The characteristics of oxidizing and reducing agents may be summarized as follows:

Oxidizing Agent
- Is reduced
- Gains electrons
- Causes oxidation

Reducing Agent
- Is oxidized
- Loses electrons
- Causes reduction

Question 8.25

Write the oxidation half-reaction, the reduction half-reaction, and the complete reaction for the formation of calcium sulfide from the elements Ca and S.

Question 8.26

Write the oxidation half-reaction, the reduction half-reaction, and the complete reaction for the formation of calcium iodide from calcium metal and I_2. Remember, the electron gain *must* equal the electron loss.

Applications of Oxidation and Reduction

Oxidation-reduction processes are important in many areas as diverse as industrial manufacturing and biochemical processes.

Corrosion

The deterioration of metals caused by an oxidation-reduction process is termed **corrosion**. Metal atoms are converted to metal ions; the structure, hence the properties, changes dramatically, and usually for the worse (Figure 8.9).

Millions of dollars are spent annually in an attempt to correct the damage resulting from corrosion. A current area of chemical research is concerned with the development of corrosion-inhibiting processes. In one type of corrosion, elemental iron is oxidized to iron(III) oxide (rust):

$$4Fe(s) + 3O_2(g) \longrightarrow 2Fe_2O_3(s)$$

At the same time that iron is oxidized, O_2 is being reduced to O^{2-} and is incorporated into the structure of iron(III) oxide. Electrons lost by iron reduce oxygen. This again shows that oxidation and reduction processes go hand in hand.

Combustion of Fossil Fuels

Burning fossil fuel is an extremely exothermic process. Energy is released to heat our homes, offices, and classrooms. The simplest fossil fuel is methane, CH_4, and its oxidation reaction is written:

$$CH_4(g) + 2O_2(g) \longrightarrow CO_2(g) + 2H_2O(g)$$

(a)

(b)

(c)

Figure 8.9
The rust (an oxide of iron) that diminishes structural strength and ruins the appearance of (a) automobiles, (b) bridges, and (c) other iron-based objects is a common example of an oxidation-reduction reaction.

A Medical Perspective

Electrochemical Reactions in the Statue of Liberty and in Dental Fillings

Throughout history, we have suffered from our ignorance of basic electrochemical principles. For example, during the Middle Ages, our chemistry ancestors (alchemists) placed an iron rod into a blue solution of copper sulfate. They noticed that bright shiny copper plated out onto an iron rod and they thought that they had changed a base metal, iron, into copper. What actually happened was the redox reaction shown in Equation 1.

$$2Fe(s) + 3Cu^{2+}(aq) \longrightarrow 2Fe^{3+}(aq) + 3Cu(s) \quad (1)$$

This misunderstanding encouraged them to embark on a futile, one-thousand-year attempt to change base metals into gold.

Over one hundred years ago, France presented the United States with the Statue of Liberty. Unfortunately, the French did not anticipate the redox reaction shown in Equation 1 when they mounted the copper skin of the statue on iron support rods. Oxygen in the atmosphere oxidized the copper skin to produce copper ions. Then, because iron is more active than copper, the displacement shown in Equation 1 aided the corrosion of the support bars. As a result of this and other reactions, the statue needed refurbishing before we celebrated its one hundredth anniversary in 1986.

Sometimes dentists also overlook possible redox reactions when placing gold caps over teeth next to teeth with amalgam fillings. The amalgam in tooth fillings is an alloy of mercury, silver, tin, and copper. Because the metals in the amalgam are more active than gold, contact between the amalgam fillings and minute numbers of gold ions results in redox reactions such as the following.*

$$3Sn(s) + 2Au^{3+}(aq) \longrightarrow 3Sn^{2+}(aq) + 2Au(s) \quad (2)$$

As a result, the dental fillings dissolve and the patients are left with a constant metallic taste in their mouths.

These examples show that like our ancestors, we continue to experience unfortunate results because of a lack of understanding of basic electrochemical principles.

Source: Ronald DeLorenzo, *Journal of Chemical Education*, May 1985, pp. 424–425.

*Equation 2 is oversimplified to illustrate more clearly the basic displacement of gold ions by metallic tin atoms. Actually, only complex ions of gold and tin can exist in aqueous solutions, not the simple cations that are shown.

For Further Understanding

Label the oxidizing agent, reducing agent, substance oxidized, and substance reduced in each equation in this perspective.

For each equation, state the substance that gains electrons and the substance that loses electrons.

Statue of Liberty redox reaction.

Methane is a hydrocarbon. The complete oxidation of any hydrocarbon (including those in gasoline, heating oil, liquid propane, and so forth) produces carbon dioxide and water. The energy released by these reactions is of paramount importance. The water and carbon dioxide are viewed as waste products, and the carbon dioxide contributes to the greenhouse effect (see An Environmental Perspective: The Greenhouse Effect and Global Warming on page 166).

Bleaching

Bleaching agents are most often oxidizing agents. Sodium hypochlorite (NaOCl) is the active ingredient in a variety of laundry products. It is an effective oxidizing agent. Products containing NaOCl are advertised for their stain-removing capabilities.

Stains are a result of colored compounds adhering to surfaces. Oxidation of these compounds produces products that are not colored or compounds that are subsequently easily removed from the surface, thus removing the stain.

Biological Processes

Respiration

There are many examples of biological oxidation-reduction reactions. For example, the electron-transport chain of aerobic respiration involves the reversible oxidation and reduction of iron atoms in cytochrome c,

$$\text{cytochrome } c \text{ (Fe}^{3+}) + e^- \longrightarrow \text{cytochrome } c \text{ (Fe}^{2+})$$

The reduced iron ion transfers an electron to an iron ion in another protein, called cytochrome c oxidase, according to the following reaction:

$$\text{cytochrome } c \text{ (Fe}^{2+}) + \text{cytochrome } c \text{ oxidase (Fe}^{3+})$$
$$\longrightarrow$$
$$\text{cytochrome } c \text{ (Fe}^{3+}) + \text{cytochrome } c \text{ oxidase (Fe}^{2+})$$

Cytochrome c oxidase eventually passes four electrons to O_2, the final electron acceptor of the chain:

$$O_2 + 4e^- + 4H^+ \longrightarrow 2H_2O$$

See Chapters 21 and 22 for the details of these energy-harvesting cellular oxidation-reduction reactions.

Metabolism

When ethanol is metabolized in the liver, it is oxidized to acetaldehyde (the molecule partially responsible for hangovers). Continued oxidation of acetaldehyde produces acetic acid, which is eventually oxidized to CO_2 and H_2O. These reactions, summarized as follows, are catalyzed by liver enzymes.

$$CH_3CH_2\text{—}OH \longrightarrow \underset{\text{Acetaldehyde}}{CH_3\overset{\overset{\displaystyle O}{\|}}{C}\text{—}H} \longrightarrow \underset{\text{Acetic acid}}{CH_3\overset{\overset{\displaystyle O}{\|}}{C}\text{—}OH} \longrightarrow CO_2 + H_2O$$

Ethanol

It is more difficult to recognize these reactions as oxidations because neither the product nor the reactant carries a charge. In previous examples we looked for an increase in positive charge as an indication that an oxidation had occurred. A decrease in positive charge (or increased negative charge) would signify reduction.

Alternative descriptions of oxidation and reduction are useful in identifying these reactions.

Oxidation is the *gain* of oxygen or *loss* of hydrogen.

Reduction is the *loss* of oxygen or *gain* of hydrogen.

In the conversion of ethanol to acetaldehyde, ethanol has six hydrogen atoms per molecule; the product acetaldehyde has four hydrogen atoms per molecule. This represents a loss of two hydrogen atoms per molecule. Therefore, ethanol has been oxidized to acetaldehyde, based on the interpretation of the above-mentioned rules.

This strategy is most useful for recognizing oxidation and reduction of *organic compounds* and organic compounds of biological interest, *biochemical compounds*. Organic compounds and their structures and reactivity are the focus of Chapters 10 through 15 and biochemical compounds are described in Chapters 16 through 23.

Voltaic Cells

When zinc metal is dipped into a copper(II) sulfate solution, zinc atoms are oxidized to zinc ions and copper(II) ions are reduced to copper metal, which deposits on the surface of the zinc metal (Figure 8.10). This reaction is summarized as follows:

 LEARNING GOAL

Figure 8.10
The spontaneous reaction of zinc metal and Cu^{2+} ions is the basis of the cell depicted in Figure 8.11.

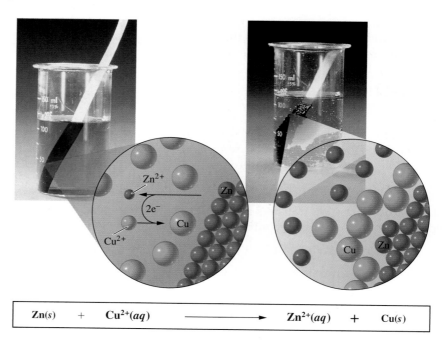

$$Zn(s) + Cu^{2+}(aq) \longrightarrow Zn^{2+}(aq) + Cu(s)$$

$$\overbrace{Zn(s) + Cu^{2+}(aq) \longrightarrow Zn^{2+}(aq) + Cu(s)}^{\text{Oxidation/e}^- \text{ loss}}_{\text{Reduction/e}^- \text{ gain}}$$

In the reduction of aqueous copper(II) ions by zinc metal, electrons flow from the zinc rod directly to copper(II) ions in the solution. If electron transfer from the zinc rod to the copper ions in solution could be directed through an external electrical circuit, this spontaneous oxidation-reduction reaction could be used to produce an electrical current that could perform some useful function.

Recall that solutions of ionic salts are good conductors of electricity (Chapter 6).

However, when zinc metal in one container is connected by a copper wire with a copper(II) sulfate solution in a separate container, no current flows through the wire. A complete, or continuous circuit is necessary for current to flow. To complete the circuit, we connect the two containers with a tube filled with a solution of an electrolyte such as potassium chloride. This tube is described as a *salt bridge*.

Current now flows through the external circuit (Figure 8.11). The device shown in Figure 8.11 is an example of a *voltaic cell*. A **voltaic cell** is an *electrochemical* cell that converts stored *chemical* energy into *electrical* energy.

This cell consists of two *half-cells*. The oxidation half-reaction occurs in one half-cell and the reduction half-reaction occurs in the other half-cell. The sum of the two half-cell reactions is the overall oxidation-reduction reaction that describes the cell. The electrode at which oxidation occurs is called the **anode,** and the electrode at which reduction occurs is the **cathode.** In the device shown in Figure 8.11, the zinc metal is the anode. At this electrode the zinc atoms are oxidized to zinc ions:

$$\text{Anode half-reaction: } Zn(s) \longrightarrow Zn^{2+}(aq) + 2e^-$$

Electrons released at the anode travel through the external circuit to the cathode (the copper rod) where they are transferred to copper(II) ions in the solution. Copper(II) ions are reduced to copper atoms that deposit on the copper metal surface, the cathode:

$$\text{Cathode half-reaction: } Cu^{2+}(aq) + 2e^- \longrightarrow Cu(s)$$

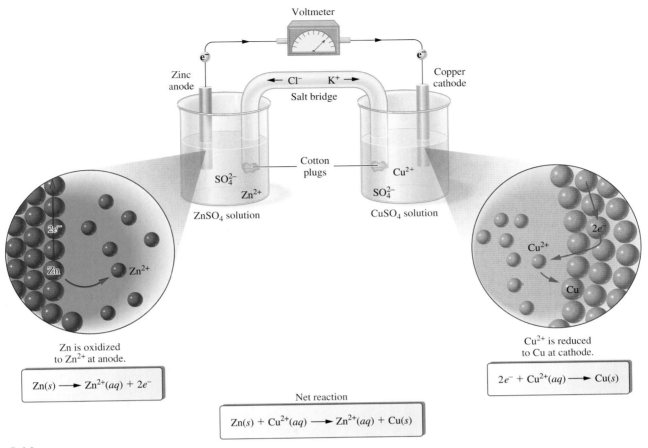

Figure 8.11
A voltaic cell generating electrical current by the reaction:

$$Zn(s) + Cu^{2+}(aq) \longrightarrow Zn^{2+}(aq) + Cu(s)$$

Each electrode consists of the pure metal, zinc or copper. Zinc is oxidized, releasing electrons that flow to the copper, reducing Cu^{2+} to Cu. The salt bridge completes the circuit and the voltmeter displays the voltage (or chemical potential) associated with the reaction.

The sum of these half-cell reactions is the cell reaction:

$$Zn(s) + Cu^{2+}(aq) \longrightarrow Zn^{2+}(aq) + Cu(s)$$

Voltaic cells are found in many aspects of our life, as convenient and reliable sources of electrical energy, the battery. Batteries convert stored chemical energy to an electrical current to power a wide array of different commercial appliances: radios, portable televisions and computers, flashlights, a host of other useful devices.

Technology has made modern batteries smaller, safer, and more dependable than our crudely constructed copper-zinc voltaic cell. In fact, the silver cell (Figure 8.12) is sufficiently safe and nontoxic that it can be implanted in the human body as a part of a pacemaker circuit that is used to improve heart rhythm. A rather futuristic potential application of voltaic cells is noted in A Medical Perspective: Turning the Human Body into a Battery on page 270.

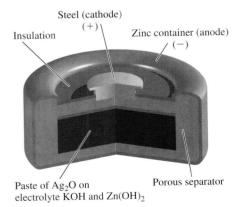

Figure 8.12
A silver battery used in cameras, heart pacemakers, and hearing aids. This battery is small, stable, and nontoxic (hence implantable in the human body).

Electrolysis

Electrolysis reactions use electrical energy to cause nonspontaneous oxidation- reduction reactions to occur. They are the reverse of voltaic cells. One common application is the rechargeable battery. When it is being used to power a device, such

A Medical Perspective

Turning the Human Body into a Battery

The heart has its own natural pacemaker that sends nerve impulses (pulses of electrical current) throughout the heart approximately seventy-two times per minute. These electrical pulses cause your heart muscles to contract (beat), which pumps blood through the body. The fibers that carry the nerve impulses can be damaged by disease, drugs, heart attacks, and surgery. When these heart fibers are damaged, the heart may run too slowly, stop temporarily, or stop altogether. To correct this condition, artificial heart pacemakers (see figure below) are surgically inserted in the human body. A pacemaker (pacer) is a battery-driven device that sends an electrical current (pulse) to the heart about seventy-two times per minute. Over 300,000 Americans are now wearing artificial pacemakers with an additional 30,000 pacemakers installed each year.

Yearly operations used to be necessary to replace the pacemaker's batteries. Today, pacemakers use improved batteries that last much longer, but even these must be replaced eventually.

It would be very desirable to develop a permanent battery to run pacemakers. Some scientists began working on ways of converting the human body itself into a battery (voltaic cell) to power artificial pacemakers.

Several methods for using the human body as a voltaic cell have been suggested. One of these is to insert platinum and zinc

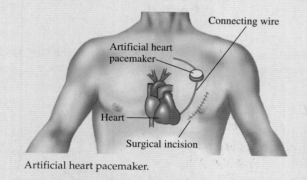

Artificial heart pacemaker.

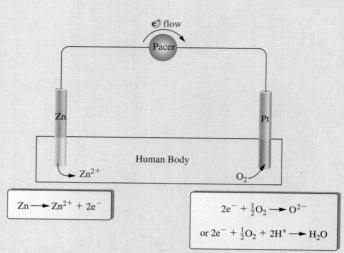

The "body battery."

electrodes into the human body as diagrammed in the figure above. The pacemaker and the electrodes would be worn internally. This "body battery" could easily generate the small amount of current (5×10^{-5} ampere) that is required by most pacemakers. This "body battery" has been tested on animals for periods exceeding four months without noticeable problems.

Source: Ronald DeLorenzo, *Problem Solving in General Chemistry,* 2nd ed., Wm. C. Brown, Publishers, Dubuque, Iowa, 1993, pp. 336–338.

For Further Understanding

What are some criteria that must be considered when choosing the electrode material?

Combine the two half-reactions for the "body battery" to yield a complete oxidation-reduction equation.

LEARNING GOAL

as a laptop computer, it behaves as a voltaic cell. After some time, the chemical reaction approaches completion and the voltaic cell "runs down." The cell reaction is reversible and the battery is plugged into a battery charger. The charger is really an external source of electrical energy that reverses the chemical reaction in the battery, bringing it back to its original state. The cell has been operated as an electrolytic cell. Removal of the charging device turns the cell back into a voltaic device, ready to spontaneously react to produce electrical current once again.

The relationship between a voltaic cell and an electrolytic cell is illustrated in Figure 8.13.

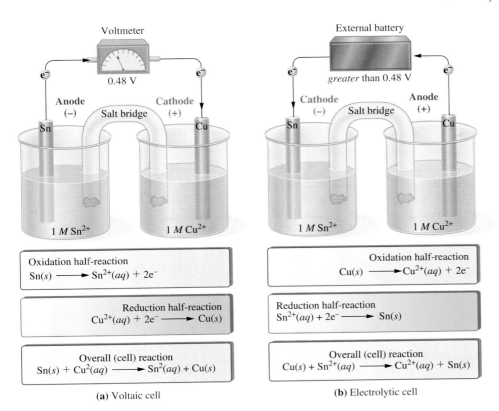

Figure 8.13
(a) A voltaic cell is converted to (b) an electrolytic cell by attaching a battery with a voltage sufficiently large to reverse the reaction. This process underlies commercially available rechargeable batteries.

SUMMARY

8.1 Acids and Bases

One of the earliest definitions of acids and bases is the *Arrhenius theory*. According to this theory, an *acid* dissociates to form hydrogen ions, H^+, and a *base* dissociates to form hydroxide ions, OH^-. The *Brønsted-Lowry theory* defines an acid as a proton (H^+) donor and a base as a proton acceptor.

Water, the solvent in many acid-base reactions, is *amphiprotic*. It has both acid and base properties.

The strength of acids and bases in water depends on their degree of dissociation, the extent to which they react with the solvent, water. Acids and bases are strong when the reaction with water is virtually 100% complete and weak when the reaction with water is much less than 100% complete.

Weak acids and weak bases dissolve in water principally in the molecular form. Only a small percentage of the molecules dissociate to form the *hydronium* ion or *hydroxide* ion.

Aqueous solutions of acids and bases are electrolytes. The dissociation of the acid or base produces ions, which conduct an electrical current. Strong acids and bases are strong electrolytes. Weak acids and bases are weak electrolytes.

Although pure water is virtually 100% molecular, a small number of water molecules do ionize. This process occurs by the transfer of a proton from one water molecule to another, producing a hydronium ion and a hydroxide ion. This process is the *autoionization*, or self-ionization, of water.

Pure water at room temperature has a hydronium ion concentration of 1.0×10^{-7} M. One hydroxide ion is produced for each hydronium ion. Therefore, the hydroxide ion concentration is also 1.0×10^{-7} M. The product of hydronium and hydroxide ion concentration (1.0×10^{-14}) is the *ion product for water*.

8.2 pH: A Measurement Scale for Acids and Bases

The *pH scale* correlates the hydronium ion concentration with a number, the pH, that serves as a useful indicator of the degree of acidity or basicity of a solution. The pH of a solution is defined as the negative logarithm of the molar concentration of the hydronium ion (pH = $-\log [H_3O^+]$).

8.3 Reactions Between Acids and Bases

The reaction of an acid with a base to produce a salt and water is referred to as *neutralization*. Neutralization requires equal numbers of moles of H_3O^+ and OH^- to produce a neutral solution (no excess acid or base). A neutralization reaction may be used to determine the concentration of an unknown acid or base solution. The technique of *titration* involves the addition of measured amounts of a *standard solution* (one whose concentration is known) from a *buret* to neutralize the second, unknown solution. The *equivalence point* is signaled by an *indicator*.

8.4 Acid-Base Buffers

A *buffer solution* contains components that enable the solution to resist large changes in pH when acids or bases are added. The basis of buffer action is an equilibrium between either a weak acid and its salt or a weak base and its salt.

A buffer solution follows LeChatelier's principle, which states that an equilibrium system, when stressed, will shift its equilibrium to alleviate that stress.

Buffering against base is a function of the concentration of the weak acid for an acidic buffer. Buffering against acid is dependent on the concentration of the anion of the salt.

A buffer solution can be described by an equilibrium-constant expression. The equilibrium-constant expression for an acidic system can be rearranged and solved for $[H_3O^+]$. In that way, the pH of a buffer solution can be obtained, if the composition of the solution is known. Alternatively, the *Henderson-Hasselbalch equation*, derived from the equilibrium constant expression, may be used to calculate the pH of a buffer solution.

8.5 Oxidation-Reduction Processes

Oxidation is defined as a loss of electrons, loss of hydrogen atoms, or gain of oxygen atoms. *Reduction* is defined as a gain of electrons, gain of hydrogen atoms, or loss of oxygen atoms.

Oxidation and reduction are complementary processes. The oxidation half-reaction produces an electron that is the reactant for the reduction half-reaction. The combination of two half-reactions, one oxidation and one reduction, produces the complete reaction.

The *reducing agent* releases electrons for the reduction of a second substance to occur. The *oxidizing agent* accepts electrons, causing the oxidation of a second substance to take place.

A *voltaic cell* is an electrochemical cell that converts chemical energy into electrical energy. *Electrolysis* is the opposite of a battery. It converts electrical energy into chemical potential energy.

KEY TERMS

acid (8.1)
amphiprotic (8.1)
anode (8.5)
Arrhenius theory (8.1)
autoionization (8.1)
base (8.1)
Brønsted-Lowry theory (8.1)
buffer capacity (8.4)
buffer solution (8.4)
buret (8.3)
cathode (8.5)
conjugate acid (8.1)
conjugate acid-base pair (8.1)
conjugate base (8.1)
corrosion (8.5)
electrolysis (8.5)
equivalence point (8.3)
Henderson-Hasselbalch equation (8.4)
hydronium ion (8.1)
indicator (8.3)
ion product for water (8.1)
neutralization (8.3)
oxidation (8.5)
oxidizing agent (8.5)
pH scale (8.2)
polyprotic substance (8.3)
reducing agent (8.5)
reduction (8.5)
standard solution (8.3)
titration (8.3)
voltaic cell (8.5)

QUESTIONS AND PROBLEMS

Acids and Bases

Foundations

8.27 a. Define an acid according to the Arrhenius theory.
      b. Define an acid according to the Brønsted-Lowry theory.
8.28 a. Define a base according to the Arrhenius theory.
      b. Define a base according to the Brønsted-Lowry theory.
8.29 What are the essential differences between the Arrhenius and Brønsted-Lowry theories?
8.30 Why is ammonia described as a Brønsted-Lowry base and not an Arrhenius base?

Applications

8.31 Write an equation for the reaction of each of the following with water:
      a. HNO_2
      b. HCN
8.32 Write an equation for the reaction of each of the following with water:
      a. HNO_3
      b. $HCOOH$
8.33 Select the conjugate acid-base pairs for each reaction in Question 8.31.
8.34 Select the conjugate acid-base pairs for each reaction in Question 8.32.
8.35 Label each of the following as a strong or weak acid (consult Figure 8.2, if necessary):
      a. H_2SO_3
      b. H_2CO_3
      c. H_3PO_4
8.36 Label each of the following as a strong or weak base (consult Figure 8.2, if necessary):
      a. KOH
      b. CN^-
      c. SO_4^{2-}
8.37 Identify the conjugate acid-base pairs in each of the following chemical equations:
      a. $NH_4^+ (aq) + CN^- (aq) \rightleftharpoons NH_3 (aq) + HCN(aq)$
      b. $CO_3^{2-} (aq) + HCl (aq) \rightleftharpoons HCO_3^- (aq) + Cl^- (aq)$
8.38 Identify the conjugate acid-base pairs in each of the following chemical equations:
      a. $HCOOH (aq) + NH_3 (aq) \rightleftharpoons HCOO^- (aq) + NH_4^+ (aq)$
      b. $HCl (aq) + OH^- (aq) \rightleftharpoons H_2O (l) + Cl^- (aq)$
8.39 Distinguish between the terms acid-base *strength* and acid-base *concentration*.
8.40 Of the diagrams shown here, which one represents:
      a. a concentrated strong acid
      b. a dilute strong acid
      c. a concentrated weak acid
      d. a dilute weak acid

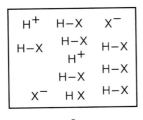

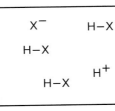

8.41 Classify each of the following as a Brønsted acid, Brønsted base, or both:
   a. H_3O^+
   b. OH^-
   c. H_2O

8.42 Classify each of the following as a Brønsted acid, Brønsted base, or both:
   a. NH_4^+
   b. NH_3

8.43 Classify each of the following as a Brønsted acid, Brønsted base, or both:
   a. H_2CO_3
   b. HCO_3^-
   c. CO_3^{2-}

8.44 Classify each of the following as a Brønsted acid, Brønsted base, or both:
   a. H_2SO_4
   b. HSO_4^-
   c. SO_4^{2-}

8.45 Write the formula of the conjugate acid of CN^-.
8.46 Write the formula of the conjugate acid of Br^-.
8.47 Write the formula of the conjugate base of HI.
8.48 Write the formula of the conjugate base of HCOOH.

pH of Acid and Base Solutions

8.49 Calculate the $[H_3O^+]$ of an aqueous solution that is:
   a. 1.0×10^{-7} M in OH^-
   b. 1.0×10^{-3} M in OH^-

8.50 Calculate the $[H_3O^+]$ of an aqueous solution that is:
   a. 1.0×10^{-9} M in OH^-
   b. 1.0×10^{-5} M in OH^-

8.51 Label each solution in Problem 8.49 as acidic, basic, or neutral.
8.52 Label each solution in Problem 8.50 as acidic, basic, or neutral.
8.53 Calculate the pH of a solution that has
   a. $[H_3O^+] = 1.0 \times 10^{-7}$
   b. $[OH^-] = 1.0 \times 10^{-9}$

8.54 Calculate the pH of a solution that has:
   a. $[H_3O^+] = 1.0 \times 10^{-10}$
   b. $[OH^-] = 1.0 \times 10^{-5}$

8.55 Calculate both $[H_3O^+]$ and $[OH^-]$ for a solution that is:
   a. pH = 1.00
   b. pH = 9.00

8.56 Calculate both $[H_3O^+]$ and $[OH^-]$ for a solution that is:
   a. pH = 5.00
   b. pH = 7.20

8.57 Calculate both $[H_3O^+]$ and $[OH^-]$ for a solution that is:
   a. pH = 1.30
   b. pH = 9.70

8.58 Calculate both $[H_3O^+]$ and $[OH^-]$ for a solution that is:
   a. pH = 5.50
   b. pH = 7.00

8.59 What is a neutralization reaction?
8.60 Describe the purpose of a titration.
8.61 The pH of urine may vary between 4.5 and 8.2. Determine the H_3O^+ concentration and OH^- concentration if the measured pH is:
   a. 6.00
   b. 5.20
   c. 7.80

8.62 The hydronium ion concentration in blood of three different patients was:

Patient	$[H_3O^+]$
A	5.0×10^{-8}
B	3.1×10^{-8}
C	3.2×10^{-8}

   What is the pH of each patient's blood? If the normal range is 7.30–7.50, which, if any, of these patients have an abnormal blood pH?

8.63 Determine how many times more acidic a solution is at:
   a. pH 2 relative to pH 4
   b. pH 7 relative to pH 11
   c. pH 2 relative to pH 12

8.64 Determine how many times more basic a solution is at:
   a. pH 6 relative to pH 4
   b. pH 10 relative to pH 9
   c. pH 11 relative to pH 6

8.65 What is the H_3O^+ concentration of a solution with a pH of:
   a. 5.0
   b. 12.0
   c. 5.5

8.66 What is the OH^- concentration of each solution in Question 8.65?
8.67 Calculate the pH of a solution with a H_3O^+ concentration of:
   a. 1.0×10^{-6} M
   b. 1.0×10^{-8} M
   c. 5.6×10^{-4} M

8.68 What is the OH^- concentration of each solution in Question 8.67?
8.69 Calculate the pH of a solution that has $[H_3O^+] = 7.5 \times 10^{-4}$ M.
8.70 Calculate the pH of a solution that has $[H_3O^+] = 6.6 \times 10^{-5}$ M.
8.71 Calculate the pH of a solution that has $[OH^-] = 5.5 \times 10^{-4}$ M.
8.72 Calculate the pH of a solution that has $[OH^-] = 6.7 \times 10^{-9}$ M.

Buffer Solutions

Foundations

8.73 Which of the following are capable of forming a buffer solution?
   a. NH_3 and NH_4Cl
   b. HNO_3 and KNO_3

8.74 Which of the following are capable of forming a buffer solution?
   a. HBr and $MgCl_2$
   b. H_2CO_3 and $NaHCO_3$

8.75 Define:
   a. buffer solution
   b. acidosis (refer to A Medical Perspective: Control of Blood pH on page 262)

8.76 Define:
   a. alkalosis (refer to A Medical Perspective: Control of Blood pH on page 262)
   b. standard solution

Applications

8.77 For the equilibrium situation involving acetic acid,

$$CH_3COOH(aq) + H_2O(l) \rightleftharpoons CH_3COO^-(aq) + H_3O^+(aq)$$

explain the equilibrium shift occurring for the following changes:
a. A strong acid is added to the solution.
b. The solution is diluted with water.

8.78 For the equilibrium situation involving acetic acid,

$$CH_3COOH(aq) + H_2O(l) \rightleftharpoons CH_3COO^-(aq) + H_3O^+(aq)$$

explain the equilibrium shift occurring for the following changes:
a. A strong base is added to the solution.
b. More acetic acid is added to the solution.

8.79 What is $[H_3O^+]$ for a buffer solution that is 0.200 M in acid and 0.500 M in the corresponding salt if the weak acid $K_a = 5.80 \times 10^{-7}$?

8.80 What is the pH of the solution described in Question 8.79?

8.81 What does K_a tell us about acid strength?

8.82 What does K_b tell us about base strength?

8.83 Calculate the pH of a buffer system containing 1.0 M CH_3COOH and 1.0 M CH_3COONa. (K_a of acetic acid, CH_3COOH, is 1.8×10^{-5})

8.84 Calculate the pH of a buffer system containing 1.0 M NH_3 and 1.0 M NH_4Cl. (K_a of NH_4^+, the acid in this system, is 5.6×10^{-10})

8.85 The pH of blood plasma is 7.40. The principal buffer system is HCO_3^-/H_2CO_3. Calculate the ratio $[HCO_3^-]/[H_2CO_3]$ in blood plasma. (K_a of H_2CO_3, carbonic acid, is 4.5×10^{-7})

8.86 The pH of blood plasma from a patient was found to be 7.6, a life-threatening situation. Calculate the ratio $[HCO_3^-]/[H_2CO_3]$ in this sample of blood plasma. (K_a of H_2CO_3, carbonic acid, is 4.5×10^{-7})

Oxidation-Reduction Reactions

8.87 Define:
a. oxidation
b. oxidizing agent

8.88 Define:
a. reduction
b. reducing agent

8.89 During an oxidation process in an oxidation-reduction reaction, does the species oxidized gain or lose electrons?

8.90 During an oxidation-reduction reaction, is the oxidizing agent oxidized or reduced?

8.91 During an oxidation-reduction reaction, is the reducing agent oxidized or reduced?

8.92 Do metals tend to be good oxidizing agents or good reducing agents?

8.93 In the following reaction, identify the oxidized species, reduced species, oxidizing agent, and reducing agent:

$$Cl_2(aq) + 2KI(aq) \longrightarrow 2KCl(aq) + I_2(aq)$$

8.94 In the following reaction, identify the oxidized species, reduced species, oxidizing agent, and reducing agent:

$$Zn(s) + Cu^{2+}(aq) \longrightarrow Zn^{2+}(aq) + Cu(s)$$

8.95 Write the oxidation and reduction half-reactions for the equation in Question 8.93.

8.96 Write the oxidation and reduction half-reactions for the equation in Question 8.94.

8.97 Explain the relationship between oxidation-reduction and voltaic cells.

8.98 Compare and contrast a battery and electrolysis.

8.99 Describe one application of voltaic cells.

8.100 Describe one application of electrolytic cells.

CRITICAL THINKING PROBLEMS

1. Acid rain is a threat to our environment because it can increase the concentration of toxic metal ions, such as Cd^{2+} and Cr^{3+}, in rivers and streams. If cadmium and chromium are present in sediment as $Cd(OH)_2$ and $Cr(OH)_3$, write reactions that demonstrate the effect of acid rain. Use the library or internet to find the properties of cadmium and chromium responsible for their environmental impact.

2. Aluminum carbonate is more soluble in acidic solution, forming aluminum cations. Write a reaction (or series of reactions) that explains this observation.

3. Carbon dioxide reacts with the hydroxide ion to produce the bicarbonate anion. Write the Lewis dot structures for each reactant and product. Label each as a Brønsted acid or base. Explain the reaction using the Brønsted theory. Why would the Arrhenius theory provide an inadequate description of this reaction?

4. Maalox is an antacid composed of $Mg(OH)_2$ and $Al(OH)_3$. Explain the origin of the trade name Maalox. Write chemical reactions that demonstrate the antacid activity of Maalox.

5. Acid rain has been described as a regional problem, whereas the greenhouse effect is a global problem. Do you agree with this statement? Why or why not?

GENERAL CHEMISTRY 9

The Nucleus, Radioactivity, and Nuclear Medicine

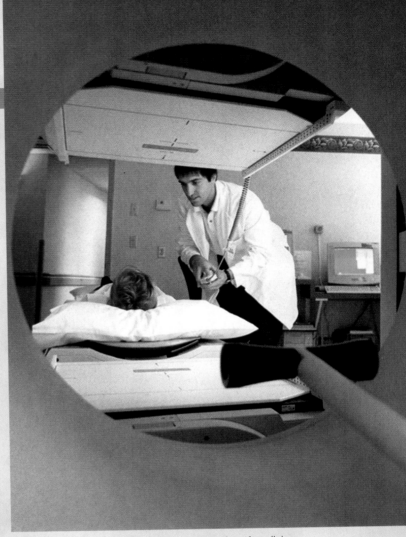

Nuclear technology has revolutionized the practice of medicine.

Learning Goals

1. Enumerate the characteristics of alpha, beta, and gamma radiation.
2. Write balanced equations for common nuclear processes.
3. Calculate the amount of radioactive substance remaining after a specified number of half-lives.
4. Describe the various ways in which nuclear energy may be used to generate electricity: fission, fusion, and the breeder reactor.
5. Explain the process of radiocarbon dating.
6. Cite several examples of the use of radioactive isotopes in medicine.
7. Describe the use of ionizing radiation in cancer therapy.
8. Discuss the preparation and use of radioisotopes in diagnostic imaging studies.
9. Explain the difference between natural and artificial radioactivity.
10. Describe the characteristics of radioactive materials that relate to radiation exposure and safety.
11. Be familiar with common techniques for the detection of radioactivity.
12. Know the common units in which radiation intensity is represented: the curie, roentgen, rad, and rem.

Outline

Chemistry Connection:
An Extraordinary Woman in Science

9.1 Natural Radioactivity
9.2 Writing a Balanced Nuclear Equation
9.3 Properties of Radioisotopes
9.4 Nuclear Power
9.5 Radiocarbon Dating
9.6 Medical Applications of Radioactivity

An Environmental Perspective:
Nuclear Waste Disposal

9.7 Biological Effects of Radiation

A Medical Perspective:
Magnetic Resonance Imaging

9.8 Measurement of Radiation

An Environmental Perspective:
Radon and Indoor Air Pollution

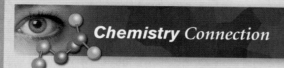

Chemistry Connection

An Extraordinary Woman in Science

The path to a successful career in science, or any other field for that matter, is seldom smooth or straight. That was certainly true for Madame Marie Sklodowska Curie. Her lifelong ambition was to raise a family and do something interesting for a career. This was a lofty goal for a nineteenth-century woman.

The political climate in Poland, coupled with the prevailing attitudes toward women and careers, especially careers in science, certainly did not make it any easier for Mme. Curie. To support herself and her sister, she toiled at menial jobs until moving to Paris to resume her studies.

It was in Paris that she met her future husband and fellow researcher, Pierre Curie. Working with crude equipment in a laboratory that was primitive, even by the standards of the time, she and Pierre made a most revolutionary discovery only two years after Henri Becquerel discovered radioactivity. Radioactivity, the emission of energy from certain substances, was released from *inside* the atom and was independent of the molecular form of the substance. The absolute proof of this assertion came only after the Curies processed over one *ton* of a material (pitchblende) to isolate less than a gram of pure radium. The difficult conditions under which this feat was accomplished are perhaps best stated by Sharon Bertsch McGrayne in her book *Nobel Prize Women in Science* (Birch Lane Press, New York, p. 23):

The only space large enough at the school was an abandoned dissection shed. The shack was stifling hot in summer and freezing cold in winter. It had no ventilation system for removing poisonous fumes, and its roof leaked. A chemist accustomed to Germany's modern laboratories called it "a cross between a stable and a potato cellar and, if I had not seen the work table with the chemical apparatus, I would have thought it a practical joke." This ramshackle shed became the symbol of the Marie Curie legend.

The pale green glow emanating from the radium was beautiful to behold. Mme. Curie would go to the shed in the middle of the night to bask in the light of her accomplishment. She did not realize that this wonderful accomplishment would, in time, be responsible for her death.

Mme. Curie received not one, but two Nobel Prizes, one in physics and one in chemistry. She was the first woman in France to earn the rank of professor.

As you study this chapter, the contributions of Mme. Curie, Pierre Curie, and the others of that time will become even more clear. Ironically, the field of medicine has been a major beneficiary of advances in nuclear and radiochemistry, despite the toxic properties of those same radioactive materials.

Introduction

Chapter 2 describes the electronic structure of atoms.

Our discussion of the atom and atomic structure revealed a nucleus containing protons and neutrons surrounded by electrons. Until now, we have treated the nucleus as simply a region of positive charge in the center of the atom. The focus of our interest has been the electrons and their arrangement around the nucleus. Electron arrangement is an essential part of a discussion of bonding or chemical change.

In this chapter we consider the nucleus and nuclear properties. The behavior of nuclei may have as great an effect on our everyday lives as any of the thousands of synthetic compounds developed over the past several decades. Examples of nuclear technology range from everyday items (smoke detectors) to sophisticated instruments for medical diagnosis and treatment and electrical power generation (nuclear power plants).

Beginning in 1896 with Becquerel's discovery of radiation emitted from uranium ore, the technology arising from this and related findings has produced both risks and benefits. Although early discoveries of radioactivity and its properties expanded our fundamental knowledge and brought fame to the investigators, it was not accomplished without a price. Several early investigators died prematurely of cancer and other diseases caused by the radiation they studied.

Even today, the existence of nuclear energy and its associated technology is a mixed blessing. On one side, the horrors of Nagasaki and Hiroshima, the fear of nuclear war, and potential contamination of populated areas resulting from the peaceful application of nuclear energy are critical problems facing society. Conversely, hundreds of thousands of lives have been saved because of the early detection of disease such as cancer by diagnosis based on the interaction of radiation and the body and the cure of cancer using techniques such as cobalt-60 treatment. Furthermore, nuclear energy is an alternative energy source, providing an opportunity for us to compensate for the depletion of oil reserves.

9.1 Natural Radioactivity

Radioactivity is the process by which some atoms emit energy and particles. The energy and particles are termed *radiation*. Nuclear radiation occurs as a result of an alteration in nuclear composition or structure. This process occurs in a nucleus that is unstable and hence radioactive. Radioactivity is a nuclear event: *matter and energy released during this process come from the nucleus.*

We shall designate the nucleus using *nuclear symbols*, analogous to the *atomic symbols* that were introduced in Section 2.1. The nuclear symbols consist of the *symbol* of the element, the *atomic number* (the number of protons in the nucleus), and the *mass number*, which is defined as the sum of neutrons and protons in the nucleus.

With the use of nuclear symbols, the fluorine nucleus is represented as

$$\text{Mass number} \longrightarrow {}^{19}_{9}\text{F} \longleftarrow \text{Symbol of the element}$$
$$\text{Atomic number (or nuclear charge)} \nearrow$$

Be careful not to confuse the mass number (a simple count of the neutrons and protons) with the atomic mass, which includes the contribution of electrons and is a true *mass* figure.

This symbol is equivalent to writing *fluorine-19*. This alternative representation is frequently used to denote specific isotopes of elements.

Not all nuclei are unstable. Only unstable nuclei undergo change and produce radioactivity, the process of radioactive decay. Recall that different atoms of the same element having different masses exist as *isotopes*. One isotope of an element may be radioactive, whereas others of the same element may be quite stable. It is important to distinguish between the terms *isotope* and **nuclide.** The term *isotope* refers to any atoms that have the same atomic number but different mass number. The term **nuclide** refers to any atom characterized by an atomic number and a mass number.

Isotopes are introduced in Section 2.1.

Many elements in the periodic table occur in nature as mixtures of isotopes. Two common examples include carbon (Figure 9.1),

$${}^{12}_{6}\text{C} \qquad {}^{13}_{6}\text{C} \qquad {}^{14}_{6}\text{C}$$
Carbon-12   Carbon-13   Carbon-14

and hydrogen,

$${}^{1}_{1}\text{H} \qquad {}^{2}_{1}\text{H} \qquad {}^{3}_{1}\text{H}$$
Hydrogen-1   Hydrogen-2   Hydrogen-3
Protium   Deuterium   Tritium
      (symbol D)   (symbol T)

Protium is a stable isotope and makes up more than 99.9% of naturally occurring hydrogen. Deuterium (D) can be isolated from hydrogen; it can form compounds such as "heavy water," D_2O. Heavy water is a potential source of deuterium for fusion processes. Tritium (T) is rare and unstable, hence radioactive.

Figure 9.1
Three isotopes of carbon. Each nucleus contains the same number of protons. Only the number of neutrons is different; hence, each isotope has a different mass.

Carbon-12 has
6 protons and
6 neutrons

Carbon-13 has
6 protons and
7 neutrons

Carbon-14 has
6 protons and
8 neutrons

In writing the symbols for a nuclear process, it is essential to indicate the particular isotope involved. This is why the mass number and atomic number are used. These values tell us the number of neutrons in the species, hence the isotope's identity.

Three types of natural radiation emitted by unstable nuclei are *alpha particles*, *beta particles*, and *gamma rays*.

Alpha Particles

An **alpha particle** (α) contains two protons and two neutrons. An alpha particle is identical to the nucleus of the helium atom (He) or a *helium ion* (He^{2+}), which also contains two protons (atomic number = 2) and two neutrons (mass number − atomic number = 2). Having no electrons to counterbalance the nuclear charge, the alpha particle may be symbolized as

$$^{4}_{2}He^{2+} \quad \text{or} \quad ^{4}_{2}He \quad \text{or} \quad \alpha$$

Alpha particles have a relatively large mass compared to other nuclear particles. Consequently, alpha particles emitted by radioisotopes are relatively slow-moving particles (approximately 10% of the speed of light), and they are stopped by barriers as thin as a few pages of this book.

Beta Particles

The **beta particle** (β), in contrast, is a fast-moving electron traveling at approximately 90% of the speed of light as it leaves the nucleus. It is formed in the nucleus by the conversion of a neutron into a proton. The beta particle is represented as

$$^{0}_{-1}e \quad \text{or} \quad ^{0}_{-1}\beta \quad \text{or} \quad \beta$$

The subscript −1 is written in the same position as the atomic number and, like the atomic number (number of protons), indicates the charge of the particle.

Beta particles are smaller and faster than alpha particles. They are more penetrating and are stopped only by more dense materials such as wood, metal, or several layers of clothing.

Gamma Rays

Gamma rays (γ) are the most energetic part of the electromagnetic spectrum (see Section 2.2), and result from nuclear processes; in contrast, alpha radiation and beta radiation are matter. Because electromagnetic radiation has no protons, neutrons, or electrons, the symbol for a gamma ray is simply

$$\gamma$$

Alpha, beta, and gamma radiation have widespread use in the field of medicine.

LEARNING GOAL 1

Other radiation particles, such as neutrinos and deuterons, will not be discussed here.

TABLE 9.1   A Summary of the Major Properties of Alpha, Beta, and Gamma Radiation

Name and Symbol	Identity	Charge	Mass (amu)	Velocity	Penetration
Alpha (α)	Helium nucleus	+2	4.0026	5–10% of the speed of light	Low
Beta (β)	Electron	−1	0.000549	Up to 90% of the speed of light	Medium
Gamma (γ)	Radiant energy	0	0	Speed of light	High

Gamma radiation is highly energetic and is the most penetrating form of nuclear radiation. Barriers of lead, concrete, or, more often, a combination of the two are required for protection from this type of radiation.

Properties of Alpha, Beta, and Gamma Radiation

Important properties of alpha, beta, and gamma radiation are summarized in Table 9.1.

Alpha, beta, and gamma radiation are collectively termed *ionizing radiation*. **Ionizing radiation** produces a trail of ions throughout the material that it penetrates. The ionization process changes the chemical composition of the material. When the material is living tissue, radiation-induced illness may result (Section 9.7).

The penetrating power of alpha radiation is very low. Damage to internal organs from this form of radiation is negligible except when an alpha particle emitter is actually ingested. Beta particles have much higher velocities than alpha particles; still, they have limited penetrating power. They cause skin and eye damage and, to a lesser extent, damage to internal organs. Shielding is required when working with beta emitters. Pregnant women must take special precautions. The great penetrating power and high energy of gamma radiation make it particularly difficult to shield. Hence, it can damage internal organs.

Anyone working with any type of radiation must take precautions. Radiation safety is required, monitored, and enforced in the United States under provisions of the Occupational Safety and Health Act (OSHA).

1 LEARNING GOAL

Question 9.1

Gamma radiation is a form of *electromagnetic radiation*. Provide examples of other forms of electromagnetic radiation.

Question 9.2

How does the energy of gamma radiation compare with that of other regions of the electromagnetic spectrum?

9.2 Writing a Balanced Nuclear Equation

Nuclear equations represent nuclear change in much the same way as chemical equations represent chemical change.

A **nuclear equation** can be used to represent the process of radioactive decay. In radioactive decay a *nuclide* breaks down, producing a *new nuclide, smaller particles, and/or energy*. The concept of mass balance, required when writing chemical equations, is also essential for nuclear equations. When writing a balanced equation, remember that:

2 LEARNING GOAL

Alpha Decay

Consider the decay of one isotope of uranium, $^{238}_{92}U$, into thorium and an alpha particle. Because an alpha particle is lost in this process, this decay is called *alpha decay*.
Examine the balanced equation for this nuclear reaction:

$$^{238}_{92}U \longrightarrow {}^{234}_{90}Th + {}^{4}_{2}He$$

Uranium-238   Thorium-234   Helium-4

The sum of the mass numbers on the right (234 + 4 = 238) are equal to the mass number on the left. The atomic numbers on the right (90 + 2 = 92) are equal to the atomic number on the left.

Beta Decay

Beta decay is illustrated by the decay of one of the less-abundant nitrogen isotopes, $^{16}_{7}N$. Upon decomposition, nitrogen-16 produces oxygen-16 and a beta particle. Conceptually, a neutron = proton + electron. In beta decay, one neutron in nitrogen-16 is converted to a proton and the electron, the beta particle, is released. The reaction is represented as

$$^{16}_{7}N \longrightarrow {}^{16}_{8}O + {}^{0}_{-1}e$$

or

$$^{16}_{7}N \longrightarrow {}^{16}_{8}O + \beta$$

Note that the mass number of the beta particle is zero, because the electron includes no protons or neutrons. Sixteen nuclear particles are accounted for on both sides of the reaction arrow. Note also that the product nuclide has the same mass number as the parent nuclide but the atomic number has *increased* by one unit.
The atomic number on the left (+7) is counterbalanced by [8 + (−1)] or (+7) on the right. Therefore the equation is correctly balanced.

Position Emission

The decay of carbon-11 to a stable isotope, boron-11, is one example of position emission.

$$^{11}_{6}C \longrightarrow {}^{11}_{5}B + {}^{0}_{1}e$$

or

$$^{11}_{6}C \longrightarrow {}^{11}_{5}B + {}^{0}_{1}\beta$$

A **position** has the same mass as an electron, or beta particle, but opposite (+) charge. In contrast to beta emission, the product nuclide has the same mass number as the parent nuclide, but the atomic number has *decreased* by one unit.
The atomic number on the left (+6) is counterbalanced by [5 + (+1)] or (+6) on the right. Therefore, the equation is correctly balanced.

Gamma Production

If *gamma radiation* were the only product of nuclear decay, there would be no measurable change in the mass or identity of the radioactive nuclei. This is so because the gamma emitter has simply gone to a lower energy state. An example of an isotope that decays in this way is technetium-99m. It is described as a **metastable isotope,** meaning that it is unstable and increases its stability through gamma decay without change in the mass or charge of the isotope. The letter *m* is used to denote a metastable isotope. The decay equation for $^{99m}_{43}Tc$ is

$$^{99m}_{43}Tc \longrightarrow {}^{99}_{43}Tc + \gamma$$

More often, gamma radiation is produced along with other products. For example, iodine-131 decays as follows:

$$^{131}_{53}I \longrightarrow {}^{131}_{54}Xe + {}^{0}_{-1}\beta + \gamma$$

Iodine-131   Xenon-131   Beta particle   Gamma ray

This reaction may also be represented as

$$^{131}_{53}I \longrightarrow {}^{131}_{54}Xe + {}^{0}_{-1}e + \gamma$$

An isotope of xenon, a beta particle, and gamma radiation are produced.

Predicting Products of Nuclear Decay

It is possible to use a nuclear equation to predict one of the products of a nuclear reaction if the others are known. Consider the following example, in which we represent the unknown product as ?:

$$^{40}_{19}K \longrightarrow ? + {}^{0}_{-1}e$$

Step 1. The mass number of this isotope of potassium is 40. Therefore the sum of the mass number of the products must also be 40, and ? must have a mass number of 40.

Step 2. Likewise, the atomic number on the left is 19, and the sum of the unknown atomic number plus the charge of the beta particle (-1) must equal 19.

Step 3. The unknown atomic number must be 20, because $[20 + (-1) = 19]$. The unknown is

$$^{40}_{20}?$$

If we consult the periodic table, the element that has atomic number 20 is calcium; therefore $? = {}^{40}_{20}Ca$.

EXAMPLE 9.1

Predicting the Products of Radioactive Decay

Determine the identity of the unknown product of the alpha decay of curium-245:

$$^{245}_{96}Cm \longrightarrow {}^{4}_{2}He + ?$$

Solution

Step 1. The mass number of the curium isotope is 245. Therefore the sum of the mass numbers of the products must also be 245, and ? must have a mass number of 241.

Step 2. Likewise, the atomic number on the left is 96, and the sum of the unknown atomic number plus the atomic number of the alpha particle (2) must equal 96.

Step 3. The unknown atomic number must be 94, because $[94 + 2 = 96]$. The unknown is

$$^{241}_{94}?$$

Referring to the periodic table, we find that the element that has atomic number 94 is plutonium; therefore $? = {}^{241}_{94}Pu$.

Question 9.3

Complete each of the following nuclear equations:

a. $^{85}_{36}Kr \longrightarrow ? + ^{0}_{-1}e$
b. $? \longrightarrow ^{4}_{2}He + ^{222}_{86}Rn$

Question 9.4

Complete each of the following nuclear equations:

a. $^{239}_{92}U \longrightarrow ? + ^{0}_{-1}e$
b. $^{11}_{5}B \longrightarrow ^{7}_{3}Li + ?$

9.3 Properties of Radioisotopes

Why are some isotopes radioactive but others are not? Do all radioactive isotopes decay at the same rate? Are all radioactive materials equally hazardous? We address these and other questions in this section.

Nuclear Structure and Stability

A measure of nuclear stability is the **binding energy** of the nucleus. The binding energy of the nucleus is the energy required to break up a nucleus into its component protons and neutrons. This must be very large, because identically charged protons in the nucleus exert extreme repulsive forces on one another. These forces must be overcome if the nucleus is to be stable. When a nuclide decays, some energy is released because the products are more stable than the parent nuclide. This released energy is the source of the high-energy radiation emitted and the basis for all nuclear technology.

Why are some isotopes more stable than others? The answer to this question is not completely clear. Evidence obtained so far points to several important factors that describe stable nuclei:

- Nuclear stability correlates with the ratio of neutrons to protons in the isotope. For example, for light atoms a neutron:proton ratio of 1 characterizes a stable atom.
- Nuclei with large numbers of protons (84 or more) tend to be unstable.
- Naturally occurring isotopes containing 2, 8, 20, 50, 82, or 126 protons or neutrons are stable. These *magic numbers* seem to indicate the presence of energy levels in the nucleus, analogous to electronic energy levels in the atom.
- Isotopes with even numbers of protons or neutrons are generally more stable than those with odd numbers of protons or neutrons.
- All isotopes (except hydrogen-1) with more protons than neutrons are unstable. However, the reverse is not true.

Half-Life

LEARNING GOAL

The **half-life ($t_{1/2}$)** is the time required for one-half of a given quantity of a substance to undergo change. Not all radioactive isotopes decay at the same rate. The rate of nuclear decay is generally represented in terms of the half-life of the isotope. Each isotope has its own characteristic half-life that may be as short as a few millionths of a second or as long as billions of years. Half-lives of some naturally occurring and synthetic isotopes are given in Table 9.2.

TABLE 9.2 Half-Lives of Selected Radioisotopes

Name	Symbol	Half-Life
Carbon-14	$^{14}_{6}C$	5730 years
Cobalt-60	$^{60}_{27}Co$	5.3 years
Hydrogen-3	$^{3}_{1}H$	12.3 years
Iodine-131	$^{131}_{53}I$	8.1 days
Iron-59	$^{59}_{26}Fe$	45 days
Molybdenum-99	$^{99}_{42}Mo$	67 hours
Sodium-24	$^{24}_{11}Na$	15 hours
Strontium-90	$^{90}_{38}Sr$	28 years
Technetium-99m	$^{99m}_{43}Tc$	6 hours
Uranium-235	$^{235}_{92}U$	710 million years

The stability of an isotope is indicated by the isotope's half-life. Isotopes with short half-lives decay rapidly; they are very unstable. This is not meant to imply that substances with long half-lives are less hazardous. Often, just the reverse is true.

Imagine that we begin with 100 mg of a radioactive isotope that has a half-life of 24 hours. After one half-life, or 24 hours, 1/2 of 100 mg will have decayed to other products, and 50 mg remain. After two half-lives (48 hours), 1/2 of the remaining material has decayed, leaving 25 mg, and so forth:

$$100 \text{ mg} \xrightarrow[\text{Half-life (24 h)}]{\text{One}} 50 \text{ mg} \xrightarrow[\text{Half-life (48 h total)}]{\text{A Second}} 25 \text{ mg} \longrightarrow \text{etc.}$$

Decay of a radioisotope that has a reasonably short $t_{1/2}$ is experimentally determined by following its activity as a function of time. Graphing the results produces a radioactive decay curve as shown in Figure 9.2.

The mass of any radioactive substance remaining after a period may be calculated with a knowledge of the initial mass and the half-life of the isotope, following the scheme just outlined. The general equation for this process is:

$$m_f = m_i(.5)^n$$

Refer to the discussion of radiation exposure and safety in Sections 9.7 and 9.8.

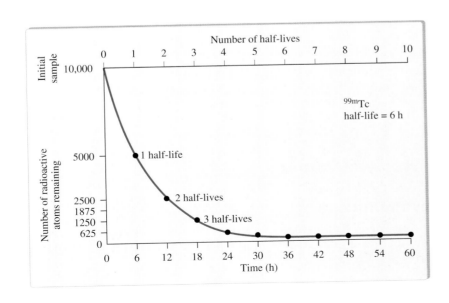

Figure 9.2
The decay curve for the medically useful radioisotope technetium-99m. Note that the number of radioactive atoms remaining—hence the radioactivity—approaches zero.

where m_f = final or remaining mass
 m_i = initial mass
 n = number of half-lives

EXAMPLE 9.2 Predicting the Extent of Radioactive Decay

A 50.0-mg supply of iodine-131, used in hospitals in the treatment of hyperthyroidism, was stored for 32.4 days. If the half-life of iodine-131 is 8.1 days, how many milligrams remain?

Solution

First calculate n, the number of half-lives elapsed using the half-life as a conversion factor:

$$n = 32.4 \text{ days} \times \frac{1 \text{ half-life}}{8.1 \text{ days}} = 4.0 \text{ half-lives}$$

Then calculate the amount remaining:

$$50.0 \text{ mg} \xrightarrow{\text{first half-life}} 25.0 \text{ mg} \xrightarrow{\text{second half-life}} 12.5 \text{ mg} \xrightarrow{\text{third half-life}} 6.25 \text{ mg} \xrightarrow{\text{fourth half-life}} 3.13 \text{ mg}$$

Hence, 3.13 mg of iodine-131 remain after 32.4 days.

An Alternate Strategy

Use the equation

$$m_f = m_i(.5)^n$$
$$m_f = 50.0 \text{ mg}(.5)^4$$
$$m_f = 3.13 \text{ mg of iodine-131 remain after 32.4 days.}$$

Note that both strategies produce the same answer.

Question 9.5

A 100.0-ng sample of sodium-24 was stored in a lead-lined cabinet for 2.5 days. How much sodium-24 remained? See Table 9.2 for the half-life of sodium-24.

Question 9.6

If a patient is administered 10 ng of technetium-99m, how much will remain one day later, assuming that no technetium has been eliminated by any other process? See Table 9.2 for the half-life of technetium-99m.

9.4 Nuclear Power

Energy Production

LEARNING GOAL 4

Einstein predicted that a small amount of nuclear mass corresponds to a very large amount of energy that is released when the nucleus breaks apart. Einstein's equation is

$$E = mc^2$$

An Environmental Perspective

Nuclear Waste Disposal

Nuclear waste arises from a variety of sources. A major source is the spent fuel from nuclear power plants. Medical laboratories generate significant amounts of low-level waste from tracers and therapy. Even household items with limited lifetimes, such as certain types of smoke detectors, use a tiny amount of radioactive material.

Virtually everyone is aware, through television and newspapers, of the problems of solid waste (nonnuclear) disposal that our society faces. For the most part, this material will degrade in some reasonable amount of time. Still, we are disposing of trash and garbage at a rate that far exceeds nature's ability to recycle it.

Now imagine the problem with nuclear waste. We cannot alter the rate at which it decays. This is defined by the half-life. We can't heat it, stir it, or add a catalyst to speed up the process as we can with chemical reactions. Furthermore, the half-lives of many nuclear waste products are very long: plutonium, for example, has a half-life in excess of 24,000 years. Ten half-lives represents the approximate time required for the radioactivity of a substance to reach background levels. So we are talking about a *very* long storage time.

Where on earth can something so very hazardous be contained and stored with reasonable assurance that it will lie undisturbed for a quarter of a million years? Perhaps this is a rhetorical question. Scientists, engineers, and politicians have debated this question for almost fifty years. As yet, no permanent disposal site has been agreed upon. Most agree that the best solution is burial in a stable rock formation, but there is no firm agreement on the location. Fear of earthquakes, which may release large quantities of radioactive materials into the underground water system, is the most serious consideration. Such a disaster could render large sections of the country unfit for habitation.

Many argue for the continuation of temporary storage sites with the hope that the progress of science and technology will, in the years ahead, provide a safer and more satisfactory long-term solution.

A photograph of the earth, taken from the moon, clearly illustrates the limits of resources and the limits to waste disposal.

The nuclear waste problem, important for its own sake, also affects the development of future societal uses of nuclear chemistry. Before we can fully enjoy its benefits, we must learn to use and dispose of it safely.

For Further Understanding

Summarize the major arguments supporting expanded use of nuclear power for electrical energy.

Enumerate the characteristics of an "ideal" solution to the nuclear waste problem.

in which

$$E = \text{energy}$$

$$m = \text{mass}$$

$$c = \text{speed of light}$$

This kinetic energy, when rapidly released, is the basis for the greatest instruments of destruction developed by humankind, nuclear bombs. However, when heat energy is released in a controlled fashion, as in a nuclear power plant, the heat energy converts liquid water into steam. The steam, in turn, drives an electrical generator, producing electricity.

Nuclear Fission

Fission (splitting) occurs when a heavy nuclear particle is split into smaller nuclei by a smaller nuclear particle (such as a neutron). This splitting process is accompanied by the release of large amounts of energy.

A nuclear power plant uses a fissionable material (capable of undergoing fission), such as uranium-235, as fuel. The energy released by the fission process in the nuclear core heats water in an adjoining chamber, producing steam. The high pressure of the steam drives a turbine, which converts this heat energy into electricity using an electric power generator. The energy transformation may be summarized as follows:

nuclear energy → heat energy → mechanical energy → electrical energy
Nuclear reactor — Steam — Turbine — Electricity

The fission reaction, once initiated, is self-perpetuating. For example, neutrons are used to initiate the reaction:

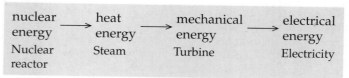

$$^1_0 n + ^{235}_{92}U \longrightarrow ^{236}_{92}U \longrightarrow ^{92}_{36}Kr + ^{141}_{56}Ba + 3\,^1_0 n + \text{energy}$$

Fuel — Unstable — Products of reaction

Note that three neutrons are released as product for each single reacting neutron. Each of the three neutrons produced is available to initiate another fission process. Nine neutrons are released from this process. These, in turn, react with other nuclei. The fission process continues and intensifies, producing very large amounts of energy (Figure 9.3). This process of intensification is referred to as a **chain reaction.**

To maintain control over the process and to prevent dangerous overheating, rods fabricated from cadmium or boron are inserted into the core. These rods,

Figure 9.3
The fission of uranium-235 producing a chain reaction. Note that the number of available neutrons, which "trigger" the decomposition of the fissionable nuclei to release energy, increases at each step in the "chain." In this way the reaction builds in intensity. Control rods stabilize (or limit) the extent of the chain reaction to a safe level.

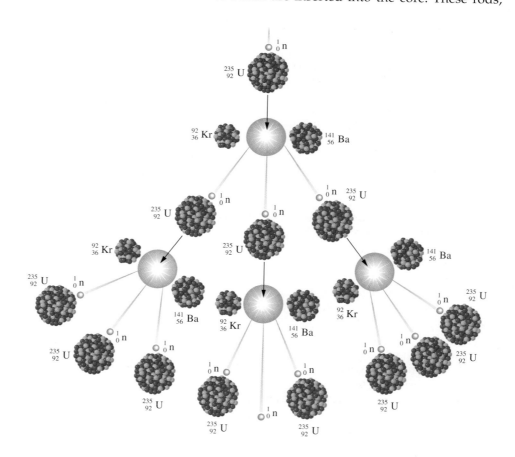

9.4 Nuclear Power

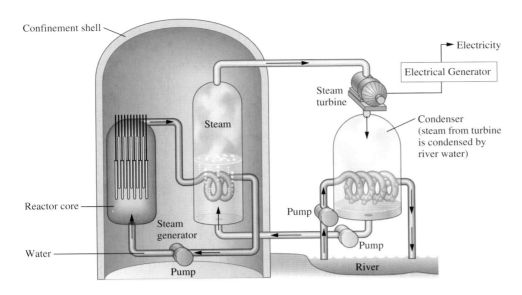

Figure 9.4
A representation of the "energy zones" of a nuclear reactor. Heat produced by the reactor core is carried by water in a second zone to a boiler. Water in the boiler (third zone) is converted to steam, which drives a turbine to convert heat energy to electrical energy. The isolation of these zones from each other allows heat energy transfer without actual physical mixing. This minimizes the transport of radioactive material into the environment.

which are controlled by the reactor's main operating system, absorb free neutrons as needed, thereby moderating the reaction.

A nuclear fission reactor may be represented as a series of energy transfer zones, as depicted in Figure 9.4. A view of the core of a fission reactor is shown in Figure 9.5.

Nuclear Fusion

Fusion (meaning *to join together*) results from the combination of two small nuclei to form a larger nucleus with the concurrent release of large amounts of energy. The best example of a fusion reactor is the sun. Continuous fusion processes furnish our solar system with light and heat.

An example of a fusion reaction is the combination of two isotopes of hydrogen, deuterium (^2_1H) and tritium (^3_1H), to produce helium, a neutron, and energy:

$$^2_1H + ^3_1H \longrightarrow ^4_2He + ^1_0n + \text{energy}$$

Although fusion is capable of producing tremendous amounts of energy, no commercially successful fusion plant exists in the United States. Safety concerns relating to problems of containment of the reaction, resulting directly from the technological problems associated with containing high temperatures (millions of degrees) and pressures required to sustain a fusion process, have slowed the development of fusion reactors.

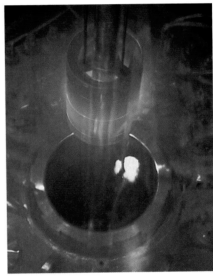

Figure 9.5
The core of a nuclear reactor located at Oak Ridge National Laboratories in Tennessee.

Breeder Reactors

A **breeder reactor** is a variation of a fission reactor that literally manufactures its own fuel. A perceived shortage of fissionable isotopes makes the breeder an attractive alternative to conventional fission reactors. A breeder reactor uses $^{238}_{92}U$, which is abundant but nonfissionable. In a series of steps, the uranium-238 is converted to plutonium-239, which *is* fissionable and undergoes a fission chain reaction, producing energy. The attractiveness of a reactor that makes its own fuel from abundant starting materials is offset by the high cost of the system, potential environmental damage, and fear of plutonium proliferation. Plutonium can be readily used to manufacture nuclear bombs. Currently only France and Japan operate breeder reactors for electrical power generation.

9.5 Radiocarbon Dating

Natural radioactivity is useful in establishing the approximate age of objects of archaeological, anthropological, or historical interest. **Radiocarbon dating** is the estimation of the age of objects through measurement of isotopic ratios of carbon.

Radiocarbon dating is based on the measurement of the relative amounts (or ratio) of $^{14}_{6}C$ and $^{12}_{6}C$ present in an object. The $^{14}_{6}C$ is formed in the upper atmosphere by the bombardment of $^{14}_{7}N$ by high-speed neutrons (cosmic radiation):

$$^{14}_{7}N + ^{1}_{0}n \longrightarrow ^{14}_{6}C + ^{1}_{1}H$$

The carbon-14, along with the more abundant carbon-12, is converted into living plant material by the process of photosynthesis. Carbon proceeds up the food chain as the plants are consumed by animals, including humans. When a plant or animal dies, the uptake of both carbon-14 and carbon-12 ceases. However, the amount of carbon-14 slowly decreases because carbon-14 is radioactive ($t_{1/2}$ = 5730 years). Carbon-14 decay produces nitrogen:

$$^{14}_{6}C \longrightarrow ^{14}_{7}N + ^{0}_{-1}e$$

When an artifact is found and studied, the relative amounts of carbon-14 and carbon-12 are determined. By using suitable equations involving the $t_{1/2}$ of carbon-14, it is possible to approximate the age of the artifact.

This technique has been widely used to increase our knowledge about the history of the earth, to establish the age of objects (Figure 9.6), and even to detect art forgeries. Early paintings were made with inks fabricated from vegetable dyes (plant material that, while alive, metabolized carbon).

The carbon-14 dating technique is limited to objects that are less than fifty thousand years old, or approximately nine half-lives, which is a practical upper limit. Older objects that have geological or archaeological significance may be dated using naturally occurring isotopes having much longer half-lives.

Examples of useful dating isotopes are listed in Table 9.3.

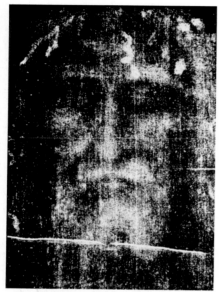

Figure 9.6
Radiocarbon dating was used in the authentication study of the Shroud of Turin. It is a minimally destructive technique and is valuable in estimating the age of historical artifacts.

TABLE 9.3 Isotopes Useful in Radioactive Dating			
Isotope	Half-Life (years)	Upper Limit (years)	Dating Applications
Carbon-14	5730	5×10^4	Charcoal, organic material, artwork
Tritium ($^{3}_{1}H$)	12.3	1×10^2	Aged wines, artwork
Potassium-40	1.3×10^9	Age of earth (4×10^9)	Rocks, planetary material
Rhenium-187	4.3×10^{10}	Age of earth (4×10^9)	Meteorites
Uranium-238	4.5×10^9	Age of earth (4×10^9)	Rocks, earth's crust

9.6 Medical Applications of Radioactivity

The use of radiation in the treatment of various forms of cancer, as well as the newer area of **nuclear medicine,** the use of radioisotopes in diagnosis, has become widespread in the past quarter century. Let's look at the properties of radiation that make it an indispensable tool in modern medical care.

Cancer Therapy Using Radiation

When high-energy radiation, such as gamma radiation, passes through a cell, it may collide with one of the molecules in the cell and cause it to lose one or more electrons, causing a series of events that result in the production of ion pairs. For this reason, such radiation is termed *ionizing radiation* (Section 9.1).

Ions produced in this way are highly energetic. Consequently they may damage biological molecules and cause changes in cellular biochemical processes. Interaction of ionizing radiation with intracellular water produces free electrons and other particles that can damage DNA. This may result in diminished or altered cell function or, in extreme cases, the death of the cell.

An organ that is cancerous is composed of both healthy cells and malignant cells. Tumor cells are more susceptible to the effects of gamma radiation than normal cells because they are undergoing cell division more frequently. Therefore exposure of the tumor area to carefully targeted and controlled dosages of high-energy gamma radiation from cobalt-60 (a high-energy gamma ray source) will kill a higher percentage of abnormal cells than normal cells. If the dosage is administered correctly, a sufficient number of malignant cells will die, destroying the tumor, and enough normal cells will survive to maintain the function of the affected organ.

Gamma radiation can cure cancer. Paradoxically, the exposure of healthy cells to gamma radiation can actually cause cancer. For this reason, radiation therapy for cancer is a treatment that requires unusual care and sophistication.

Nuclear Medicine

The diagnosis of a host of biochemical irregularities or diseases of the human body has been made routine through the use of radioactive tracers. Medical **tracers** are small amounts of radioactive substances used as probes to study internal organs. Medical techniques involving tracers are **nuclear imaging** procedures.

A small amount of the tracer, an isotope of an element that is known to be attracted to the organ of interest, is administered to the patient. For a variety of reasons, such as ease of administration of the isotope to the patient and targeting the organ of interest, the isotope is often a part of a larger molecule or ion. Because the isotope is radioactive, its path may be followed by using suitable detection devices. A "picture" of the organ is obtained, often far more detailed than is possible with conventional X-rays. Such techniques are noninvasive; that is, surgery is not required to investigate the condition of the internal organ, eliminating the risk associated with an operation.

The radioactive isotope of an element chosen for tracer studies has chemical behavior similar to any other isotope of the same element. For example, iodine-127, the most abundant nonradioactive isotope of iodine, is used by the body in the synthesis of thyroid hormones and tends to concentrate in the thyroid gland. Both radioactive iodine-131 and iodine-127 behave in the same way, making it possible to use iodine-131 to study the thyroid. The rate of uptake of the radioactive isotope gives valuable information regarding underactivity or overactivity (hypoactive or hyperactive thyroid).

Isotopes with short half-lives are preferred for tracer studies. These isotopes emit their radiation in a more concentrated burst (short half-life materials have greater activity), facilitating their detection. If the radioactive decay is easily detected, the method is more sensitive and thus capable of providing more information. Furthermore, an isotope with a short half-life decays to background more rapidly. This is a mechanism for removal of the radioactivity from the body. If the radioactive element is also rapidly metabolized and excreted, this is obviously beneficial as well.

The following examples illustrate the use of imaging procedures for diagnosis of disease.

TABLE 9.4 Isotopes Commonly Used in Nuclear Medicine

Area of Body	Isotope	Use
Blood	Red blood cells tagged with chromium-51	Determine blood volume in body
Bone	*Technetium-99m, barium-131	Allow early detection of the extent of bone tumors and active sites of rheumatoid arthritis
Brain	*Technetium-99m	Detect and locate brain tumors and stroke
Coronary artery	Thallium-201	Determine the presence and location of obstructions in coronary arteries
Heart	*Technetium-99m	Determine cardiac output, size, and shape
Kidney	*Technetium-99m	Determine renal function and location of cysts; a common follow-up procedure for kidney transplant patients
Liver-spleen	*Technetium-99m	Determine size and shape of liver and spleen; location of tumors
Lung	Xenon-133	Determine whether lung fills properly; locate region of reduced ventilation and tumors
Thyroid	Iodine-131	Determine rate of iodine uptake by thyroid

*The destination of this isotope is determined by the identity of the compound in which it is incorporated.

- *Bone disease and injury.* The most widely used isotope for bone studies is technetium-99m, which is incorporated into a variety of ions and molecules that direct the isotope to the tissue being investigated. Technetium compounds containing phosphate are preferentially adsorbed on the surface of bone. New bone formation (common to virtually all bone injuries) increases the incorporation of the technetium compound. As a result, an enhanced image appears at the site of the injury. Bone tumors behave in a similar fashion.
- *Cardiovascular diseases.* Thallium-201 is used in the diagnosis of coronary artery disease. The isotope is administered intravenously and delivered to the heart muscle in proportion to the blood flow. Areas of restricted flow are observed as having lower levels of radioactivity, indicating some type of blockage.
- *Pulmonary disease.* Xenon is one of the noble gases. Radioactive xenon-133 may be inhaled by the patient. The radioactive isotope will be transported from the lungs and distributed through the circulatory system. Monitoring the distribution, as well as the reverse process, the removal of the isotope from the body (exhalation), can provide evidence of obstructive pulmonary disease, such as cancer or emphysema.

Examples of useful isotopes and the organ(s) in which they tend to concentrate are summarized in Table 9.4.

For many years, imaging with radioactive tracers was used exclusively for diagnosis. Recent applications have expanded to other areas of medicine. Imaging is now used extensively to guide surgery, assist in planning radiation therapy, and support the technique of angioplasty.

Question 9.7

Technetium-99m is used in diagnostic imaging studies involving the brain. What fraction of the radioisotope remains after 12 hours have elapsed? See Table 9.2 for the half-life of technetium-99m.

> **Question 9.8**
>
> Barium-131 is a radioisotope used to study bone formation. A patient ingested barium-131. How much time will elapse until only one-fourth of the barium-131 remains, assuming that none of the isotope is eliminated from the body through normal processes? The half-life of barium-131 is 11.6 minutes.

Making Isotopes for Medical Applications

LEARNING GOAL 8

LEARNING GOAL 9

In early experiments with radioactivity, the radioactive isotopes were naturally occurring. For this reason the radioactivity produced by these unstable isotopes is described as **natural radioactivity.** If, on the other hand, a normally stable, nonradioactive nucleus is made radioactive, the resulting radioactivity is termed **artificial radioactivity.** The stable nucleus is made unstable by the introduction of "extra" protons, neutrons, or both.

The process of forming radioactive substances is often accomplished in the core of a **nuclear reactor,** in which an abundance of small nuclear particles, particularly neutrons, is available. Alternatively, extremely high-velocity charged particles (such as alpha and beta particles) may be produced in **particle accelerators,** such as a cyclotron. Accelerators are extremely large and use magnetic and electric fields to "push and pull" charged particles toward their target at very high speeds. A portion of the accelerator at the Brookhaven National Laboratory is shown in Figure 9.7.

Many isotopes that are useful in medicine are produced by particle bombardment. A few examples include the following:

- Gold-198, used as a tracer in the liver, is prepared by neutron bombardment.

$$^{197}_{79}\text{Au} + ^{1}_{0}\text{n} \longrightarrow ^{198}_{79}\text{Au}$$

- Gallium-67, used in the diagnosis of Hodgkin's disease, is prepared by proton bombardment.

$$^{66}_{30}\text{Zn} + ^{1}_{1}\text{p} \longrightarrow ^{67}_{31}\text{Ga}$$

Figure 9.7
A portion of a linear accelerator located at Brookhaven National Laboratory in New York. Particles can be accelerated at velocities close to the speed of light and accurately strike small "target" nuclei. At such facilities, rare isotopes can be synthesized and their properties studied.

A Medical Perspective

Magnetic Resonance Imaging

The Nobel prize in physics was awarded to Otto Stern in 1943 and to Isidor Rabi in 1944. They discovered that certain atomic nuclei have a property known as spin, analogous to the spin associated with electrons which we discussed in Chapter 2. The spin of electrons is responsible for the magnetic properties of atoms. Spinning nuclei behave as tiny magnets, producing magnetic fields as well.

One very important aspect of this phenomenon is the fact that the atoms in close proximity to the spinning nuclei (its chemical environment) exert an effect on the nuclear spin. In effect, measurable differences in spin are indicators of their surroundings. This relationship has been exhaustively studied for one atom in particular, hydrogen, and magnetic resonance techniques have become useful tools for the study of molecules containing hydrogen.

Human organs and tissue are made up of compounds containing hydrogen atoms. In the 1970s and 1980s the experimental technique was extended beyond tiny laboratory samples of pure compounds to the most complex sample possible—the human body. The result of these experiments is termed *magnetic resonance imaging (MRI)*.

Dr. Paul Barnett of the Greater Baltimore Medical Center studies images obtained using MRI.

MRI is noninvasive to the body, requires no use of radioactive substances, and is quick, safe, and painless. A person is placed in a cavity surrounded by a magnetic field, and an image (based on the extent of radio frequency energy absorption) is generated, stored, and sorted in a computer. Differences between normal and malignant tissue, atherosclerotic thickening of an aortal wall, and a host of other problems may be seen clearly in the final image.

Advances in MRI technology have provided medical practitioners with a powerful tool in diagnostic medicine. This is but one more example of basic science leading to technological advancement.

For Further Understanding

Why is hydrogen a useful atom to study in biological systems?

Why would MRI provide minimal information about bone tissue?

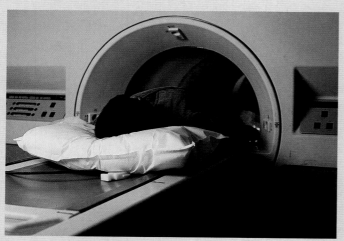

A patient entering an MRI scanner.

Some medically useful isotopes, with short half-lives, must be prepared near the site of the clinical test. Preparation and shipment from a reactor site would waste time and result in an isotopic solution that had already undergone significant decay, resulting in diminished activity.

A common example is technetium-99m. It has a half-life of only six hours. It is prepared in a small generator, often housed in a hospital's radiology laboratory (Figure 9.8). The generator contains radioactive molybdate ion (MoO_4^{2-}). Molybdenum-99 is more stable than technetium-99m; it has a half-life of 67 hours.

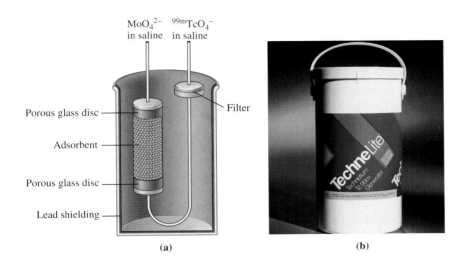

Figure 9.8
Preparation of technetium-99m.
(a) A diagram depicting the conversion of $^{99}MoO_4^{2-}$ to $^{99m}TcO_4^-$ through radioactive decay. The radioactive pertechnetate ion is periodically removed from the generator in saline solution and used in tracer studies.
(b) A photograph of a commercially available technetium-99m generator suitable for use in a hospital laboratory.

The molybdenum in molybdate ion decays according to the following nuclear equation:

$$^{99}_{42}Mo \longrightarrow {}^{99m}_{43}Tc + {}^{0}_{-1}e$$

Chemically, radioactive molybdate MoO_4^{2-} converts to radioactive pertechnetate ion (TcO_4^-). The radioactive TcO_4^- is removed from the generator when needed. It is administered to the patient as an aqueous salt solution that has an osmotic pressure identical to that of human blood.

9.7 Biological Effects of Radiation

It is necessary to use suitable precautions in working with radioactive substances. The chosen protocol is based on an understanding of the effects of radiation, dosage levels and "tolerable levels," the way in which radiation is detected and measured, and the basic precepts of radiation safety.

Radiation Exposure and Safety

In working with radioactive materials, the following factors must be considered.

 **LEARNING GOAL**

The Magnitude of the Half-Life

In considering safety, isotopes with short half-lives have, at the same time, one major disadvantage and one major advantage.

On one hand, short-half-life radioisotopes produce a larger amount of radioactivity per unit time than a long-half-life substance. For example, consider equal amounts of hypothetical isotopes that produce alpha particles. One has a half-life of ten days; the other has a half-life of one hundred days. After one half-life, each substance will produce exactly the same number of alpha particles. However, the first substance generates the alpha particles in only one-tenth of the time, hence emits ten times as much radiation per unit time. Equal exposure times will result in a higher level of radiation exposure for substances with short half-lives, and lower levels for substances with long half-lives.

On the other hand, materials with short half-lives (weeks, days, or less) may be safer to work with, especially if an accident occurs. Over time (depending on the magnitude of the half-life) radioactive isotopes will decay to **background radiation** levels. This is the level of radiation attributable to our surroundings on a day-to-day basis.

Higher levels of exposure in a short time produce clearer images.

Virtually all matter is composed of both radioactive and nonradioactive isotopes. Small amounts of radioactive material in the air, water, soil, and so forth make up a part of the background levels. Cosmic rays from outer space continually bombard us with radiation, contributing to the total background. Owing to the inevitability of background radiation, there can be no situation on earth where we observe zero radiation levels.

An isotope with a short half-life, for example 5.0 min, may decay to background in as few as ten half-lives

$$10 \text{ half-lives} \times \frac{5.0 \text{ min}}{1 \text{ half-life}} = 50 \text{ min}$$

See An Environmental Perspective: Nuclear Waste Disposal on page 285.

A spill of such material could be treated by waiting ten half-lives, perhaps by going to lunch. When you return to the laboratory, the material that was spilled will be no more radioactive than the floor itself. An accident with plutonium-239, which has a half-life of 24,000 years, would be quite a different matter! After fifty minutes, virtually all of the plutonium-239 would still remain. Long-half-life isotopes, by-products of nuclear technology, pose the greatest problems for safe disposal. Finding a site that will remain undisturbed "forever" is quite a formidable task.

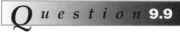

Describe the advantage of using isotopes with short half-lives for tracer applications in a medical laboratory.

Can you think of any disadvantage associated with the use of isotopes described in Question 9.9? Explain.

Shielding

Alpha and beta particles, being relatively low in penetrating power, require low-level **shielding.** A lab coat and gloves are generally sufficient protection from this low-penetration radiation. On the other hand, shielding made of lead, concrete, or both is required for gamma rays (and X-rays, which are also high-energy radiation). Extensive manipulation of gamma emitters is often accomplished in laboratory and industrial settings by using robotic control: computer-controlled mechanical devices that can be programmed to perform virtually all manipulations normally carried out by humans.

Distance from the Radioactive Source

Radiation intensity varies *inversely* with the *square* of the distance from the source. Doubling the distance from the source *decreases* the intensity by a factor of four (2^2). Again, the use of robot manipulators is advantageous, allowing a greater distance between the operator and the radioactive source.

Time of Exposure

The effects of radiation are cumulative. Generally, potential damage is directly proportional to the time of exposure. Workers exposed to moderately high levels of radiation on the job may be limited in the time that they can perform that task. For example, workers involved in the cleanup of the Three Mile Island nuclear plant, incapacitated in 1979, observed strict limits on the amount of time that they could be involved in the cleanup activities.

An Environmental Perspective

Radon and Indoor Air Pollution

Marie and Pierre Curie first discovered that air in contact with radium compounds became radioactive. Later experiments by Ernest Rutherford and others isolated the radioactive substance from the air. This substance was an isotope of the noble gas radon (Rn).

We now know that radium (Ra) produces radon by spontaneous decay:

$$^{226}_{88}Ra \longrightarrow {}^{4}_{2}He + {}^{222}_{86}Rn$$

Radium in trace quantities is found in the soil and rock and is unequally distributed in the soil. The decay product, radon, is emitted from the soil to the surrounding atmosphere. Radon is also found in higher concentrations when uranium is found in the soil. This is not surprising, because radium is formed as a part of the stepwise decay of uranium.

If someone constructs a building over soil or rock that has a high radium content (or uses stone with a high radium content to build the foundation!), the radon gas can percolate through the basement and accumulate in the house. Couple this with the need to build more energy-efficient, well-insulated dwellings, and the radon levels in buildings in some regions of the country can become quite high.

Radon itself is radioactive; however, its radiation is not the major problem. Because it is a gas and chemically inert, it is rapidly exhaled after breathing. However, radon decays to polonium:

$$^{222}_{86}Rn \longrightarrow {}^{4}_{2}He + {}^{218}_{84}Po$$

This polonium isotope is radioactive and is a nonvolatile heavy metal that can attach itself to bronchial or lung tissue, emitting hazardous radiation and producing other isotopes that are also radioactive.

In the United States, homes are now being tested and monitored for radon. In many states, proof of acceptable levels of radon is a condition of sale of the property. Studies continue to attempt to find reasonable solutions to the radon problem. Current recommendations include sealing cracks and openings in basements, increasing ventilation, and evaluating sites before construction of buildings. Debate continues within the scientific community regarding a safe and attainable indoor air quality standard for radon.

For Further Understanding

Why is indoor radon more hazardous than outdoor radon?

Polonium-218 has a very long half-life. Explain why this constitutes a potential health problem.

Types of Radiation Emitted

Alpha and beta emitters are generally less hazardous than gamma emitters, owing to differences in energy and penetrating power that require less shielding. However, ingestion or inhalation of an alpha emitter or beta emitter can, over time, cause serious tissue damage; the radioactive substance is in direct contact with sensitive tissue. An Environmental Perspective: Radon and Indoor Air Pollution (above) expands on this problem.

Waste Disposal

Virtually all applications of nuclear chemistry create radioactive waste and, along with it, the problems of safe handling and disposal. Most disposal sites, at present, are considered temporary, until a long-term safe solution can be found. Figure 9.9 conveys a sense of the enormity of the problem. Also, An Environmental Perspective: Nuclear Waste Disposal on page 285, examines this problem in more detail.

9.8 Measurement of Radiation

The changes that take place when radiation interacts with matter (such as photographic film) provide the basis of operation for various radiation detection devices.

The principal detection methods involve the use of either photographic film to create an image of the location of the radioactive substance or a counter that allows the measurement of intensity of radiation emitted from some source by converting the radiation energy to an electrical signal.

Figure 9.9
Photograph of the construction of one-million-gallon capacity storage tanks for radioactive waste. Located in Hanford, Washington, they are now covered with 6–8 ft of earth.

Nuclear Imaging

LEARNING GOAL

This approach is often used in nuclear medicine. An isotope is administered to a patient, perhaps iodine-131, which is used to study the thyroid gland, and the isotope begins to concentrate in the organ of interest. Nuclear images (photographs) of that region of the body are taken at periodic intervals using a special type of film. The emission of radiation from the radioactive substance creates the image, in much the same way as light causes the formation of images on conventional film in a camera. Upon development of the series of photographs, a record of the organ's uptake of the isotope over time enables the radiologist to assess the condition of the organ.

Computer Imaging

The coupling of rapid developments in the technology of television and computers, resulting in the marriage of these two devices, has brought about a versatile alternative to photographic imaging.

A specialized television camera, sensitive to emitted radiation from a radioactive substance administered to a patient, develops a continuous and instantaneous record of the voyage of the isotope throughout the body. The signal, transmitted to the computer, is stored, sorted, and portrayed on a monitor. Advantages include increased sensitivity, allowing a lower dose of the isotope, speed through elimination of the developing step, and versatility of application, limited perhaps only by the creativity of the medical practitioners.

A particular type of computer imaging, useful in diagnostic medicine, is the CT scanner. The CT scanner measures the interaction of X-rays with biological tissue, gathering huge amounts of data and processing the data to produce detailed information, all in a relatively short time. Such a device may be less hazardous than conventional X-ray techniques because it generates more useful information per unit of radiation. It often produces a superior image. A photograph of a CT scanner is shown in Figure 9.10, and an image of a damaged spinal bone, taken by a CT scanner, is shown in Figure 9.11.

CT represents *Computer-aided Tomography:* the computer reconstructs a series of measured images of tissue density (tomography). Small differences in tissue density may indicate the presence of a tumor.

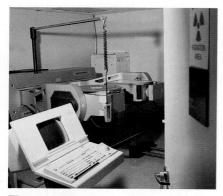

Figure 9.10
An imaging laboratory at the Greater Baltimore Medical Center.

The Geiger Counter

A Geiger counter is an instrument that detects ionizing radiation. Ions, produced by radiation passing through a tube filled with an ionizable gas, can conduct an electrical current between two electrodes. This current flow can be measured and is proportional to the level of radiation (Figure 9.12). Such devices, which were routinely used in laboratory and industrial monitoring, have been largely replaced by more sophisticated devices, often used in conjunction with a computer.

Film Badges

A common sight in any hospital or medical laboratory or any laboratory that routinely uses radioisotopes is the film badge worn by all staff members exposed in any way to low-level radioactivity.

A film badge is merely a piece of photographic film that is sensitive to energies corresponding to radioactive emissions. It is shielded from light, which would interfere, and mounted in a clip-on plastic holder that can be worn throughout the workday. The badges are periodically collected and developed. The degree of darkening is proportional to the amount of radiation to which the worker has been exposed, just as a conventional camera produces images on film in proportion to the amount of light that it "sees."

Proper record keeping thus allows the laboratory using radioactive substances to maintain an ongoing history of each individual's exposure and, at the same time, promptly pinpoint any hazards that might otherwise go unnoticed.

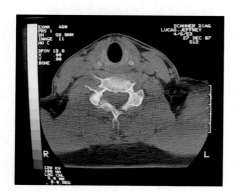

Figure 9.11
Damage observed in a spinal bone on a CT scan image.

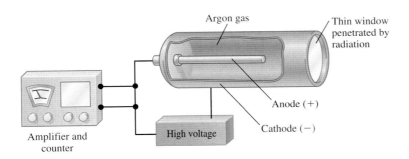

Figure 9.12
The design of a Geiger counter used for the measurement of radioactivity.

Units of Radiation Measurement

The amount of radiation emitted by a source or received by an individual is reported in a variety of ways, using units that describe different aspects of radiation. The *curie* and the *roentgen* describe the intensity of the emitted radiation, whereas the *rad* and the *rem* describe the biological effects of radiation.

The Curie

The **curie** is a measure of the amount of radioactivity in a radioactive source. The curie is independent of the nature of the radiation (alpha, beta, or gamma) and its effect on biological tissue. A curie is defined as the amount of radioactive material that produces 3.7×10^{10} atomic disintegrations per second.

 LEARNING GOAL

The Roentgen

The **roentgen** is a measure of very high energy ionizing radiation (X-ray and gamma ray) only. The roentgen is defined as the amount of radiation needed to produce 2×10^9 ion pairs when passing through one cm^3 of air at 0°C. The roentgen is a measure of radiation's interaction with air and gives no information about the effect on biological tissue.

The Rad

The **rad**, or *radiation absorbed dosage*, provides more meaningful information than either of the previous units of measure. It takes into account the nature of the absorbing material. It is defined as the dosage of radiation able to transfer 2.4×10^{-3} cal of energy to one kg of matter.

The Rem

The **rem**, or *roentgen equivalent for man*, describes the biological damage caused by the absorption of different kinds of radiation by the human body. The *rem* is obtained by multiplication of the *rad* by a factor called the *relative biological effect (RBE)*. The RBE is a function of the type of radiation (alpha, beta, or gamma). Although a beta particle is more penetrating than an alpha particle, an alpha particle is approximately ten times more damaging to biological tissue. As a result, the RBE is ten for alpha particles and one for beta particles. Relative yearly radiation dosages received by Americans are shown in Figure 9.13.

The **lethal dose (LD$_{50}$)** of radiation is defined as the acute dosage of radiation that would be fatal for 50% of the exposed population within 30 days. An estimated lethal dose is 500 rems. Some biological effects, however, may be detectable at a level as low as 25 rem.

Figure 9.13
Relative yearly radiation dosages for individuals in the continental United States. Red, yellow, and green shading indicates higher levels of background radiation. Blue shading indicates regions of lower background exposure.

Question 9.11

From a clinical standpoint, what advantages does expressing radiation in rems have over the use of other radiation units?

Question 9.12

Is the roentgen unit used in the measurement of alpha particle radiation? Why or why not?

SUMMARY

9.1 Natural Radioactivity

Radioactivity is the process by which atoms emit energetic, ionizing particles or rays. These particles or rays are termed radiation. Nuclear radiation occurs because the nucleus is unstable, hence radioactive. Nuclear symbols consist of the elemental symbol, the atomic number, and the mass number.

Not all *nuclides* are unstable. Only unstable nuclides undergo change and produce radioactivity in the process of radioactive decay. Three types of natural radiation emitted by unstable nuclei are *alpha particles, beta particles,* and *gamma rays*. This radiation is collectively termed *ionizing radiation*.

9.2 Writing a Balanced Nuclear Equation

A *nuclear equation* represents a nuclear process such as radioactive decay. The total of the mass numbers on each side of the reaction arrow must be identical, and the sum of the atomic numbers of the reactants must equal the sum of the atomic numbers of the products. Nuclear equations can be used to predict products of nuclear reactions.

9.3 Properties of Radioisotopes

The *binding energy* of the nucleus is a measure of nuclear stability. When an isotope decays, energy is released. Nuclear stability correlates with the ratio of neutrons to protons in the isotope. Nuclei with large numbers of protons tend to be unstable, and isotopes containing 2, 8, 20, 50, 82, or 126 protons or neutrons (magic numbers) are stable. Also, isotopes with even numbers of protons or neutrons are generally more stable than those with odd numbers of protons or neutrons.

The *half-life*, $t_{1/2}$, is the time required for one-half of a given quantity of a substance to undergo change. Each isotope has its own characteristic half-life. The degree of stability of an isotope is indicated by the isotope's half-life. Isotopes with short half-lives decay rapidly; they are very unstable.

9.4 Nuclear Power

Einstein predicted that a small amount of nuclear mass would convert to a very large amount of energy when the nucleus breaks apart. Fission reactors are used to generate electrical power. Technological problems with *fusion* and *breeder* reactors have prevented their commercialization in the United States.

9.5 Radiocarbon Dating

Radiocarbon dating is based on the measurement of the relative amounts of carbon-12 and carbon-14 present in an object. The ratio of the masses of these isotopes changes slowly over time, making it useful in determining the age of objects containing carbon.

9.6 Medical Applications of Radioactivity

The use of radiation in the treatment of various forms of cancer, and in the newer area of *nuclear medicine,* has become widespread in the past quarter century.

Ionizing radiation causes changes in cellular biochemical processes that may damage or kill the cell. A cancerous organ is composed of both healthy and malignant cells. Exposure of the tumor area to controlled dosages of high-energy gamma radiation from cobalt-60 will kill a higher percentage of abnormal cells than normal cells and is a valuable cancer therapy.

The diagnosis of a host of biochemical irregularities or diseases of the human body has been made routine through the use of radioactive tracers. *Tracers* are small amounts of radioactive substances used as probes to study internal organs. Because the isotope is radioactive, its path may be followed by using suitable detection devices. A "picture" of the organ is obtained, far more detailed than is possible with conventional X-rays.

The radioactivity produced by unstable isotopes is described as *natural radioactivity*. A normally stable, nonradioactive nucleus can be made radioactive, and this is termed *artificial radioactivity* (the process produces synthetic isotopes). Synthetic isotopes are often used in clinical situations. Isotopic synthesis may be carried out in the core of a *nuclear reactor* or in a *particle accelerator*. Short-lived isotopes, such as technetium-99m, are often produced directly at the site of the clinical testing.

9.7 Biological Effects of Radiation

Safety considerations are based on the magnitude of the *half-life, shielding,* distance from the radioactive source, time of exposure, and type of radiation emitted. We are never entirely free of the effects of radioactivity. *Background radiation* is normal radiation attributable to our surroundings.

Virtually all applications of nuclear chemistry create radioactive waste and, along with it, the problems of safe handling and disposal. Most disposal sites are considered temporary, until a long-term safe solution can be found.

9.8 Measurement of Radiation

The changes that take place when radiation interacts with matter provide the basis for various radiation detection devices. Photographic imaging, computer imaging, the Geiger counter, and film badges represent the most frequently used devices for detecting and measuring radiation.

Commonly used radiation units include the *curie*, a measure of the amount of radioactivity in a radioactive source; the *roentgen*, a measure of high-energy radiation (X-ray and gamma ray); the *rad* (radiation absorbed dosage), which takes into account the nature of the absorbing material; and the *rem* (roentgen equivalent for man), which describes the biological damage caused by the absorption of different kinds of radiation by the human body. The *lethal dose* of radiation, LD_{50}, is defined as the dose that would be fatal for 50% of the exposed population within thirty days.

KEY TERMS

alpha particle (9.1)
artificial radioactivity (9.6)
background radiation (9.7)
beta particle (9.1)
binding energy (9.3)
breeder reactor (9.4)
chain reaction (9.4)
curie (9.8)
fission (9.4)
fusion (9.4)
gamma ray (9.1)
half-life ($t_{1/2}$) (9.3)
ionizing radiation (9.1)
lethal dose (LD_{50}) (9.8)
metastable isotope (9.2)
natural radioactivity (9.6)
nuclear equation (9.2)
nuclear imaging (9.6)
nuclear medicine (9.6)
nuclear reactor (9.6)
nuclide (9.1)
particle accelerator (9.6)
positron (9.2)
rad (9.8)
radioactivity (9.1)
radiocarbon dating (9.5)
rem (9.8)
roentgen (9.8)
shielding (9.7)
tracer (9.6)

QUESTIONS AND PROBLEMS

Natural Radioactivity

Fundamentals

9.13 Describe the meaning of the term *natural radioactivity*.
9.14 What is background radiation?
9.15 What is the composition of an alpha particle?
9.16 What is alpha decay?
9.17 What is the composition of a beta particle?
9.18 What is the composition of a positron?
9.19 In what way do beta particles and positrons differ?
9.20 In what way are beta particles and positrons similar?
9.21 What are the major differences between alpha and beta particles?
9.22 What are the major differences between alpha particles and gamma radiation?
9.23 How do nuclear reactions and chemical reactions differ?
9.24 We can control the rate of chemical reactions. Can we control the rate of natural radiation?

Applications

9.25 Write the nuclear symbol for an alpha particle.
9.26 Write the nuclear symbol for a beta particle.
9.27 Write the nuclear symbol for uranium-235.
9.28 How many protons and neutrons are contained in the nucleus of uranium-235?
9.29 How many protons and neutrons are contained in each of the three isotopes of hydrogen?
9.30 How many protons and neutrons are contained in each of the three isotopes of carbon?
9.31 Write the nuclear symbol for nitrogen-15.
9.32 Write the nuclear symbol for carbon-14.
9.33 Compare and contrast the three major types of radiation produced by nuclear decay.
9.34 Rank the three major types of radiation in order of size, speed, and penetrating power.
9.35 How does an α particle differ from a helium atom?
9.36 What is the major difference between β and δ radiation?

Writing a Balanced Nuclear Equation

Fundamentals

9.37 Write a nuclear reaction to represent cobalt-60 decaying to nickel-60 plus a beta particle plus a gamma ray.
9.38 Write a nuclear reaction to represent radium-226 decaying to radon-222 plus an alpha particle.
9.39 Complete the following nuclear reaction:

$$^{23}_{11}Na + ^{2}_{1}H \longrightarrow ? + ^{1}_{1}H$$

9.40 Complete the following nuclear reaction:

$$^{238}_{92}U + ^{14}_{7}N \longrightarrow ? + 6^{1}_{0}n$$

9.41 Complete the following nuclear reaction:

$$^{24}_{10}Ne \longrightarrow \beta + ?$$

9.42 Complete the following nuclear reaction:

$$^{190}_{78}Pt \longrightarrow \alpha + ?$$

9.43 Complete the following nuclear reaction:

$$? \longrightarrow ^{140}_{56}Ba + ^{0}_{-1}e$$

9.44 Complete the following nuclear reaction:

$$? \longrightarrow ^{214}_{90}Th + ^{4}_{2}He$$

Applications

9.45 Element 107 was synthesized by bombarding bismuth-209 with chromium-54. Write the equation for this process if one product is a neutron.
9.46 Element 109 was synthesized by bombarding bismuth-209 with iron-58. Write the equation for this process if one product is a neutron.
9.47 Write a balanced nuclear equation for beta emission by magnesium-27.

9.48 Write a balanced nuclear equation for alpha decay of bismuth-212.
9.49 Write a balanced nuclear equation for position emission by nitrogen-12.
9.50 Write a balanced nuclear equation for the formation of polonium-206 through alpha decay.

Properties of Radioisotopes
Fundamentals
9.51 What is the difference between natural radioactivity and artificial radioactivity?
9.52 Is the fission of uranium-235 an example of natural or artificial radioactivity?
9.53 Summarize the major characteristics of nuclei for which we predict a high degree of stability.
9.54 Explain why the binding energy of a nucleus is expected to be large.

Applications
9.55 Would you predict oxygen-20 to be stable? Explain your reasoning.
9.56 Would you predict cobalt-59 to be stable? Explain your reasoning.
9.57 Would you predict chromium-48 to be stable? Explain your reasoning.
9.58 Would you predict lithium-9 to be stable? Explain your reasoning.
9.59 If 3.2 mg of the radioisotope iodine-131 is administered to a patient, how much will remain in the body after 24 days, assuming that no iodine has been eliminated from the body by any other process? (See Table 9.2 for the half-life of iodine-131.)
9.60 A patient receives 9.0 ng of a radioisotope with a half-life of 12 hours. How much will remain in the body after 2.0 days, assuming that radioactive decay is the only path for removal of the isotope from the body?
9.61 A sample containing 1.00×10^2 mg of iron-59 is stored for 135 days. What mass of iron-59 will remain at the end of the storage period? (See Table 9.2 for the half-life of iron-59.)
9.62 An instrument for cancer treatment containing a cobalt-60 source was manufactured in 1978. In 1995 it was removed from service and, in error, was buried in a landfill with the source still in place. What percentage of its initial radioactivity will remain in the year 2010? (See Table 9.2 for the half-life of cobalt-60.)
9.63 The half-life of molybdenum-99 is 67 hr. A 200 μg quantity decays, over time, to 25 μg. How much time has elapsed?
9.64 The half-life of strontium-87 is 2.8 hr. What percentage of this isotope will remain after 8 hours and 24 minutes?

Nuclear Power
Fundamentals
9.65 Which type of nuclear process splits nuclei to release energy?
9.66 Which type of nuclear process combines small nuclei to release energy?
9.67 a. Describe the process of fission.
     b. How is this reaction useful as the basis for the production of electrical energy?
9.68 a. Describe the process of fusion.
     b. How could this process be used for the production of electrical energy?

Applications
9.69 Write a balanced nuclear equation for a fusion reaction.
9.70 What are the major disadvantages of a fission reactor for electrical energy production?
9.71 What is meant by the term *breeder reactor?*
9.72 What are the potential advantages and disadvantages of breeder reactors?
9.73 Describe what is meant by the term *chain reaction.*
9.74 Why are cadmium rods used in a fission reaction?
9.75 What is the greatest barrier to development of fusion reactors?
9.76 What type of nuclear reaction fuels our solar system?

Radiocarbon Dating
9.77 Describe the process used to determine the age of the wooden coffin of King Tut.
9.78 What property of carbon enables us to assess the age of a painting?

Medical Applications of Radioactivity
9.79 The isotope indium-111 is used in medical laboratories as a label for blood platelets. To prepare indium-111, silver-108 is bombarded with an alpha particle, forming an intermediate isotope of indium. Write a nuclear equation for the process, and identify the intermediate isotope of indium.
9.80 Radioactive molybdenum-99 is used to produce the tracer isotope, technetium-99m. Write a nuclear equation for the formation of molybdenum-99 from stable molybdenum-98 bombarded with neutrons.
9.81 Describe an application of each of the following isotopes:
    a. technetium-99m
    b. xenon-133
9.82 Describe an application of each of the following isotopes:
    a. iodine-131
    b. thallium-201
9.83 Why is radiation therapy an effective treatment for certain types of cancer?
9.84 Describe how medically useful isotopes may be prepared.
9.85 What is the source of background radiation?
9.86 Why do high-altitude jet flights increase a person's exposure to background radiation?

Answer questions 9.87 through 9.94 based on the assumption that you are employed by a clinical laboratory that prepares radioactive isotopes for medical diagnostic tests. Consider α, β, positron, and γ emission.

9.87 What would be the effect on your level of radiation exposure if you increase your average distance from a radioactive source?
9.88 Would wearing gloves have any significant effect? Why?
9.89 Would limiting your time of exposure have a positive effect? Why?
9.90 Would wearing a lab apron lined with thin sheets of lead have a positive effect? Why?
9.91 Would the use of robotic manipulation of samples have an effect? Why?
9.92 Would the use of concrete rather than wood paneling help to protect workers in other parts of the clinic? Why?
9.93 Would the thickness of the concrete in question 9.92 be an important consideration? Why?
9.94 Suggest a protocol for radioactive waste disposal.

Measurement of Radiation
9.95 X-ray technicians often wear badges containing photographic film. How is this film used to indicate exposure to X-rays?
9.96 Why would a Geiger counter be preferred to film for assessing the immediate danger resulting from a spill of some solution containing a radioisotope?
9.97 What is meant by the term *relative biological effect?*
9.98 What is meant by the term *lethal dose* of radiation?

9.99 Define each of the following units:
   a. curie
   b. roentgen

9.100 Define each of the following radiation units:
   a. rad
   b. rem

CRITICAL THINKING PROBLEMS

1. Isotopes used as radioactive tracers have chemical properties that are similar to those of a nonradioactive isotope of the same element. Explain why this is a critical consideration in their use.
2. A chemist proposes a research project to discover a catalyst that will speed up the decay of radioactive isotopes that are waste products of a medical laboratory. Such a discovery would be a potential solution to the problem of nuclear waste disposal. Critique this proposal.
3. A controversial solution to the disposal of nuclear waste involves burial in sealed chambers far below the earth's surface. Describe potential pros and cons of this approach.
4. What type of radioactive decay is favored if the number of protons in the nucleus is much greater than the number of neutrons? Explain.
5. If the proton-to-neutron ratio in question 4 (above) were reversed, what radioactive decay process would be favored? Explain.
6. Radioactive isotopes are often used as "tracers" to follow an atom through a chemical reaction, and the following is an example. Acetic acid reacts with methyl alcohol by eliminating a molecule of water to form methyl acetate. Explain how you would use the radioactive isotope oxygen-18 to show whether the oxygen atom in the water product comes from the —OH of the acid or the —OH of the alcohol.

$$H_3C-\underset{\underset{Acetic\ acid}{}}{\overset{\overset{O}{\|}}{C}}-OH + \underset{Methyl\ alcohol}{HOCH_3} \longrightarrow H_3C-\underset{\underset{Methyl\ acetate}{}}{\overset{\overset{O}{\|}}{C}}-O-CH_3 + H_2O$$

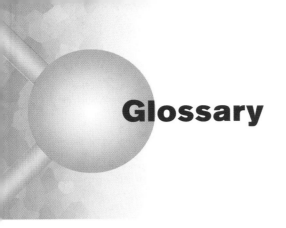

Glossary

A

absolute specificity (19.5) the property of an enzyme that allows it to bind and catalyze the reaction of only one substrate

accuracy (1.4) the nearness of an experimental value to the true value

acetal (13.4) the family of organic compounds formed via the reaction of two molecules of alcohol with an aldehyde in the presence of an acid catalyst; acetals have the following general structure:

$$R^1-\underset{\underset{H}{|}}{\overset{\overset{OR^2}{|}}{C}}-OR^3$$

acetyl coenzyme A (acetyl CoA) (14.4, 22.2) a molecule composed of coenzyme A and an acetyl group; the intermediate that provides acetyl groups for complete oxidation by aerobic respiration

acid (8.1) a substance that behaves as a proton donor

acid anhydride (14.3) the product formed by the combination of an acid chloride and a carboxylate ion; structurally they are two carboxylic acids with a water molecule removed:

$$(Ar)\ R-\overset{\overset{O}{\|}}{C}-O-\overset{\overset{O}{\|}}{C}-R\ (Ar)$$

acid-base reaction (4.3) reaction that involves the transfer of a hydrogen ion (H⁺) from one reactant to another

acid chloride (14.3) member of the family of organic compounds with the general formula

$$(Ar)\ R-\overset{\overset{O}{\|}}{C}-Cl$$

activated complex (7.3) the arrangement of atoms at the top of the potential energy barrier as a reaction proceeds

activation energy (7.3) the threshold energy that must be overcome to produce a chemical reaction

active site (19.4) the cleft in the surface of an enzyme that is the site of substrate binding

active transport (17.6) the movement of molecules across a membrane against a concentration gradient

acyl carrier protein (ACP) (23.4) the protein that forms a thioester linkage with fatty acids during fatty acid synthesis

acyl group (14: Intro, 15.3) the functional group found in carboxylic acid derivatives that contains the carbonyl group attached to one alkyl or aryl group:

$$(Ar)\ R-\overset{\overset{O}{\|}}{C}-$$

addition polymer (11.5) polymers prepared by the sequential addition of a monomer

addition reaction (11.5, 13.4) a reaction in which two molecules add together to form a new molecule; often involves the addition of one molecule to a double or triple bond in an unsaturated molecule; e.g., the addition of alcohol to an aldehyde or ketone to form a hemiacetal or hemiketal

adenosine triphosphate (ATP) (15.4, 21.1) a nucleotide composed of the purine adenine, the sugar ribose, and three phosphoryl groups; the primary energy storage and transport molecule used by the cells in cellular metabolism

adipocyte (23.1) a fat cell

adipose tissue (23.1) fatty tissue that stores most of the body lipids

aerobic respiration (22.3) the oxygen-requiring degradation of food molecules and production of ATP

alcohol (12.1) an organic compound that contains a hydroxyl group (—OH) attached to an alkyl group

aldehyde (13.1) a class of organic molecules characterized by a carbonyl group; the carbonyl carbon is bonded to a hydrogen atom and to another hydrogen or an alkyl or aryl group. Aldehydes have the following general structure:

$$(Ar)-\overset{\overset{O}{\|}}{C}-H \qquad R-\overset{\overset{O}{\|}}{C}-H$$

aldol condensation (13.4) a reaction in which aldehydes or ketones react to form a larger molecule

aldose (16.2) a sugar that contains an aldehyde (carbonyl) group

aliphatic hydrocarbon (10.1) any member of the alkanes, alkenes, and alkynes or the substituted alkanes, alkenes, and alkynes

alkali metal (2.4) an element within Group IA (1) of the periodic table

alkaline earth metal (2.4) an element within Group IIA (2) of the periodic table

alkaloid (15.2) a class of naturally occurring compounds that contain one or more nitrogen heterocyclic rings; many of the alkaloids have medicinal and other physiological effects

alkane (10.2) a hydrocarbon that contains only carbon and hydrogen and is bonded together through carbon-hydrogen and carbon-carbon single bonds; a saturated hydrocarbon with the general molecular formula C_nH_{2n+2}

alkene (11.1) a hydrocarbon that contains one or more carbon-carbon double bonds; an unsaturated hydrocarbon with the general formula C_nH_{2n}

alkyl group (10.2) a hydrocarbon group that results from the removal of one hydrogen from the original hydrocarbon (e.g., methyl, —CH₃; ethyl, —CH₂CH₃)

alkyl halide (10.5) a substituted hydrocarbon with the general structure R—X, in which R— represents any alkyl group and X = a halogen (F—, Cl—, Br—, or I—)

alkylammonium ion (15.1) the ion formed when the lone pair of electrons of the nitrogen atom of an amine is shared with a proton (H⁺) from a water molecule

alkyne (11.1) a hydrocarbon that contains one or more carbon-carbon triple bonds; an unsaturated hydrocarbon with the general formula C_nH_{2n-2}

allosteric enzyme (19.9) an enzyme that has an effector binding site and an active site; effector binding changes the shape of the active site, rendering it either active or inactive

alpha particle (9.1) a particle consisting of two protons and two neutrons; the alpha particle is identical to a helium nucleus

amide bond (15.3) the bond between the carbonyl carbon of a carboxylic acid and the amino nitrogen of an amine

G-1

amides (15.3) the family of organic compounds formed by the reaction between a carboxylic acid derivative and an amine and characterized by the amide group

amines (15.1) the family of organic molecules with the general formula RNH_2, R_2NH, or R_3N (R— can represent either an alkyl or aryl group); they may be viewed as substituted ammonia molecules in which one or more of the ammonia hydrogens has been substituted by a more complex organic group

α-amino acid (18.2) the subunits of proteins composed of an α-carbon bonded to a carboxylate group, a protonated amino group, a hydrogen atom, and a variable R group

aminoacyl group (15.4) the functional group that is characteristic of an amino acid; the aminoacyl group has the following general structure:

$$H_3\overset{+}{N}-\underset{R}{\underset{|}{C}}\overset{H}{\underset{|}{-}}\overset{O}{\underset{\|}{C}}-$$

aminoacyl tRNA (20.6) the transfer RNA covalently linked to the correct amino acid

aminoacyl tRNA binding site of ribosome (A-site) (20.6) a pocket on the surface of a ribosome that holds the aminoacyl tRNA during translation

aminoacyl tRNA synthetase (20.6) an enzyme that recognizes one tRNA and covalently links the appropriate amino acid to it

amorphous solid (5.3) a solid with no organized, regular structure

amphibolic pathway (22.9) a metabolic pathway that functions in both anabolism and catabolism

amphiprotic (8.1) a substance that can behave either as a Brønsted acid or a Brønsted base

amylopectin (16.6) a highly branched form of amylose; the branches are attached to the C-6 hydroxyl by α(1 → 6) glycosidic linkage; a component of starch

amylose (16.6) a linear polymer of α-D-glucose molecules bonded in α(1 → 4) glycosidic linkage that is a major component of starch; a polysaccharide storage form

anabolism (21.1, 22.9) all of the cellular energy-requiring biosynthetic pathways

anaerobic threshold (21.4) the point at which the level of lactate in the exercising muscle inhibits glycolysis and the muscle, deprived of energy, ceases to function

analgesic (15.2) any drug that acts as a painkiller, e.g., aspirin, acetaminophen

anaplerotic reaction (22.9) a reaction that replenishes a substrate needed for a biochemical pathway

anesthetic (15.2) a drug that causes a lack of sensation in part of the body (local anesthetic) or causes unconsciousness (general anesthetic)

angular structure (3.4) a planar molecule with bond angles other than 180°

anion (2.1) a negatively charged atom or group of atoms

anode (8.5) the positively charged electrode in an electrical cell

anomers (16.4) isomers of cyclic monosaccharides that differ from one another in the arrangement of bonds around the hemiacetal carbon

antibodies (18.1) immunoglobulins; specific glycoproteins produced by cells of the immune system in response to invasion by infectious agents

anticodon (20.4) a sequence of three ribonucleotides on a tRNA that are complementary to a codon on the mRNA; codon-anticodon binding results in delivery of the correct amino acid to the site of protein synthesis

antigen (18.1) any substance that is able to stimulate the immune system; generally a protein or large carbohydrate

antiparallel strands (20.2) a term describing the polarities of the two strands of the DNA double helix; on one strand the sugar-phosphate backbone advances in the 5' → 3' direction; on the opposite, complementary strand the sugar-phosphate backbone advances in the 3' → 5' direction

apoenzyme (19.7) the protein portion of an enzyme that requires a cofactor to function in catalysis

aqueous solution (6.1) any solution in which the solvent is water

arachidonic acid (17.2) a fatty acid derived from linoleic acid; the precursor of the prostaglandins

aromatic hydrocarbon (10.1, 11.6) an organic compound that contains the benzene ring or a derivative of the benzene ring

Arrhenius theory (8.1) a theory that describes an acid as a substance that dissociates to produce H^+ and a base as a substance that dissociates to produce OH^-

artificial radioactivity (9.6) radiation that results from the conversion of a stable nucleus to another, unstable nucleus

atherosclerosis (17.4) deposition of excess plasma cholesterol and other lipids and proteins on the walls of arteries, resulting in decreased artery diameter and increased blood pressure

atom (2.1) the smallest unit of an element that retains the properties of that element

atomic mass (2.1) the mass of an atom expressed in atomic mass units

atomic mass unit (4.1) 1/12 of the mass of a ^{12}C atom, equivalent to 1.661×10^{-24} g

atomic number (2.1) the number of protons in the nucleus of an atom; it is a characteristic identifier of an element

atomic orbital (2.3, 2.5) a specific region of space where an electron may be found

ATP synthase (22.6) a multiprotein complex within the inner mitochondrial membrane that uses the energy of the proton (H^+) gradient to produce ATP

autoionization (8.1) also known as *self-ionization*, the reaction of a substance, such as water, with itself to produce a positive and a negative ion

Avogadro's law (5.1) a law that states that the volume is directly proportional to the number of moles of gas particles, assuming that the pressure and temperature are constant

Avogadro's number (4.1) 6.022×10^{23} particles of matter contained in 1 mol of a substance

axial atom (10.4) an atom that lies above or below a cycloalkane ring

B

background radiation (9.7) the radiation that emanates from natural sources

barometer (5.1) a device for measuring pressure

base (8.1) a substance that behaves as a proton acceptor

base pair (20.2) a hydrogen-bonded pair of bases within the DNA double helix; the standard base pairs always involve a purine and a pyrimidine; in particular, adenine always base pairs with thymine and cytosine with guanine

Benedict's reagent (16.4) a buffered solution of Cu^{2+} ions that can be used to test for reducing sugars or to distinguish between aldehydes and ketones

Benedict's test (13.4) a test used to determine the presence of reducing sugars or to distinguish between aldehydes and ketones; it requires a buffered solution of Cu^{2+} ions that are reduced to Cu^+, which precipitates as brick-red Cu_2O

beta particle (9.1) an electron formed in the nucleus by the conversion of a neutron into a proton

bile (23.1) micelles of lecithin, cholesterol, bile salts, protein, inorganic ions, and bile pigments that aid in lipid digestion by emulsifying fat droplets

binding energy (9.3) the energy required to break down the nucleus into its component parts

bioinformatics (20.10) an interdisciplinary field that uses computer information sciences and DNA technology to devise methods for understanding, analyzing, and applying DNA sequence information

boat conformation (10.4) a form of a six-member cycloalkane that resembles a rowboat. It is less stable than the chair conformation because the hydrogen atoms are not perfectly staggered

boiling point (3.3) the temperature at which the vapor pressure of a liquid is equal to the atmospheric pressure

bond energy (3.4) the amount of energy necessary to break a chemical bond

Boyle's law (5.1) a law stating that the volume of a gas varies inversely with the pressure exerted if the temperature and number of moles of gas are constant

breeder reactor (9.4)   a nuclear reactor that produces its own fuel in the process of providing electrical energy

Brønsted-Lowry theory (8.1)   a theory that describes an acid as a proton donor and a base as a proton acceptor

buffer capacity (8.4)   a measure of the ability of a solution to resist large changes in pH when a strong acid or strong base is added

buffer solution (8.4)   a solution containing a weak acid or base and its salt (the conjugate base or acid) that is resistant to large changes in pH upon addition of strong acids or bases

buret (8.3)   a device calibrated to deliver accurately known volumes of liquid, as in a titration

C

C-terminal amino acid (18.3)   the amino acid in a peptide that has a free $\alpha\text{-}CO_2^-$ group; the last amino acid in a peptide

calorimetry (7.2)   the measurement of heat energy changes during a chemical reaction

cap structure (20.4)   a 7-methylguanosine unit covalently bonded to the 5' end of a mRNA by a 5'–5' triphosphate bridge

carbinol carbon (12.4)   that carbon in an alcohol to which the hydroxyl group is attached

carbohydrate (16.1)   generally sugars and polymers of sugars; the primary source of energy for the cell

carbonyl group (13: Intro)   the functional group that contains a carbon-oxygen double bond: —C=O; the functional group found in aldehydes and ketones

carboxyl group (14.1)   the —COOH functional group; the functional group found in carboxylic acids

carboxylic acid (14.1)   a member of the family of organic compounds that contain the —COOH functional group

carboxylic acid derivative (14.2)   any of several families of organic compounds, including the esters and amides, that are derived from carboxylic acids and have the general formula

$$\text{(Ar)}-\overset{\overset{\displaystyle O}{\|}}{C}-Z \qquad R-\overset{\overset{\displaystyle O}{\|}}{C}-Z$$

Z = —OR or OAr for the esters, and Z = —NH$_2$ for the amides

carcinogen (20.7)   any chemical or physical agent that causes mutations in the DNA that lead to uncontrolled cell growth or cancer

catabolism (21.1, 22.9)   the degradation of fuel molecules and production of ATP for cellular functions

catalyst (7.3)   any substance that increases the rate of a chemical reaction (by lowering the activation energy of the reaction) and that is not destroyed in the course of the reaction

cathode (8.5)   the negatively charged electrode in an electrical cell

cathode rays (2.2)   a stream of electrons that is given off by the cathode (negative electrode) in a cathode ray tube

cation (2.1)   a positively charged atom or group of atoms

cellulose (16.6)   a polymer of β-D-glucose linked by β(1 → 4) glycosidic bonds

central dogma (20.4)   a statement of the directional transfer of the genetic information in cells: DNA → RNA → Protein

chain reaction (9.4)   the process in a fission reactor that involves neutron production and causes subsequent reactions accompanied by the production of more neutrons in a continuing process

chair conformation (10.4)   the most energetically favorable conformation for a six-member cycloalkane; so-called for its resemblance to a lawn chair

Charles's law (5.1)   a law stating that the volume of a gas is directly proportional to the temperature of the gas, assuming that the pressure and number of moles of the gas are constant

chemical bond (3.1)   the attractive force holding two atomic nuclei together in a chemical compound

chemical equation (4.3)   a record of chemical change, showing the conversion of reactants to products

chemical formula (4.2)   the representation of a compound or ion in which elemental symbols represent types of atoms and subscripts show the relative numbers of atoms

chemical property (1.2)   characteristics of a substance that relate to the substance's participation in a chemical reaction

chemical reaction (1.2)   a process in which atoms are rearranged to produce new combinations

chemistry (1.1)   the study of matter and the changes that matter undergoes

chiral carbon (16.3)   a carbon atom bonded to four different atoms or groups of atoms

chiral molecule (16.3)   molecule capable of existing in mirror-image forms

cholesterol (17.4)   a twenty-seven-carbon steroid ring structure that serves as the precursor of the steroid hormones

chromosome (20.2)   a piece of DNA that carries all the genetic instructions, or genes, of an organism

chylomicron (17.5, 23.1)   a plasma lipoprotein (aggregate of protein and triglycerides) that carries triglycerides from the intestine to all body tissues via the bloodstream

***cis-trans* isomers** (10.3)   isomers that differ from one another in the placement of substituents on a double bond or ring

citric acid cycle (22.4)   a cyclic biochemical pathway that is the final stage of degradation of carbohydrates, fats, and amino acids. It results in the complete oxidation of acetyl groups derived from these dietary fuels

cloning vector (20.8)   a DNA molecule that can carry a cloned DNA fragment into a cell and that has a replication origin that allows the DNA to be replicated abundantly within the host cell

coagulation (18.10)   the process by which proteins in solution are denatured and aggregate with one another to produce a solid

codon (20.4)   a group of three ribonucleotides on the mRNA that specifies the addition of a specific amino acid onto the growing peptide chain

coenzyme (19.7)   an organic group required by some enzymes; it generally serves as a donor or acceptor of electrons or a functional group in a reaction

coenzyme A (22.2)   a molecule derived from ATP and the vitamin pantothenic acid; coenzyme A functions in the transfer of acetyl groups in lipid and carbohydrate metabolism

cofactor (19.7)   an inorganic group, usually a metal ion, that must be bound to an apoenzyme to maintain the correct configuration of the active site

colipase (23.1)   a protein that aids in lipid digestion by binding to the surface of lipid droplets and facilitating binding of pancreatic lipase

colligative property (6.4)   property of a solution that is dependent only on the concentration of solute particles

colloidal suspension (6.1)   a heterogeneous mixture of solute particles in a solvent; distribution of solute particles is not uniform because of the size of the particles

combination reaction (4.3)   a reaction in which two substances join to form another substance

combined gas law (5.1)   an equation that describes the behavior of a gas when volume, pressure, and temperature may change simultaneously

combustion (10.5)   the oxidation of hydrocarbons by burning in the presence of air to produce carbon dioxide and water

competitive inhibitor (19.10)   a structural analog; a molecule that has a structure very similar to the natural substrate of an enzyme, competes with the natural substrate for binding to the enzyme active site, and inhibits the reaction

complementary strands (20.2)   the opposite strands of the double helix are hydrogen-bonded to one another such that adenine and thymine or guanine and cytosine are always paired

complete protein (18.11)   a protein source that contains all the essential and nonessential amino acids

complex lipid (17.5)   a lipid bonded to other types of molecules

compound (1.2)   a substance that is characterized by constant composition and that can be chemically broken down into elements

concentration (1.5, 6.2) a measure of the quantity of a substance contained in a specified volume of solution

condensation (5.2) the conversion of a gas to a liquid

condensation polymer (14.2) a polymer, which is a large molecule formed by combination of many small molecules (monomers) that results from joining of monomers in a reaction that forms a small molecule, such as water or an alcohol

condensed formula (10.2) a structural formula showing all of the atoms in a molecule and placing them in a sequential arrangement that details which atoms are bonded to each other; the bonds themselves are not shown

conformations, conformers (10.4) discrete, distinct isomeric structures that may be converted, one to the other, by rotation about the bonds in the molecule

conjugate acid (8.1) substance that has one more proton than the base from which it is derived

conjugate acid-base pair (8.1) two species related to each other through the gain or loss of a proton

conjugate base (8.1) substance that has one less proton than the acid from which it is derived

constitutional isomers (10.2) two molecules having the same molecular formulas, but different chemical structures

Cori Cycle (21.6) a metabolic pathway in which the lactate produced by working muscle is taken up by cells in the liver and converted back to glucose by gluconeogenesis

corrosion (8.5) the unwanted oxidation of a metal

covalent bond (3.1) a pair of electrons shared between two atoms

covalent solid (5.3) a collection of atoms held together by covalent bonds

crenation (6.4) the shrinkage of red blood cells caused by water loss to the surrounding medium

cristae (22.1) the folds of the inner membrane of the mitochondria

crystal lattice (3.2) a unit of a solid characterized by a regular arrangement of components

crystalline solid (5.3) a solid having a regular repeating atomic structure

curie (9.8) the quantity of radioactive material that produces 3.7×10^{10} nuclear disintegrations per second

cycloalkane (10.3) a cyclic alkane; a saturated hydrocarbon that has the general formula C_nH_{2n}

D

Dalton's law (5.1) also called the law of partial pressures; states that the total pressure exerted by a gas mixture is the sum of the partial pressures of the component gases

data (1.3) a group of facts resulting from an experiment

decomposition reaction (4.3) the breakdown of a substance into two or more substances

defense proteins (18.1) proteins that defend the body against infectious diseases. Antibodies are defense proteins

degenerate code (20.5) a term used to describe the fact that several triplet codons may be used to specify a single amino acid in the genetic code

dehydration (of alcohols) (12.5) a reaction that involves the loss of a water molecule, in this case the loss of water from an alcohol and the simultaneous formation of an alkene

deletion mutation (20.7) a mutation that results in the loss of one or more nucleotides from a DNA sequence

denaturation (18.10) the process by which the organized structure of a protein is disrupted, resulting in a completely disorganized, nonfunctional form of the protein

density (1.5) mass per unit volume of a substance

deoxyribonucleic acid (DNA) (20.1) the nucleic acid molecule that carries all of the genetic information of an organism; the DNA molecule is a double helix composed of two strands, each of which is composed of phosphate groups, deoxyribose, and the nitrogenous bases thymine, cytosine, adenine, and guanine

deoxyribonucleotide (20.1) a nucleoside phosphate or nucleotide composed of a nitrogenous base in β-N-glycosidic linkage to the 1' carbon of the sugar 2'-deoxyribose and with one, two, or three phosphoryl groups esterified at the hydroxyl of the 5' carbon

diabetes mellitus (23.3) a disease caused by the production of insufficient levels of insulin and characterized by the appearance of very high levels of glucose in the blood and urine

dialysis (6.6) the removal of waste material via transport across a membrane

diglyceride (17.3) the product of esterification of glycerol at two positions

dipole-dipole interactions (5.2) attractive forces between polar molecules

disaccharide (16.1) a sugar composed of two monosaccharides joined through an oxygen atom bridge

dissociation (3.3) production of positive and negative ions when an ionic compound dissolves in water

disulfide (12.9) an organic compound that contains a disulfide group (—S—S—)

DNA polymerase III (20.3) the enzyme that catalyzes the polymerization of daughter DNA strands using the parental strand as a template

double bond (3.4) a bond in which two pairs of electrons are shared by two atoms

double helix (20.2) the spiral staircase-like structure of the DNA molecule characterized by two sugar-phosphate backbones wound around the outside and nitrogenous bases extending into the center

double-replacement reaction (4.3) a chemical change in which cations and anions "exchange partners"

dynamic equilibrium (7.4) the state that exists when the rate of change in the concentration of products and reactants is equal, resulting in no net concentration change

E

eicosanoid (17.2) any of the derivatives of twenty-carbon fatty acids, including the prostaglandins, leukotrienes, and thromboxanes

electrolysis (8.5) an electrochemical process that uses electrical energy to cause nonspontaneous oxidation-reduction reactions to occur

electrolyte (3.3, 6.1) a material that dissolves in water to produce a solution that conducts an electrical current

electrolytic solution (3.3) a solution composed of an electrolytic solute dissolved in water

electromagnetic radiation (2.3) energy that is propagated as waves at the speed of light

electromagnetic spectrum (2.3) the complete range of electromagnetic waves

electron (2.1) a negatively charged particle outside of the nucleus of an atom

electron affinity (2.7) the energy released when an electron is added to an isolated atom

electron configuration (2.5) the arrangement of electrons around a nucleus of an atom, ion, or a collection of nuclei of a molecule

electron density (2.3) the probability of finding the electron in a particular location

electron transport system (22.6) the series of electron transport proteins embedded in the inner mitochondrial membrane that accept high-energy electrons from NADH and $FADH_2$ and transfer them in stepwise fashion to molecular oxygen (O_2)

electronegativity (3.1) a measure of the tendency of an atom in a molecule to attract shared electrons

element (1.2) a substance that cannot be decomposed into simpler substances by chemical or physical means

elimination reaction (12.5) a reaction in which a molecule loses atoms or ions from its structure

elongation factor (20.6) proteins that facilitate the elongation phase of translation

emulsifying agent (17.3) a bipolar molecule that aids in the suspension of fats in water

enantiomers (16.3) stereoisomers that are nonsuperimposable mirror images of one another

endothermic reaction (7.1) a chemical or physical change in which energy is absorbed

energy (1.1) the capacity to do work

energy level (2.3) one of numerous atomic regions where electrons may be found

enthalpy (7.1) a term that represents heat energy

entropy (7.1) a measure of randomness or disorder

enzyme (18.1, 19: Intro) a protein that serves as a biological catalyst

enzyme specificity (19.5) the ability of an enzyme to bind to only one, or a very few, substrates and thus catalyze only a single reaction

enzyme-substrate complex (19.4) a molecular aggregate formed when the substrate binds to the active site of the enzyme

equatorial atom (10.4) an atom that lies in the plane of a cycloalkane ring

equilibrium reaction (7.4) a reaction that is reversible and the rates of the forward and reverse reactions are equal

equivalence point (8.3) the situation in which reactants have been mixed in the molar ratio corresponding to the balanced equation

equivalent (6.3) the number of grams of an ion corresponding to Avogadro's number of electrical charges

error (1.4) the difference between the true value and the experimental value for data or results

essential amino acid (18.11) an amino acid that cannot be synthesized by the body and must therefore be supplied by the diet

essential fatty acids (17.2) the fatty acids linolenic and linoleic acids that must be supplied in the diet because they cannot be synthesized by the body

ester (14.2) a carboxylic acid derivative formed by the reaction of a carboxylic acid and an alcohol. Esters have the following general formula:

$$\underset{R-C-OR}{\overset{O}{\parallel}} \quad \underset{R-C-O(Ar)}{\overset{O}{\parallel}} \quad \underset{(Ar)-C-O(Ar)}{\overset{O}{\parallel}}$$

esterification (17.2) the formation of an ester in the reaction of a carboxylic acid and an alcohol

ether (12.8) an organic compound that contains two alkyl and/or aryl groups attached to an oxygen atom; R—O—R, Ar—O—R, and Ar—O—Ar

eukaryote (20.2) an organism having cells containing a true nucleus enclosed by a nuclear membrane and having a variety of membrane-bound organelles that segregate different cellular functions into different compartments

evaporation (5.2) the conversion of a liquid to a gas below the boiling point of the liquid

exon (20.4) protein-coding sequences of a gene found on the final mature mRNA

exothermic reaction (7.1) a chemical or physical change that releases energy

extensive property (1.2) a property of a substance that depends on the quantity of the substance

F

F_0F_1 complex (22.6) an alternative term for the ATP synthase, the multiprotein complex in the inner mitochondrial membrane that uses the energy of the proton gradient to produce ATP

facilitated diffusion (17.6) movement of a solute across a membrane from an area of high concentration to an area of low concentration through a transmembrane protein, or permease

fatty acid (14.1, 17.2) any member of the family of continuous-chain carboxylic acids that generally contain four to twenty carbon atoms; the most concentrated source of energy used by the cell

feedback inhibition (19.9) the process whereby excess product of a biosynthetic pathway turns off the entire pathway for its own synthesis

fermentation (12.3, 21.4) anaerobic (in the absence of oxygen) catabolic reactions that occur with no net oxidation. Pyruvate or an organic compound produced from pyruvate is reduced as NADH is oxidized

fibrous protein (18.5) a protein composed of peptides arranged in long sheets or fibers

Fischer Projection (16.3) a two-dimensional drawing of a molecule, which shows a chiral carbon at the intersection of two lines and horizontal lines representing bonds projecting out of the page and vertical lines representing bonds that project into the page

fission (9.4) the splitting of heavy nuclei into lighter nuclei accompanied by the release of large quantities of energy

fluid mosaic model (17.6) the model of membrane structure that describes the fluid nature of the lipid bilayer and the presence of numerous proteins embedded within the membrane

formula (3.2) the representation of the fundamental compound unit using chemical symbols and numerical subscripts

formula unit (4.2) the smallest collection of atoms from which the formula of a compound can be established

formula weight (4.2) the mass of a formula unit of a compound relative to a standard (carbon-12)

free energy (7.1) the combined contribution of entropy and enthalpy for a chemical reaction

fructose (16.4) a ketohexose that is also called levulose and fruit sugar; the sweetest of all sugars, abundant in honey and fruits

fuel value (7.2) the amount of energy derived from a given mass of material

functional group (10.1) an atom (or group of atoms and their bonds) that imparts specific chemical and physical properties to a molecule

fusion (9.4) the joining of light nuclei to form heavier nuclei, accompanied by the release of large amounts of energy

G

galactose (16.4) an aldohexose that is a component of lactose (milk sugar)

galactosemia (16.5) a human genetic disease caused by the inability to convert galactose to a phosphorylated form of glucose (glucose-1-phosphate) that can be used in cellular metabolic reactions

gamma ray (9.1) a high-energy emission from nuclear processes, traveling at the speed of light; the high-energy region of the electromagnetic spectrum

gaseous state (1.2) a physical state of matter characterized by a lack of fixed shape or volume and ease of compressibility

genome (20.2) the complete set of genetic information in all the chromosomes of an organism

geometric isomer (10.3, 11.3) an isomer that differs from another isomer in the placement of substituents on a double bond or a ring

globular protein (18.6) a protein composed of polypeptide chains that are tightly folded into a compact spherical shape

glucagon (21.7, 23.6) a peptide hormone synthesized by the α-cells of the islets of Langerhans in the pancreas and secreted in response to low blood glucose levels; glucagon promotes glycogenolysis and gluconeogenesis and thereby increases the concentration of blood glucose

gluconeogenesis (21.6) the synthesis of glucose from noncarbohydrate precursors

glucose (16.4) an aldohexose, the most abundant monosaccharide; it is a component of many disaccharides, such as lactose and sucrose, and of polysaccharides, such as cellulose, starch, and glycogen

glyceraldehyde (16.3) an aldotriose that is the simplest carbohydrate; phosphorylated forms of glyceraldehyde are important intermediates in cellular metabolic reactions

glyceride (17.3) a lipid that contains glycerol

glycogen (16.6, 21.7) a long, branched polymer of glucose stored in liver and muscles of animals; it consists of a linear backbone of α-D-glucose in α(1 → 4) linkage, with numerous short branches attached to the C-6 hydroxyl group by α(1 → 6) linkage

glycogenesis (21.7) the metabolic pathway that results in the addition of glucose to growing glycogen polymers when blood glucose levels are high

glycogen granule (21.7) a core of glycogen surrounded by enzymes responsible for glycogen synthesis and degradation

glycogenolysis (21.7) the biochemical pathway that results in the removal of glucose molecules from glycogen polymers when blood glucose levels are low

glycolysis (21.3) the enzymatic pathway that converts a glucose molecule into two molecules of pyruvate; this anaerobic process generates a net energy yield of two molecules of ATP and two molecules of NADH

glycoprotein (18.7) a protein bonded to sugar groups

glycosidic bond (16.1) the bond between the hydroxyl group of the C-1 carbon of one sugar and a hydroxyl group of another sugar

group (2.4) any one of eighteen vertical columns of elements; often referred to as a family

group specificity (19.5) an enzyme that catalyzes reactions involving similar substrate molecules having the same functional groups

guanosine triphosphate (GTP) (21.6) a nucleotide composed of the purine guanosine, the sugar ribose, and three phosphoryl groups

H

half-life ($t_{1/2}$) (9.3) the length of time required for one-half of the initial mass of an isotope to decay to products

halogen (2.4) an element found in Group VIIA (17) of the periodic table

halogenation (10.5, 11.5) a reaction in which one of the C—H bonds of a hydrocarbon is replaced with a C—X bond (X = Br or Cl generally)

Haworth projection (16.4) a means of representing the orientation of substituent groups around a cyclic sugar molecule

α-helix (18.5) a right-handed coiled secondary structure maintained by hydrogen bonds between the amide hydrogen of one amino acid and the carbonyl oxygen of an amino acid four residues away

heme group (18.9) the chemical group found in hemoglobin and myoglobin that is responsible for the ability to carry oxygen

hemiacetal (13.4, 16.4) the family of organic compounds formed via the reaction of one molecule of alcohol with an aldehyde in the presence of an acid catalyst; hemiacetals have the following general structure:

$$R^1-\underset{\underset{H}{|}}{\overset{\overset{OH}{|}}{C}}-OR^2$$

hemiketal (13.4, 16.4) the family of organic compounds formed via the reaction of one molecule of alcohol with a ketone in the presence of an acid catalyst; hemiketals have the following general structure:

$$R^1-\underset{\underset{R^2}{|}}{\overset{\overset{OH}{|}}{C}}-OR^3$$

hemoglobin (18.9) the major protein component of red blood cells; the function of this red, iron-containing protein is transport of oxygen

hemolysis (6.4) the rupture of red blood cells resulting from movement of water from the surrounding medium into the cell

Henderson-Hasselbalch equation (8.4) an equation for calculating the pH of a buffer system:

$$pH = pKa + \log \frac{[\text{conjugate base}]}{[\text{weak acid}]}$$

Henry's law (6.1) a law stating that the number of moles of a gas dissolved in a liquid at a given temperature is proportional to the partial pressure of the gas

heterocyclic amine (15.2) a heterocyclic compound that contains nitrogen in at least one position in the ring skeleton

heterocyclic aromatic compound (11.7) cyclic aromatic compound having at least one atom other than carbon in the structure of the aromatic ring

heterogeneous mixture (1.2) a mixture of two or more substances characterized by nonuniform composition

hexose (16.2) a six-carbon monosaccharide

high-density lipoprotein (HDL) (17.5) a plasma lipoprotein that transports cholesterol from peripheral tissue to the liver

holoenzyme (19.7) an active enzyme consisting of an apoenzyme bound to a cofactor

homogeneous mixture (1.2) a mixture of two or more substances characterized by uniform composition

hybridization (20.8) a technique for identifying DNA or RNA sequences that is based on specific hydrogen bonding between a radioactive probe and complementary DNA or RNA sequences

hydrate (4.2) any substance that has water molecules incorporated in its structure

hydration (11.5, 12.5) a reaction in which water is added to a molecule, e.g., the addition of water to an alkene to form an alcohol

hydrocarbon (10.1) a compound composed solely of the elements carbon and hydrogen

hydrogen bonding (5.2) the attractive force between a hydrogen atom covalently bonded to a small, highly electronegative atom and another atom containing an unshared pair of electrons

hydrogenation (11.5, 13.4, 17.2) a reaction in which hydrogen (H_2) is added to a double or a triple bond

hydrohalogenation (11.5) the addition of a hydrohalogen (HCl, HBr, or HI) to an unsaturated bond

hydrolase (19.1) an enzyme that catalyzes hydrolysis reactions

hydrolysis (14.2) a chemical change that involves the reaction of a molecule with water; the process by which molecules are broken into their constituents by addition of water

hydronium ion (8.1) a protonated water molecule, H_3O^+

hydrophilic amino acid (18.1) "water loving"; a polar or ionic amino acid that has a high affinity for water

hydrophobic amino acid (18.2) "water fearing"; a nonpolar amino acid that prefers contact with other nonpolar amino acids over contact with water

hydroxyl group (12.1) the —OH functional group that is characteristic of alcohols

hyperammonemia (22.8) a genetic defect in one of the enzymes of the urea cycle that results in toxic or even fatal elevation of the concentration of ammonium ions in the body

hyperglycemia (21.7) blood glucose levels that are higher than normal

hypertonic solution (6.4, 17.6) the more concentrated solution of two separated by a semipermeable membrane

hypoglycemia (21.7) blood glucose levels that are lower than normal

hypothesis (1.1) an attempt to explain observations in a commonsense way

hypotonic solution (6.4, 17.6) the more dilute solution of two separated by a semipermeable membrane

I

ideal gas (5.1) a gas in which the particles do not interact and the volume of the individual gas particles is assumed to be negligible

ideal gas law (5.1) a law stating that for an ideal gas the product of pressure and volume is proportional to the product of the number of moles of the gas and its temperature; the proportionality constant for an ideal gas is symbolized R

incomplete protein (18.11) a protein source that does not contain all the essential and nonessential amino acids

indicator (8.3) a solute that shows some condition of a solution (such as acidity or basicity) by its color

induced fit model (19.4) the theory of enzyme-substrate binding that assumes that the enzyme is a flexible molecule and that both the substrate and the enzyme change their shapes to accommodate one another as the enzyme-substrate complex forms

initiation factors (20.6) proteins that are required for formation of the translation initiation complex, which is composed of the large and small ribosomal subunits, the mRNA, and the initiator tRNA, methionyl tRNA

inner mitochondrial membrane (22.1) the highly folded, impermeable membrane within the mitochondrion that is the location of the electron transport system and ATP synthase

insertion mutation (20.7) a mutation that results in the addition of one or more nucleotides to a DNA sequence

insulin (21.7, 23.6) a hormone released from the pancreas in response to high blood glucose levels; insulin stimulates glycogenesis, fat storage, and cellular uptake and storage of glucose from the blood

intensive property (1.2) a property of a substance that is independent of the quantity of the substance

intermembrane space (22.1) the region between the outer and inner mitochondrial membranes, which is the location of the proton (H^+) reservoir that drives ATP synthesis

intermolecular force (3.5) any attractive force that occurs between molecules

intramolecular force (3.5) any attractive force that occurs within molecules

intron (20.4) a noncoding sequence within a eukaryotic gene that must be removed from the primary transcript to produce a functional mRNA

ion (2.1)   an electrically charged particle formed by the gain or loss of electrons

ionic bonding (3.1)   an electrostatic attractive force between ions resulting from electron transfer

ionic solid (5.3)   a solid composed of positive and negative ions in a regular three-dimensional crystalline arrangement

ionization energy (2.7)   the energy needed to remove an electron from an atom in the gas phase

ionizing radiation (9.1)   radiation that is sufficiently high in energy to cause ion formation upon impact with an atom

ion pair (3.1)   the simplest formula unit for an ionic compound

ion product for water (8.1)   the product of the hydronium and hydroxide ion concentrations in pure water at a specified temperature; at 25°C, it has a value of 1.0×10^{-14}

irreversible enzyme inhibitor (19.10)   a chemical that binds strongly to the R groups of an amino acid in the active site and eliminates enzyme activity

isoelectric point (18.10)   a situation in which a protein has an equal number of positive and negative charges and therefore has an overall net charge of zero

isoelectronic (2.6)   atoms, ions, and molecules containing the same number of electrons

isomerase (19.1)   an enzyme that catalyzes the conversion of one isomer to another

isotonic solution (6.4, 17.6)   a solution that has the same solute concentration as another solution with which it is being compared; a solution that has the same osmotic pressure as a solution existing within a cell

isotope (2.1)   atom of the same element that differs in mass because it contains different numbers of neutrons

I.U.P.A.C. Nomenclature System (10.2)   the International Union of Pure and Applied Chemistry (I.U.P.A.C.) standard, universal system for the nomenclature of organic compounds

K

α-keratin (18.5)   a member of the family of fibrous proteins that form the covering of most land animals; major components of fur, skin, beaks, and nails

ketal (13.4)   the family of organic compounds formed via the reaction of two molecules of alcohol with a ketone in the presence of an acid catalyst; ketals have the following general structure:

$$R^1\!-\!\underset{\underset{R^2}{|}}{\overset{\overset{OR^3}{|}}{C}}\!-\!OR^4$$

ketoacidosis (23.3)   a drop in the pH of the blood caused by elevated levels of ketone bodies

ketone (13.1)   a family of organic molecules characterized by a carbonyl group; the carbonyl carbon is bonded to two alkyl groups, two aryl groups, or one alkyl and one aryl group; ketones have the following general structures:

$$\underset{R-C-R}{\overset{O}{\|}} \quad \underset{R-C-(Ar)}{\overset{O}{\|}} \quad \underset{(Ar)-C-(Ar)}{\overset{O}{\|}}$$

ketone bodies (23.3)   acetone, acetoacetone, and β-hydroxybutyrate produced from fatty acids in the liver via acetyl CoA

ketose (16.2)   a sugar that contains a ketone (carbonyl) group

ketosis (23.3)   an abnormal rise in the level of ketone bodies in the blood

kinetic energy (1.5)   the energy resulting from motion of an object [kinetic energy = 1/2 (mass)(velocity)^2]

kinetic-molecular theory (5.1)   the fundamental model of particle behavior in the gas phase

kinetics (7.3)   the study of rates of chemical reactions

L

lactose (16.5)   a disaccharide composed of β-D-galactose and either α- or β-D-glucose in β(1 → 4) glycosidic linkage; milk sugar

lactose intolerance (16.5)   the inability to produce the digestive enzyme lactase, which degrades lactose to galactose and glucose

lagging strand (20.3)   in DNA replication, the strand that is synthesized discontinuously from numerous RNA primers

law (1.1)   a summary of a large quantity of information

law of conservation of mass (4.3)   a law stating that, in chemical change, matter cannot be created or destroyed

leading strand (20.3)   in DNA replication, the strand that is synthesized continuously from a single RNA primer

LeChatelier's principle (7.4)   a law stating that when a system at equilibrium is disturbed, the equilibrium shifts in the direction that minimizes the disturbance

lethal dose (LD$_{50}$) (9.8)   the quantity of toxic material (such as radiation) that causes the death of 50% of a population of an organism

Lewis symbol (3.1)   representation of an atom or ion using the atomic symbol (for the nucleus and core electrons) and dots to represent valence electrons

ligase (19.1)   an enzyme that catalyzes the joining of two molecules

linear structure (3.4)   the structure of a molecule in which the bond angles about the central atom(s) is (are) 180°

line formula (10.2)   the simplest representation of a molecule in which it is assumed that there is a carbon atom at any location where two or more lines intersect, there is a carbon at the end of any line, and each carbon is bonded to the correct number of hydrogen atoms

linkage specificity (19.5)   the property of an enzyme that allows it to catalyze reactions involving only one kind of bond in the substrate molecule

lipase (23.1)   an enzyme that hydrolyzes the ester linkage between glycerol and the fatty acids of triglycerides

lipid (17.1)   a member of the group of biological molecules of varying composition that are classified together on the basis of their solubility in nonpolar solvents

liquid state (1.2)   a physical state of matter characterized by a fixed volume and the absence of a fixed shape

lock-and-key model (19.4)   the theory of enzyme-substrate binding that depicts enzymes as inflexible molecules; the substrate fits into the rigid active site in the same way a key fits into a lock

London forces (5.2)   weak attractive forces between molecules that result from short-lived dipoles that occur because of the continuous movement of electrons in the molecules

lone pair (3.4)   an electron pair that is not involved in bonding

low-density lipoprotein (LDL) (17.5)   a plasma lipoprotein that carries cholesterol to peripheral tissues and helps to regulate cholesterol levels in those tissues

lyase (19.1)   an enzyme that catalyzes a reaction involving double bonds

M

maltose (16.5)   a disaccharide composed of α-D-glucose and a second glucose molecule in α(1 → 4) glycosidic linkage

Markovnikov's rule (11.5)   the rule stating that a hydrogen atom, adding to a carbon-carbon double bond, will add to the carbon having the larger number of hydrogens attached to it

mass (1.5)   a quantity of matter

mass number (2.1)   the sum of the number of protons and neutrons in an atom

matrix space (22.1)   the region of the mitochondrion within the inner membrane; the location of the enzymes that carry out the reactions of the citric acid cycle and β-oxidation of fatty acids

matter (1.1)   the material component of the universe

melting point (3.3, 5.3)   the temperature at which a solid converts to a liquid

messenger RNA (20.4)   an RNA species produced by transcription and that specifies the amino acid sequence for a protein

metal (2.4)   an element located on the left side of the periodic table (left of the "staircase" boundary)

metallic bond (5.3)   a bond that results from the orbital overlap of metal atoms

metallic solid (5.3)   a solid composed of metal atoms held together by metallic bonds

metalloid (2.4) an element along the "staircase" boundary between metals and nonmetals; metalloids exhibit both metallic and nonmetallic properties

metastable isotope (9.2) an isotope that will give up some energy to produce a more stable form of the same isotope

micelle (23.1) an aggregation of molecules having nonpolar and polar regions; the nonpolar regions of the molecules aggregate, leaving the polar regions facing the surrounding water

mitochondria (22.1) the cellular "power plants" in which the reactions of the citric acid cycle, the electron transport system, and ATP synthase function to produce ATP

mixture (1.2) a material composed of two or more substances

molality (6.4) the number of moles of solute per kilogram of solvent

molar mass (4.1) the mass in grams of 1 mol of a substance

molar volume (5.1) the volume occupied by 1 mol of a substance

molarity (6.3) the number of moles of solute per liter of solution

mole (4.1) the amount of substance containing Avogadro's number of particles

molecular formula (10.2) a formula that provides the atoms and number of each type of atom in a molecule but gives no information regarding the bonding pattern involved in the structure of the molecule

molecular solid (5.3) a solid in which the molecules are held together by dipole-dipole and London forces (van der Waals forces)

molecule (3.2) a unit in which the atoms of two or more elements are held together by chemical bonds

monatomic ion (3.2) an ion formed by electron gain or loss from a single atom

monoglyceride (17.3) the product of the esterification of glycerol at one position

monomer (11.5) the individual molecules from which a polymer is formed

monosaccharide (16.1) the simplest type of carbohydrate consisting of a single saccharide unit

movement protein (18.1) a protein involved in any aspect of movement in an organism, for instance actin and myosin in muscle tissue and flagellin that composes bacterial flagella

mutagen (20.7) any chemical or physical agent that causes changes in the nucleotide sequence of a gene

mutation (20.7) any change in the nucleotide sequence of a gene

myoglobin (18.9) the oxygen storage protein found in muscle

N

N-terminal amino acid (18.3) the amino acid in a peptide that has a free α-N^+H_3 group; the first amino acid of a peptide

natural radioactivity (9.6) the spontaneous decay of a nucleus to produce high-energy particles or rays

negative allosterism (19.9) effector binding inactivates the active site of an allosteric enzyme

neurotransmitter (15.5) a chemical that carries a message, or signal, from a nerve cell to a target cell

neutral glyceride (17.3) the product of the esterification of glycerol at one, two, or three positions

neutralization (8.3) the reaction between an acid and a base

neutron (2.1) an uncharged particle, with the same mass as the proton, in the nucleus of an atom

nicotinamide adenine dinucleotide (NAD^+) (21.3) a molecule synthesized from the vitamin niacin and the nucleotide ATP and that serves as a carrier of hydride anions; a coenzyme that is an oxidizing agent used in a variety of metabolic processes

noble gas (2.4) elements in Group VIIIA (18) of the periodic table

nomenclature (3.2) a system for naming chemical compounds

nonelectrolyte (3.3, 6.1) a substance that, when dissolved in water, produces a solution that does not conduct an electrical current

nonessential amino acid (18.11) any amino acid that can be synthesized by the body

nonmetal (2.4) an element located on the right side of the periodic table (right of the "staircase" boundary)

nonreducing sugar (16.5) a sugar that cannot be oxidized by Benedict's or Tollens' reagent

normal boiling point (5.2) the temperature at which a substance will boil at 1 atm of pressure

nuclear equation (9.2) a balanced equation accounting for the products and reactants in a nuclear reaction

nuclear imaging (9.6) the generation of images of components of the body (organs, tissues) using techniques based on the measurement of radiation

nuclear medicine (9.6) a field of medicine that uses radioisotopes for diagnostic and therapeutic purposes

nuclear reactor (9.6) a device for conversion of nuclear energy into electrical energy

nucleosome (20.2) the first level of chromosome structure consisting of a strand of DNA wrapped around a small disk of histone proteins

nucleotide (20.1, 21.1) a molecule composed of a nitrogenous base, a five-carbon sugar, and one, two, or three phosphoryl groups

nucleus (2.1) the small, dense center of positive charge in the atom

nuclide (9.1) any atom characterized by an atomic number and a mass number

nutrient protein (18.1) a protein that serves as a source of amino acids for embryos or infants

nutritional Calorie (7.2) equivalent to one kilocalorie (1000 calories); also known as a large Calorie

O

octet rule (2.6) a rule predicting that atoms form the most stable molecules or ions when they are surrounded by eight electrons in their highest occupied energy level

oligosaccharide (16.1) an intermediate-sized carbohydrate composed of from three to ten monosaccharides

order of the reaction (7.3) the exponent of each concentration term in the rate equation

osmolarity (6.4) molarity of particles in solution; this value is used for osmotic pressure calculations

osmosis (6.4, 17.6) net flow of a solvent across a semipermeable membrane in response to a concentration gradient

osmotic pressure (6.4, 17.6) the net force with which water enters a solution through a semipermeable membrane; alternatively, the pressure required to stop net transfer of solvent across a semipermeable membrane

outer mitochondrial membrane (22.1) the membrane that surrounds the mitochondrion and separates it from the contents of the cytoplasm; it is highly permeable to small "food" molecules

β-oxidation (23.2) the biochemical pathway that results in the oxidation of fatty acids and the production of acetyl CoA

oxidation (8.5, 12.6, 13.4, 14.1) a loss of electrons; in organic compounds it may be recognized as a loss of hydrogen atoms or the gain of oxygen

oxidation-reduction reaction (4.3, 9.5) also called redox reaction, a reaction involving the transfer of one or more electrons from one reactant to another

oxidative deamination (22.7) an oxidation-reduction reaction in which NAD^+ is reduced and the amino acid is deaminated

oxidative phosphorylation (21.3, 22.6) production of ATP using the energy of electrons harvested during biological oxidation-reduction reactions

oxidizing agent (8.5) a substance that oxidizes, or removes electrons from, another substance; the oxidizing agent is reduced in the process

oxidoreductase (19.1) an enzyme that catalyzes an oxidation-reduction reaction

P

pancreatic serine proteases (19.11) a family of proteolytic enzymes, including trypsin, chymotrypsin, and elastase, that arose by divergent evolution

parent compound or parent chain (10.2) in the I.U.P.A.C. Nomenclature System the parent compound is the longest carbon-carbon chain containing the principal functional group in the molecule that is being named

partial pressure (5.1) the pressure exerted by one component of a gas mixture

particle accelerator (9.6) a device for production of high-energy nuclear particles based on the interaction of charged particles with magnetic and electrical fields

passive transport (17.6) the net movement of a solute from an area of high concentration to an area of low concentration

pentose (16.2) a five-carbon monosaccharide

pentose phosphate pathway (21.5) an alternative pathway for glucose degradation that provides the cell with reducing power in the form of NADPH

peptide bond (15.4, 18.3) the amide bond between two amino acids in a peptide chain

peptidyl tRNA binding site of ribosome (P-site) (20.6) a pocket on the surface of the ribosome that holds the tRNA bound to the growing peptide chain

percent yield (4.5) the ratio of the actual and theoretical yields of a chemical reaction multiplied by 100%

period (2.4) any one of seven horizontal rows of elements in the periodic table

periodic law (2.4) a law stating that properties of elements are periodic functions of their atomic numbers (Note that Mendeleev's original statement was based on atomic masses.)

peripheral membrane protein (17.6) a protein bound to either the inner or the outer surface of a membrane

phenol (12.7) an organic compound that contains a hydroxyl group (—OH) attached to a benzene ring

phenyl group (11.6) a benzene ring that has had a hydrogen atom removed, C_6H_5—

pH optimum (19.8) the pH at which an enzyme catalyzes the reaction at maximum efficiency

phosphatidate (17.3) a molecule of glycerol with fatty acids esterified to C-1 and C-2 of glycerol and a free phosphoryl group esterified at C-3

phosphoester (14.4) the product of the reaction between phosphoric acid and an alcohol

phosphoglyceride (17.3) a molecule with fatty acids esterified at the C-1 and C-2 positions of glycerol and a phosphoryl group esterified at the C-3 position

phospholipid (17.3) a lipid containing a phosphoryl group

phosphopantetheine (23.4) the portion of coenzyme A and the acyl carrier protein that is derived from the vitamin pantothenic acid

phosphoric anhydride (14.4) the bond formed when two phosphate groups react with one another and a water molecule is lost

pH scale (8.2) a numerical representation of acidity or basicity of a solution; pH = −log $[H_3O^+]$

physical change (1.2) a change in the form of a substance but not in its chemical composition; no chemical bonds are broken in a physical change

physical property (1.2) a characteristic of a substance that can be observed without the substance undergoing change (examples include color, density, melting and boiling points)

plasma lipoprotein (17.5) a complex composed of lipid and protein that is responsible for the transport of lipids throughout the body

β-pleated sheet (18.5) a common secondary structure of a peptide chain that resembles the pleats of an Oriental fan

point mutation (20.7) the substitution of one nucleotide pair for another within a gene

polar covalent bonding (3.4) a covalent bond in which the electrons are not equally shared

polar covalent molecule (3.4) a molecule that has a permanent electric dipole moment resulting from an unsymmetrical electron distribution; a dipolar molecule

poly(A) tail (20.4) a tract of 100–200 adenosine monophosphate units covalently attached to the 3′ end of eukaryotic messenger RNA molecules

polyatomic ion (3.2) an ion containing a number of atoms

polymer (11.5) a very large molecule formed by the combination of many small molecules (called monomers) (e.g., polyamides, nylons)

polyprotic substance (8.3) a substance that can accept or donate more than one proton per molecule

polysaccharide (16.1) a large, complex carbohydrate composed of long chains of monosaccharides

polysome (20.6) complexes of many ribosomes all simultaneously translating a single mRNA

positive allosterism (19.9) effector binding activates the active site of an allosteric enzyme

positron (9.2) particle that has the same mass as an electron but opposite (+) charge

post-transcriptional modification (20.4) alterations of the primary transcripts produced in eukaryotic cells; these include addition of a poly(A) tail to the 3′ end of the mRNA, addition of the cap structure to the 5′ end of the mRNA, and RNA splicing

potential energy (1.5) stored energy or energy caused by position or composition

precipitate (6.1) an insoluble substance formed and separated from a solution

precision (1.4) the degree of agreement among replicate measurements of the same quantity

pressure (5.1) a force per unit area

primary (1°) alcohol (12.4) an alcohol with the general formula RCH_2OH

primary (1°) amine (15.1) an amine with the general formula RNH_2

primary (1°) carbon (10.2) a carbon atom that is bonded to only one other carbon atom

primary structure (of a protein) (18.4) the linear sequence of amino acids in a protein chain determined by the genetic information of the gene for each protein

primary transcript (20.4) the RNA product of transcription in eukaryotic cells, before post-transcriptional modifications are carried out

product (4.3, 19.2) the chemical species that results from a chemical reaction and that appears on the right side of a chemical equation

proenzyme (19.9) the inactive form of a proteolytic enzyme

prokaryote (20.2) an organism with simple cellular structure in which there is no true nucleus enclosed by a nuclear membrane and there are no true membrane-bound organelles in the cytoplasm

promoter (20.4) the sequence of nucleotides immediately before a gene that is recognized by the RNA polymerase and signals the start point and direction of transcription

properties (1.2) characteristics of matter

prostaglandins (17.2) a family of hormonelike substances derived from the twenty-carbon fatty acid, arachidonic acid; produced by many cells of the body, they regulate many body functions

prosthetic group (18.7) the nonprotein portion of a protein that is essential to the biological activity of the protein; often a complex organic compound

protein (18: Intro) a macromolecule whose primary structure is a linear sequence of α-amino acids and whose final structure results from folding of the chain into a specific three-dimensional structure; proteins serve as catalysts, structural components, and nutritional elements for the cell

protein modification (19.9) a means of enzyme regulation in which a chemical group is covalently added to or removed from a protein. The chemical modification either turns the enzyme on or turns it off

proteolytic enzyme (19.11) an enzyme that hydrolyzes the peptide bonds between amino acids in a protein chain

proton (2.1) a positively charged particle in the nucleus of an atom

pure substance (1.2) a substance with constant composition

purine (20.1) a family of nitrogenous bases (heterocyclic amines) that are components of DNA and RNA and consist of a six-sided ring fused to a five-sided ring; the common purines in nucleic acids are adenine and guanine

pyridoxal phosphate (22.7) a coenzyme derived from vitamin B_6 that is required for all transamination reactions

pyrimidine (20.1) a family of nitrogenous bases (heterocyclic amines) that are components of nucleic acids and consist of a single six-sided ring; the common pyrimidines of DNA are cytosine and thymine; the common pyrimidines of RNA are cytosine and uracil

pyrimidine dimer (20.7) UV-light induced covalent bonding of two adjacent pyrimidine bases in a strand of DNA

pyruvate dehydrogenase complex (22.2) a complex of all the enzymes and coenzymes required for the synthesis of CO_2 and acetyl CoA from pyruvate

Q

quantization (2.3) a characteristic that energy can occur only in discrete units called quanta

quaternary ammonium salt (15.1) an amine salt with the general formula $R_4N^+A^-$ (in which R— can be an alkyl or aryl group or a hydrogen atom and A^- can be any anion)

quaternary (4°) carbon (10.2) a carbon atom that is bonded to four other carbon atoms

quaternary structure (of a protein) (18.7) aggregation of more than one folded peptide chain to yield a functional protein

R

rad (9.8) abbreviation for *radiation absorbed dose*, the absorption of 2.4×10^{-3} calories of energy per kilogram of absorbing tissue

radioactivity (9.1) the process by which atoms emit high-energy particles or rays; the spontaneous decomposition of a nucleus to produce a different nucleus

radiocarbon dating (9.5) the estimation of the age of objects through measurement of isotopic ratios of carbon

Raoult's law (6.4) a law stating that the vapor pressure of a component is equal to its mole fraction times the vapor pressure of the pure component

rate constant (7.3) the proportionality constant that relates the rate of a reaction and the concentration of reactants

rate equation (7.3) expresses the rate of a reaction in terms of reactant concentration and a rate constant

rate of chemical reaction (7.3) the change in concentration of a reactant or product per unit time

reactant (1.2, 4.3) starting material for a chemical reaction, appearing on the left side of a chemical equation

reducing agent (8.5) a substance that reduces, or donates electrons to, another substance; the reducing agent is itself oxidized in the process

reducing sugar (16.4) a sugar that can be oxidized by Benedict's or Tollens' reagents; includes all monosaccharides and most disaccharides

reduction (8.5, 12.6) the gain of electrons; in organic compounds it may be recognized by a gain of hydrogen or loss of oxygen

regulatory proteins (18.1) proteins that control cell functions such as metabolism and reproduction

release factor (20.6) a protein that binds to the termination codon in the empty A-site of the ribosome and causes the peptidyl transferase to hydrolyze the bond between the peptide and the peptidyl tRNA

rem (9.8) abbreviation for *roentgen equivalent for man*, the product of rad and RBE

replication fork (20.3) the point at which new nucleotides are added to the growing daughter DNA strand

replication origin (20.3) the region of a DNA molecule where DNA replication always begins

representative element (2.4) member of the groups of the periodic table designated as A

resonance (3.4) a condition that occurs when more than one valid Lewis structure can be written for a particular molecule

resonance form (3.4) one of a number of valid Lewis structures for a particular molecule

resonance hybrid (3.4) a description of the bonding in a molecule resulting from a superimposition of all valid Lewis structures (resonance forms)

restriction enzyme (20.8) a bacterial enzyme that recognizes specific nucleotide sequences on a DNA molecule and cuts the sugar-phosphate backbone of the DNA at or near that site

result (1.3) the outcome of a designed experiment, often determined from individual bits of data

reversible, competitive enzyme inhibitor (19.10) a chemical that resembles the structure and charge distribution of the natural substrate and competes with it for the active site of an enzyme

reversible, noncompetitive enzyme inhibitor (19.10) a chemical that binds weakly to an amino acid R group of an enzyme and inhibits activity; when the inhibitor dissociates, the enzyme is restored to its active form

reversible reaction (7.4) a reaction that will proceed in either direction, reactants to products or products to reactants

ribonucleic acid (RNA) (20.1) single-stranded nucleic acid molecules that are composed of phosphoryl groups, ribose, and the nitrogenous bases uracil, cytosine, adenine, and guanine

ribonucleotide (20.1) a ribonucleoside phosphate or nucleotide composed of a nitrogenous base in β-*N*-glycosidic linkage to the 1′ carbon of the sugar ribose and with one, two, or three phosphoryl groups esterified at the hydroxyl of the 5′ carbon of the ribose

ribose (16.4) a five-carbon monosaccharide that is a component of RNA and many coenzymes

ribosomal RNA (rRNA) (20.4) the RNA species that are structural and functional components of the small and large ribosomal subunits

ribosome (20.6) an organelle composed of a large and a small subunit, each of which is made up of ribosomal RNA and proteins; the platform on which translation occurs and that carries the enzymatic activity that forms peptide bonds

RNA polymerase (20.4) the enzyme that catalyzes the synthesis of RNA molecules using DNA as the template

RNA splicing (20.4) removal of portions of the primary transcript that do not encode protein sequences

röentgen (9.8) the dose of radiation producing 2.1×10^9 ions in 1 cm^3 of air at 0°C and 1 atm of pressure

S

saccharide (16.1) a sugar molecule

saponification (14.2, 17.2) a reaction in which a soap is produced; more generally, the hydrolysis of an ester by an aqueous base

saturated fatty acid (17.2) a long-chain monocarboxylic acid in which each carbon of the chain is bonded to the maximum number of hydrogen atoms

saturated hydrocarbon (10.1) an alkane; a hydrocarbon that contains only carbon and hydrogen bonded together through carbon-hydrogen and carbon-carbon single bonds

saturated solution (6.1) one in which undissolved solute is in equilibrium with the solution

scientific method (1.1) the process of studying our surroundings that is based on experimentation

scientific notation (1.4) a system used to represent numbers as powers of ten

secondary (2°) alcohol (12.4) an alcohol with the general formula R_2CHOH

secondary (2°) amine (15.1) an amine with the general formula R_2NH

secondary (2°) carbon (10.2) a carbon atom that is bonded to two other carbon atoms

secondary structure (of a protein) (18.5) folding of the primary structure of a protein into an α-helix or a β-pleated sheet; folding is maintained by hydrogen bonds between the amide hydrogen and the carbonyl oxygen of the peptide bond

semiconservative DNA replication (20.3) DNA polymerase "reads" each parental strand of DNA and produces a complementary daughter strand; thus, all newly synthesized DNA molecules consist of one parental and one daughter strand

semipermeable membrane (6.4, 17.6) a membrane permeable to the solvent but not the solute; a material that allows the transport of certain substances from one side of the membrane to the other

shielding (9.7) material used to provide protection from radiation

sickle cell anemia (18.9) a human genetic disease resulting from inheriting mutant hemoglobin genes from both parents

significant figures (1.4) all digits in a number known with certainty and the first uncertain digit

silent mutation (20.7) a mutation that changes the sequence of the DNA but does not alter the amino acid sequence of the protein encoded by the DNA

single bond (3.4) a bond in which one pair of electrons is shared by two atoms

single-replacement reaction (4.3) also called substitution reaction, one in which one atom in a molecule is displaced by another

soap (14.2) any of a variety of the alkali metal salts of fatty acids

solid state (1.2) a physical state of matter characterized by its rigidity and fixed volume and shape

solubility (3.5, 6.1) the amount of a substance that will dissolve in a given volume of solvent at a specified temperature

solute (6.1) a component of a solution that is present in lesser quantity than the solvent

solution (6.1) a homogeneous (uniform) mixture of two or more substances

solvent (6.1) the solution component that is present in the largest quantity

specific gravity (1.5) the ratio of the density of a substance to the density of water at 4°C or any specified temperature

specific heat (7.2) the quantity of heat (calories) required to raise the temperature of 1 g of a substance one degree Celsius

spectroscopy (2.3) the measurement of intensity and energy of electromagnetic radiation

speed of light (2.3) 2.99×10^8 m/s in a vacuum

sphingolipid (17.4) a phospholipid that is derived from the amino alcohol sphingosine rather than from glycerol

sphingomyelin (17.4) a sphingolipid found in abundance in the myelin sheath that surrounds and insulates cells of the central nervous system

standard solution (8.3) a solution whose concentration is accurately known

standard temperature and pressure (STP) (5.1) defined as 273 K and 1 atm

stereochemical specificity (19.5) the property of an enzyme that allows it to catalyze reactions involving only one enantiomer of the substrate

stereochemistry (16.3) the study of the spatial arrangement of atoms in a molecule

stereoisomers (10.3, 16.3) a pair of molecules having the same structural formula and bonding pattern but differing in the arrangement of the atoms in space

steroid (17.4) a lipid derived from cholesterol and composed of one five-sided ring and three six-sided rings; the steroids include sex hormones and anti-inflammatory compounds

structural analog (19.10) a chemical having a structure and charge distribution very similar to those of a natural enzyme substrate

structural formula (10.2) a formula showing all of the atoms in a molecule and exhibiting all bonds as lines

structural isomers (10.2) molecules having the same molecular formula but different chemical structures

structural protein (18.1) a protein that provides mechanical support for large plants and animals

sublevel (2.5) a set of equal-energy orbitals within a principal energy level

substituted hydrocarbon (10.1) a hydrocarbon in which one or more hydrogen atoms is replaced by another atom or group of atoms

substitution reaction (10.5, 11.6) a reaction that results in the replacement of one group for another

substrate (19.1) the reactant in a chemical reaction that binds to an enzyme active site and is converted to product

substrate-level phosphorylation (21.3) the production of ATP by the transfer of a phosphoryl group from the substrate of a reaction to ADP

sucrose (16.5) a disaccharide composed of α-D-glucose and β-D-fructose in (α1 → β2) glycosidic linkage; table sugar

supersaturated solution (6.1) a solution that is more concentrated than a saturated solution (Note that such a solution is not at equilibrium.)

surface tension (5.2) a measure of the strength of the attractive forces at the surface of a liquid

surfactant (5.2) a substance that decreases the surface tension of a liquid

surroundings (7.1) the universe outside of the system

suspension (6.1) a heterogeneous mixture of particles; the suspended particles are larger than those found in a colloidal suspension

system (7.1) the process under study

T

temperature (1.5) a measure of the relative "hotness" or "coldness" of an object

temperature optimum (19.8) the temperature at which an enzyme functions optimally and the rate of reaction is maximal

terminal electron acceptor (22.6) the final electron acceptor in an electron transport system that removes the low-energy electrons from the system; in aerobic organisms the terminal electron acceptor is molecular oxygen

termination codon (20.6) a triplet of ribonucleotides with no corresponding anticodon on a tRNA; as a result, translation will end, because there is no amino acid to transfer to the peptide chain

terpene (17.4) the general term for lipids that are synthesized from isoprene units; the terpenes include steroids, bile salts, lipid-soluble vitamins, and chlorophyll

tertiary (3°) alcohol (12.4) an alcohol with the general formula R_3COH

tertiary (3°) amine (15.1) an amine with the general formula R_3N

tertiary (3°) carbon (10.2) a carbon atom that is bonded to three other carbon atoms

tertiary structure (of a protein) (18.6) the globular, three-dimensional structure of a protein that results from folding the regions of secondary structure; this folding occurs spontaneously as a result of interactions of the side chains or R groups of the amino acids

tetrahedral structure (3.4) a molecule consisting of four groups attached to a central atom that occupy the four corners of an imagined regular tetrahedron

tetrose (16.2) a four-carbon monosaccharide

theoretical yield (4.5) the maximum amount of product that can be produced from a given amount of reactant

theory (1.1) a hypothesis supported by extensive testing that explains and predicts facts

thermodynamics (7.1) the branch of science that deals with the relationship between energies of systems, work, and heat

thioester (14.4) the product of a reaction between a thiol and a carboxylic acid

thiol (12.9) an organic compound that contains a thiol group (—SH)

titration (8.3) the process of adding a solution from a buret to a sample until a reaction is complete, at which time the volume is accurately measured and the concentration of the sample is calculated

Tollens' test (13.4) a test reagent (silver nitrate in ammonium hydroxide) used to distinguish aldehydes and ketones; also called the Tollens' silver mirror test

tracer (9.6) a radioisotope that is rapidly and selectively transmitted to the part of the body for which diagnosis is desired

transaminase (22.7) an enzyme that catalyzes the transfer of an amino group from one molecule to another

transamination (22.7) a reaction in which an amino group is transferred from one molecule to another

transcription (20.4) the synthesis of RNA from a DNA template

transferase (19.1) an enzyme that catalyzes the transfer of a functional group from one molecule to another

transfer RNA (tRNA) (15.4, 20.4) small RNAs that bind to a specific amino acid at the 3′ end and mediate its addition at the appropriate site in a growing peptide chain; accomplished by recognition of the correct codon on the mRNA by the complementary anticodon on the tRNA

transition element (2.4) any element located between Groups IIA (2) and IIIA (13) in the long periods of the periodic table

transition state (19.6) the unstable intermediate in catalysis in which the enzyme has altered the form of the substrate so that it now shares properties of both the substrate and the product

translation (20.4) the synthesis of a protein from the genetic code carried on the mRNA

translocation (20.6) movement of the ribosome along the mRNA during translation

transmembrane protein (17.6) a protein that is embedded within a membrane and crosses the lipid bilayer, protruding from the membrane both inside and outside the cell

transport protein (18.1) a protein that transports materials across the cell membrane or throughout the body

triglyceride (17.3, 23.1) triacylglycerol; a molecule composed of glycerol esterified to three fatty acids

trigonal pyramidal molecule (3.4) a nonplanar structure involving three groups bonded to a central atom in which each group is equidistant from the central atom

triose (16.2) a three-carbon monosaccharide

triple bond (3.4) a bond in which three pairs of electrons are shared by two atoms

U

uncertainty (1.4) the degree of doubt in a single measurement

unit (1.3) a determinate quantity (of length, time, etc.) that has been adopted as a standard of measurement

unsaturated fatty acid (17.2) a long-chain monocarboxylic acid having at least one carbon-to-carbon double bond

unsaturated hydrocarbon (10.1, 11: Intro) a hydrocarbon containing at least one multiple (double or triple) bond

urea cycle (22.8) a cyclic series of reactions that detoxifies ammonium ions by incorporating them into urea, which is excreted from the body

uridine triphosphate (UTP) (21.7) a nucleotide composed of the pyrimidine uracil, the sugar ribose, and three phosphoryl groups and that serves as a carrier of glucose-1-phosphate in glycogenesis

V

valence electron (2.5) electron in the outermost shell (principal quantum level) of an atom

valence shell electron pair repulsion theory (VSEPR) (3.4) a model that predicts molecular geometry using the premise that electron pairs will arrange themselves as far apart as possible, to minimize electron repulsion

van der Waals forces (5.2) a general term for intermolecular forces that include dipole-dipole and London forces

vapor pressure of a liquid (5.2) the pressure exerted by the vapor at the surface of a liquid at equilibrium

very low density lipoprotein (VLDL) (17.5) a plasma lipoprotein that binds triglycerides synthesized by the liver and carries them to adipose tissue for storage

viscosity (5.2) a measure of the resistance to flow of a substance at constant temperature

vitamin (19.7) an organic substance that is required in the diet in small amounts; water-soluble vitamins are used in the synthesis of coenzymes required for the function of cellular enzymes; lipid-soluble vitamins are involved in calcium metabolism, vision, and blood clotting

voltaic cell (8.5) an electrochemical cell that converts chemical energy into electrical energy

W

wax (17.4) a collection of lipids that are generally considered to be esters of long-chain alcohols

weight (1.5) the force exerted on an object by gravity

weight/volume percent [% (W/V)] (6.2) the concentration of a solution expressed as a ratio of grams of solute to milliliters of solution multiplied by 100%

weight/weight percent [% (W/W)] (6.2) the concentration of a solution expressed as a ratio of mass of solute to mass of solution multiplied by 100%

Z

Zaitsev's rule (12.5) states that in an elimination reaction, the alkene with the greatest number of alkyl groups on the double-bonded carbon (the more highly substituted alkene) is the major product of the reaction

Answers to Odd-Numbered Problems

Chapter 1

1.1 a. Physical property
     b. Chemical property
     c. Physical property
     d. Physical property
     e. Physical property
1.3 a. Pure substance
     b. Heterogeneous mixture
     c. Homogeneous mixture
     d. Pure substance
1.5 a. 1.0×10^3 mL
     b. 1.0×10^6 µL
     c. 1.0×10^{-3} kL
     d. 1.0×10^2 cL
     e. 1.0×10^{-1} daL
1.7 a. 1.3×10^{-2} m
     b. 0.71 L
     c. 2.00 oz
     d. 1.5×10^{-4} m²
1.9 a. Three
     b. Three
     c. Four
     d. Two
     e. Three
1.11 a. 2.4×10^{-3}
     b. 1.80×10^{-2}
     c. 2.24×10^2
1.13 a. 8.09
     b. 5.9
     c. 20.19
1.15 a. 51
     b. 8.0×10^1
     c. 1.6×10^2
1.17 a. 61.4
     b. 6.17
     c. 6.65×10^{-2}
1.19 a. 0°C
     b. 273 K
1.21 23.7 g
1.23 a. Chemistry is the study of matter and the changes that matter undergoes.
     b. Matter is the material component of the universe.
     c. Energy is the ability to do work.
1.25 a. Potential energy is stored energy, or energy due to position or composition.
     b. Kinetic energy is the energy resulting from motion of an object.
     c. Data are a group of facts resulting from an experiment.
1.27 a. Gram (or kilogram)
     b. Liter
     c. Meter
1.29 Mass is an independent quantity while weight is dependent on gravity.
1.31 Density is mass per volume. Specific gravity is the ratio of the density of a substance to the density of water at 4°C.
1.33 The scientific method is an organized way of doing science.
1.35 A theory.
1.37 A physical property is a characteristic of a substance that can be observed without the substance undergoing a change in chemical composition.
1.39 Chemical properties of matter include flammability and toxicity.
1.41 A pure substance has constant composition with only a single substance whereas a mixture is composed of two or more substances.
1.43 Mixtures are composed of two or more substances. A homogeneous mixture has uniform composition while a heterogeneous mixture has non-uniform composition.
1.45 a. Chemical reaction
     b. Physical change
     c. Physical change
1.47 a. Physical property
     b. Chemical property
1.49 a. Pure substance
     b. Pure substance
     c. Mixture
1.51 a. Homogeneous
     b. Homogeneous
     c. Homogeneous
1.53 A gas is made up of particles that are widely separated. A gas will expand to fill any container and it has no definite shape or volume.
1.55 a. Extensive property
     b. Extensive property
     c. Intensive property
1.57 An element is a pure substance that cannot be changed into a simpler form of matter by any chemical reaction. An atom is the smallest unit of an element that retains the properties of that element.
1.59 a. Iron, oxygen, carbon are just a few of the more than 100 possible elements
     b. Sodium chloride, water, sucrose, ethyl alcohol
1.61 a. 32 oz
     b. 1.0×10^{-3} t
     c. 9.1×10^2 g
     d. 9.1×10^5 mg
     e. 9.1×10^1 da

AP-1

1.63
a. 6.6×10^{-3} lb
b. 1.1×10^{-1} oz
c. 3.0×10^{-3} kg
d. 3.0×10^{2} cg
e. 3.0×10^{3} mg

1.65
a. 10.0°C
b. 283.2 K

1.67
a. 293.2 K
b. 68.0°F

1.69 4 L

1.71 101°F

1.73 5 cm is shorter than 5 in.

1.75 5.0 μg is smaller that 5.0 mg.

1.77
a. 3
b. 3
c. 3
d. 4
e. 4
f. 3

1.79
a. 3.87×10^{-3}
b. 5.20×10^{-2}
c. 2.62×10^{-3}
d. 2.43×10^{1}
e. 2.40×10^{2}
f. 2.41×10^{0}

1.81
a. Precision is a measure of the agreement of replicate results.
b. Accuracy is the degree of agreement between the true value and measured value.

1.83
a. 1.5×10^{4}
b. 2.41×10^{-1}
c. 5.99
d. 1139.42
e. 7.21×10^{3}

1.85
a. 1.23×10^{1}
b. 5.69×10^{-2}
c. -1.527×10^{3}
d. 7.89×10^{-7}
e. 9.2×10^{7}
f. 5.280×10^{-3}
g. 1.279×10^{0}
h. -5.3177×10^{2}

1.87
a. 3,240
b. 0.000150
c. 0.4579
d. −683,000
e. −0.0821
f. 299,790,000
g. 1.50
h. 602,200,000,000,000,000,000,000

1.89 6.00 g/mL

1.91 1.08×10^{3} g

1.93 teak

1.95 0.789

1.97 Lead has the lowest density and platinum has the greatest density.

1.99 12.6 mL

Chapter 2

2.1
a. 16 protons, 16 electrons, 16 neutrons
b. 11 protons, 11 electrons, 12 neutrons

2.3 20.18 amu

2.5 Electron density is the probability that an electron will be found in a particular region of an atomic orbital.

2.7
a. Zr (zirconium)
b. 22.99
c. Cr (chromium)
d. Bi (bismuth)

2.9
a. Helium, atomic number = 2, mass = 4.00 amu
b. Fluorine, atomic number = 9, mass = 19.00 amu
c. Manganese, atomic number = 25, mass = 54.94 amu

2.11
a. Total electrons = 11, valence electrons = 1
b. Total electrons = 12, valence electrons = 2
c. Total electrons = 16, valence electrons = 6
d. Total electrons = 17, valence electrons = 7
e. Total electrons = 18, valence electrons = 8

2.13
a. Sulfur: $1s^2, 2s^2, 2p^6, 3s^2, 3p^4$
b. Calcium: $1s^2, 2s^2, 2p^6, 3s^2, 3p^6, 4s^2$

2.15
a. [Ne] $3s^2, 3p^4$
b. [Ar] $4s^2$

2.17
a. Ca^{2+} and Ar are isoelectronic
b. Sr^{2+} and Kr are isoelectronic
c. S^{2-} and Ar are isoelectronic
d. Mg^{2+} and Ne are isoelectronic
e. P^{3-} and Ar are isoelectronic

2.19
a. (Smallest) F, N, Be (largest)
b. (Lowest) Be, N, F (highest)
c. (Lowest) Be, N, F, (highest)

2.21
a. 8 protons, 8 electrons, 8 neutrons.
b. 16 neutrons.

2.23

Particle	Mass	Charge
a. electron	5.4×10^{-4} amu	−1
b. proton	1.00 amu	+1
c. neutron	1.00 amu	0

2.25
a. An ion is a charged atom or group of atoms formed by the loss or gain of electrons.
b. A loss of electrons by a neutral species results in a cation.
c. A gain of electrons by a neutral species results in an anion.

2.27 From the periodic table, all isotopes of Rn have 86 protons. Isotopes differ in the number of neutrons.

2.29
a. 34
b. 46

2.31
a. $^{1}_{1}H$
b. $^{14}_{6}C$

2.33

	Atomic Symbol	# Protons	# Neutrons	# Electrons	Charge
a.	$^{23}_{11}Na$	11	12	11	0
b.	$^{32}_{16}S^{2-}$	16	16	18	2−
c.	$^{16}_{8}O$	8	8	8	0
d.	$^{24}_{12}Mg^{2+}$	12	12	10	2+
e.	$^{39}_{19}K^{+}$	19	20	18	1+

2.35
a. Neutrons
b. Protons
c. Protons, neutrons
d. Ion
e. Nucleus, negative

2.37
- All matter consists of tiny particles called atoms.
- Atoms cannot be created, divided, destroyed, or converted to any other type of atom.
- All atoms of a particular element have identical properties.
- Atoms of different elements have different properties.
- Atoms combine in simple whole-number ratios.
- Chemical change involves joining, separating, or rearranging atoms.

2.39
a. Chadwick—demonstrated the existence of the neutron in 1932.
b. Goldstein—identified positive charge in the atom.

2.41 a. Dalton—developed the Law of Multiple Proportions; determined the relative atomic weights of the elements known at that time; developed the first scientific atomic theory.
   b. Crookes—developed the cathode ray tube and discovered "cathode rays;" characterized electron properties.
2.43 Our understanding of the nucleus is based on the gold foil experiment performed by Geiger and interpreted by Rutherford. In this experiment, Geiger bombarded a piece of gold foil with alpha particles, and observed that some alpha particles passed straight through the foil, others were deflected and some simply bounced back. This led Rutherford to propose that the atom consisted of a small, dense nucleus (alpha particles bounced back), surrounded by a cloud of electrons (some alpha particles were deflected). The size of the nucleus is small when compared to the volume of the atom (alpha particles were able to pass through the foil).
2.45 Crookes used the cathode ray tube. He observed particles emitted by the cathode and traveling toward the anode. This ray was deflected by an electric field. Thomson measured the curvature of the ray influenced by the electric and magnetic fields. This measurement provided the mass to charge ratio of the negative particle. Thomson also gave the particle the name, electron.
2.47 A cathode ray is the negatively charged particle formed in a cathode ray tube.
2.49 Radiowave ↑
   Microwave |
   Infrared | Increasing
   Visible | Wavelength
   Ultraviolet |
   X-ray |
   Gamma ray |
2.51 Infrared radiation has greater energy than microwave radiation.
2.53 Spectroscopy is the measurement of intensity and energy of electromagnetic radiation.
2.55 According to Bohr, Planck, and others, electrons exist only in certain allowed regions, quantum levels, outside of the nucleus.
2.57 • Electrons are found in orbits at discrete distances from the nucleus.
   • The orbits are quantized—they are of discrete energies.
   • Electrons can only be found in these orbits, never in between (they are able to jump instantaneously from orbit to orbit).
   • Electrons can undergo transitions—if an electron absorbs energy, it will jump to a higher orbit; when the electron falls back to a lower orbit, it will release energy.
2.59 Bohr's atomic model was the first to successfully account for electronic properties of atoms, specifically, the interaction of atoms and light (spectroscopy).
2.61 a. Sodium
   b. Potassium
   c. Magnesium
2.63 Group IA (or 1) is known collectively as the alkali metals and consists of lithium, sodium, potassium, rubidium, cesium, and francium.
2.65 Group VIIA (or 17) is known collectively as the halogens and consists of fluorine, chlorine, bromine, iodine, and astatine.
2.67 a. True
   b. True
2.69 a. Na, Ni, Al
   b. Na, Al
   c. Na, Ni, Al
   d. Ar

2.71 a. One
   b. One
   c. Three
   d. Seven
   e. Zero (or eight)
   f. Zero (or two)
2.73 A principal energy level is designated $n = 1, 2, 3$, and so forth. It is similar to Bohr's orbits in concept. A sublevel is a part of a principal energy level and is designated s, p, d, and f.
2.75 The s orbital represents the probability of finding an electron in a region of space surrounding the nucleus.
2.77 Three p orbitals (p_x, p_y, p_z) can exist in a given principal energy level.
2.79 A $3p$ orbital is a higher energy orbital than a $2p$ orbital because it is a part of a higher energy principal energy level.
2.81 $2\ e^-$ for $n = 1$
   $8\ e^-$ for $n = 2$
   $18\ e^-$ for $n = 3$
2.83 a. $3p$ orbital
   b. $3s$ orbital
   c. $3d$ orbital
   d. $4s$ orbital
   e. $3d$ orbital
   f. $3p$ orbital
2.85 a. Not possible
   b. Possible
   c. Not possible
   d. Not possible
2.87 a. Li^+
   b. O^{2-}
   c. Ca^{2+}
   d. Br^-
   e. S^{2-}
   f. Al^{3+}
2.89 a. Isoelectronic
   b. Isoelectronic
2.91 a. Na^+
   b. S^{2-}
   c. Cl^-
2.93 a. $1s^2, 2s^2, 2p^6, 3s^2, 3p^6$
   b. $1s^2, 2s^2, 2p^6$
2.95 a. (Smallest) F, O, N (Largest)
   b. (Smallest) Li, K, Cs (Largest)
   c. (Smallest) Cl, Br, I (Largest)
2.97 a. (Smallest) O, N, F (Largest)
   b. (Smallest) Cs, K, Li (Largest)
   c. (Smallest) I, Br, Cl (Largest)
2.99 A positive ion is always smaller than its parent atom because the positive charge of the nucleus is shared among fewer electrons in the ion. As a result, each electron is pulled closer to the nucleus and the volume of the ion decreases.
2.101 The fluoride ion has a completed octet of electrons and an electron configuration resembling its nearest noble gas.

Chapter 3

3.1 a. LiBr
   b. $CaBr_2$
   c. Ca_3N_2
3.3 a. Potassium cyanide
   b. Magnesium sulfide
   c. Magnesium acetate
3.5 a. $CaCO_3$
   b. $NaHCO_3$
   c. Cu_2SO_4

3.7 a. Diboron trioxide
     b. Nitrogen oxide
     c. Iodine chloride
     d. Phosphorus trichloride

3.9 a. P_2O_5
     b. SiO_2

3.11 a. H:Ö:H
     b. H:C:H with H above and H below

3.13 a. $[H:Ö:H \text{ with H above}]^+$
     b. $[:Ö:H]^-$

3.15 a. $\left[:\ddot{O}-C(=O)-\ddot{O}-H\right]^- \longleftrightarrow \left[:\ddot{O}=C(-\ddot{O}:)-\ddot{O}-H\right]^-$
     b. $\left[:\ddot{O}:P(:\ddot{O}:)(:\ddot{O}:):\ddot{O}:\right]^{3-}$

3.17 a. The bonded nuclei are closer together when a double bond exists, in comparison to a single bond.
     b. The bond strength increases as the bond order increases. Therefore, a double bond is stronger than a single bond.

3.19 a. $\left[:\ddot{O}:Se::\ddot{O} \longleftrightarrow :\ddot{O}::Se:\ddot{O}:\right]$

3.21 a. H:P:H with H above (and structural formula with P bonded to 3 H's)
     b. H:Si:H with H above and H below (and structural formula with Si bonded to 4 H's)

3.23 a. Oxygen is more electronegative than sulfur; the bond is polar. The electrons are pulled toward the oxygen atom.
     b. Nitrogen is more electronegative than carbon; the bond is polar. The electrons are pulled toward the nitrogen atom.
     c. There is no electronegativity difference between two identical atoms; the bond is nonpolar.
     d. Chlorine is more electronegative than iodine; the bond is polar. The electrons are pulled toward the chlorine atom.

3.25 a. Nonpolar
     b. Polar
     c. Polar
     d. Nonpolar

3.27 a. H_2O
     b. CO
     c. NH_3
     d. ICl

3.29 a. Ionic
     b. Covalent
     c. Covalent
     d. Covalent

3.31 a. Covalent
     b. Covalent
     c. Covalent
     d. Ionic

3.33 a. $Li\cdot + :\ddot{Br}\cdot \longrightarrow Li^+ + :\ddot{Br}:^-$
     b. $\cdot Mg\cdot + 2:\ddot{Cl}\cdot \longrightarrow Mg^{2+} + 2:\ddot{Cl}:^-$

3.35 a. $:\ddot{S}\cdot + 2H\cdot \longrightarrow :\ddot{S}:H$ with H below
     b. $\cdot\ddot{P}\cdot + 3H\cdot \longrightarrow H:\ddot{P}:H$ with H below

3.37 He has two valence electrons (electron configuration $1s^2$) and a complete N=1 level. It has a stable electron configuration, with no tendency to gain or lose electrons, and satisfies the octet rule (2 e^- for period 1). Hence, it is nonreactive.
He:

3.39 a. Sodium ion
     b. Copper(I) ion (or cuprous ion)
     c. Magnesium ion
     d. Iron(II) ion (or ferrous ion)
     e. Iron(III) ion (or ferric ion)

3.41 a. Sulfide ion
     b. Chloride ion
     c. Carbonate ion

3.43 a. K^+
     b. Br^-

3.45 a. SO_4^{2-}
     b. NO_3^-

3.47 a. NaCl
     b. $MgBr_2$

3.49 a. AgCN
     b. NH_4Cl

3.51 a. Magnesium chloride
     b. Aluminum chloride

3.53 a. Nitrogen dioxide
     b. Sulfur trioxide

3.55 a. Al_2O_3
     b. Li_2S

3.57 a. CO or CO_2
     b. SO_2 or SO_3

3.59 a. $NaNO_3$
     b. $Mg(NO_3)_2$

3.61 a. NH_4I
     b. $(NH_4)_2SO_4$

3.63 a. Copper(II) sulfide
     b. Copper(II) sulfate

3.65 a. Sodium hypochlorite
     b. Sodium chlorite

3.67 Ionic solid state compounds exist in regular, repeating, three-dimensional structures; the crystal lattice. The crystal lattice is made up of positive and negative ions. Solid state covalent compounds are made up of molecules which may be arranged in a regular crystalline pattern or in an irregular (amorphous) structure.

3.69 The boiling points of ionic solids are generally much higher than those of covalent solids.

3.71 KCl would be expected to exist as a solid at room temperature; it is an ionic compound, and ionic compounds are characterized by high melting points.

3.73 Water will have a higher boiling point. Water is a polar molecule with strong intermolecular attractive forces, whereas carbon tetrachloride is a nonpolar molecule with weak intermolecular attractive forces. More energy, hence, a higher temperature is required to overcome the attractive forces among the water molecules.

3.75 a. $H\cdot$
     b. $He:$
     c. $\cdot\dot{C}\cdot$
     d. $\cdot\dot{N}\cdot$

3.77 a. Li^+
     b. Mg^{2+}
     c. $:\ddot{Cl}:^-$
     d. $:\ddot{P}:^{3-}$

3.79 a. Cl:N:Cl with Cl below (N bonded to three Cl, lone pair on N)

    b. H:C:O:H with H above and below C (methanol Lewis structure)

    c. :S::C::S:

3.81 a. Cl:N:Cl with Cl below
    Pyramidal,
    polar
    water soluble

    b. H:C:O:H with H above and below C
    Tetrahedral around C,
    angular around O
    polar
    water soluble

    c. :S::C::S:
    Linear
    nonpolar
    not water soluble

3.83 Resonance can occur when more than one valid Lewis structure can be written for a molecule. Each individual structure which can be drawn is a resonance form. The true nature of the structure for the molecule is the resonance hybrid, which consists of the "average" of the resonance forms.

3.85 [H:O: / H:C:C:O: / H]⁻ ⟷ [H:O: / H:C:C::O / H]⁻

3.87 H:C:C:O:H with H H above and H H below

3.89 H:C:C:C:H with H O H above and H H below

3.91 a. Polar covalent
    b. Polar covalent
    c. Ionic
    d. Ionic
    e. Ionic

3.93 a. [:C≡N:]⁻
    b. [:Si≡P:]⁻
    c), d), and e) are ionic compounds.

3.95 A molecule containing no polar bonds *must* be nonpolar. A molecule containing polar bonds may or may not itself be polar. It depends upon the number and arrangement of the bonds.

3.97 Polar compounds have strong intermolecular attractive forces. Higher temperatures are needed to overcome these forces and convert the solid to a liquid; hence, we predict higher melting points for polar compounds when compared to non-polar compounds.

3.99 Yes

Chapter 4

4.1 26.98 g Al/mol Al

4.3 a. 1.51×10^{24} oxygen atoms
    b. 3.01×10^{24} oxygen atoms

4.5 14.0 g He

4.7 a. 17.04 g/mol.
    b. 180.18 g/mol.
    c. 237.95 g/mol.

4.9 a. DR
    b. SR
    c. DR
    d. D

4.11 a. $KCl(aq) + AgNO_3(aq) \rightarrow KNO_3(aq) + AgCl(s)$
     A precipitation reaction occurs.
     b. $CH_3COOK(aq) + AgNO_3(aq) \rightarrow$ no reaction
     No precipitation reaction occurs.

4.13 a. $4Fe(s) + 3O_2(g) \rightarrow 2Fe_2O_3(s)$
     b. $2C_6H_6(l) + 15O_2(g) \rightarrow 12CO_2(g) + 6H_2O(g)$

4.15 a. 90.1 g H_2O
     b. 0.590 mol LiCl

4.17 a. 3 mol O_2
     b. 96.00 g O_2

4.19 a. $4Fe(s) + 3O_2(g) \rightarrow 2Fe_2O_3(s)$
     b. 3.50 g Fe

4.21 a. 132.0 g SnF_2
     b. 3.79 % yield

4.23 Examples of other packaging units include a *ream* of paper (500 sheets of paper), a six-pack of soft drinks, a case of canned goods (24 cans), to name a few.

4.25 a. 28.09 g.
     b. 107.9 g.

4.27 39.95 g.

4.29 40.36 g Ne

4.31 4.00 g He/mol He

4.33 a. 5.00 mol He
     b. 1.7 mol Na
     c. 4.2×10^{-2} mol Cl_2

4.35 1.62×10^3 g Ag

4.37 A molecule is a single unit comprised of atoms joined by covalent bonds. An ion-pair is composed of positive and negatively charged ions joined by electrostatic attraction, the ionic bond. The ion pairs, unlike the molecule, do not form single units; the electrostatic charge is directed to other ions in a crystal lattice, as well.

4.39 a. 58.44 g/mol
     b. 142.04 g/mol
     c. 357.49 g/mol

4.41 32.00 g/mol O_2

4.43 249.70 grams

4.45 a. 0.257 mol NaCl
     b. 0.106 mol Na_2SO_4

4.47 a. 18.02 g H_2O
     b. 116.9 g NaCl

4.49 a. 40.0 g He
     b. 2.02×10^2 g H

4.51 a. 2.43 g Mg
     b. 10.0 g $CaCO_3$

4.53 a. 4.00 g NaOH
     b. 9.81 g H_2SO_4

4.55 a. 0.420 mol KBr
     b. 0.415 mol $MgSO_4$

4.57 a. 6.57×10^{-1} mol CS_2
b. 2.14×10^{-1} mol $Al_2(CO_3)_3$

4.59 The ultimate basis for a correct chemical equation is the law of conservation of mass. No mass may be gained or lost in a chemical reaction, and the chemical equation must reflect this fact.

4.61 The subscript tells us the number of atoms or ions contained in one unit of the compound.

4.63 a. $MgCO_3 (s) \xrightarrow{\Delta} MgO (s) + CO_2 (g)$
b. $Zn(s) + CuSO_4(aq) \rightarrow ZnSO_4(aq) + Cu(s)$

4.65 $2NaOH(aq) + FeCl_2(aq) \rightarrow Fe(OH)_2(s) + 2NaCl(aq)$

4.67 Heat is necessary for the reaction to occur.

4.69 If we change the subscript we change the identity of the compound.

4.71 A reactant is the starting material for a chemical reaction.

4.73 A product is the chemical species that results from a chemical reaction.

4.75 a. $2C_2H_6(g) + 7O_2(g) \rightarrow 4CO_2(g) + 6H_2O(g)$
b. $6K_2O(s) + P_4O_{10}(s) \rightarrow 4K_3PO_4(s)$
c. $MgBr_2(aq) + H_2SO_4(aq) \rightarrow 2HBr(g) + MgSO_4(aq)$

4.77 a. $Ca(s) + F_2(g) \rightarrow CaF_2(s)$
b. $2Mg(s) + O_2(g) \rightarrow 2MgO(s)$
c. $3H_2(g) + N_2(g) \rightarrow 2NH_3(g)$

4.79 a. $2C_4H_{10}(g) + 13O_2(g) \rightarrow 10H_2O(g) + 8CO_2(g)$
b. $Au_2S_3(s) + 3H_2(g) \rightarrow 2Au(s) + 3H_2S(g)$
c. $Al(OH)_3(s) + 3HCl(aq) \rightarrow AlCl_3(aq) + 3H_2O(l)$
d. $(NH_4)_2Cr_2O_7(s) \rightarrow Cr_2O_3(s) + N_2(g) + 4H_2O(g)$
e. $C_2H_5OH(l) + 3O_2(g) \rightarrow 2CO_2(g) + 3H_2O(g)$

4.81 a. $N_2(g) + 3H_2(g) \rightarrow 2NH_3(g)$
b. $HCl(aq) + NaOH(aq) \rightarrow NaCl(aq) + H_2O(l)$

4.83 a. $C_6H_{12}O_6(s) + 6O_2(g) \rightarrow 6H_2O(l) + 6CO_2(g)$
b. $Na_2CO_3(s) \xrightarrow{\Delta} Na_2O(s) + CO_2(g)$

4.85 50.3 g B_2O_3

4.87 104 g $CrCl_3$

4.89 a. $N_2(g) + 3H_2(g) \rightarrow 2NH_3(g)$
b. Three moles of H_2 will react with one mole of N_2.
c. One mole of N_2 will produce two moles of the product NH_3.
d. 1.50 mol H_2
e. 17.0 g NH_3

4.91 a. 149.21 g/mol.
b. 1.20×10^{24} O atoms
c. 32.00 g O
d. 10.7 g O

4.93 7.39 g O_2

4.95 6.14×10^4 g O_2

4.97 70.6 g $C_{10}H_{22}$

4.99 9.13×10^2 g N_2

4.101 92.6%

4.103 6.85×10^2 g N_2

Chapter 5

5.1 a. 0.954 atm
b. 0.382 atm
c. 0.730 atm

5.3 a. 38 atm
b. 25 atm

5.5 a. 3.76 L
b. 3.41 L
c. 2.75 L

5.7 0.200 atm

5.9 4.46 mol H_2

5.11 9.00 L

5.13 0.223 mol N_2

5.15 In all cases, gas particles are much further apart than similar particles in the liquid or solid state. In most cases, particles in the liquid state are, on average, farther apart than those in the solid state. Water is the exception; liquid water's molecules are closer together than they are in the solid state.

5.17 Pressure is a force/unit area. Gas particles are in continuous, random motion. Collisions with the walls of the container results in a force (mass × acceleration) on the walls of the container. The sum of these collisional forces constitutes the pressure exerted by the gas.

5.19 Gases are easily compressed simply because there is a great deal of space between particles; they can be pushed closer together (compressed) because the space is available.

5.21 Gas particles are in continuous, random motion. They are free (minimal attractive forces between particles) to roam, up to the boundary of their container.

5.23 Gases exhibit more ideal behavior at low pressures. At low pressures, gas particles are more widely separated and therefore the attractive forces between particles are less. The ideal gas model assumes negligible attractive forces between gas particles.

5.25 The kinetic molecular theory states that the average kinetic energy of the gas particles increases as the temperature increases. Kinetic energy is proportional to (velocity)^2. Therefore, as the temperature increases the gas particle velocity increases and the rate of mixing increases as well.

5.27 The volume of the balloon is directly proportional to the pressure the gas exerts on the inside surface of the balloon. As the balloon cools, its pressure drops, and the balloon contracts. (Recall that the speed, hence the force exerted by the molecules decreases as the temperature decreases.)

5.29 Boyle's law states that the volume of a gas varies inversely with the gas pressure if the temperature and the number of moles of gas are held constant.

5.31 Volume will decrease according to Boyle's law. Volume is inversely proportional to the pressure exerted on the gas.

5.33 1 atm

5.35 5 L-atm

5.37 5.23 atm

5.39 Charles's law states that the volume of a gas varies directly with the absolute temperature if pressure and number of moles of gas are constant.

5.41 The Kelvin scale is the only scale that is directly proportional to molecular motion, and it is the motion that determines the physical properties of gases.

5.43 No. The volume is proportional to the temperature in K, not Celsius.

5.45 0.96 L

5.47 1.51 L

5.49 • Volume and temperature are *directly* proportional; increasing T *increases* V.
• Volume and pressure are *inversely* proportional; decreasing P *increases* V.
Therefore, both variables work together to *increase* the volume.

5.51 $V_f = \dfrac{P_i V_i T_f}{P_f T_i}$

5.53 1.82×10^{-2} L

5.55 Avogadro's law states that equal volumes of any ideal gas contain the same number of moles if measured at constant temperature and pressure.

5.57 6.00 L

5.59 No. One mole of an ideal gas will occupy exactly 22.4 L; however, there is no completely ideal gas and careful measurement will show a different volume.

5.61 Standard temperature is 273K.
5.63 0.80 mol
5.65 22.4 L
5.67 0.276 mol
5.69 5.94×10^{-2} L
5.71 22.4 L
5.73 9.08×10^3 L
5.75 172°C
5.77 Dalton's law states that the total pressure of a mixture of gases is the sum of the partial pressures of the component gases.
5.79 0.74 atm
5.81 Intermolecular forces in liquids are considerably stronger than intermolecular forces in gases. Particles are, on average, much closer together in liquids and the strength of attraction is inversely proportional to the distance of separation.
5.83 The vapor pressure of a liquid increases as the temperature of the liquid increases.
5.85 Evaporation is the conversion of a liquid to a gas at a temperature lower than the boiling point of the liquid. Condensation is the conversion of a gas to a liquid at a temperature lower than the boiling point of the liquid.
5.87 Viscosity is the resistance to flow caused by intermolecular attractive forces. Complex molecules may become entangled and not slide smoothly across one another.
5.89 All molecules exhibit London forces.
5.91 Only methanol exhibits hydrogen bonding. Methanol has an oxygen atom bonded to a hydrogen atom, a necessary condition for hydrogen bonding.
5.93 Solids are essentially incompressible because the average distance of separation among particles in the solid state is small. There is literally no space for the particles to crowd closer together.
5.95 a. High melting temperature, brittle
 b. High melting temperature, hard
5.97 Beryllium.
5.99 Mercury.

Chapter 6

6.1 A chemical analysis must be performed in order to determine the identity of all components, a qualitative analysis. If only one component is found, it is a pure substance; two or more components indicates a true solution.
6.3 After the container of soft drink is opened, CO_2 diffuses into the surrounding atmosphere; consequently the partial pressure of CO_2 over the soft drink decreases and the equilibrium
$$CO_2 (g) \leftrightarrow CO_2 (aq)$$
shifts to the left, lowering the concentration of CO_2 in the soft drink.
6.5 16.7% NaCl
6.7 7.50% KCl
6.9 2.56×10^{-2} % oxygen
6.11 20.0 % oxygen
6.13 2.00×10^5 ppm
6.15 0.125 mol HCl
6.17 To prepare the solution, dilute 1.7×10^{-2} L of 12 M HCl with sufficient water to produce 1.0×10^2 mL of total solution.
6.19 Pure water
6.21 1.0×10^{-2} osmol
6.23 0.24 atm
6.25 Polar carbon monoxide is more soluble in water.
6.27 1.54×10^{-2} mol Na$^+$/L

6.29 a. 2.00% NaCl
 b. 6.60% $C_6H_{12}O_6$
6.31 a. 5.00% ethanol
 b. 10.0% ethanol
6.33 a. 21.0% NaCl
 b. 3.75% NaCl
6.35 a. 2.25 g NaCl
 b. 3.13 g $NaC_2H_3O_2$
6.37 19.5% KNO_3
6.39 1.00 g sugar
6.41 0.04% (w/w) solution is more concentrated.
6.43 2.0×10^{-3} ppt
6.45 a. 0.342 M NaCl
 b. 0.367 M $C_6H_{12}O_6$
6.47 a. 1.46 g NaCl
 b. 9.00 g $C_6H_{12}O_6$
6.49 0.146 M $C_{12}H_{22}O_{11}$
6.51 5.00×10^{-2} L
6.53 20.0 M
6.55 0.900 M
6.57 158 g glucose
6.59 0.0500 M
6.61 A colligative property is a solution property that depends on the concentration of solute particles rather than the identity of the particles.
6.63 Salt is an ionic substance that dissociates in water to produce positive and negative ions. These ions (or particles) lower the freezing point of water. If the concentration of salt particles is large, the freezing point may be depressed below the surrounding temperature, and the ice would melt.
6.65 Chemical properties depend on the identity of the substance, whereas colligative properties depend on concentration, not identity.
6.67 Raoult's law states that when a solute is added to a solvent, the vapor pressure of the solvent decreases in proportion to the concentration of the solute.
6.69 One mole of $CaCl_2$ produces three moles of particles in solution whereas one mole of NaCl produces two moles of particles in solution. Therefore, a one molar $CaCl_2$ solution contains a greater number of particles than a one molar NaCl solution and will produce a greater freezing-point depression.
6.71 Sucrose
6.73 Sucrose
6.75 24 atm
6.77 A → B
6.79 No net flow
6.81 Hypertonic
6.83 Hypotonic
6.85 Water is often termed the "universal solvent" because it is a polar molecule and will dissolve, at least to some extent, most ionic and polar covalent compounds. The majority of our body mass is water and this water is an important part of the nutrient transport system due to its solvent properties. This is true in other animals and plants as well. Because of its ability to hydrogen bond, water has a high boiling point and a low vapor pressure. Also, water is abundant and easily purified.
6.87 The shelf life is a function of the stability of the ammonia-water solution. The ammonia can react with the water to convert to the extremely soluble and stable ammonium ion. Also, ammonia and water are polar molecules. Polar interactions, particularly hydrogen bonding, are strong and contribute to the long-term solution stability.

6.89 [structures: H₂O, Na⁺, and H₂O with lone pairs]

6.91 Polar; like dissolves like (H₂O is polar)

6.93 H—N : δ⁻ δ⁺ H—O : δ⁻ (with H atoms)

6.95 In dialysis, sodium ions move from a region of high concentration to a region of low concentration. If we wish to remove (transport) sodium ions from the blood, they can move to a region of lower concentration, the dialysis solution.

6.97 Elevated concentrations of sodium ion in the blood may cause confusion, stupor, or coma.

6.99 Elevated concentrations of sodium ion in the blood may occur whenever large amounts of water are lost. Diarrhea, diabetes, and certain high-protein diets are particularly problematic.

Chapter 7

7.1 a. Exothermic
b. Exothermic
c. Exothermic

7.3 He(g)

7.5 $\Delta G = (+) - T(-)$
ΔG must always be positive.

7.7 13°C

7.9 2.7×10^3 J

7.11 $\dfrac{2.1 \times 10^2 \text{ nutritional Cal}}{\text{candy bar}}$

7.13 Heat energy produced by the friction of striking the match provides the activation energy necessary for this combustion process.

7.15 If the enzyme catalyzed a process needed to sustain life, the substance interfering with that enzyme would be classified as a poison.

7.17 a. rate = $k[N_2]^n[O_2]^{n'}$
b. rate = $k[C_4H_6]^n$

7.19 At rush hour, approximately the same number of passengers enter and exit the train at any given stop. Throughout the trip, the number of passengers on the train may be essentially unchanged, but the identity of the individual passengers is continually changing.

7.21 Measure the concentrations of products and reactants at a series of times until no further concentration change is observed.

7.23 a. $K_{eq} = \dfrac{[N_2][O_2]^2}{[NO_2]^2}$
b. $K_{eq} = [H_2]^2[O_2]$

7.25 Product formation

7.27 8.2×10^{-2}

7.29 a. A would decrease
b. A would increase
c. A would decrease
d. A would remain the same

7.31 joule

7.33 An exothermic reaction is one in which energy is released during chemical change.

7.35 A fuel must release heat in the combustion (oxidation) process.

7.37 The temperature of the water (or solution) is measured in a calorimeter. If the reaction being studied is exothermic, release energy heats the water and the temperature increases. In an endothermic reaction, heat flows from the water to the reaction and the water temperature decreases.

7.39 Double-walled containers, used in calorimeters, provide a small airspace between the part of the calorimeter (inside wall) containing the sample solution and the outside wall, contacting the surroundings. This makes heat transfer more difficult.

7.41 Free energy is the combined contribution of entropy and enthalpy for a chemical reaction.

7.43 The first law of thermodynamics, the law of conservation of energy, states that the energy of the universe is constant.

7.45 Enthalpy is a measure of heat energy.

7.47 1.20×10^3 cal

7.49 5.02×10^3 J

7.51 a. Entropy increases.
b. Entropy increases.

7.53 $\Delta G = (-) - T(+)$
ΔG must always be negative and the process is always spontaneous.

7.55 Isopropyl alcohol quickly evaporates (liquid → gas) after being applied to the skin. Conversion of a liquid to a gas requires heat energy. The heat energy is supplied by the skin. When this heat is lost, the skin temperature drops.

7.57 Decomposition of leaves and twigs to produce soil.

7.59 The activated complex is the arrangement of reactants in an unstable transition state as a chemical reaction proceeds. The activated complex must form in order to convert reactants to products.

7.61 The rate of a reaction is the change in concentration of a reactant or product per unit time. The rate constant is the proportionality constant that relates rate and concentration. The order is the exponent of each concentration term in the rate equation.

7.63 Increase

7.65 A catalyst increases the rate of a reaction without itself undergoing change.

7.67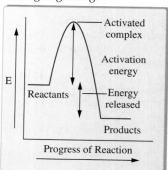
Non-catalyzed reaction
Higher activation energy

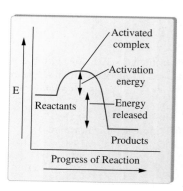
Catalyzed reaction
Lower activation energy

7.69 Enzymes are biological catalysts. The enzyme lysozyme catalyzes a process that results in the destruction of the cell walls of many harmful bacteria. This helps to prevent disease in organisms. The breakdown of foods to produce material for construction and repair of body tissue, as well as energy, is catalyzed by a variety of enzymes. For example, amylase begins the hydrolysis of starch in the mouth.

7.71 An increase in concentration of reactants means that there are more molecules in a certain volume. The probability of collision is enhanced because they travel a shorter distance before meeting another molecule. The rate is proportional to the number of collisions per unit time.

7.73 Rate = k $[N_2O_4]$

7.75 A catalyst speeds up a chemical reaction by facilitating the formation of the activated complex, thus lowering the activation energy, the energy barrier for the reaction.

7.77 Products

7.79 Ice and water at 0°C

7.81 Le Chaltelier's principle states that when a system at equilibrium is disturbed, the equilibrium shifts in the direction that minimizes the disturbance.

7.83 A dynamic equilibrium has fixed concentrations of all reactants and products - these concentrations do not change with time. However, the process is dynamic because products and reactants are continuously being formed and consumed. The concentrations do not change because the rates of production and consumption are equal.

7.85 $K_{eq} = \dfrac{[NO_2]^2}{[N_2O_4]}$

7.87 A physical equilibrium describes physical change; examples include the equilibrium between ice and water, or the equilibrium vapor pressure of a liquid. A chemical equilibrium describes chemical change; examples include the reactions shown in questions 7.93 and 7.94.

7.89 a. Equilibrium shifts to the left.
 b. No change
 c. No change

7.91 a. False
 b. False

7.93 a. PCl_3 increases
 b. PCl_3 decreases
 c. PCl_3 decreases
 d. PCl_3 decreases
 e. PCl_3 remains the same

7.95 Decrease

7.97 $K_{eq} = \dfrac{[CO][H_2]}{[H_2O]}$

7.99 False

7.101 Removing the cap allows CO_2 to escape into the atmosphere. This corresponds to the removal of product (CO_2):
$$CO_2 (l) \rightleftharpoons CO_2 (g)$$
The equilibrium shifts to the right, dissolved CO_2 is lost, and the beverage goes "flat".

7.103 $K_{eq} = \dfrac{[NH_3]^2}{[N_2][H_2]^3}$

Chapter 8

8.1 a. $HF(aq) + H_2O (l) \rightleftharpoons H_3O^+ (aq) + F^- (aq)$
 b. $NH_3(aq) + H_2O (l) \rightleftharpoons NH_4^+(aq) + OH^-(aq)$

8.3 a. HF and F^-; H_2O and H_3O^+
 b. NH_3 and NH_4^+ ; H_2O and OH^-

8.5 a. NH_4^+
 b. H_2SO_4

8.7 1.0×10^{-11} M

8.9 12.00

8.11 3.2×10^{-9} M

8.13 0.1000 M NaOH

8.15 $CO_2 + H_2O \rightleftharpoons H_2CO_3 \rightleftharpoons H_3O^+ + HCO_3^-$
An increase in the partial pressure of CO_2 is a stress on the left side of the equilibrium. The equilibrium will shift to the right in an effort to decrease the concentration of CO_2. This will cause the molar concentration of H_2CO_3 to increase.

8.17 In Question 8.15, the equilibrium shifts to the right. Therefore the molar concentration of H_3O^+ should increase.
 In Question 8.16, the equilibrium shifts to the left. Therefore the molar concentration of H_3O^+ should decrease.

8.19 4.87

8.21 4.74

8.23 4.87

8.25 $Ca \rightarrow Ca^{2+} + 2\,e^-$ (oxidation ½ reaction)
 $S + 2\,e^- \rightarrow S^{2-}$ (reduction ½ reaction)
 $Ca + S \rightarrow CaS$ (complete reaction)

8.27 a. An Arrhenius acid is a substance that dissociates, producing hydrogen ions.
 b. A Brønsted-Lowry acid is a substance that behaves as a proton donor.

8.29 The Brønsted-Lowry theory provides a broader view of acid-base theory than does the Arrhenius theory. Brønsted-Lowry emphasizes the role of the solvent in the dissociation process.

8.31 a. $HNO_2 (aq) + H_2O (l) \rightleftharpoons H_3O^+ (aq) + NO_2^- (aq)$
 b. $HCN (aq) + H_2O (l) \rightleftharpoons H_3O^+ (aq) + CN^- (aq)$

8.33 a. HNO_2 and NO_2^-; H_2O and H_3O^+
 b. HCN and CN^-; H_2O and H_3O^+

8.35 a. Weak
 b. Weak
 c. Weak

8.37 a. CN^- and HCN; NH_3 and NH_4^+
 b. CO_3^{2-} and HCO_3^-; Cl^- and HCl

8.39 Concentration refers to the quantity of acid or base contained in a specified volume of solvent. Strength refers to the degree of dissociation of the acid or base.

8.41 a. Bronsted acid
 b. Bronsted base
 c. Both

8.43 a. Bronsted acid
 b. Both
 c. Bronsted base

8.45 HCN

8.47 I^-

8.49 a. 1.0×10^{-7} M
 b. 1.0×10^{-11} M

8.51 a. Neutral
 b. Basic

8.53 a. 7.00
 b. 5.00

8.55 a. $[H_3O^+] = 1.0 \times 10^{-1}$ M
 $[OH^-] = 1.0 \times 10^{-13}$ M
 b. $[H_3O^+] = 1.0 \times 10^{-9}$ M
 $[OH^-] = 1.0 \times 10^{-5}$ M

8.57 a. $[H_3O^+] = 5.0 \times 10^{-2}$ M
 $[OH^-] = 2.0 \times 10^{-13}$ M
 b. $[H_3O^+] = 2.0 \times 10^{-10}$ M
 $[OH^-] = 5.0 \times 10^{-5}$ M

8.59 A neutralization reaction is one in which an acid and a base react to produce water and a salt (a "neutral" solution).

8.61 a. $[H_3O^+] = 1.0 \times 10^{-6}$ M
$[OH^-] = 1.0 \times 10^{-8}$ M
b. $[H_3O^+] = 6.3 \times 10^{-6}$ M
$[OH^-] = 1.6 \times 10^{-9}$ M
c. $[H_3O^+] = 1.6 \times 10^{-8}$ M
$[OH^-] = 6.3 \times 10^{-7}$ M
8.63 a. 1×10^2
b. 1×10^4
c. 1×10^{10}
8.65 a. $[H_3O^+] = 1 \times 10^{-5}$
b. $[H_3O^+] = 1 \times 10^{-12}$
c. $[H_3O^+] = 3.2 \times 10^{-6}$
8.67 a. pH = 6.00
b. pH = 8.00
c. pH = 3.25
8.69 pH = 3.12
8.71 pH = 10.74
8.73 a. NH_3 and NH_4Cl can form a buffer solution.
b. HNO_3 and KNO_3 cannot form a buffer solution.
8.75 a. A buffer solution contains components (a weak acid and its salt or a weak base and its salt) that enable the solution to resist large changes in pH when acids or bases are added.
b. Acidosis is a medical condition characterized by higher-than-normal levels of CO_2 in the blood and lower-than-normal blood pH.
8.77 a. Addition of strong acid is equivalent to adding H_3O^+. This is a stress on the right side of the equilibrium and the equilibrium will shift to the left. Consequently the $[CH_3COOH]$ increases.
b. Water, in this case, is a solvent and does not appear in the equilibrium expression. Hence, it does not alter the position of the equilibrium.
8.79 $[H_3O^+] = 2.32 \times 10^{-7}$ M
8.81 The very fact that an acid has a K_a value means that it is a weak acid. The smaller the value of K_a, the weaker the acid.
8.83 pH = 4.74
8.85 $11.2 = \dfrac{[HCO_3^-]}{[H_2CO_3]}$
8.87 a. Oxidation is the loss of electrons, loss of hydrogen atoms, or gain of oxygen atoms.
b. An oxidizing agent removes electrons from another substance. In doing so the oxidizing agent becomes reduced.
8.89 The species oxidized *loses* electrons.
8.91 During an oxidation-reduction reaction the species *oxidized* is the reducing agent.
8.93 Cl_2 + $2KI$ → $2KCl + I_2$
substance reduced substance oxidized
oxidizing agent reducing agent
8.95 $2KI \rightarrow 2K^+ + 2e^- + I_2$ (oxidation ½ reaction)
$Cl_2 + 2e^- \rightarrow 2Cl^-$ (reduction ½ reaction)
8.97 An oxidation-reduction reaction must take place to produce electron flow in a voltaic cell.
8.99 Storage battery.

Chapter 9

9.1 X-ray, ultraviolet, visible, infra-red, microwave, and radiowave.
9.3 a. $^{85}_{36}Kr \rightarrow ^{85}_{37}Rb + ^{0}_{-1}e$
b. $^{226}_{88}Ra \rightarrow ^{4}_{2}He + ^{222}_{86}Rn$
9.5 6.3 ng of sodium-24 remain after 2.5 days
9.7 1/4 of the radioisotope remains after 2 half-lives
9.9 Isotopes with short half-lives release their radiation rapidly. There is much more radiation per unit time observed with short half-life substances; hence, the signal is stronger and the sensitivity of the procedure is enhanced.
9.11 The rem takes into account the relative biological effect of the radiation in addition to the quantity of radiation. This provides a more meaningful estimate of potential radiation damage to human tissue.
9.13 Natural radioactivity is the spontaneous decay of a nucleus to produce high-energy particles or rays.
9.15 Two protons and two neutrons
9.17 An electron with a –1 charge.
9.19 A positron has a positive charge and a beta particle has a negative charge.
9.21 • charge, α = +2, β = –1
• mass, α = 4 amu, β = 0.000549 amu
• velocity, α = 10% of C, β = 90% of C
9.23 Chemical reactions involve joining, separating and rearranging atoms; valence electrons are critically involved. Nuclear reactions involve only changes in nuclear composition.
9.25 $^{4}_{2}He$
9.27 $^{235}_{92}U$
9.29 $^{1}_{1}H$ 1 – 1 = 0 neutrons
$^{2}_{1}H$ 2 – 1 = 1 neutron
$^{3}_{1}H$ 3 – 1 = 2 neutrons
9.31 $^{15}_{7}N$
9.33 Alpha and beta particles are matter; gamma radiation is pure energy. Alpha particles are large and relatively slow moving. They are the least energetic and least penetrating. Gamma radiation moves at the speed of light, highly energetic, and most penetrating.
9.35 A helium atom has two electrons; an α particle has no electrons.
9.37 $^{60}_{27}Co \rightarrow ^{60}_{28}Ni + ^{0}_{-1}\beta + \gamma$
9.39 $^{23}_{11}Na + ^{2}_{1}H \rightarrow ^{24}_{11}Na + ^{1}_{1}H$
9.41 $^{24}_{10}Ne \rightarrow \beta + ^{24}_{11}Na$
9.43 $^{140}_{55}Cs \rightarrow ^{140}_{56}Ba + ^{0}_{-1}e$
9.45 $^{209}_{83}Bi + ^{54}_{24}Cr \rightarrow ^{262}_{107}Bh + ^{1}_{0}n$
9.47 $^{27}_{12}Mg \rightarrow ^{0}_{-1}e + ^{27}_{13}Al$
9.49 $^{12}_{7}N \rightarrow ^{12}_{6}C + ^{0}_{1}e$
9.51 Natural radioactivity is a spontaneous process; artificial radioactivity is nonspontaneous and results from a nuclear reaction that produces an unstable nucleus.
9.53 • Nuclei for light atoms tend to be most stable if their neutron/proton ratio is close to 1.
• Nuclei with more than 84 protons tend to be unstable.
• Isotopes with a "magic number" of protons or neutrons (2, 8, 20, 50, 82, or 126 protons or neutrons) tend to be stable.
• Isotopes with even numbers of protons or neutrons tend to be more stable.
9.55 $^{20}_{8}O$; Oxygen-20 has 20 – 8 = 12 neutrons, an n/p of 12/8, or 1.5. The n/p is probably too high for stability even though it does have a "magic number" of protons and an even number of protons and neutrons.
9.57 $^{48}_{24}Cr$; Chromium-48 has 48 – 24 = 24 neutrons, an n/p of 24/24, or 1.0. It also has an even number of protons and neutrons. It would probably be stable.
9.59 0.40 mg of iodine-131 remains
9.61 13 mg of iron-59 remains
9.63 201 hours
9.65 Fission

9.67 a. The fission process involves the breaking down of large, unstable nuclei into smaller, more stable nuclei. This process releases some of the binding energy in the form of heat and/or light.
  b. The heat generated during the fission process could be used to generate steam, which is then used to drive a turbine to create electricity.
9.69 $^{3}_{1}H + ^{1}_{1}H \rightarrow ^{4}_{2}He +$ energy
9.71 A "breeder" reactor creates the fuel which can be used by a conventional fission reactor during its fission process.
9.73 The reaction in a fission reactor that involves neutron production and causes subsequent reactions accompanied by the production of more neutrons in a continuing process.
9.75 High operating temperatures
9.77 Radiocarbon dating is a process used to determine the age of objects. The ratio of the masses of the stable isotope, carbon-12, and unstable isotope, carbon-14, is measured. Using this value and the half-life of carbon-14, the age of the coffin may be calculated.
9.79 $^{108}_{47}Ag + ^{4}_{2}He \rightarrow ^{112}_{49}In$
9.81 a. Technetium-99 m is used to study the heart (cardiac output, size, and shape), kidney (follow-up procedure for kidney transplant), and liver and spleen (size, shape, presence of tumors).
  b. Xenon-133 is used to locate regions of reduced ventilation and presence of tumors in the lung.
9.83 Radiation therapy provides sufficient energy to destroy molecules critical to the reproduction of cancer cells.
9.85 Background radiation, radiation from natural sources, is emitted by the sun as cosmic radiation, and from naturally radioactive isotopes found throughout our environment.
9.87 Level decreases
9.89 Positive effect
9.91 Positive effect
9.93 Yes
9.95 A film badge detects gamma radiation by darkening photographic film in proportion to the amount of radiation exposure over time. Badges are periodically collected and evaluated for their level of exposure. This mirrors the level of exposure of the personnel wearing the badges.
9.97 Relative biological effect is a measure of the damage to biological tissue caused by different forms of radiation.
9.99 a. The curie is the amount of radioactive material needed to produce 3.7×10^{10} atomic disintegrations per second.
  b. The roentgen is the amount of radioactive material needed to produce 2×10^{9} ion-pairs when passing through 1 cc of air at 0°C.

Chapter 10

10.1 The student could test the solubility of the substance in water and in an organic solvent, such as hexane. Solubility in hexane would suggest an organic substance; whereas solubility in water would indicate an inorganic compound. The student could also determine the melting and boiling points of the substance. If the melting and boiling points are very high, an inorganic substance would be suspected.

10.3

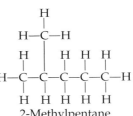

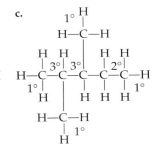

10.5

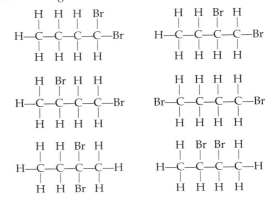

10.7 a. 2,3-Dimethylbutane
  b. 2,2-Dimethylpentane
  c. 2,2-Dimethylpropane
  d. 1,2,3-Tribromopropane

10.9 a. The straight chain isomers of molecular formula C_4H_9Br:

  b. The straight chain isomers of molecular formula $C_4H_8Br_2$:

10.11
a. *trans*-1-Bromo-2-ethylcyclobutane
b. *trans*-1, 2-Dimethylcyclopropane
c. Propylcyclohexane

10.13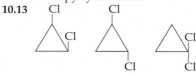

10.15 Three of the six axial hydrogen atoms of cyclohexane lie above the ring. The remaining three hydrogen atoms lie below the ring.

10.17 a. The combustion of cyclobutane:

◇ + 6O₂ ⟶ 4CO₂ + 4H₂O + heat energy

b. The monobromination of propane will produce two products as shown in the following two equations:

H H H   H Br H
| | | | | |
H—C—C—C—H + Br₂ —Light or heat→ H—C—C—C—H + HBr
| | | | | |
H H H   H H H

H H H   H H Br
| | | | | |
H—C—C—C—H + Br₂ —Light or heat→ H—C—C—C—H + HBr
| | | | | |
H H H   H H H

c. 2CH₃CH₃ + 7O₂ → 4CO₂ + 6H₂O + energy

d.
H H H H   H H H H
| | | | | | | |
H—C—C—C—C—H + Cl₂ —Heat or light→ H—C—C—C—C—H + HCl
| | | | | | | |
H H H H   H Cl H H

H H H H   H H H H
| | | | | | | |
H—C—C—C—C—H + Cl₂ —Heat or light→ H—C—C—C—C—H + HCl
| | | | | | | |
H H H H   Cl H H H

10.19 The products in the reactions in Problem 10.17b are 1-bromopropane and 2-bromopropane. The products of the reactions in Problem 10.17d are 1-chlorobutane and 2-chlorobutane.

10.21 The number of organic compounds is nearly limitless because carbon forms stable covalent bonds with other carbon atoms in a variety of different patterns. In addition, carbon can form stable bonds with other elements and functional groups, producing many families of organic compounds, including alcohols, aldehydes, ketones, esters, ethers, amines, and amides. Finally, carbon can form double or triple bonds with other carbon atoms to produce organic molecules with different properties.

10.23 The allotropes of carbon include graphite, diamond, and buckminsterfullerene.

10.25 Because ionic substances often form three-dimensional crystals made up of many positive and negative ions, they generally have much higher melting and boiling points than covalent compounds.

10.27 a. LiCl > H₂O > CH₄
b. NaCl > C₃H₈ > C₂H₆

10.29 a. LiCl would be a solid; H₂O would be a liquid; and CH₄ would be a gas.
b. NaCl would be a solid; both C₃H₈ and C₂H₆ would be gases.

10.31
a. Water-soluble inorganic compounds
b. Inorganic compounds
c. Organic compounds
d. Inorganic compounds
e. Organic compounds

10.33 a.
              H   H
              | |
          H—C—H   H—C—H
              | |
H—C—C—C—C—C—H
      | | | | |
      H H   H H   H

b.
      H Br H H
      | | | |
H—C—C—C—C—H
      | | | |
      H H Br H

10.35 a. CH₃CH₂CHCH₃
                    |
                   CH₃

b.
        CH₃   CH₃
         | |
CH₃CH₂CHCH₂CH₂CHCH₃
         |
        CH₃

10.37 a., b., c. (structural diagrams)

10.39 a., b., c. (structural diagrams, c contains Br)

10.41 Structure b is not possible because there are five bonds to carbon-2. Structure d is not possible because there are five bonds to carbon-3. Structure e is not possible because there are five bonds to carbon-3. Structure f is not possible because there are five bonds to carbon-3.

10.43 a.
            H
            |
         H—C—H
            |
H H   H H H
| | | | |
H—C—C—C—C—C—H
| | | | |
H H   H H
         |
         H—C—H
            |
            H

b.
H H H H H
| | | | |
H—C—C—C—C—C—H
| | | | |
H H H H H

10.45
a.
                    H
                    |
              H—C—H
               H |
               | H—C—H
         H—C—H |
            | H—C—H
H H | H H H |
| | | | | | |
H—C—C—C—C—C—C—C—C—H
| | | | | | |
H H | H H H H
        H

b.
             H
             |
        H—C—H
          |
     H |
     | H H
H—C—C—C—C—H
     | | |
     | H H
     H—C—H
          |
          H

10.47 An alcohol

H H
| |
H—C—C—OH
| |
H H

An aldehyde

H O
| ‖
H—C—C—H
|
H

A ketone

$$\text{H}-\overset{\overset{\text{H}}{|}}{\underset{\underset{\text{H}}{|}}{\text{C}}}-\overset{\overset{\text{O}}{\|}}{\text{C}}-\overset{\overset{\text{H}}{|}}{\underset{\underset{\text{H}}{|}}{\text{C}}}-\text{H}$$

A carboxylic acid

$$\text{H}-\overset{\overset{\text{H}}{|}}{\underset{\underset{\text{H}}{|}}{\text{C}}}-\overset{\overset{\text{O}}{\|}}{\text{C}}-\text{OH}$$

An amine

$$\text{H}-\overset{\overset{\text{H}}{|}}{\underset{\underset{\text{H}}{|}}{\text{C}}}-\overset{\overset{\text{H}}{|}}{\underset{\underset{\text{H}}{|}}{\text{C}}}-\text{N}\overset{\text{H}}{\underset{\text{H}}{}}$$

10.49 a. C_nH_{2n+2}
b. C_nH_{2n-2}
c. C_nH_{2n}
d. C_nH_{2n}
e. C_nH_{2n-2}

10.51 Alkanes have only carbon-to-carbon and carbon-to-hydrogen single bonds, as in the molecule ethane:

$$\text{H}-\overset{\overset{\text{H}}{|}}{\underset{\underset{\text{H}}{|}}{\text{C}}}-\overset{\overset{\text{H}}{|}}{\underset{\underset{\text{H}}{|}}{\text{C}}}-\text{H}$$

Alkenes have at least one carbon-to-carbon double bond, as in the molecule ethene:

$$\overset{\text{H}}{\underset{\text{H}}{}}\text{C}=\text{C}\overset{\text{H}}{\underset{\text{H}}{}}$$

Alkynes have at least one carbon-to-carbon triple bond, as in the molecule ethyne:

$$\text{H}-\text{C}\equiv\text{C}-\text{H}$$

10.53 a. A carboxylic acid:

$$\text{CH}_3\text{CH}_2-\overset{\overset{\text{O}}{\|}}{\text{C}}-\text{OH}$$

b. An amine: $CH_3CH_2CH_2-NH_2$
c. An alcohol: $CH_3CH_2CH_2-OH$
d. An ether: $CH_3CH_2-O-CH_2CH_3$

10.55

Aspirin
Acetylsalicylic acid
(Carboxyl group and Ester groups labeled)

10.57 Hydrocarbons are nonpolar molecules, and hence are not soluble in water.

10.59 a. heptane > hexane > butane > ethane
b. $CH_3CH_2CH_2CH_2CH_2CH_2CH_2CH_2CH_3$ > $CH_3CH_2CH_2CH_2CH_3$ > $CH_3CH_2CH_3$

10.61 a. Heptane and hexane would be liquid at room temperature; butane and ethane would be gases.
b. $CH_3CH_2CH_2CH_2CH_2CH_2CH_2CH_2CH_3$ and $CH_3CH_2CH_2CH_2CH_3$ would be liquids at room temperature; $CH_3CH_2CH_3$ would be a gas.

10.63 Nonane: $CH_3CH_2CH_2CH_2CH_2CH_2CH_2CH_2CH_3$
Pentane: $CH_3CH_2CH_2CH_2CH_3$
Propane: $CH_3CH_2CH_3$

10.65 a.
$$\overset{\overset{\text{Br}}{|}}{\text{CH}_3\text{CHCH}_2\text{CH}_3}$$

b.
$$\text{CH}_3-\overset{\overset{\text{Cl}}{|}}{\underset{\underset{\text{CH}_3}{|}}{\text{C}}}-\text{CH}_3$$

c.
$$\text{CH}_3-\overset{\overset{\text{CH}_3}{|}}{\underset{\underset{\text{CH}_3}{|}}{\text{C}}}-\text{CH}_2\text{CH}_2\text{CH}_2\text{CH}_3$$

10.67 a. 2,2-Dibromobutane:

$$\text{H}-\overset{\overset{\text{H}}{|}}{\underset{\underset{\text{H}}{|}}{\text{C}}}-\overset{\overset{\text{Br}}{|}}{\underset{\underset{\text{Br}}{|}}{\text{C}}}-\overset{\overset{\text{H}}{|}}{\underset{\underset{\text{H}}{|}}{\text{C}}}-\overset{\overset{\text{H}}{|}}{\underset{\underset{\text{H}}{|}}{\text{C}}}-\text{H}$$

b. 2-Iododecane:

$$\text{H}-\text{C}(\text{H})_2-\text{C}(\text{H})(\text{I})-\text{C}(\text{H})_2-\text{C}(\text{H})_2-\text{C}(\text{H})_2-\text{C}(\text{H})_2-\text{C}(\text{H})_2-\text{C}(\text{H})_2-\text{C}(\text{H})_2-\text{C}(\text{H})_2-\text{H}$$

c. 1,2-Dichloropentane:

$$\text{Cl}-\overset{\overset{\text{H}}{|}}{\underset{\underset{\text{H}}{|}}{\text{C}}}-\overset{\overset{\text{Cl}}{|}}{\underset{\underset{\text{H}}{|}}{\text{C}}}-\overset{\overset{\text{H}}{|}}{\underset{\underset{\text{H}}{|}}{\text{C}}}-\overset{\overset{\text{H}}{|}}{\underset{\underset{\text{H}}{|}}{\text{C}}}-\overset{\overset{\text{H}}{|}}{\underset{\underset{\text{H}}{|}}{\text{C}}}-\text{H}$$

d. 1-Bromo-2-methylpentane:

$$\begin{array}{c} \text{H} \\ | \\ \text{H}-\text{C}-\text{H} \\ \\ \text{H}-\text{C}-\text{C}-\text{C}-\text{C}-\text{H} \\ | \quad | \quad | \quad | \\ \text{Br} \; \text{H} \; \text{H} \; \text{H} \; \text{H} \end{array}$$

10.69 a. 3-Methylpentane c. 1-Bromoheptane
b. 2,5-Dimethylhexane d. 1-Chloro-3-methylbutane

10.71 a. 2-Chloropropane d. 1-Chloro-2-methylpropane
b. 2-Iodobutane e. 2-Iodo-2-methylpropane
c. 2,2-Dibromopropane

10.73 a. 2-Chlorohexane c. 3-Chloropentane
b. 1,4-Dibromobutane d. 2-Methylheptane

10.75 a. The first pair of molecules are constitutional isomers: hexane and 2-methylpentane.
b. The second pair of molecules are identical. Both are heptane.

10.77 a. Incorrect: 3-Methylhexane
b. Incorrect: 2-Methylbutane
c. Incorrect: 3-Methylheptane
d. Correct

10.79 a.
$$CH_3CHCH_2CHCH_3$$
with CH_3 branches
The name given in the problem is correct.

b.
$$CH_3CH_2CH_2CHCH_2CH_3$$
with CH_3 branch
The correct name is 3-methylhexane.

c. $I—CH_2CH_2CH_2CH_2CH_2—I$ The name given in the problem is correct.

d. $CH_3CH_2CH_2CH_2CH_2CHCH_2CH_2CH_3$ with CH_2CH_3 branch
The correct name is 4-ethylnonane.

e.
The name given in the problem is correct.

10.81 Cycloalkanes are a family of molecules having carbon-to-carbon bonds in a ring structure.
10.83 The general formula for a cycloalkane is C_nH_{2n}.
10.85 a. Chlorocyclopropane
b. *cis*-1, 2-Dichlorocyclopropane
c. *trans*-1, 2-Dichlorocyclopropane
d. Bromocyclobutane
10.87 a. 1-Bromo-2-methylcyclobutane: b. Iodocyclopropane:
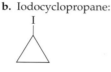

10.89 a. Incorrect—1, 2-Dibromocyclobutane
b. Incorrect—1, 2-Diethylcyclobutane
c. Correct
d. Incorrect—1, 2, 3-Trichlorocyclohexane
10.91 a.   b. (cyclobutane with two CH₃)
c.   d. (cyclohexane with CH₂CH₃ groups)

10.93 a. *cis*-1, 2-Dibromocyclopentane
b. *trans*-1, 3-Dibromocyclopentane
c. *cis*-1, 2-Dimethylcyclohexane
d. *cis*-1, 2-Dimethylcyclopropane
10.95 Conformational isomers are distinct isomeric structures that may be converted into one another by rotation about the bonds in the molecule.
10.97 In the chair conformation the hydrogen atoms, and thus the electron pairs of the C—H bonds, are farther from one another. As a result, there is less electron repulsion and the structure is more stable (more energetically favored). In the boat conformation, the electron pairs are more crowded. This causes greater electron repulsion, producing a less stable, less energetically favored conformation.
10.99 Because conformations are freely and rapidly interconverted, they cannot be separated from one another.

10.101 One conformation is more stable than the other because the electron pairs of the carbon-hydrogen bonds are farther from one another.
10.103 a. $C_3H_8 + 5O_2 \rightarrow 4H_2O + 3CO_2$
b. $C_7H_{16} + 11O_2 \rightarrow 8H_2O + 7CO_2$
c. $C_9H_{20} + 14O_2 \rightarrow 10H_2O + 9CO_2$
d. $2C_{10}H_{22} + 31O_2 \rightarrow 22H_2O + 20CO_2$
10.105 a. $8CO_2 + 10H_2O$
b. $Br—C(CH_3)_2—CH_3 + CH_3CHCH_2Br + 2\,HBr$
c. Cl_2 + light
10.107 The following molecules are all isomers of C_6H_{14}.

$CH_3CH_2CH_2CH_2CH_2CH_3$ $CH_3CHCH_2CH_2CH_3$ with CH_3
Hexane   2-Methylpentane

$CH_3CH_2CHCH_2CH_3$ with CH_3 $CH_3CHCHCH_3$ with two CH_3
3-Methylpentane   2,3-Dimethylbutane

$CH_3CCH_2CH_3$ with two CH_3
2,2-Dimethylbutane

a. 2, 3-Dimethylbutane produces only two monobrominated derivatives: 1-bromo-2, 3-dimethylbutane and 2-bromo-2, 3-dimethylbutane.
b. Hexane produces three monobrominated products: 1-bromohexane, 2-bromohexane, and 3-bromohexane. 2, 2-Dimethylbutane also produces three monobrominated products: 1-bromo-2, 2-dimethylbutane, 2-bromo-3, 3-dimethylbutane, and 1-bromo-3, 3-dimethylbutane.
c. 3-Methylpentane produces four monobrominated products: 1-bromo-3-methylpentane, 2-bromo-3-methylpentane, 3-bromo-3-methylpentane, and 1-bromo-2-ethylbutane.

10.109 The hydrocarbon is cyclooctane, having a molecular formula of C_8H_{16}.

(cyclooctane) $+ 12O_2 \longrightarrow 8CO_2 + 8H_2O$

Chapter 11

11.1 a.
$$Br—CH_2—CH_2—C\equiv C—CH_2—CH_2—H$$

b.
$$H—CH_2—C\equiv C—CH_2—H$$

11.3 a. Cl—C≡C—Cl

b.
H—C≡C—CH₂—CH₂—CH₂—CH₂—CH₂—CH₂—CH₂—I
(with all H's shown on each carbon)

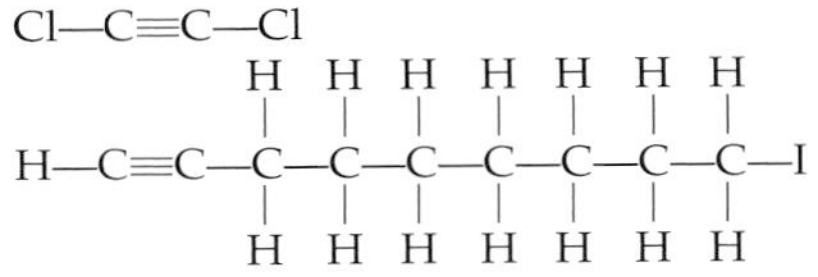

11.5 a.

cis-3-Hexene (H and H on same side, CH₃CH₂ groups on same side)

trans-3-Hexene (H and CH₂CH₃ arrangement)

b.

trans-2,3-Dibromo-2-butene

cis-2,3-Dibromo-2-butene

11.7 Molecule c can exist as *cis*- and *trans*-isomers because there are two different groups on each of the carbon atoms attached by the double bond.

11.9 a. CH₃CH₂ and CH₂CH₂CH₂CH₃ on C=C with H, H

b. CH₃ and H on C=C with H and Cl—CH₂CHCH₃

c. CH₃ and Cl on C=C with Cl and CH₃

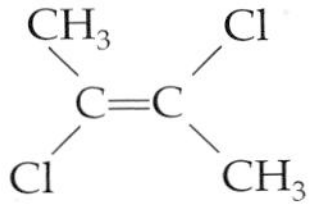

11.11 The hydrogenation of the *cis* and *trans* isomers of 2-pentene would produce the same product, pentane.

11.13 a. H₃C—C≡C—CH₃ + 2 H₂ →(Ni) butane (H—C—C—C—C—H with all H's)

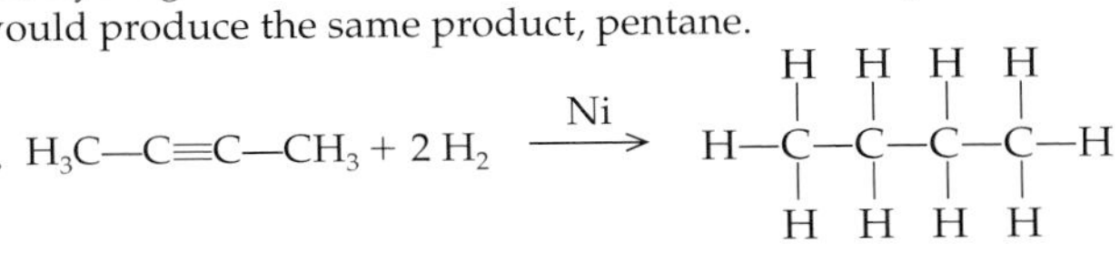

2-Butyne → Butane

b. H₃C—C≡C—CH₂CH₃ + 2 H₂ →(Ni) pentane

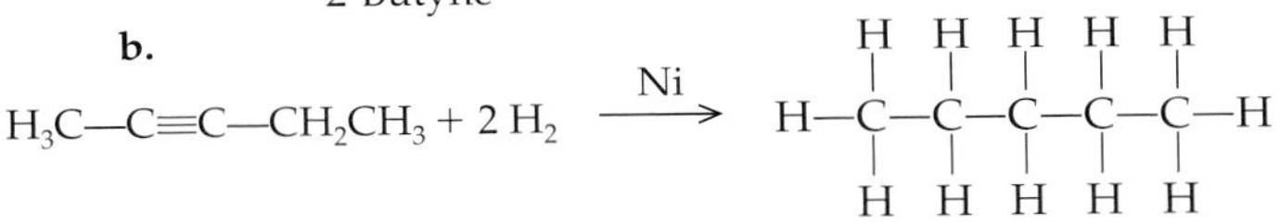

2-Pentyne → Pentane

11.15 a. CH₃CH=CH₂ + Br₂ → H—C—C—CH (with Br's on middle and end)

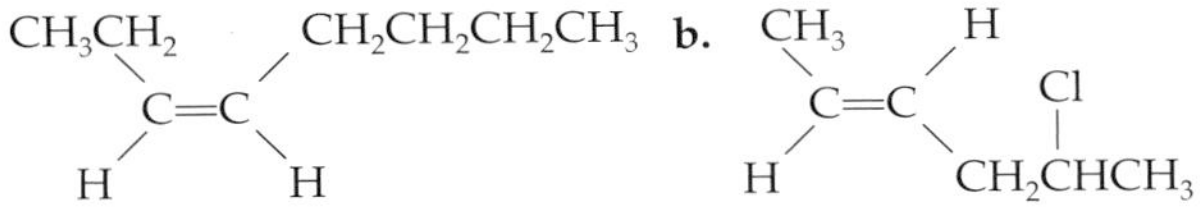

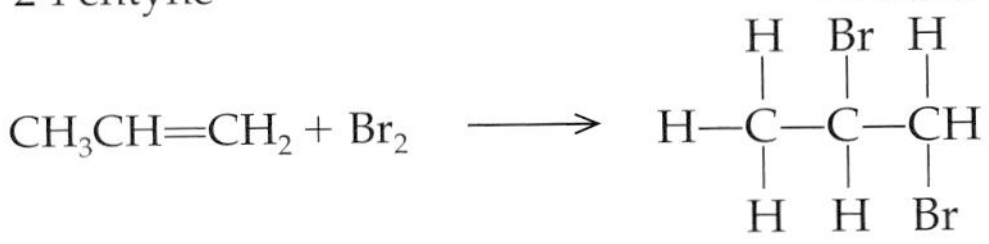

b. CH₃CH=CHCH₃ + Br₂ → H—C—C—C—C—H (with Br on C2 and C3)

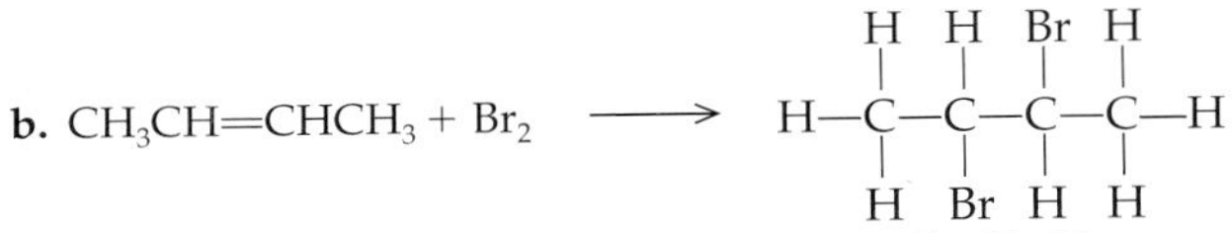

11.17 a. CH₃C≡CCH₃ + 2Cl₂ → H—C—C—C—C—H (with Cl, Cl, Cl, Cl on middle carbons)

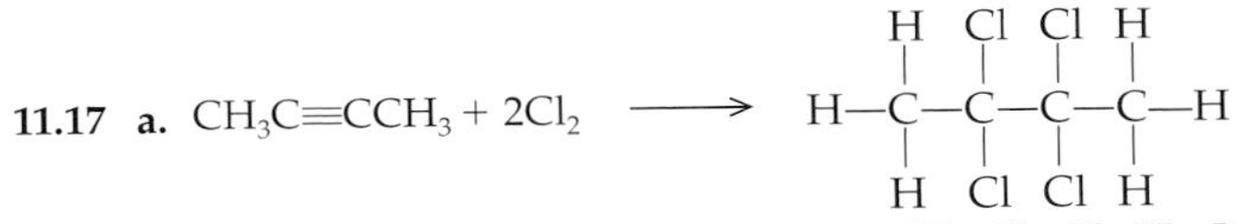

b. CH₃C≡CCH₂CH₃ + 2Cl₂ → H—C—C—C—C—C—H (with Cl's on C2 and C3)

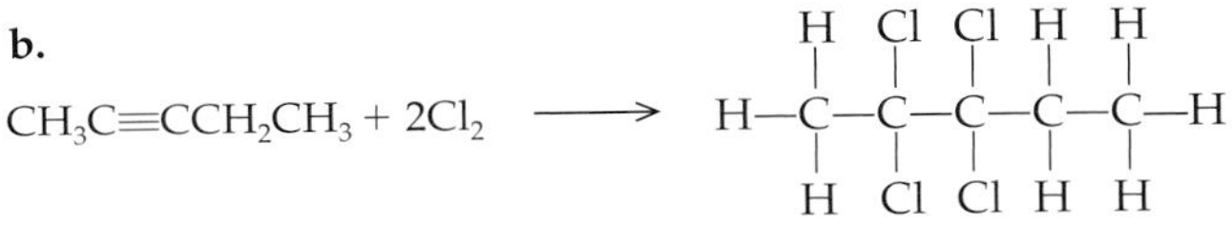

11.19 a. CH₃CH=CHCH₃ + H₂O →(H⁺) CH₃CHCH₂CH₃ (only product)
                                             |
                                             OH

b. CH₂=CHCH₂CH₂CHCH₃ + H₂O →(H⁺)
                   |
                   CH₃

CH₃CHCH₂CH₂CHCH₃ (major product)
   | |
   OH   CH₃

CH₂=CHCH₂CH₂CHCH₃ + H₂O →(H⁺)
                   |
                   CH₃

CH₂CH₂CH₂CH₂CHCH₃ (minor product)
   | |
   OH   CH₃

c. CH₃CH₂CH₂CH=CHCH₂CH₃ + H₂O →(H⁺)

CH₃CH₂CH₂CHCH₂CH₂CH₃
         |
         OH

CH₃CH₂CH₂CH=CHCH₂CH₃ + H₂O →(H⁺)

CH₃CH₂CH₂CH₂CHCH₂CH₃
             |
             OH

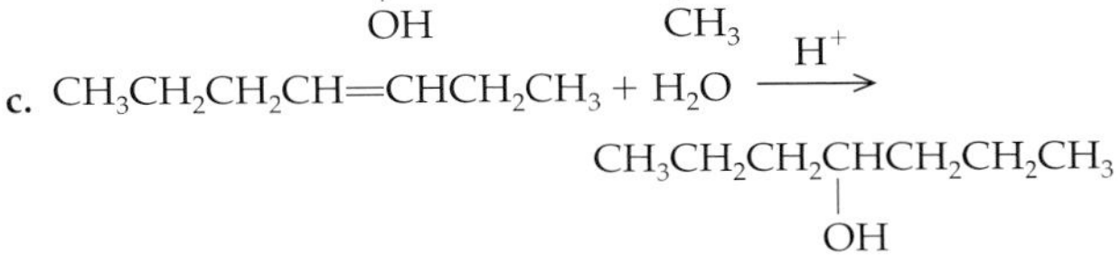

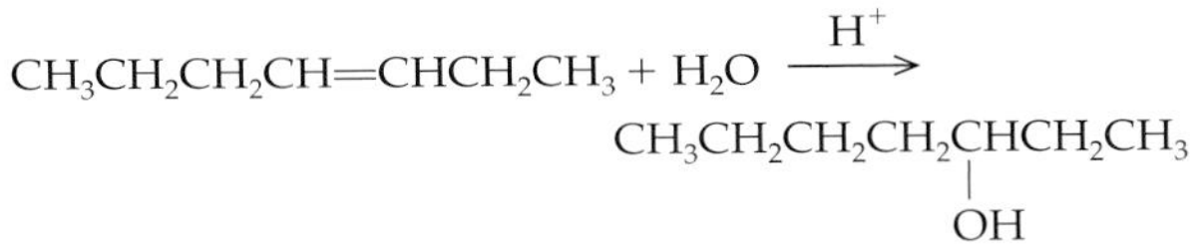

These products will be formed in approximately equal amounts.

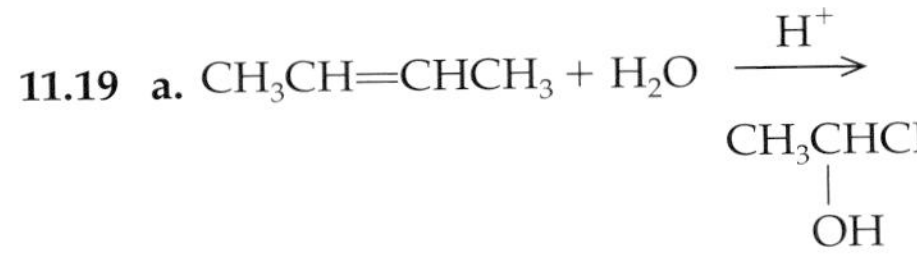

d. CH₃CHClCH=CHCHClCH₃ + H₂O →(H⁺)

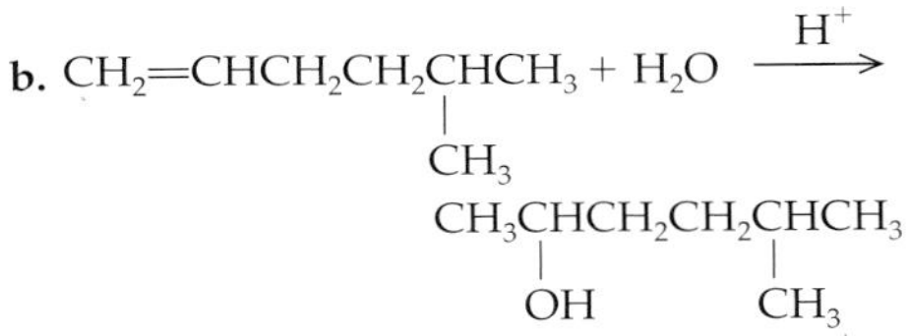

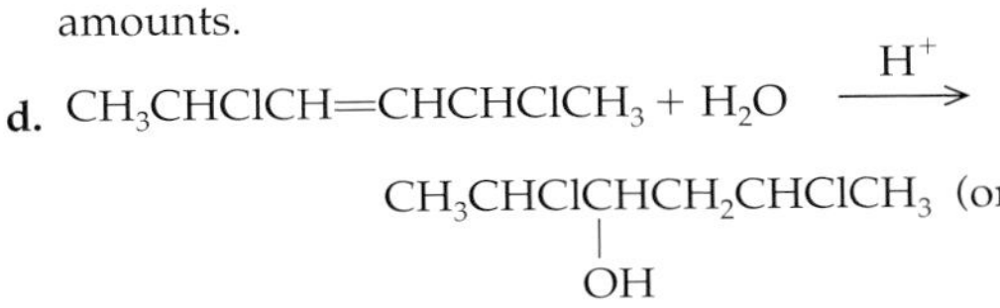

CH₃CHClCHCH₂CHClCH₃ (only product)
         |
         OH

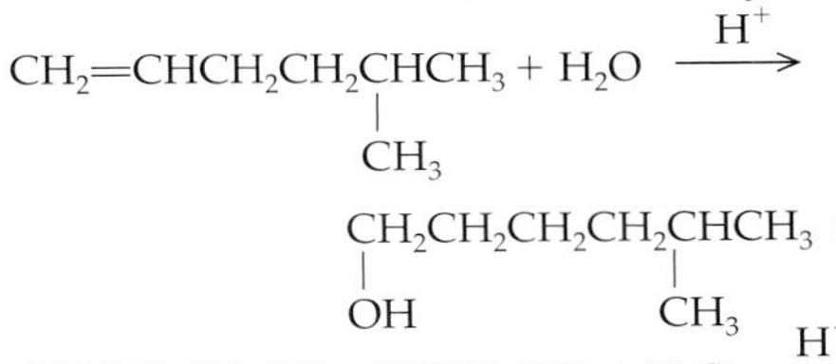

11.21 a. H₃C—C≡CH + H₂O →(H⁺) H—C—C=C—H (with OH on middle C)
                                              |
                                           H OH

→ H—C—C—C—H (ketone, C=O on middle)

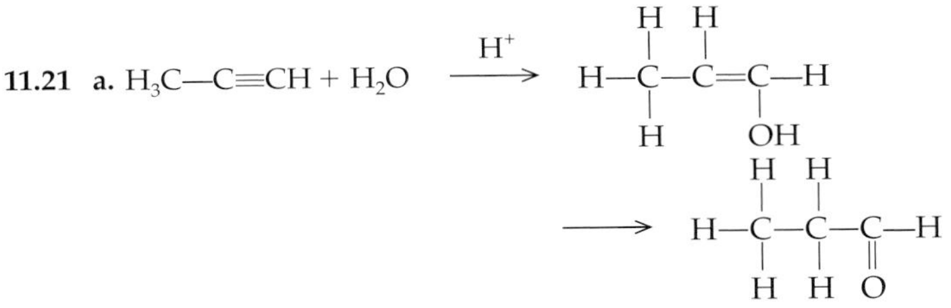

Or

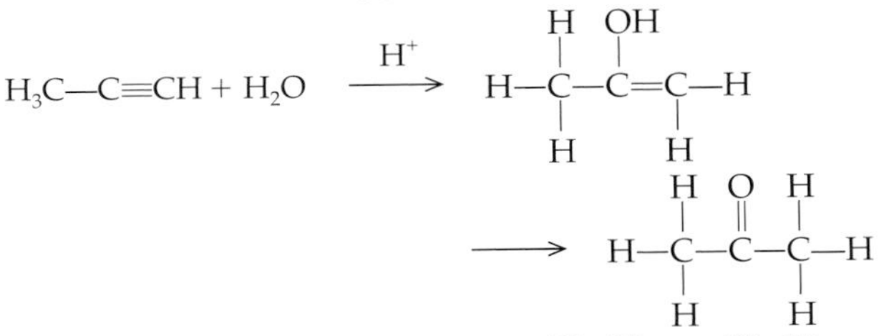

H₃C—C≡CH + H₂O →(H⁺) H—C—C=C—H (with OH on terminal C)
                         |
                         H   H

→ H—C—C—C—H (aldehyde, terminal C=O)

b. H₃C—C≡CCH₂CH₃ + H₂O →(H⁺) H—C—C=C—C—C—H (with OH)

→ H—C—C—C—C—C—H (ketone with C=O)

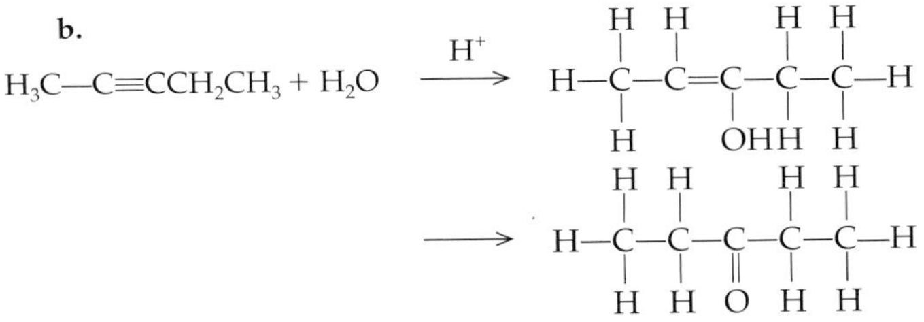

Or

H₃C—C≡CCH₂CH₃ + H₂O →(H⁺) H—C—C=C—C—C—H (with OH on other C)

→ H—C—C—C—C—C—H (ketone with C=O on other position)

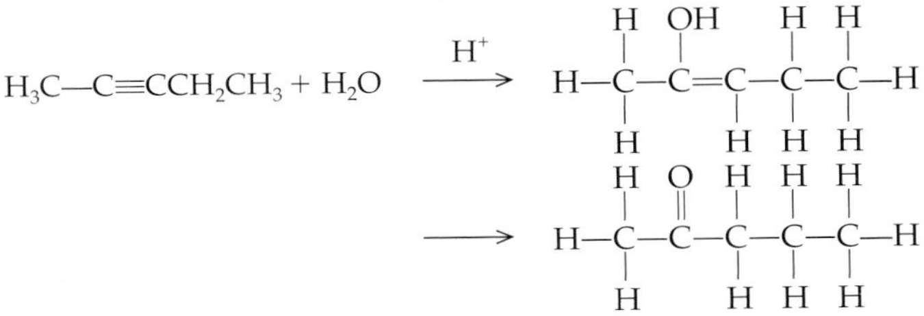

11.23 a. Reactant—*cis*-2-butene; Only product—butane
b. Reactant—1-butene; Major product—2-butanol
c. Reactant—2-butene; Only product—2,3-dichlorobutane
d. Reactant—1-pentene; Major product—2-bromopentane

11.25
a. 1,3,5-trichlorobenzene
b. 2-methylphenol (o-cresol)
c. 2,4-dibromophenol
d. 1,4-dinitrobenzene
e. 2-nitroaniline
f. 3-nitrotoluene

11.27 The longer the carbon chain of an alkene, the higher the boiling point.

11.29 The general formula for an alkane is C_nH_{2n+2}.
The general formula for an alkene is C_nH_{2n}.
The general formula for an alkyne is C_nH_{2n-2}.

11.31 Ethene is a planar molecule. All of the bond angles are 120°.

11.33 In alkanes, such as ethane, the four bonds around each carbon atom have tetrahedral geometry. The bond angles are 109.5°. In alkenes, such as ethene, each carbon is bonded by two single bonds and one double bond. The molecule is planar and each bond angle is approximately 120°.

11.35 Ethyne is a linear molecule. All of the bond angles are 180°.

11.37 In alkanes, such as ethane, the four bonds around each carbon atom have tetrahedral geometry. The bond angles are 109.5°. In alkenes, such as ethene, each carbon is bonded by two single bonds and one double bond. The molecule is planar and each bond angle is approximately 120°. In alkynes, such as ethyne, each carbon is bonded by one single bond and one triple bond. The molecule is linear and the bond angles are 180°.

11.39 a. 2-Pentyne > Propyne > Ethyne
b. 3-Decene > 2-Butene > Ethene

11.41 Identify the longest carbon chain containing the carbon-to-carbon double or triple bond. Replace the *–ane* suffix of the alkane name with *–ene* for an alkene or *-yne* for an alkyne. Number the chain to give the lowest number to the first of the two carbons involved in the double or triple bond. Determine the name and carbon number of each substituent group and place that information as a prefix in front of the name of the parent compound.

11.43 Geometric isomers of alkenes differ from one another in the placement of substituents attached to each of the carbon atoms of the double bond. Of the pair of geometric isomers, the *cis*-isomer, is the one in which identical groups are on the same side of the double bond.

11.45 a. (CH₃)(CH₂CH₂CH₃)C=C(CH₃)(H)
b. (CH₃CH₂)(H)C=C(H)(CH₂CH₂CH₃)
c. (CH₂)(CH₂CH₃)C=C(H)(H)
d. (CH₃)(CH₃CCl)C=C(CH₂CH₂CH₃)(H) ... (H)(H)
e. (CH₃CH)(CH₃)... C=C ... (H)(CH—CHCH₂CH₃)(Br)(CH₃)

11.47 a. 3-Methyl-1-pentene
b. 7-Bromo-1-heptene
c. 5-Bromo-3-heptene
d. 1-*t*-Butyl-4-methylcyclohexene

11.49 a. 1,3,5-Trifluoropentane: $CH_2F-CH_2-CHF-CH_2-CH_2F$
b. *cis*-2-Octene: (H₃C)(H)C=C(H)(CH₂CH₂CH₂CH₂CH₃)
c. Dipropylacetylene: $CH_3CH_2CH_2-C\equiv C-CH_2CH_2CH_3$

11.51 a. 2,3-Dibromobutane could not exist as *cis* and *trans* isomers.
b. (CH₃)(H)C=C(CH₂CH₂CH₂CH₃)(H) *cis*-2-Heptene
(CH₃)(H)C=C(H)(CH₂CH₂CH₂CH₃) *trans*-2-Heptene
c. (CH₃)(Br)C=C(CH₃)(Br) *cis*-2,3-Dibromo-2-butene
(CH₃)(Br)C=C(Br)(CH₃) *trans*-2,3-Dibromo-2-butene
d. Propene cannot exist as *cis* and *trans* isomers.

11.53 Alkenes b and c would not exhibit *cis-trans* isomerism.

11.55 Alkenes b and d can exist as both *cis*- and *trans*- isomers.

11.57 a. 1,5-Nonadiene
b. 1,4,7-Nonatriene
c. 2,5-Octadiene
d. 4-Methyl-2,5-heptadiene

11.59 R₂C=CR₂ + H₂ →(Pt, Pd, or Ni / heat or pressure)→ R−CH−CH−R with R groups

11.61 R₂C=CR₂ + X₂ → R−CX−CX−R

11.63 R₂C=CR₂ + H₂O →(H⁺)→ R−CH−C(OH)−R

11.65 The primary difference between complete hydrogenation of an alkene and an alkyne is that 2 moles of H₂ are required for the complete hydrogenation of an alkyne.

11.67 Addition of bromine (Br₂) to an alkene results in a color change from red to colorless. If equimolar quantities of Br₂ are added to hexene, the reaction mixture will change from red to colorless. This color change will not occur if cyclohexane is used.

11.69
a. H_2
b. H_2O
c. HBr
d. $19O_2 \rightarrow 12CO_2 + 14H_2O$
e. Cl_2
f. (cyclopentene)

11.71 a.

$H_3C-C\equiv C-CH_3 + 2H_2 \xrightarrow{\text{Pt, Pd, or Ni}}_{\text{heat or pressure}} H_3C-\underset{H}{\underset{|}{\overset{H}{\overset{|}{C}}}}-\underset{H}{\underset{|}{\overset{H}{\overset{|}{C}}}}-CH_3$

2-Butyne

b.

$CH_3CH_2-C\equiv C-CH_3 + 2X_2 \longrightarrow CH_3CH_2-\underset{X}{\underset{|}{\overset{X}{\overset{|}{C}}}}-\underset{X}{\underset{|}{\overset{X}{\overset{|}{C}}}}-CH_3$

2-Pentyne

11.73 $CH_2=CHCH_2CH_2CH_3$, $CH_3CH=CHCH_2CH_3$,

$CH_3\underset{CH_3}{\underset{|}{C}}=CHCH_3$, $CH_2=\underset{CH_3}{\underset{|}{C}}CH_2CH_3$, $CH_2=CH\underset{CH_3}{\underset{|}{C}}HCH_3$

11.75 a.

$CH_3\underset{Br}{\underset{|}{C}}HCH_2CH_3$

b.

$CH_3CH_2-\underset{CH_3}{\underset{|}{\overset{I}{\overset{|}{C}}}}-CH_2CH_2CH_3 + CH_3\underset{I}{\underset{|}{C}}H\underset{CH_3}{\underset{|}{C}}HCH_2CH_2CH_3$

(major product)   (minor product)

c. (chlorocyclopentane)

11.77 A polymer is a macromolecule composed of repeating structural units called *monomers*.

11.79

$n \; \underset{F}{\overset{F}{C}}=\underset{F}{\overset{F}{C}} \longrightarrow \left[\underset{F}{\underset{|}{\overset{F}{\overset{|}{C}}}}-\underset{F}{\underset{|}{\overset{F}{\overset{|}{C}}}}\right]_n$

Tetrafluoroethene   Teflon

11.81 a.

$CH_3\underset{H}{\overset{H}{C}}=CCH_2CH_3 + H_2O \xrightarrow{H^+} CH_3CHCH_2CH_2CH_3$
                                                                       OH

2-Pentene

$CH_3\overset{H}{\underset{H}{C}}=CCH_2CH_3 + H_2O \xrightarrow{H^+} CH_3CH_2CHCH_2CH_3$
                                                                        OH

These products will be formed in approximately equal amounts.

b. $CH_2=\underset{Br}{\underset{|}{C}}-H + H_2O \xrightarrow{H^+} CH_2CHCH_3$ (major product)
                                                               Br   OH

3-Bromo-1-propene

$CH_2\underset{Br}{\underset{|}{C}}H=C-H + H_2O \xrightarrow{H^+} CH_2CH_2CHOH$ (minor product)
                                                                      Br

c. 3,4-Dimethylcyclohexene + $H_2O \xrightarrow{H^+}$ (two products: one with OH, CH_3, CH_3; the other HO, CH_3, CH_3)

These products will be formed in approximately equal amounts.

11.83 a. $CH_2=CH\underset{CH_3}{\underset{|}{C}}HCH_2CH_3 + H_2O \xrightarrow{H^+} CH_2CH_2CH_2\underset{CH_3}{\underset{|}{C}}HCH_3$
                                                                        OH

(This is the minor product of this reaction.)

b. $CH_3\overset{H}{\underset{H}{C}}=CCH_2CH_2CH_3 + HBr \longrightarrow CH_3CH_2\underset{Br}{\underset{|}{C}}HCH_2CH_2CH_3$

OR

$CH_3CH_2\overset{H}{\underset{H}{C}}=CCH_2CH_3 + HBr \longrightarrow CH_3CH_2\underset{Br}{\underset{|}{C}}HCH_2CH_2CH_3$

c. (methylcyclohexene) + HBr $\longrightarrow$ (bromo methylcyclohexane)

d. (ethylcyclopentene) + $H_2O \xrightarrow{H^+}$ (ethyl-hydroxycyclopentane)

11.85 a.
$CH_2=CHCH_2CH=CHCH_3 + 2H_2 \xrightarrow[\text{heat}]{\text{Pt}} CH_3(CH_2)_4CH_3$

1,4-Hexadiene   Hexane

b.
$CH_3CH=CHCH=CHCH=CHCH_3 + 3H_2 \xrightarrow[\text{heat}]{\text{Ni}} CH_3(CH_2)_6CH_3$

2,4,6-Octatriene   Octane

c. 1,3-Cyclohexadiene + $2H_2 \xrightarrow[\text{Pressure}]{\text{Pd}}$ Cyclohexane

d. 1,3,5-Cyclooctatriene + $3H_2 \xrightarrow[\text{heat}]{\text{Ni}}$ Cyclooctane

11.87 The term aromatic hydrocarbon was first used as a term to describe the pleasant-smelling resins of tropical trees.

11.89 Resonance hybrids are molecules for which more than one valid Lewis structure can be written.

11.91 a. (benzene with CH_3 and two Br substituents)
b. (benzene with CH_2CH_3, CH_2CH_3, CH_2CH_3 substituents)

11.93
a. [structure: phenol with CH₃ at meta position — OH on benzene, CH₃ meta]
b. [structure: benzene with CH₂CH₂CH₃]
c. [structure: 1,3,5-trinitrobenzene with NO₂ groups — O₂N, NO₂, NO₂]
d. [structure: benzene with CH₃ and Cl]

c. CH₃CHCH₃ attached to benzene
d. benzene with CH₃ (top) and Br (ortho), Cl (para)

11.95 Kekulé proposed that single and double carbon-carbon bonds alternate around the benzene ring. To explain why benzene does not react like other unsaturated compounds, he proposed that the double and single bonds shift positions rapidly.

11.97 An addition reaction involves addition of a molecule to a double or triple bond in an unsaturated molecule. In a substitution reaction, one chemical group replaces another.

11.99 benzene + Cl₂ →(FeCl₃) chlorobenzene + HCl

11.101 Pyrimidine structure

11.103 Purine structure

Chapter 12

12.1
a. 4-Methyl-1-pentanol
b. 4-Methyl-2-hexanol
c. 1, 2, 3-Propanetriol
d. 4-Chloro-3-methyl-1-hexanol

12.3
a. Primary
b. Secondary
c. Tertiary
d. Aromatic (phenol)
e. Secondary

12.5
a. $CH_3CH=CH_2 + H_2O \xrightarrow{H^+} CH_3CH(OH)CH_3 + CH_3CH_2CH_2OH$
   major product   minor product
b. $CH_2=CH_2 + H_2O \xrightarrow{H^+} CH_3CH_2OH$
c. $CH_3CH_2CH=CHCH_2CH_3 + H_2O \xrightarrow{H^+} CH_3CH_2CH_2CH(OH)CH_2CH_3$

12.7
a. The major product is a secondary alcohol (2-propanol) and the minor product is a primary alcohol (1-propanol).
b. The product, ethanol, is a primary alcohol.
c. The product, 3-hexanol, is a secondary alcohol.

12.9 The following equation represents the reduction of the ketone, butanone. This reaction requires a catalyst.

$$CH_3CH_2\overset{O}{\overset{\|}{C}}CH_3 + H_2 \rightarrow CH_3CH_2CH(OH)CH_3$$

12.11 The following equation represents the reduction of the aldehyde, butanal. This reaction requires a catalyst. The product is 1-butanol.

$$CH_3CH_2CH_2\overset{O}{\overset{\|}{C}}-H + H_2 \rightarrow CH_3CH_2CH_2CH_2OH$$

12.13
a. Ethanol
b. 2-Propanol (major product), 1-Propanol (minor product)
c. 2-Butanol

12.15
a. 2-Butanol is the major product. 1-Butanol is the minor product.
b. 2-Methyl-2-propanol is the major product. 2-Methyl-1-propanol is the minor product.

12.17
a. $CH_3CH=CH_2$
b. $CH_3CH=CHCH_3 + CH_3CH_2CH=CH_2$

12.19
a. $\underset{CH_3}{\overset{CH_3}{CH_3\overset{|}{C}CH_2CH_2OH}} \rightarrow \underset{CH_3}{\overset{CH_3}{CH_3\overset{|}{C}CH_2\overset{O}{\overset{\|}{C}}-H}}$

b. $CH_3CH_2OH \rightarrow CH_3\overset{O}{\overset{\|}{C}}-H$

12.21
a. $CH_3\underset{OH}{\overset{|}{CH}}CH_2CH_3 \rightarrow CH_3\overset{O}{\overset{\|}{C}}CH_2CH_3$
b. $CH_3\underset{OH}{\overset{|}{CH}}CH_2CH_2CH_3 \rightarrow CH_3\overset{O}{\overset{\|}{C}}CH_2CH_2CH_3$

12.23
a. 1-Ethoxypropane
b. 1-Methoxypropane

12.25
a. Ethyl propyl ether
b. Methyl propyl ether

12.27
$CH_3CH_2OH + CH_3CH_2OH \xrightarrow{H^+} CH_3CH_2-O-CH_2CH_3 + H_2O$
Ethanol   Diethyl ether   Water

12.29 The longer the hydrocarbon tail of an alcohol becomes, the less water soluble it will be.

12.31 a < d < c < b

12.33
a. CH_3CH_2OH
b. $CH_3CH_2CH_2CH_2OH$
c. $CH_3\underset{OH}{\overset{|}{CH}}CH_3$

12.35 The I.U.P.A.C. rules for the nomenclature of alcohols require you to name the parent compound, that is the longest continuous carbon chain bonded to the –OH group. Replace the –e ending of the parent alkane with –ol of the alcohol. Number the parent chain so that the carbon bearing the hydroxyl group has the lowest possible number. Name and number all other substituents. If there is more than one hydroxyl group, the –ol ending will be modified to reflect the number. If there are two –OH groups, the suffix –diol is used; if it has three –OH groups, the suffix –triol is used, etc.

12.37
  a. 1-Heptanol
  b. 2-Propanol
  c. 2,2-Dimethylpropanol

12.39
  a. 3-Hexanol:

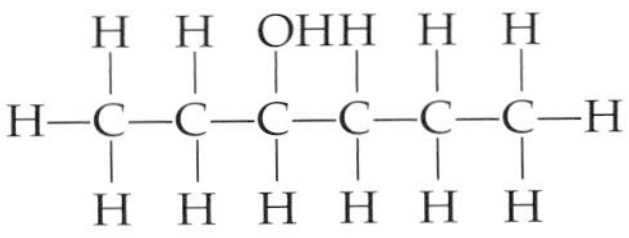

  H H OH H H H
  | | | | | |
  H—C—C—C—C—C—C—H
  | | | | | |
  H H H H H H

  b. 1,2,3-Pentanetriol:

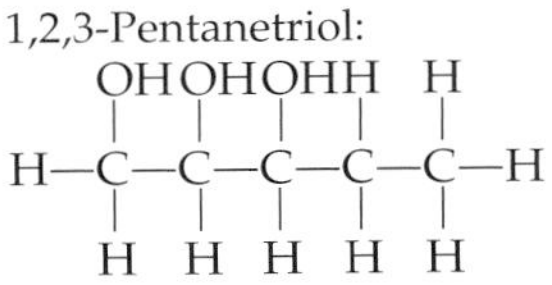

  OH OH OH H H
  | | | | |
  H—C—C—C—C—C—H
  | | | | |
  H H H H H

  c. 2-Methyl-2-pentanol:

  H OH H H
  | | | |
  H—C—C—C—C—H
  | | | |
  H H H H
  |
  H—C—H
  |
  H

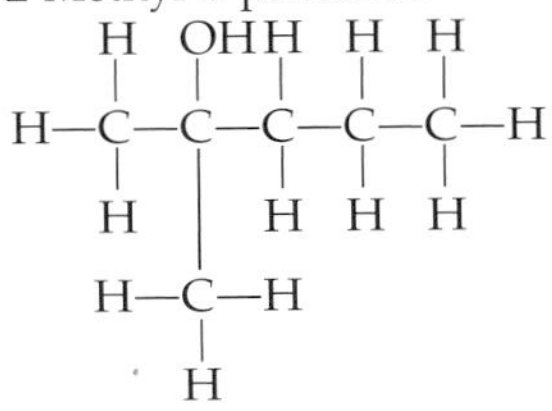

12.41
  a. Cyclopentanol
  b. Cyclooctanol
  c. 3-Methylcyclohexanol

12.43
  a. Methyl alcohol
  b. Ethyl alcohol
  c. Ethylene glycol
  d. Propyl alcohol

12.45
  a. 4-Methyl-2-hexanol 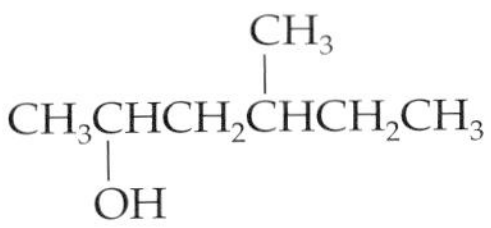
   $CH_3CHCH_2CHCH_2CH_3$ with CH_3 and OH substituents
  b. Isobutyl alcohol 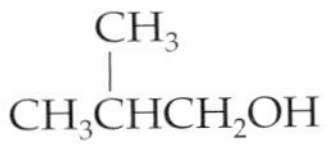
   CH_3CHCH_2OH with CH_3
  c. 1,5-Pentanediol $CH_2CH_2CH_2CH_2CH_2$ with OH on each end
  d. 2-Nonanol $CH_3CHCH_2CH_2CH_2CH_2CH_2CH_2CH_3$ with OH
  e. 1,3,5-Cyclohexanetriol

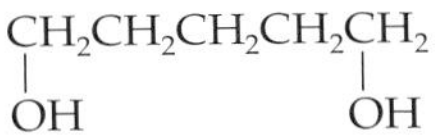

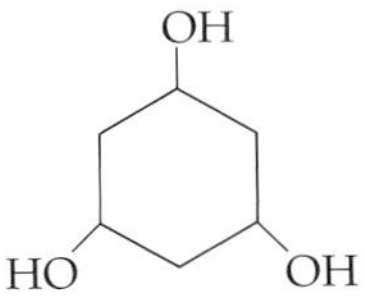

12.47 Denatured alcohol is 100% ethanol to which benzene or methanol is added. The additive makes the ethanol unfit to drink and prevents illegal use of pure ethanol.

12.49 Fermentation is the anaerobic degradation of sugar that involves no net oxidation. The alcohol fermentation, carried out by yeast, produces ethanol and carbon dioxide.

12.51 When the ethanol concentration in a fermentation reaches 12–13%, the yeast producing the ethanol are killed by it. To produce a liquor of higher alcohol concentration, the product of the original fermentation must be distilled.

12.53 The carbinol carbon is the one to which the hydroxyl group is bonded.

12.55
  a. Primary
  b. Secondary
  c. Tertiary
  d. Tertiary
  e. Tertiary

12.57
  a. Tertiary
  b. Secondary
  c. Primary
  d. Tertiary

12.59 Alkene + H_2O $\xrightarrow{H^+}$ Alcohol

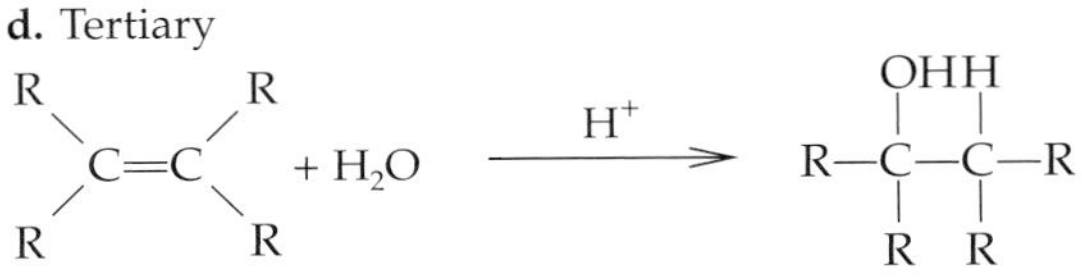

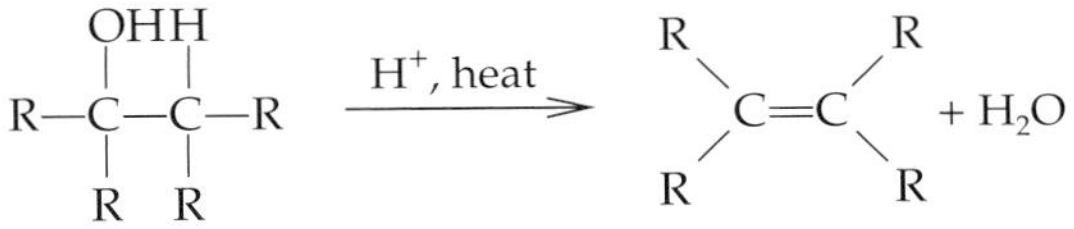

12.61 Alcohol $\xrightarrow{H^+, \text{heat}}$ Alkene + H_2O

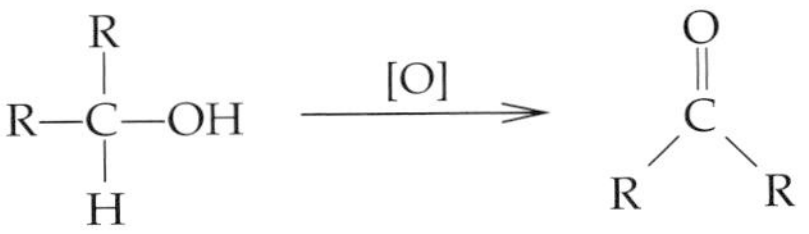

12.63 Secondary alcohol $\xrightarrow{[O]}$ Ketone

12.65
  a. 2-Pentanol (major product), 1-pentanol (minor product)
  b. 2-Pentanol and 3-pentanol
  c. 3-Methyl-2-butanol (major product), 3-methyl-1-butanol (minor product)
  d. 3,3-Dimethyl-2-butanol (major product), 3,3-dimethyl-1-butanol (minor product)

12.67
  a. $CH_3CH=CHCH_2CH_2CH_3 + H_2O \xrightarrow{H^+}$
  2-Hexene → 2-Hexanol or 3-Hexanol
  These products will be formed in approximately equal amounts.

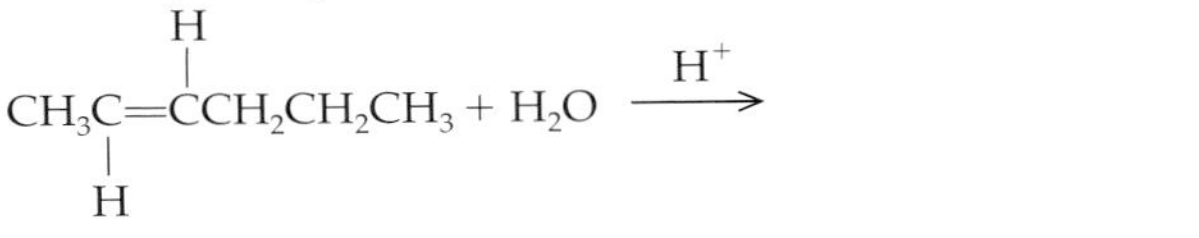

  b. Cyclopentene + H_2O $\xrightarrow{H^+}$ Cyclopentanol

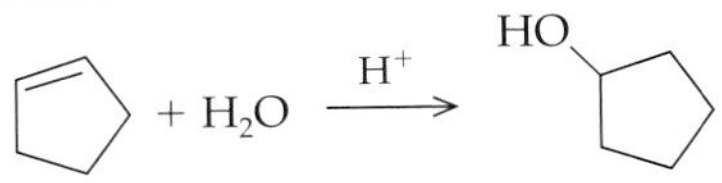

  c. $CH_2=CHCH_2CH_2CH_2CH_2CH_2CH_3 + H_2O \xrightarrow{H^+}$
  1-Octene → 2-Octanol (major product) or 1-Octanol (minor product)

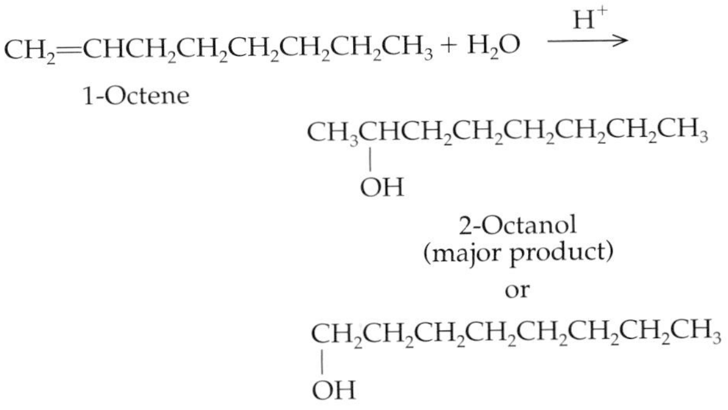

d.

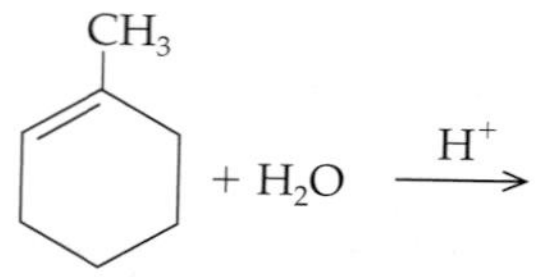

1-Methylcyclohexene + H$_2$O $\xrightarrow{H^+}$

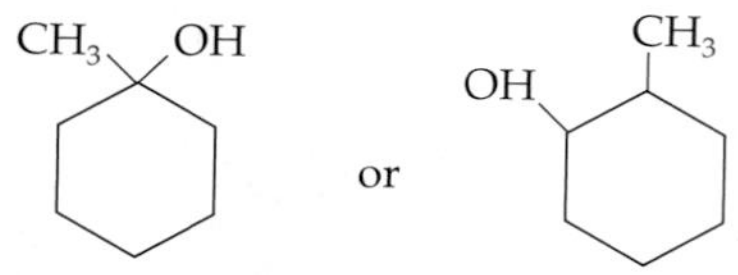

1-Methylcyclohexanol (major product) or 2-Methylcyclohexanol (minor product)

12.69
a. 2-Butanone
b. N.R.
c. Cyclohexanone
d. N.R.

12.71
a. 3-Pentanone
b. Propanal (Upon further oxidation, propanoic acid would be formed.)
c. 4-Methyl-2-pentanone
d. N.R.
e. 3-Phenylpropanal (Upon further oxidation, 3-phenylpropanoic acid will be formed.)

12.73

CH$_3$CH$_2$OH $\xrightarrow{\text{liver enzymes}}$ CH$_3$—C(=O)—H

Ethanol → Ethanal

The product, ethanal, is responsible for the symptoms of a hangover.

12.75 The reaction in which a water molecule is added to 1-butene is a hydration reaction.

CH$_3$CH$_2$CH=CH$_2$ + H$_2$O $\xrightarrow{H^+}$ CH$_3$CH$_2$CHOHCH$_3$

1-Butene → 2-Butanol

12.77

CH$_3$CH=CH$_2$ $\xrightarrow{H_2O, H^+}$ CH$_3$—CHOH—CH$_3$ $\xrightarrow{[O]}$ CH$_3$—C(=O)—CH$_3$

Propene (propylene) → 2-Propanol (isopropanol) → Propanone (acetone)

12.79 (cholesterol structure)

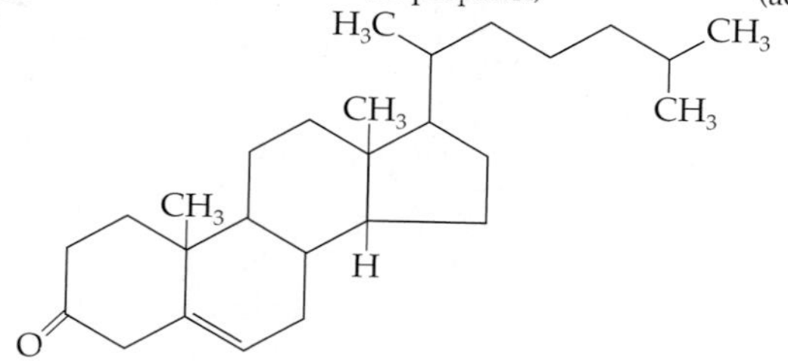

12.81 Oxidation is a loss of electrons, whereas reduction is a gain of electrons.

12.83

CH$_3$CH$_2$CH$_3$ < CH$_3$CH$_2$CH$_2$OH < CH$_3$CH$_2$C(=O)—H < CH$_3$CH$_2$C(=O)—OH

12.85 Phenols are compounds with an —OH attached to a benzene ring.

12.87 Picric acid: 2,4,6,-Trinitrotoluene:

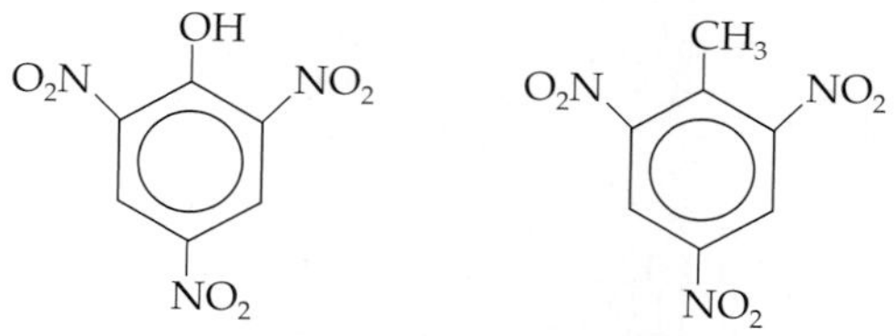

Picric acid is water-soluble because of the polar hydroxyl group that can form hydrogen bonds with water.

12.89 Hexachlorophene, hexylresorcinol, and o-phenylphenol are phenol compounds used as antiseptics or disinfectants.

12.91 Alcohols of molecular formula C$_4$H$_{10}$O

CH$_3$CH$_2$CH$_2$CH$_2$OH,   CH$_3$CHOHCH$_2$CH$_3$,

CH$_3$CH(CH$_3$)CH$_2$OH,   CH$_3$—C(OH)(CH$_3$)—CH$_3$

Ethers of molecular formula C$_4$H$_{10}$O
CH$_3$—O—CH$_2$CH$_2$CH$_3$ CH$_3$CH$_2$—O—CH$_2$CH$_3$
CH$_3$—O—CH(CH$_3$)CH$_3$

12.93 Penthrane: 2, 2-Dichloro-1, 1-difluoro-1-methoxyethane
Enthrane: 2-Chloro-1-(difluoromethoxy)-1, 1, 2-trifluoroethane

12.95
a. CH$_3$CH$_2$—O—CH$_2$CH$_3$ + H$_2$O
b. CH$_3$CH$_2$—O—CH$_2$CH$_3$ + CH$_3$—O—CH$_3$ + CH$_3$—O—CH$_2$CH$_3$ + H$_2$O
c. CH$_3$—O—CH$_3$ + CH$_3$—O—CH(CH$_3$) + CH$_3$CH(CH$_3$)—O—CH(CH$_3$)CH$_3$ + H$_2$O
d. 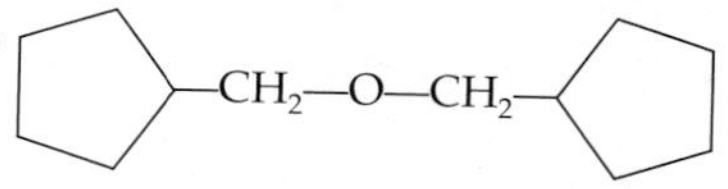 cyclopentyl—CH$_2$—O—CH$_2$—cyclopentyl

12.97
a. 2-Ethoxypentane
b. 2-Methoxybutane
c. 1-Ethoxybutane
d. Methoxycyclopentane

12.99 Cystine:

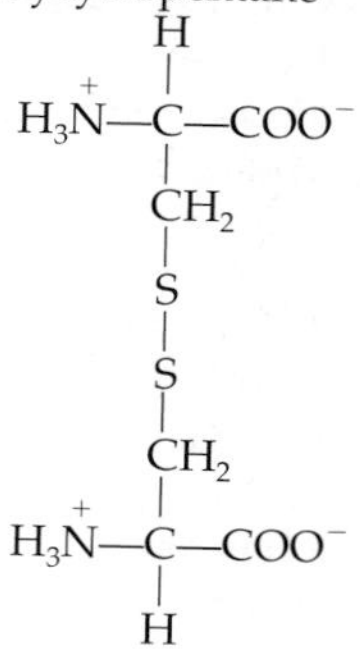

H$_3$N$^+$—CH(COO$^-$)—CH$_2$—S—S—CH$_2$—CH(COO$^-$)—NH$_3^+$

12.101
a. 1-Propanethiol
b. 2-Butanethiol
c. 2-Methyl-2-butanethiol
d. 1,4-Cyclohexanedithiol

Chapter 13

13.1 a. $CH_3-\overset{O}{\underset{\|}{C}}-CH_3$ b. $CH_3\overset{OH}{\underset{|}{C}H}CH_2CH_2CH_3$

13.3 a. $CH_3CH_2\overset{O}{\underset{\|}{C}}-OH$ b. $CH_3\overset{O}{\underset{\|}{C}}-OH$

13.5 a. Butanal
b. 2,4-Dimethylpentanal

13.7 a. I.U.P.A.C.: 3,4-Dimethylpentanal
Common: β,γ-Dimethylvaleraldehyde
b. I.U.P.A.C.: 2-Ethylpentanal
Common: α-Ethylvaleraldehyde

13.9 a. 3-Methylnonanal:

$CH_3CH_2CH_2CH_2CH_2CH_2\overset{}{\underset{|}{C}H}CH_2\overset{O}{\underset{\|}{C}}-H$
         CH_3

b. β-Bromovaleraldehyde:

$CH_3CH_2\overset{}{\underset{|}{C}H}CH_2\overset{O}{\underset{\|}{C}}-H$
     Br

13.11 a. 3-Iodobutanone
b. 4-Methyl-2-octanone

13.13 a. Methyl isopropyl ketone
(I.U.P.A.C. name: 3-Methyl-2-butanone):

$CH_3-\overset{O}{\underset{\|}{C}}-\overset{}{\underset{|}{C}H}CH_3$
            CH_3

b. 4-Heptanone:

$CH_3CH_2CH_2-\overset{O}{\underset{\|}{C}}-CH_2CH_2CH_3$

13.15 $CH_3-\overset{O}{\underset{\|}{C}}-H$

13.17 $CH_3CH_2CH_2OH \xrightarrow{H_2Cr_2O_7} CH_3CH_2\overset{O}{\underset{\|}{C}}H$
   1-Propanol   Propanal

13.19 $CH_3\overset{O}{\underset{\|}{C}}-H + Ag(NH_3)_2^+ \longrightarrow CH_3\overset{O}{\underset{\|}{C}}-O^- + Ag^0$
Ethanal   Silver ammonia   Ethanoate   Silver
              complex   anion   metal

13.21 $CH_3-\overset{O}{\underset{\|}{C}}-CH_3 + H_2 \xrightarrow{Ni} CH_3\overset{OH}{\underset{|}{C}H}CH_3$
      Propanone   2-Propanol

13.23 a. Reduction
b. Reduction
c. Reduction
d. Oxidation
e. Reduction

13.25 a. Hemiacetal
b. Ketal
c. Acetal
d. Hemiketal

13.27
$H-\overset{H}{\underset{H}{C}}-\overset{H}{\underset{H}{C}}-\overset{O}{\underset{\|}{C}}-H \rightleftharpoons H-\overset{H}{\underset{H}{C}}-\overset{OH}{\underset{}{C}}=\overset{}{\underset{H}{C}}-H$

Propanal   Propanal
Keto form   Enol form

13.29 $2CH_3CH_2\overset{O}{\underset{\|}{C}}-H \xrightarrow{OH^-} CH_3CH_2\overset{OH}{\underset{|}{C}H}\overset{}{\underset{|}{C}H}\overset{O}{\underset{\|}{C}}-H$
                                                CH_3

Propanal   3-Hydroxy-2-methylpentanal

13.31 As the carbon chain length increases, the compounds become less polar and more hydrocarbonlike. As a result, their solubility in water decreases.

13.33 A good solvent should dissolve a wide range of compounds. Simple ketones are considered to be universal solvents because they have both a polar carbonyl group and nonpolar side chains. As a result, they dissolve organic compounds and are also miscible in water.

13.35
$CH_3-\overset{O}{\underset{\|}{C}}-H \cdots H-\overset{O}{\underset{\|}{C}}-CH_3$
         (with hydrogen bonding between the two molecules)

13.37 Alcohols have higher boiling points than aldehydes or ketones of comparable molecular weights because alcohol molecules can form intermolecular hydrogen bonds with one another. Aldehydes and ketones cannot form intermolecular hydrogen bonds.

13.39 To name an aldehyde using the I.U.P.A.C. nomenclature system, identify and name the longest carbon chain containing the carbonyl group. Replace the final -e of the alkane name with -al. Number and name all substituents as usual. Remember that the carbonyl carbon is always carbon-1 and does not need to be numbered in the name of the compound.

13.41 The common names of aldehydes are derived from the same Latin roots as the corresponding carboxylic acids. For instance, methanal is formaldehyde; ethanal is acetaldehyde; propanal is propionaldehyde, *etc*.

Substituted aldehydes are named as derivatives of the straight-chain parent compound. Greek letters are used to indicate the position of substituents. The carbon nearest the carbonyl group is the α-carbon, the next is the β-carbon, and so on.

13.43 a. $H-\overset{O}{\underset{\|}{C}}-H$ b. $H-\overset{H}{\underset{Br}{C}}-\overset{H}{\underset{Br}{C}}-\overset{H}{\underset{H}{C}}-\overset{H}{\underset{H}{C}}-\overset{H}{\underset{H}{C}}-\overset{H}{\underset{H}{C}}-\overset{O}{\underset{\|}{C}}-H$

13.45 a. 3-Chloro-2-pentanone

$H-\overset{H}{\underset{H}{C}}-\overset{H}{\underset{Cl}{C}}-\overset{O}{\underset{\|}{C}}-\overset{H}{\underset{H}{C}}-\overset{H}{\underset{H}{C}}-H$

b. Benzaldehyde

13.47 a. Butanone
b. 2-Ethylhexanal
13.49 a. 3-Nitrobenzaldehyde
b. 3,4-Dihydroxycyclopentanone
13.51 a. 3-Bromobutanal
b. 2-Chloro-2-methyl-4-heptanone
13.53 a. 4,6-Dimethyl-3-heptanone
b. 3,3-Dimethylcyclopentanone
13.55 a. Acetone
b. Ethyl methyl ketone
c. Acetaldehyde
d. Propionaldehyde
e. Methyl isopropyl ketone

13.57 a.
$$CH_3CHCH_2\overset{O}{\underset{\|}{C}}-H$$
$$\underset{OH}{}$$

b.
$$CH_3CH_2CH_2\overset{O}{\underset{\|}{CH}}-H$$
$$\underset{CH_3}{}$$

c.
$$CH_3CH_2\underset{Br}{CH}CH_2CH_2\overset{O}{\underset{\|}{C}}-H$$

d.
$$CH_3CH_2\underset{I}{CH}CH_2\overset{O}{\underset{\|}{C}}-H$$

e.
$$CH_3CH_2CH_2CH_2\underset{OH}{\overset{CH_3}{CH}}\overset{O}{\underset{\|}{CHC}}-H$$

13.59 Acetone is a good solvent because it can dissolve a wide range of compounds. It has both a polar carbonyl group and nonpolar side chains. As a result, it dissolves organic compounds and is also miscible in water.

13.61 The liver

13.63 In organic molecules, oxidation may be recognized as a gain of oxygen or a loss of hydrogen. An aldehyde may be oxidized to form a carboxylic acid as in the following example in which ethanal is oxidized to produce ethanoic acid.

$$H_3C-\overset{O}{\underset{\|}{C}}-H \xrightarrow{[O]} H_3C-\overset{O}{\underset{\|}{C}}-OH$$
Ethanal → Ethanoic acid

13.65 Addition reactions of aldehydes or ketones are those in which a second molecule is added to the double bond of the carbonyl group. An example is the addition of the alcohol ethanol to the aldehyde ethanal.

$$H_3C-\overset{O}{\underset{\|}{C}}-H + CH_3CH_2OH \xrightarrow{H^+} H_3C-\underset{H}{\overset{OH}{C}}-OCH_2CH_3$$
Ethanal, Ethanol, Hemiacetal

13.67 a.
$$CH_3CH_2\underset{OH}{CH}CH_3 \xrightarrow{[O]} CH_3CH_2\overset{O}{\underset{\|}{C}}CH_3$$
2-Butanol → Butanone

b.
$$CH_3\underset{CH_3}{CH}CH_2OH \xrightarrow{[O]} CH_3\underset{CH_3}{CH}-\overset{O}{\underset{\|}{C}}-H$$
2-Methyl-1-propanol → Methylpropanal

Note that methylpropanal can be further oxidized to methylpropanoic acid.

c.
Cyclopentanol $\xrightarrow{[O]}$ Cyclopentanone

13.69
$$R-CH_2OH \xrightarrow{[O]} R-\overset{O}{\underset{\|}{C}}-H \xrightarrow{[O]} R-\overset{O}{\underset{\|}{C}}-OH$$
Primary alcohol, Aldehyde, Carboxylic acid

13.71 a. Reduction reaction
$$CH_3-\overset{O}{\underset{\|}{C}}-H \longrightarrow CH_3CH_2OH$$
Ethanal → Ethanol

b. Reduction reaction
Cyclohexanone → Cyclohexanol

c. Oxidation reaction
$$CH_3\underset{OH}{CH}CH_3 \longrightarrow CH_3-\overset{O}{\underset{\|}{C}}-CH_3$$
2-Propanol → Propanone

13.73 Only (c) 3-methylbutanal and (f) acetaldehyde would give a positive Tollens' test.

13.75 a.
$$CH_3-\overset{O}{\underset{\|}{C}}-CH_3 + CH_3CH_2OH \xrightarrow{H^+} CH_3-\underset{OCH_2CH_3}{\overset{OH}{C}}-CH_3$$

b.
$$CH_3-\overset{O}{\underset{\|}{C}}-H + CH_3CH_2OH \xrightarrow{H^+} CH_3-\underset{OCH_2CH_3}{\overset{OH}{C}}-H$$

13.77 Hemiacetal
13.79 Acetal
13.81 a.
$$CH_3-\overset{O}{\underset{\|}{C}}-CH_3 + 2\,CH_3OH \xrightarrow{H^+} CH_3-\underset{OCH_3}{\overset{OCH_3}{C}}-CH_3 + H_2O$$

b.
$$CH_3-\overset{O}{\underset{\|}{C}}-H + 2\,CH_3OH \xrightarrow{H^+} CH_3-\underset{OCH_3}{\overset{OCH_3}{C}}-H + H_2O$$

13.83 a.
$$H-\overset{O}{\underset{\|}{C}}-OH$$
b.
$$CH_3-\overset{O}{\underset{\|}{C}}-OH$$

13.85 a. Methanal
b. Propanal

13.87 a. False
b. True
c. False
d. False

13.89
$$2\,CH_3-\overset{O}{\underset{\|}{C}}-H \xrightarrow{OH^-} CH_3\underset{OH}{CH}CH_2-\overset{O}{\underset{\|}{C}}-H$$
Ethanal → 3-Hydroxybutanal

13.91

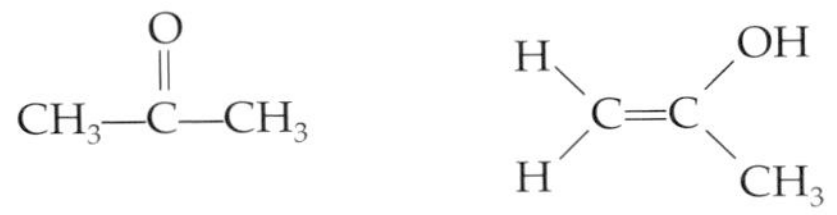

$CH_3-\overset{\overset{O}{\|}}{C}-CH_3$

Keto form of Propanone

$\underset{H}{\overset{H}{>}}C=C\underset{CH_3}{\overset{OH}{<}}$

Enol form of Propanone

13.93 a. 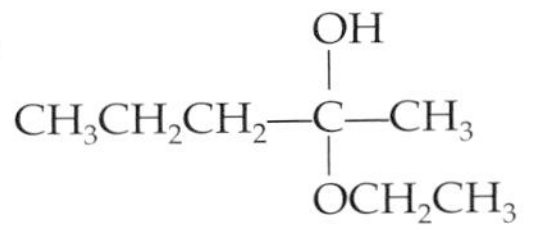 $CH_3CH_2CH_2-\underset{OCH_2CH_3}{\overset{OH}{\underset{|}{C}}}-CH_3$

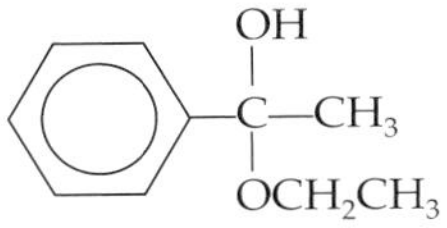

b. Ph$-\underset{OCH_2CH_3}{\overset{OH}{\underset{|}{C}}}-CH_3$

c. cyclopentyl with OH and OCH$_2$CH$_3$

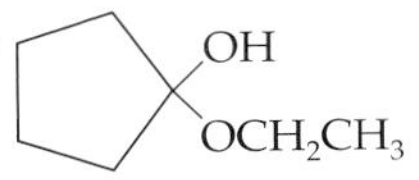

13.95 (1) $2CH_3CH_2OH$ (2) $KMnO_4/OH^-$ (3) $CH_3CH=CH_2$

Chapter 14

14.1 a. Ketone
  b. Ketone
  c. Alkane

14.3 The carboxyl group consists of two very polar groups, the carbonyl group and the hydroxyl group. Thus, carboxylic acids are very polar, in addition to which, they can hydrogen bond to one another. Aldehydes are polar, as a result of the carbonyl group, but cannot hydrogen bond to one another. As a result, carboxylic acids have higher boiling points than aldehydes of the same carbon chain length.

14.5 a. 2,4-Dimethylpentanoic acid
  b. 2,4-Dichlorobutanoic acid

14.7 a. 2,3-Dihydroxybutanoic acid:
$CH_3\underset{OH}{\overset{OH}{\underset{|}{C}H}}\overset{}{C}H-\overset{O}{\overset{\|}{C}}-OH$

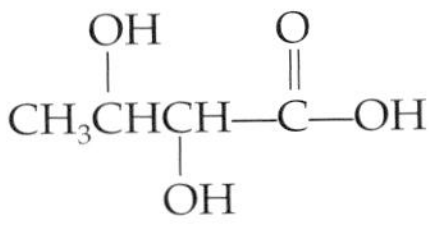

  b. 2-Bromo-3-chloro-4-methylhexanoic acid:
$CH_3CH_2\underset{CH_3}{\overset{Cl}{\underset{|}{C}H}}\underset{Br}{\overset{}{\underset{|}{C}H}}CH-\overset{O}{\overset{\|}{C}}-OH$

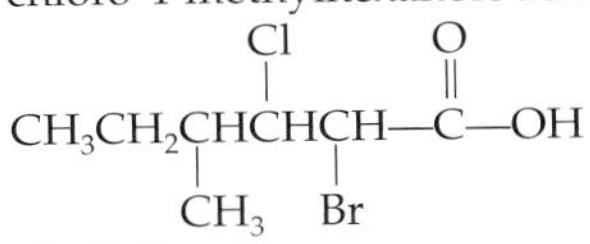

14.9 a. α,γ-Dimethylvaleric acid
  b. α,γ-Dichlorobutyric acid

14.11 a. 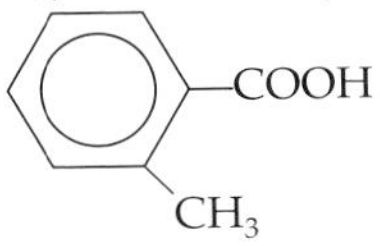 benzene ring with –COOH and –CH$_3$ (ortho)
  b. 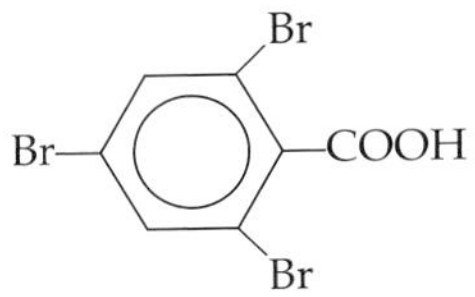 benzene ring with Br, Br, –COOH, Br
  c. 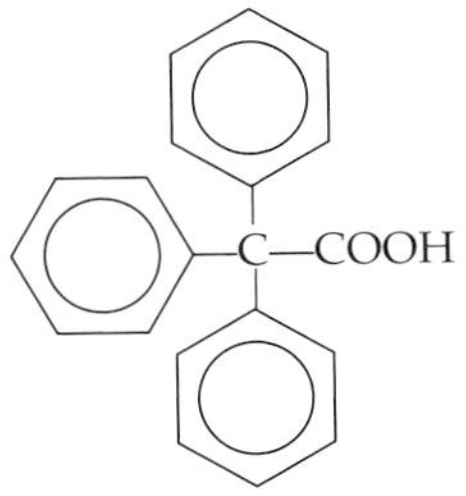 triphenyl-C–COOH

14.13 a. 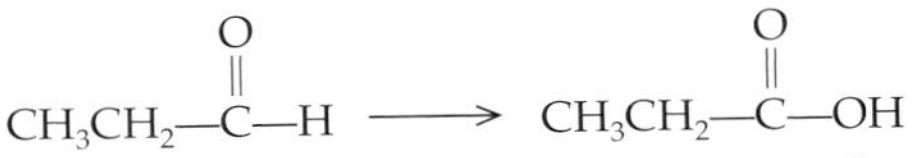 $CH_3CH_2-\overset{O}{\overset{\|}{C}}-H \longrightarrow CH_3CH_2-\overset{O}{\overset{\|}{C}}-OH$
Propanal would be the first oxidation product. However, it would quickly be oxidized further to propanoic acid.
  b. 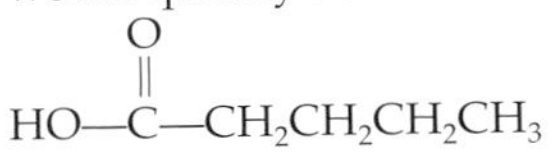 $HO-\overset{O}{\overset{\|}{C}}-CH_2CH_2CH_3$

14.15 a. 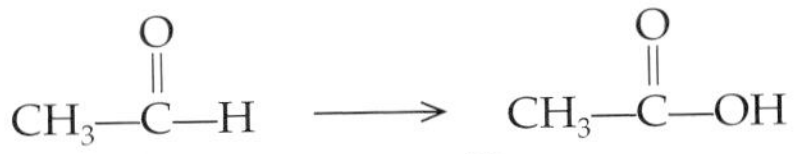 $CH_3-\overset{O}{\overset{\|}{C}}-H \longrightarrow CH_3-\overset{O}{\overset{\|}{C}}-OH$
  b. 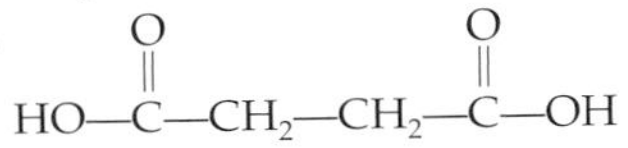 $HO-\overset{O}{\overset{\|}{C}}-CH_2-CH_2-\overset{O}{\overset{\|}{C}}-OH$

14.17 a. Potassium propanoate
  b. Barium butanoate

14.19 a. Propyl butanoate (propyl butyrate)
  b. Ethyl butanoate (ethyl butyrate)

14.21 a. The following reaction between 1-butanol and ethanoic acid produces butyl ethanoate. It requires a trace of acid and heat. It is also reversible.

$CH_3CH_2CH_2CH_2OH + CH_3COOH \leftrightarrow CH_3\overset{O}{\overset{\|}{C}}-OCH_2CH_2CH_2CH_3$

  b. The following reaction between ethanol and propanoic acid produces ethyl propanoate. It requires a trace of acid and heat. It is also reversible.

$CH_3CH_2OH + CH_3CH_2COOH \leftrightarrow CH_3CH_2\overset{O}{\overset{\|}{C}}-OCH_2CH_3$

14.23 a. $CH_3COOH + CH_3CH_2CH_2OH$
Ethanoic acid 1-Propanol
  b. $CH_3CH_2CH_2CH_2CH_2COO^-K^+ + CH_3CH_2CH_2OH$
Potassium hexanoate 1-Propanol
  c. $CH_3CH_2CH_2CH_2COO^-Na^+ + CH_3OH$
Sodium pentanoate Methanol
  d. 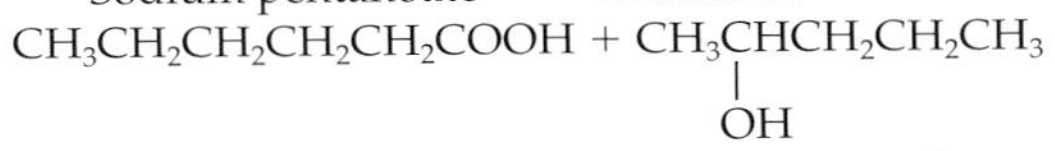 $CH_3CH_2CH_2CH_2CH_2COOH + CH_3\underset{OH}{\overset{}{\underset{|}{C}H}}CH_2CH_2CH_3$
Hexanoic acid 2-Pentanol

14.25 a. 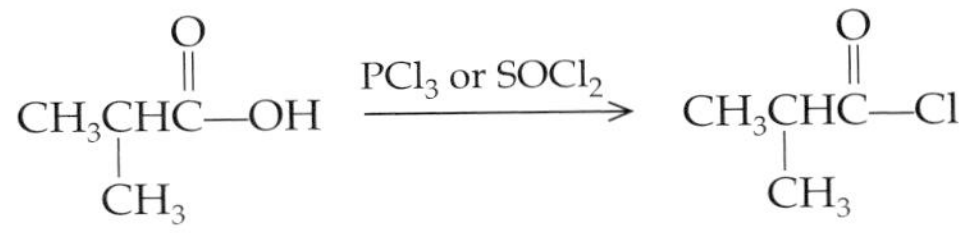 $CH_3\underset{CH_3}{\overset{}{\underset{|}{C}H}}\overset{O}{\overset{\|}{C}}-OH \xrightarrow{PCl_3 \text{ or } SOCl_2} CH_3\underset{CH_3}{\overset{}{\underset{|}{C}H}}\overset{O}{\overset{\|}{C}}-Cl$
2-Methylpropanoic acid   2-Methylpropanoyl chloride

  b. 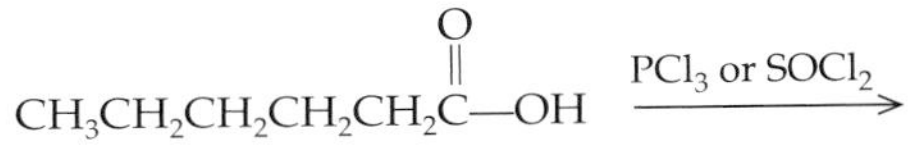 $CH_3CH_2CH_2CH_2CH_2\overset{O}{\overset{\|}{C}}-OH \xrightarrow{PCl_3 \text{ or } SOCl_2}$
Hexanoic acid 
$CH_3CH_2CH_2CH_2CH_2\overset{O}{\overset{\|}{C}}-Cl$
Hexanoyl chloride

14.27 a. 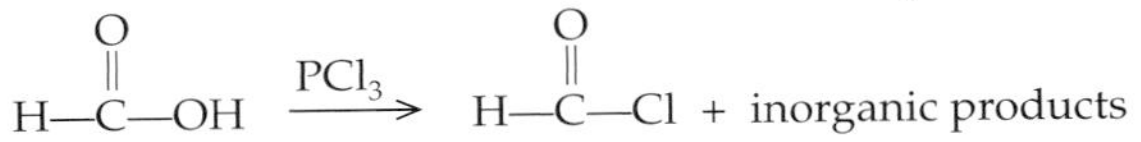 $H-\overset{O}{\overset{\|}{C}}-OH \xrightarrow{PCl_3} H-\overset{O}{\overset{\|}{C}}-Cl$ + inorganic products
Formic acid   Formyl chloride
  b.  $CH_3CH_2-\overset{O}{\overset{\|}{C}}-OH \xrightarrow{PCl_3} CH_3CH_2-\overset{O}{\overset{\|}{C}}-Cl$ + inorganic products
Propionic acid   Propionyl chloride

14.29 a. 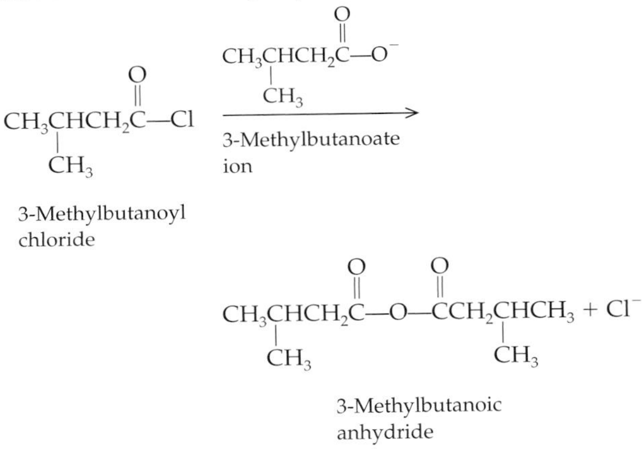 $CH_3\underset{CH_3}{\overset{}{\underset{|}{C}H}}CH_2-\overset{O}{\overset{\|}{C}}-Cl \xrightarrow{CH_3\underset{CH_3}{\overset{}{\underset{|}{C}H}}CH_2\overset{O}{\overset{\|}{C}}-O^-}$
3-Methylbutanoyl chloride   3-Methylbutanoate ion

$CH_3\underset{CH_3}{\overset{}{\underset{|}{C}H}}CH_2\overset{O}{\overset{\|}{C}}-O-\overset{O}{\overset{\|}{C}}CH_2\underset{CH_3}{\overset{}{\underset{|}{C}H}}CH_3 + Cl^-$
3-Methylbutanoic anhydride

b.

H—C(=O)—Cl →(CH₃C(=O)—O⁻, Ethanoate ion)→ H—C(=O)—O—C(=O)—CH₃ + Cl⁻

Methanoyl chloride ⟶ Ethanoic methanoic anhydride

14.31 Aldehydes are polar, as a result of the carbonyl group, but cannot hydrogen bond to one another. Alcohols are polar and can hydrogen bond as a result of the polar hydroxyl group. The carboxyl group of the carboxylic acids consists of both of these groups: the carbonyl group and the hydroxyl group. Thus, carboxylic acids are more polar than either aldehydes or alcohols, in addition to which, they can hydrogen bond to one another. As a result, carboxylic acids have higher boiling points than aldehydes or alcohols of the same carbon chain length.

14.33 a. 3-Hexanone
b. 3-Hexanone
c. Hexane

14.35 Propanoic acid > 2-Butanol > Butanal > 2-Methylbutane

14.37 a. Heptanoic acid
b. 1-Propanol
c. Pentanoic acid
d. Butanoic acid

14.39 The smaller carboxylic acids are water-soluble. They have sharp, sour tastes and unpleasant aromas.

14.41 Citric acid is found naturally in citrus fruits. It is added to foods to give them a tart flavor or to act as a food preservative and anti-oxidant. Adipic acid imparts a tart flavor to soft drinks and is a preservative.

14.43 Determine the name of the parent compound, that is the longest carbon chain containing the carboxyl group. Change the -e ending of the alkane name to -oic acid. Number the chain so that the carboxyl carbon is carbon-1. Name and number substituents in the usual way.

14.45 a. H—CH₂—CH₂—CH₂—CHBr—C(=O)—OH (shown with H H H H O / H—C—C—C—C—C—OH / H H H Br)

b. Structure with isopropyl branch and Br substituent:
H—C(H)(H)—C(H)—C(CH₃H)—C(H)(Br)—C(=O)—OH

c. Cyclohexane ring with —COOH and —Br substituents

14.47 a. I.U.P.A.C. name: Methanoic acid
Common name: Formic acid
b. I.U.P.A.C. name: 3-Methylbutanoic acid
Common name: β-Methylbutyric acid
c. I.U.P.A.C. name: Cyclopentanecarboxylic acid
Common name: Cyclovalericcarboxylic acid

14.49

Butanoic acid: H—CH₂—CH₂—CH₂—C(=O)—OH

Methylpropanoic acid: H—C(CH₃)(H)—C(=O)—OH (with methyl branch)

14.51 a. CH₃CH₂—C(CH₃)₂—CH₂CH₂COOH
b. CH₃CHCHCH₂COOH with CH₃ and Br substituents (CH₃—CH(—)—CH(Br)—CH₂—COOH)
c. Benzene ring with —COOH and two —NO₂ groups (O₂N, NO₂)
d. Cyclohexane ring with —COOH and —CH₃ (H₃C)

14.53 a. I.U.P.A.C. name: 2-Hydroxypropanoic acid
Common name: α-Hydroxypropionic acid
b. I.U.P.A.C. name: 3-Hydroxybutanoic acid
Common name: β-Hydroxybutyric acid
c. I.U.P.A.C. name: 4,4-Dimethylpentanoic acid
Common name: γ,γ-Dimethylvaleric acid
d. I.U.P.A.C. name: 3,3-Dichloropentanoic acid
Common name: β,β-Dichlorovaleric acid

14.55 In organic molecules, oxidation may be recognized as a gain of oxygen or a loss of hydrogen. An aldehyde may be oxidized to form a carboxylic acid as in the following example in which ethanal is oxidized to produce ethanoic acid.

H₃C—C(=O)—H →[O]→ H₃C—C(=O)—OH

Ethanal ⟶ Ethanoic acid

14.57 The following general equation represents the dissociation of a carboxylic acid.

R—C(=O)—OH ⇌ R—C(=O)—O⁻ + H⁺

14.59 When a strong base is added to a carboxylic acid, neutralization occurs.

14.61 Soaps are made from water, a strong base, and natural fats or oils.

14.63 a. CH₃COOH
b. CH₃CH₂CH₂—C(=O)—O—CH₃ + H₂O
c. CH₃OH

14.65 a. The oxidation of 1-pentanol yields pentanal.
b. Continued oxidation of pentanal yields pentanoic acid.

14.67 Esters are mildly polar as a result of the polar carbonyl group within the structure.

14.69 Esters are formed in the reaction of a carboxylic acid with an alcohol. The name is derived by using the alkyl or aryl portion of the alcohol I.U.P.A.C. name as the first name. The -ic acid ending of the I.U.P.A.C. name of the carboxylic acid is replaced with -ate and follows the name of the aryl or alkyl group.

14.71 a.

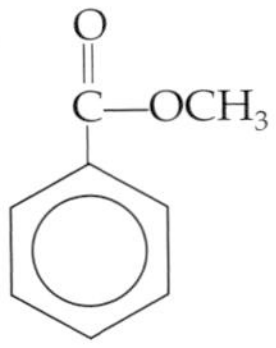

$$\text{C}_6\text{H}_5-\overset{\text{O}}{\underset{\|}{\text{C}}}-\text{OCH}_3$$

b.

$$\text{CH}_3\text{CH}_2\text{CH}_2\text{CH}_2\text{CH}_2\text{CH}_2\text{CH}_2\text{CH}_2\text{CH}_2-\overset{\text{O}}{\underset{\|}{\text{C}}}-\text{O}-\text{CH}_2\text{CH}_2\text{CH}_3$$

c.

$$\text{CH}_3\text{CH}_2-\overset{\text{O}}{\underset{\|}{\text{C}}}-\text{O}-\text{CH}_3$$

d.

$$\text{CH}_3\text{CH}_2-\overset{\text{O}}{\underset{\|}{\text{C}}}-\text{O}-\text{CH}_2\text{CH}_3$$

14.73 a. Ethyl ethanoate
b. Methyl propanoate
c. Methyl-3-methylbutanoate
d. Cyclopentyl benzoate

14.75 The following equation shows the general reaction for the preparation of an ester:

$$\text{R}-\overset{\text{O}}{\underset{\|}{\text{C}}}-\text{OH} + \text{R}-\text{OH} \underset{}{\overset{\text{H}^+,\text{ heat}}{\rightleftarrows}} \text{R}-\overset{\text{O}}{\underset{\|}{\text{C}}}-\text{OR} + \text{H}_2\text{O}$$

Carboxylic acid · Alcohol · Ester · Water

14.77 The following equation shows the general reaction for the acid-catalyzed hydrolysis of an ester:

$$\text{R}-\overset{\text{O}}{\underset{\|}{\text{C}}}-\text{OR} + \text{H}_2\text{O} \underset{}{\overset{\text{H}^+,\text{ heat}}{\rightleftarrows}} \text{R}-\overset{\text{O}}{\underset{\|}{\text{C}}}-\text{OH} + \text{R}-\text{OH}$$

Ester · Water · Carboxylic acid · Alcohol

14.79 A hydrolysis reaction is the cleavage of any bond by the addition of a water molecule.

14.81 a.

$$\text{CH}_3\text{CH}_2\text{CH}_2-\overset{\text{O}}{\underset{\|}{\text{C}}}-\text{O}-\text{CH}_2\text{CH}_3$$

b.

$$\text{CH}_3\text{CH}_2-\overset{\text{O}}{\underset{\|}{\text{C}}}-\text{OH} + \text{CH}_3\text{CH}_2\text{OH}$$

c. $\text{CH}_3\text{CH}_2\text{CH}_2\text{OH}$

d.

$$\text{CH}_3\text{CH}_2\overset{\text{Br}}{\underset{|}{\text{CH}}}\text{CH}_2-\overset{\text{O}}{\underset{\|}{\text{C}}}-\text{O}^- + \text{CH}_3\text{CH}_2\text{OH}$$

14.83 Saponification is a reaction in which a soap is produced. More generally, it is the hydrolysis of an ester in the presence of a base. The following reaction shows the base-catalyzed hydrolysis of an ester:

$$\text{CH}_3(\text{CH}_2)_{14}-\overset{\text{O}}{\underset{\|}{\text{C}}}-\text{O}-\text{CH}_3 + \text{NaOH} \longrightarrow$$

$$\text{CH}_3(\text{CH}_2)_{14}-\overset{\text{O}}{\underset{\|}{\text{C}}}-\text{O}^-\text{Na}^+ + \text{CH}_3\text{OH}$$

14.85

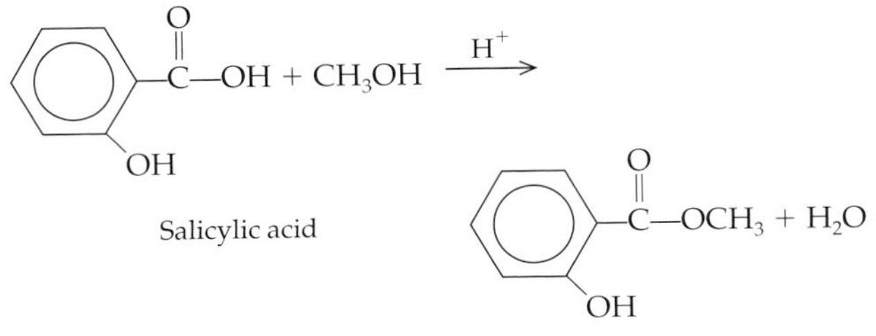

Salicylic acid $+ \text{CH}_3\text{OH} \xrightarrow{\text{H}^+}$ Methyl salicylate $+ \text{H}_2\text{O}$

14.87 Compound A is

$$\text{CH}_3\text{CH}_2\text{CH}_2\text{CH}_2-\overset{\text{O}}{\underset{\|}{\text{C}}}-\text{O}-\text{CH}_3$$

Compound B is

$$\text{CH}_3\text{CH}_2\text{CH}_2\text{CH}_2-\overset{\text{O}}{\underset{\|}{\text{C}}}-\text{OH}$$

Compound C is CH_3OH

14.89 a.

$$\text{CH}_3\text{CH}_2-\overset{\text{O}}{\underset{\|}{\text{C}}}-\text{OCH}_2\text{CH}_2\text{CH}_3 \underset{}{\overset{\text{H}^+,\text{ heat}}{\rightleftarrows}}$$
Propyl propanoate

$$\text{CH}_3\text{CH}_2-\overset{\text{O}}{\underset{\|}{\text{C}}}-\text{OH} + \text{CH}_3\text{CH}_2\text{CH}_2\text{OH}$$
Propanoic acid · 1-Propanol

b.

$$\text{H}-\overset{\text{O}}{\underset{\|}{\text{C}}}-\text{OCH}_2\text{CH}_2\text{CH}_2\text{CH}_3 \underset{}{\overset{\text{H}^+,\text{ heat}}{\rightleftarrows}}$$
Butyl methanoate

$$\text{H}-\overset{\text{O}}{\underset{\|}{\text{C}}}-\text{OH} + \text{CH}_3\text{CH}_2\text{CH}_2\text{CH}_2\text{OH}$$
Methanoic acid · 1-Butanol

c.

$$\text{H}-\overset{\text{O}}{\underset{\|}{\text{C}}}-\text{OCH}_2\text{CH}_3 \underset{}{\overset{\text{H}^+,\text{ heat}}{\rightleftarrows}}$$
Ethyl methanoate

$$\text{H}-\overset{\text{O}}{\underset{\|}{\text{C}}}-\text{OH} + \text{CH}_3\text{CH}_2\text{OH}$$
Methanoic acid · Ethanol

d.

$$\text{CH}_3\text{CH}_2\text{CH}_2\text{CH}_2-\overset{\text{O}}{\underset{\|}{\text{C}}}-\text{OCH}_3 \underset{}{\overset{\text{H}^+,\text{ heat}}{\rightleftarrows}}$$
Methyl pentanoate

$$\text{CH}_3\text{CH}_2\text{CH}_2\text{CH}_2-\overset{\text{O}}{\underset{\|}{\text{C}}}-\text{OH} + \text{CH}_3\text{OH}$$
Pentanoic acid · Ethanol

14.91 a. PCl_3, PCl_5, or SOCl_2

b.

$$\text{CH}_3-\overset{\text{O}}{\underset{\|}{\text{C}}}-\text{O}^-$$

c.

$$\text{C}_6\text{H}_{11}-\overset{\text{O}}{\underset{\|}{\text{C}}}-\text{O}^-$$

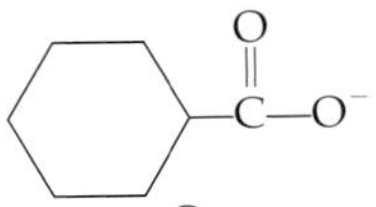

14.93 a.

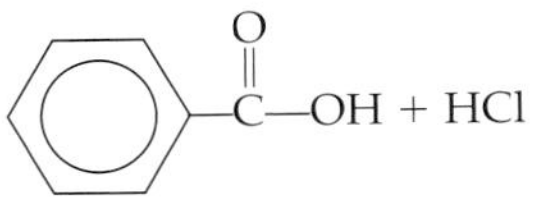

$$\text{C}_6\text{H}_5-\overset{\text{O}}{\underset{\|}{\text{C}}}-\text{OH} + \text{HCl}$$

b.

$$2\ \text{CH}_3-\overset{\text{O}}{\underset{\|}{\text{C}}}-\text{OH}$$

14.95 a.

$$\text{CH}_3(\text{CH}_2)_8-\overset{\text{O}}{\underset{\|}{\text{C}}}-\text{O}-\overset{\text{O}}{\underset{\|}{\text{C}}}-(\text{CH}_2)_8\text{CH}_3$$

b.

$$\text{CH}_3-\overset{\text{O}}{\underset{\|}{\text{C}}}-\text{O}-\overset{\text{O}}{\underset{\|}{\text{C}}}-\text{CH}_3$$

c.

$$\text{CH}_3(\text{CH}_2)_3-\overset{\text{O}}{\underset{\|}{\text{C}}}-\text{O}-\overset{\text{O}}{\underset{\|}{\text{C}}}-(\text{CH}_2)_3\text{CH}_3$$

d.

$$\text{C}_6\text{H}_5-\overset{\text{O}}{\underset{\|}{\text{C}}}-\text{Cl}$$

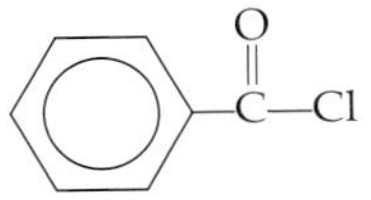

14.97 Acid chlorides are noxious, irritating chemicals. They are slightly polar and have boiling points similar to comparable aldehydes or ketones. They cannot be dissolved in water because they react violently with it.

14.99 a.

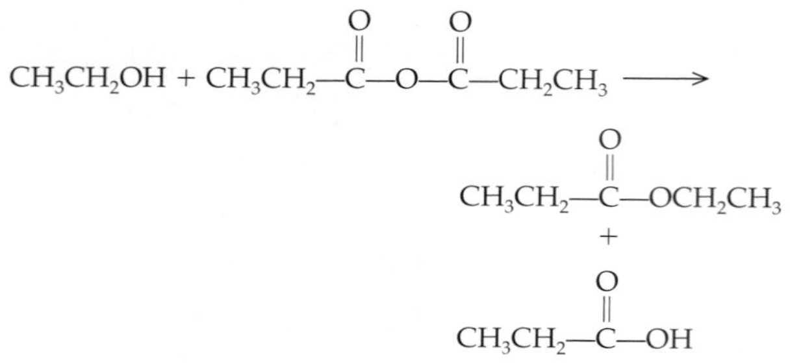

$$CH_3CH_2OH + CH_3CH_2-\overset{O}{\underset{\|}{C}}-O-\overset{O}{\underset{\|}{C}}-CH_2CH_3 \longrightarrow$$

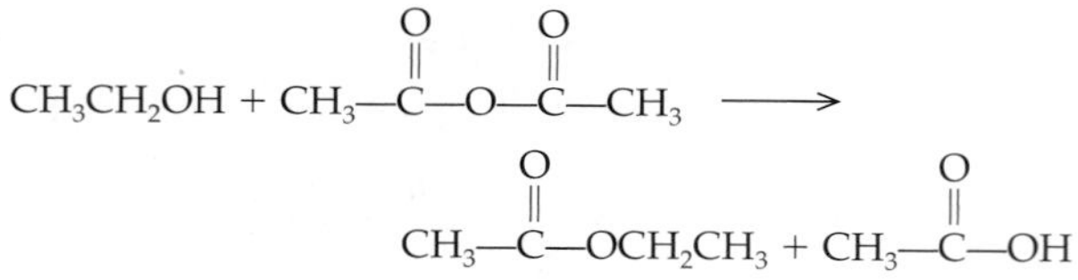

$$CH_3CH_2-\overset{O}{\underset{\|}{C}}-OCH_2CH_3$$
$$+$$
$$CH_3CH_2-\overset{O}{\underset{\|}{C}}-OH$$

b.

$$CH_3CH_2OH + CH_3-\overset{O}{\underset{\|}{C}}-O-\overset{O}{\underset{\|}{C}}-CH_3 \longrightarrow$$

$$CH_3-\overset{O}{\underset{\|}{C}}-OCH_2CH_3 + CH_3-\overset{O}{\underset{\|}{C}}-OH$$

c.

$$CH_3CH_2OH + H-\overset{O}{\underset{\|}{C}}-O-\overset{O}{\underset{\|}{C}}-H \longrightarrow$$

$$H-\overset{O}{\underset{\|}{C}}-OCH_2CH_3 + H-\overset{O}{\underset{\|}{C}}-OH$$

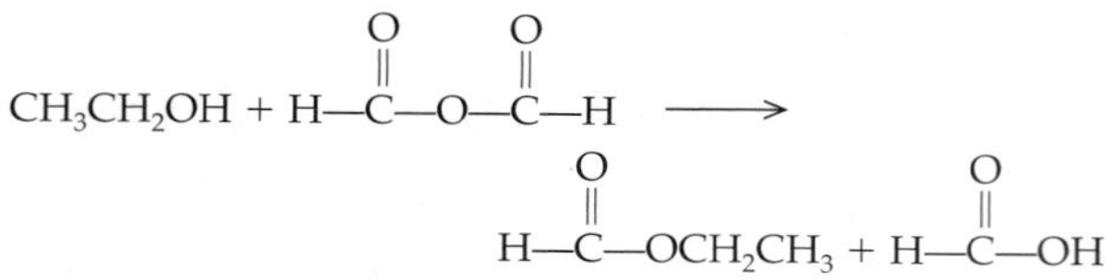

14.101 a. Monoester:

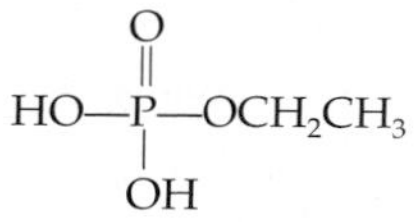

$$HO-\overset{O}{\underset{\underset{OH}{|}}{P}}-OCH_2CH_3$$

b. Diester:

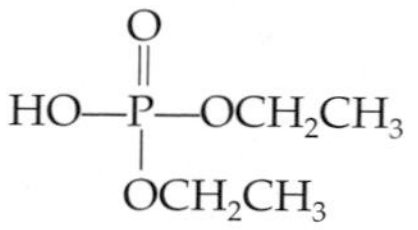

$$HO-\overset{O}{\underset{\underset{OCH_2CH_3}{|}}{P}}-OCH_2CH_3$$

c. Triester:

$$CH_3CH_2-O-\overset{O}{\underset{\underset{OCH_2CH_3}{|}}{P}}-OCH_2CH_3$$

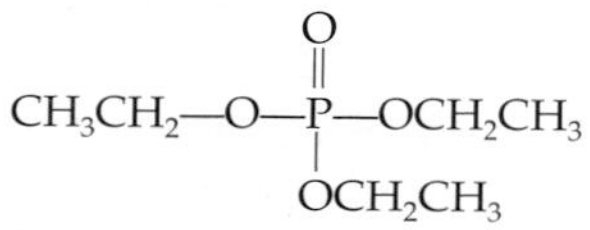

14.103 ATP is the molecule used to store the energy released in metabolic reactions. The energy is stored in the phosphoanhydride bonds between two phosphoryl groups. The energy is released when the bond is hydrolyzed. A portion of the energy can be transferred to another molecule if the phosphoryl group is transferred from ATP to the other molecule.

14.105

$$CH_3-\overset{O}{\underset{\|}{C}}\sim S-COENZYME\ A$$

The squiggle denotes a high energy bond.

14.107

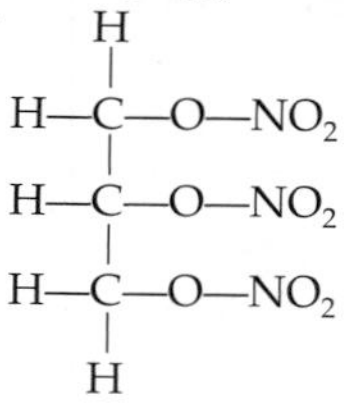

$$\begin{array}{c} H \\ | \\ H-C-O-NO_2 \\ | \\ H-C-O-NO_2 \\ | \\ H-C-O-NO_2 \\ | \\ H \end{array}$$

Chapter 15

15.1 a. Tertiary
 b. Primary
 c. Secondary

15.3 (Hydrogen bonding diagram between two methylamine-type molecules with water)

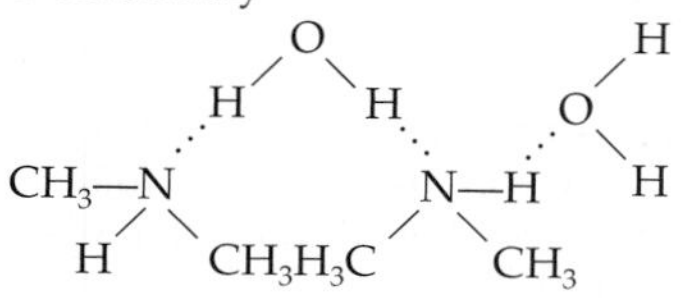

15.5 a. Methanol because the intermolecular hydrogen bonds between alcohol molecules will be stronger.
 b. Water because the intermolecular hydrogen bonds between water molecules will be stronger.
 c. Ethylamine because it has a higher molecular weight.
 d. Propylamine because propylamine molecules can form intermolecular hydrogen bonds while the nonpolar butane cannot do so.

15.7 a. Phenyl–N(H)–CH$_3$
 b. Phenyl–N(CH$_3$)–CH$_3$
 c. Phenyl–N(H)–CH$_2$CH$_3$
 d. Phenyl–N(CH$_3$)–CHCH$_3$

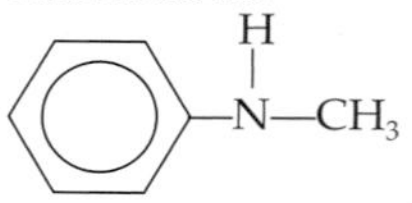

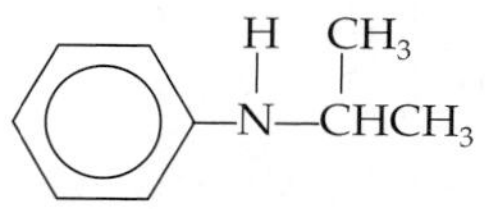

15.9 a. Propane with N-H at C2 (propan-2-amine structure)
 b. Heptane with N-H at C2
 c. Hexane with N-H branch
 d. Pentane with N-H at C2
 e. Nonane with N-H at C1, Cl at C4, I at C5
 f. Structure with N connecting two carbon chains

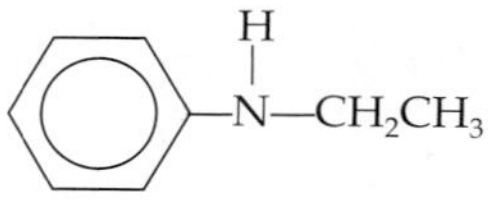

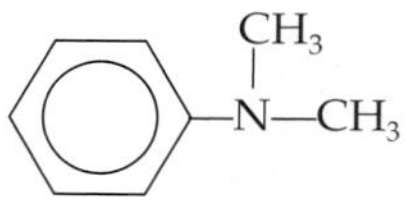

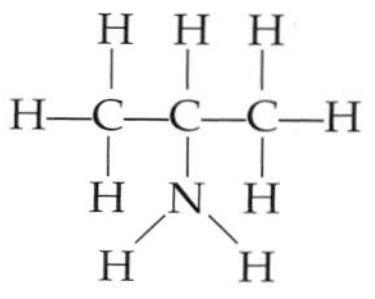

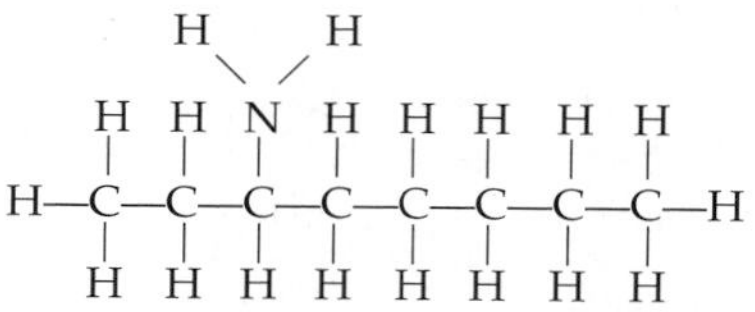

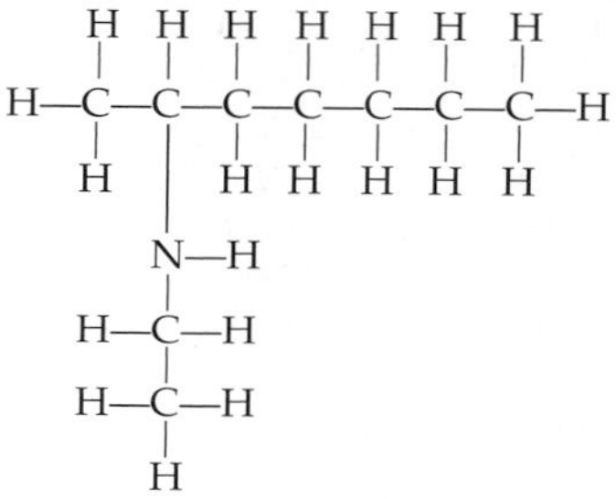

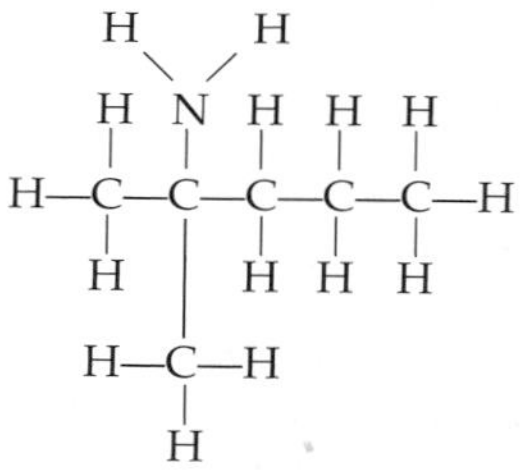

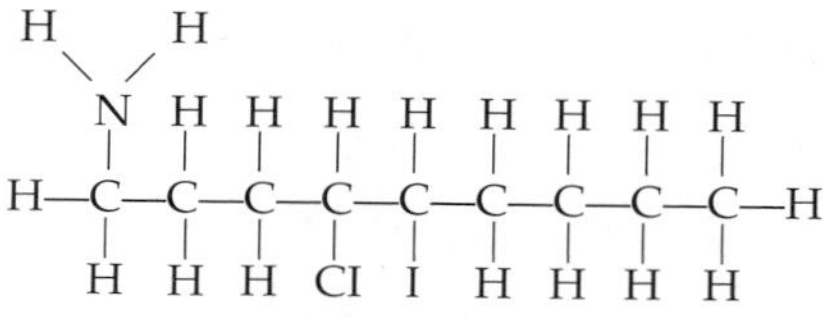

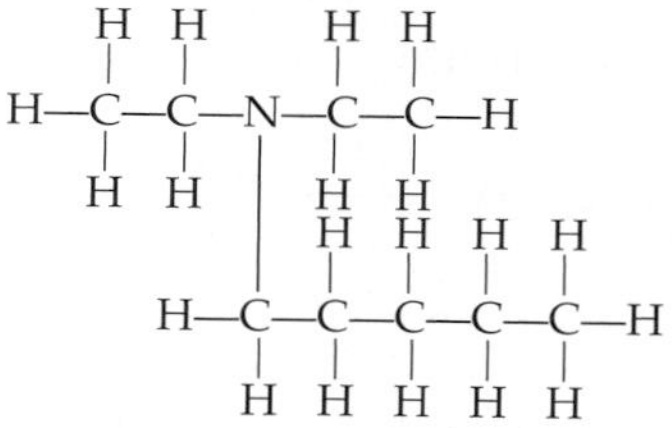

15.11 a. cyclopentyl-$NH_3^+ Br^-$

b. CH_3CH_2—$\overset{\overset{H}{|}}{\underset{\underset{H}{|}}{N^+}}$—$CH_3 + OH^-$

c. CH_3—$N^+H_3 + OH^-$

15.13 a. CH_3—NH_2

b. CH_3—$\overset{\overset{CH_3}{|}}{NH}$

15.15 The nitrogen atom is more polar than the hydrogen atom in amines; thus, the N-H bond is polar and hydrogen bonding can occur between primary or secondary amine molecules. Thus, amines have a higher boiling point than alkanes, which are nonpolar. Because nitrogen is not as electronegative as oxygen, the N-H bond is not as polar as the O-H. As a result, intermolecular hydrogen bonds between primary and secondary amine molecules are not as strong as the hydrogen bonds between alcohol molecules. Thus, alcohols have a higher boiling point.

15.17 In systematic nomenclature, primary amines are named by determining the name of the parent compound, the longest continuous carbon chain containing the amine group. The *-e* ending of the alkane chain is replaced with *-amine*. Thus, an alkane becomes an alkanamine. The parent chain is then numbered to give the carbon bearing the amine group the lowest possible number. Finally, all substituents are named and numbered and added as prefixes to the "alkanamine" name.

15.19 Amphetamines elevate blood pressure and pulse rate. They also decrease the appetite.

15.21 a. 1-Butanamine would be more soluble in water because it has a polar amine group that can form hydrogen bonds with water molecules.

b. 2-Pentanamine would be more soluble in water because it has a polar amine group that can form hydrogen bonds with water molecules.

15.23 Triethylamine molecules cannot form hydrogen bonds with one another, but 1-hexanamine molecules are able to do so.

15.25 a. 2-Butanamine
b. 3-Hexanamine
c. Cyclopentanamine
d. 2-Methyl-2-propanamine

15.27 a. CH_3CH_2—NH—CH_2CH_3
b. $CH_3CH_2CH_2CH_2NH_2$
c. $CH_3CH_2CHCH_2CH_2CH_2CH_2CH_2CH_3$ with NH_2 on the CH
d. $CH_3\overset{\overset{Br}{|}}{CH}\overset{\underset{NH_2}{|}}{CH}CH_2CH_3$
e. triphenylamine

15.29 a. $CH_3\overset{\underset{NH_2}{|}}{CH}CH_2CH_2CH_3$
b. $CH_3CH_2\overset{\overset{Br}{|}}{CH}CH_2NH_2$
c. CH_3CH_2—NH—$\overset{\underset{CH_3}{|}}{CH}CH_3$
d. cyclopentyl-NH_2

15.31
$CH_3CH_2CH_2CH_2NH_2$ — 1-Butanamine (Primary amine)

$CH_3CH_2\overset{\underset{NH_2}{|}}{CH}CH_3$ — 2-Butanamine (Primary amine)

$CH_3\overset{\underset{CH_3}{|}}{CH}CH_2NH_2$ — 2-Methyl-1-propanamine (Primary amine)

CH_3—$\overset{\overset{CH_3}{|}}{\underset{\underset{NH_2}{|}}{C}}$—$CH_3$ — 2-Methyl-2-propanamine (Primary amine)

CH_3CH_2—$\overset{\overset{CH_3}{|}}{N}$—$CH_3$ — *N,N*-Dimethylethanamine (Tertiary amine)

CH_3CH_2—NH—CH_2CH_3 — *N*-Ethylethanamine (Secondary amine)

$CH_3\overset{\underset{NH-CH_3}{|}}{CH}CH_3$ — *N*-Methyl-2-propanamine (Secondary amine)

$CH_3CH_2CH_2$—NH—CH_3 — *N*-Methyl-1-propanamine (Secondary amine)

15.33 a. Primary
b. Secondary
c. Primary
d. Tertiary

15.35 a. 4-nitrotoluene $\xrightarrow{[H]}$ 4-methylaniline (NO$_2$ → NH$_2$ on methylbenzene)

b. 2-nitrophenol $\xrightarrow{[H]}$ 2-aminophenol

c. nitrobenzene $\xrightarrow{[H]}$ aniline

d. (nitromethyl)benzene $\xrightarrow{[H]}$ benzylamine

15.37
a. H_2O
b. HBr
c. $CH_3CH_2CH_2-N^+H_3$
d. $CH_3CH_2-N^+H_2Cl^-$
   $\quad\quad\quad\quad |$
   $\quad\quad\quad\quad CH_2CH_3$

15.39 Lower molecular weight amines are soluble in water because the N—H bond is polar and can form hydrogen bonds with water molecules.

15.41 Drugs containing amine groups are generally administered as ammonium salts because the salt is more soluble in water and, hence, in body fluids.

15.43 Putrescine (1,4-Butanediamine):
$CH_2CH_2CH_2CH_2$ with NH_2 groups on each end

Cadaverine (1,5-Pentanediamine):
$CH_2CH_2CH_2CH_2CH_2$ with NH_2 groups on each end

15.45 a. Pyridine and Indole structures

b. The indole ring is found in lysergic acid diethylamide, which is a hallucinogenic drug. The pyridine ring is found in vitamin B_6, an essential water-soluble vitamin.

15.47 Morphine, codeine, quinine, and vitamin B_6

15.49 Amides have very high boiling points because the amide group consists of two very polar functional groups, the carbonyl group and the amino group. Strong intermolecular hydrogen bonding between the N-H bond of one amide and the C=O group of a second amide results in very high boiling points.

15.51 The I.U.P.A.C. names of amides are derived from the I.U.P.A.C. names of the carboxylic acids from which they are derived. The *-oic acid* ending of the carboxylic acid is replaced with the *-amide* ending.

15.53 Barbiturates are often called "downers" because they act as sedatives. They are sometimes used as anticonvulsants for epileptics and people suffering from other disorders that manifest as neurosis, anxiety, or tension.

15.55
a. I.U.P.A.C. name: Propanamide
   Common name: Propionamide
b. I.U.P.A.C. name: Pentanamide
   Common name: Valeramide
c. I.U.P.A.C. name: *N,N*-Dimethylethanamide
   Common name: *N,N*-Dimethylacetamide

15.57
a. $CH_3-\overset{O}{\overset{\|}{C}}-NH_2$
b. $CH_3CH_2-\overset{O}{\overset{\|}{C}}-NH-CH_3$
c. Phenyl$-\overset{O}{\overset{\|}{C}}-N(CH_2CH_3)_2$

d. $CH_3CH_2\overset{CH_3}{\overset{|}{C}H}CHCH_2-\overset{O}{\overset{\|}{C}}-NH_2$
   $\quad\quad\quad\quad\quad |$
   $\quad\quad\quad\quad\quad Br$

e. $CH_3-\overset{O}{\overset{\|}{C}}-\overset{CH_3}{\overset{|}{N}}-CH_3$
   $\quad\quad\quad\quad |$
   $\quad\quad\quad\quad CH_3$

15.59 *N,N*-Diethyl-*m*-toluamide:

(Structure: *m*-methylbenzene with $-C(=O)-NCH_2CH_3$ group with CH_2CH_3)

Hydrolysis of this compound would release the carboxylic acid *m*-toluic acid and the amine *N*-ethylethanamine (diethylamine).

15.61 Amides are not proton acceptors (bases) because the highly electronegative carbonyl oxygen has a strong attraction for the nitrogen lone pair of electrons. As a result they cannot "hold" a proton.

15.63 Lidocaine hydrochloride
(Structure with amide group: dimethylbenzene ring with $NH-C(=O)-CH_2-N^+(CH_2CH_3)_2 \cdot Cl^-$)

15.65 Penicillin BT
(Structure with Amide group and Carboxyl group labeled, containing $CH_3(CH_2)_3SCH_2CONH-$ group)

15.67
a. $CH_3-\overset{O}{\overset{\|}{C}}-NHCH_3 + H_3O^+ \longrightarrow$
   N-Methylethanamide
   $\quad\quad\quad\quad CH_3COOH + CH_3NH_3^+$
   $\quad\quad\quad\quad$ Ethanoic acid $\quad$ Methanamine

b. $CH_3CH_2CH_2-\overset{O}{\overset{\|}{C}}-NH-CH_3 + H_3O^+ \longrightarrow$
   N-Methylbutanamide
   $\quad\quad\quad\quad CH_3CH_2CH_2COOH + CH_3NH_3^+$
   $\quad\quad\quad\quad$ Butanoic acid $\quad$ Methanamine

c. $CH_3\overset{CH_3}{\overset{|}{C}H}CH_2-\overset{O}{\overset{\|}{C}}-NH-CH_2CH_3 + H_3O^+ \longrightarrow$
   N-Ethyl-3-methylbutanamide
   $\quad\quad\quad\quad CH_3\overset{|}{\underset{CH_3}{C}H}CH_2COOH + CH_3CH_2NH_3^+$
   $\quad\quad\quad\quad$ 3-Methylbutanoic acid $\quad$ Ethanamine

15.69 a. CH₃CH₂—C(=O)—O—C(=O)—CH₂CH₃

b. CH₃CH₂—C(=O)—NH₂ + NH₄⁺Cl⁻

c. CH₃CH₂CH₂—C(=O)—Cl + 2CH₃CH₂NH₂

15.71
H₂N—C(R)(H)—C(=O)—OH

15.73 Glycine: H₂N—CH₂—COOH   Alanine: H₂N—CH(CH₃)—COOH

15.75 H₂N—*C(H)(CH₃)—COOH

15.77 In an acyl group transfer reaction, the acyl group of an acid chloride is transferred from the Cl of the acid chloride to the N of an amine or ammonia. The product is an amide.

15.79 A chemical that carries messages or signals from a nerve to a target cell

15.81 a. Tremors, monotonous speech, loss of memory and problem-solving ability, and loss of motor function
b. Parkinson's disease
c. Schizophrenia, intense satiety sensations

15.83 In proper amounts, dopamine causes a pleasant, satisfied feeling. This feeling becomes intense as the amount of dopamine increases. Several drugs, including cocaine, heroin, amphetamines, alcohol, and nicotine increase the levels of dopamine. It is thought that the intense satiety response this brings about may contribute to addiction to these substances.

15.85 Epinephrine is a component of the flight or fight response. It stimulates glycogen breakdown to provide the body with glucose to supply the needed energy for this stress response.

15.87 The amino acid tryptophan

15.89 Perception of pain, thermoregulation, and sleep

15.91 Promotes the itchy skin rash associated with poison ivy and insect bites; the respiratory symptoms characteristic of hay fever; secretion of stomach acid

15.93 Inhibitory neurotransmitters

15.95 When acetylcholine is released from a nerve cell, it binds to receptors on the surface of muscle cells. This binding stimulates the muscle cell to contract. To stop the contraction, the acetylcholine is then broken down to choline and acetate ion. This is catalyzed by the enzyme acetylcholinesterase.

15.97 Organophosphates inactivate acetylcholinesterase by binding covalently to it. Since acetylcholine is not broken down, nerve transmission continues, resulting in muscle spasm. Pyridine aldoxime methiodide (PAM) is an antidote to organophosphate poisoning because it displaces the organophosphate, thereby allowing acetycholinesterase to function.

Chapter 16

16.1 It is currently recommended that 45–55% of the calories in the diet should be carbohydrates. Of that amount, no more than 10% should be simple sugars.

16.3 An aldose is a sugar with an aldehyde functional group. A ketose is a sugar with a ketone functional group.

16.5 a. Ketose   **d.** Aldose
b. Aldose   **e.** Ketose
c. Ketose   **f.** Aldose

16.7 a. (Fischer projection: CH₃, C=O, H—*C—OH, CH₂OH)
b. (CHO, H—*—OH, H—*—OH, HO—*—H, CH₂OH)
c. (CH₂OH, C=O, HO—*—H, H—*—OH, H—*—OH, CH₂OH)
d. (CHO, H—*—OH, CH₂OH)
e. (CH₃, C=O, H—*—OH, H—*—OH, H—*—OH, CH₂OH)
f. (CHO, H—*—OH, H—*—OH, HO—*—H, CH₂OH)

16.9 a. D-   **b.** L-   **c.** D-   **d.** D-   **e.** D-   **f.** L-

16.11
CHO
H—OH
H—OH
H—OH
CH₂OH
D-Ribose

16.13
CHO
HO—H
HO—H
HO—H
CH₂OH
L-Ribose

16.15 β-D-Galactose   α-D-Galactose (Haworth projections)

16.17 α-Amylase and β-amylase are digestive enzymes that break down the starch amylose. α-Amylase cleaves glycosidic bonds of the amylose chain at random, producing shorter polysaccharide chains. β-Amylase sequentially cleaves maltose (a disaccharide of glucose) from the reducing end of the polysaccharide chain.

16.19 A monosaccharide is the simplest sugar and consists of a single saccharide unit. A disaccharide is made up of two monosaccharides joined covalently by a glycosidic bond.

16.21 The molecular formula for a simple sugar is $(CH_2O)_n$. Typically n is an integer from 3 to 7.

16.23 Mashed potato flakes, rice, and corn starch contain amylose and amylopectin, both of which are polysaccharides. A candy bar contains sucrose, a disaccharide. Orange juice contains fructose, a monosaccharide. It may also contain sucrose if the label indicates that sugar has been added.

16.25 Four

16.27

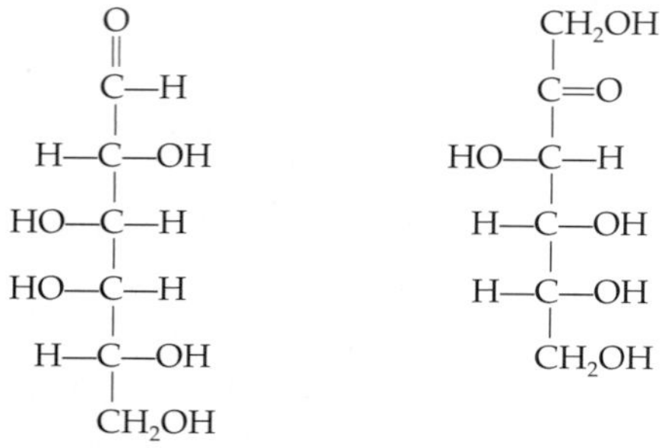

D-Galactose (An aldohexose)

D-Fructose (A ketohexose)

16.29 An *aldose* is a sugar that contains an aldehyde (carbonyl) group.

16.31 A tetrose is a sugar with a four-carbon backbone.

16.33 A ketopentose is a sugar with a five-carbon backbone and containing a ketone (carbonyl) group.

16.35
 a. β-D-Glucose is a hemiacetal.
 b. β-D-Fructose is a hemiketal.
 c. α-D-Galactose is a hemiacetal.

16.37

D-Glyceraldehyde   L-Glyceraldehyde

16.39 Stereoisomers are a pair of molecules that have the same structural formula and bonding pattern but that differ in the arrangement of the atoms in space.

16.41 A chiral carbon is one that is bonded to four different chemical groups.

16.43 A polarimeter converts monochromatic light into monochromatic plane-polarized light. This plane-polarized light is passed through a sample and into an analyzer. If the sample is optically active, it will rotate the plane of the light. The degree and angle of rotation are measured by the analyzer.

16.45 A Fischer Projection is a two-dimensional drawing of a molecule that shows a chiral carbon at the intersection of two lines. Horizontal lines at the intersection represent bonds projecting out of the page and vertical lines represent bonds that project into the page.

16.47 Dextrose is a common name used for D-glucose.

16.49 D- and L-Glyceraldehyde are a pair of enantiomers, that is, they are nonsuperimposable mirror images of one another.

16.51 a.   b.   c.

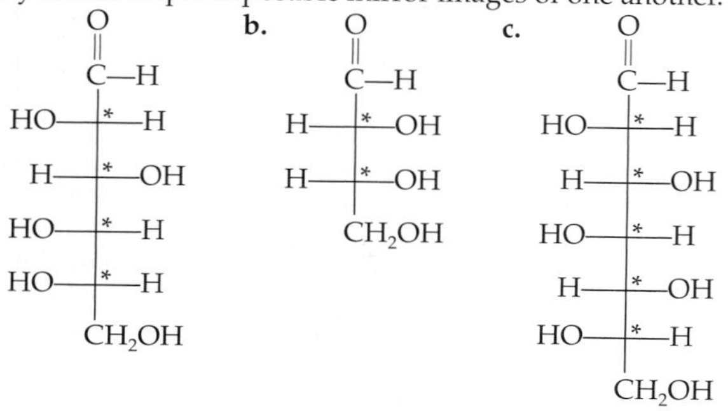

16.53 Anomers are isomers that differ in the arrangement of bonds around the hemiacetal carbon.

16.55 A hemiacetal is a member of the family of organic compounds formed in the reaction of one molecule of alcohol with an aldehyde. They have the following general structure:

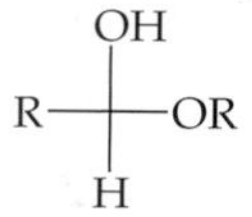

16.57 The reaction between an aldehyde and an alcohol yields a hemiacetal. Thus, when the aldehyde portion of a glucose molecule reacts with the C-5 hydroxyl group, the product is an intramolecular hemiacetal.

16.59 When the carbonyl group at C-1 of D-glucose reacts with the C-5 hydroxyl group, a new chiral carbon is created (C-1). In the α-isomer of the cyclic sugar, the C-1 hydroxyl group is below the ring; and in the β-isomer, the C-1 hydroxyl group is above the ring.

16.61 β-Maltose and α-lactose would give positive Benedict's tests. Glycogen would give only a weak reaction because there are fewer reducing ends for a given mass of the carbohydrate.

16.63 Enantiomers are stereoisomers that are nonsuperimposable mirror images of one another. For instance:

D-Glyceraldehyde   L-Glyceraldehyde

16.65 An aldehyde sugar forms an intramolecular hemiacetal when the carbonyl group of the monosaccharide reacts with a hydroxyl group on one of the other carbon atoms.

16.67 A ketal is the product formed in the reaction of two molecules of alcohol with a ketone. They have the following general structure:

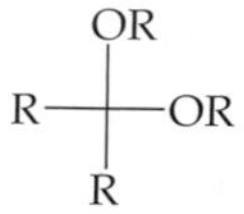

16.69 A glycosidic bond is the bond formed between the hydroxyl group of the C-1 carbon of one sugar and a hydroxyl group of another sugar.

16.71

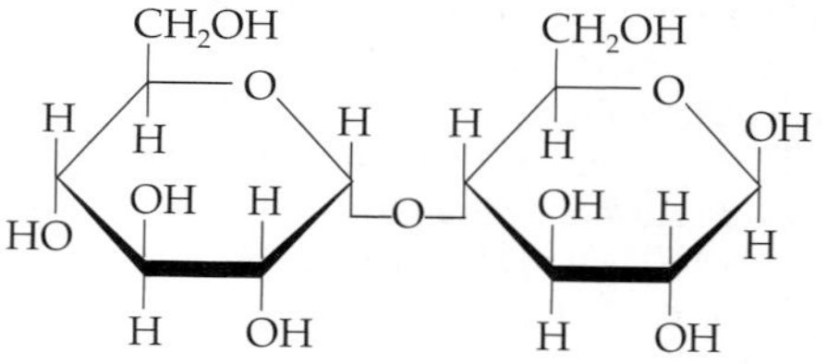

β-Maltose

16.73 Milk

16.75 Eliminating milk and milk products from the diet

16.77 Lactose intolerance is the inability to produce the enzyme lactase that hydrolyzes the milk sugar lactose into its component monosaccharides, glucose and galactose.

16.79 A polymer is a very large molecule formed by the combination of many small molecules, called monomers.
16.81 Starch
16.83 The glucose units of amylose are joined by α (1 → 4) glycosidic bonds and those of cellulose are bonded together by β (1 → 4) glycosidic bonds.
16.85 Glycogen serves as a storage molecule for glucose.
16.87 The salivary glands and the pancreas

Chapter 17

17.1
a. $CH_3(CH_2)_7CH=CH(CH_2)_7COOH$
b. $CH_3(CH_2)_{10}COOH$
c. $CH_3(CH_2)_4CH=CH-CH_2-CH=CH(CH_2)_7COOH$
d. $CH_3(CH_2)_{16}COOH$

17.3 $CH_3(CH_2)_{10}COOH + CH_3CH_2OH \xrightarrow{H^+, \text{heat}}$
Lauric acid   Ethanol
Dodecanoic acid

$CH_3(CH_2)_{10}-\overset{O}{\underset{\|}{C}}-OCH_2CH_3 + H_2O$
Ethyl dodecanoate

17.5
$CH_3CH_2-\overset{O}{\underset{\|}{C}}-OCH_2CH_2CH_2CH_3 + H_2O \xrightarrow{H^+, \text{heat}}$
Butyl propanoate
$CH_3CH_2CH_2CH_2OH + CH_3CH_2COOH$
1-Butanol   Propanoic acid

17.7
$CH_3CH_2-\overset{O}{\underset{\|}{C}}-OCH_2CH_2CH_2CH_3 + KOH \longrightarrow$
Butyl propanoate
$CH_3CH_2COO^-K^+ + CH_3CH_2CH_2CH_2OH$
Potassium propanoate   1-Butanol

17.9 $CH_3(CH_2)_5CH=CH(CH_2)_7COOH + H_2 \xrightarrow{Ni}$
cis-9-hexadecenoic acid
$CH_3(CH_2)_{14}COOH$
Hexadecanoic acid

17.11
H—C—OH
|
H—C—OH + 2 $CH_3(CH_2)_{16}COOH \longrightarrow$
|
H—C—OH
|
H

$\begin{array}{l} H-C-O-C-(CH_2)_{16}CH_3 \\ H-C-O-C-(CH_2)_{16}CH_3 \\ H-C-OH \end{array}$

17.13
a. $CH_3(CH_2)_7CH=CH(CH_2)_7-\overset{O}{\underset{\|}{C}}-O-CH_2$
    CH—OH
    CH_2—OH

$CH_3(CH_2)_7CH=CH(CH_2)_7-\overset{O}{\underset{\|}{C}}-O-CH_2$
$CH_3(CH_2)_7CH=CH(CH_2)_7-\overset{O}{\underset{\|}{C}}-O-CH$
    CH_2—OH

$CH_3(CH_2)_7CH=CH(CH_2)_7-\overset{O}{\underset{\|}{C}}-O-CH_2$
$CH_3(CH_2)_7CH=CH(CH_2)_7-\overset{O}{\underset{\|}{C}}-O-CH$
$CH_3(CH_2)_7CH=CH(CH_2)_7-\overset{O}{\underset{\|}{C}}-O-CH_2$

b. $CH_3(CH_2)_8-\overset{O}{\underset{\|}{C}}-O-CH_2$
    CH—OH
    CH_2—OH

$CH_3(CH_2)_8-\overset{O}{\underset{\|}{C}}-O-CH_2$
$CH_3(CH_2)_8-\overset{O}{\underset{\|}{C}}-O-CH$
    CH_2—OH

$CH_3(CH_2)_8-\overset{O}{\underset{\|}{C}}-O-CH_2$
$CH_3(CH_2)_8-\overset{O}{\underset{\|}{C}}-O-CH$
$CH_3(CH_2)_8-\overset{O}{\underset{\|}{C}}-O-CH_2$

17.15 Steroid nucleus (rings A, B, C, D with numbered carbons 1–17)

17.17 Receptor-mediated endocytosis

17.19 Membrane transport resembles enzyme catalysis because both processes exhibit a high degree of specificity.

17.21 Fatty acids, glycerides, nonglyceride lipids, and complex lipids

17.23 Lipid-soluble vitamins are transported into cells of the small intestine in association with dietary fat molecules. Thus, a diet low in fat reduces the amount of vitamins A, D, E, and K that enters the body.

17.25 A saturated fatty acid is one in which the hydrocarbon tail has only carbon-to-carbon single bonds. An unsaturated fatty acid has at least one carbon-to-carbon double bond.

17.27 The melting points increase.

17.29 The melting points of fatty acids increase as the length of the hydrocarbon chains increase. This is because the intermolecular attractive forces, including van der Waals forces, increase as the length of the hydrocarbon chain increases.

17.31 a. Decanoic acid
$CH_3(CH_2)_8COOH$
b. Stearic acid
$CH_3(CH_2)_{16}COOH$

17.33 a. I.U.P.A.C. name: Hexadecanoic acid
Common name: Palmitic acid
b. I.U.P.A.C. name: Dodecanoic acid
Common name: Lauric acid

17.35 a. Glycerol + 3 $CH_3(CH_2)_{12}COOH$ → tristearate triglyceride (tri-myristate) + 3 H_2O

b. Glycerol + 3 $CH_3(CH_2)_{16}COOH$ → tristearate triglyceride + 3 H_2O; reverse hydrolysis yields $3 CH_3(CH_2)_{16}C(O)OH$ + glycerol

c. $CH_3CH_2CH_2CH_2CH_2CH_2CH_2CH_2CH_2—C(=O)—OH$
$\xrightarrow{KOH}$
$CH_3CH_2CH_2CH_2CH_2CH_2CH_2CH_2CH_2—C(=O)—O^- K^+ + H_2O$

d. $CH_3(CH_2)_4CH=CHCH_2CH=CH(CH_2)_7—C(=O)—OH + 2H_2$
$\xrightarrow{Ni}$
$CH_3(CH_2)_{16}—C(=O)—OH$

17.37 The essential fatty acid linoleic acid is required for the synthesis of arachidonic acid, a precursor for the synthesis of the prostaglandins, a group of hormonelike molecules.

17.39 Aspirin effectively decreases the inflammatory response by inhibiting the synthesis of all prostaglandins. Aspirin works by inhibiting cyclooxygenase, the first enzyme in prostaglandin biosynthesis. This inhibition results from the transfer of an acetyl group from aspirin to the enzyme. Because cyclooxygenase is found in all cells, synthesis of all prostaglandins is inhibited.

17.41 Smooth muscle contraction, enhancement of fever and swelling associated with the inflammatory response, bronchial dilation, inhibition of secretion of acid into the stomach

17.43 A glyceride is a lipid ester that contains the glycerol molecule and from 1 to 3 fatty acids.

17.45 An emulsifying agent is a molecule that aids in the suspension of triglycerides in water. They are amphipathic molecules, such as lecithin, that serve as bridges holding together the highly polar water molecules and the nonpolar triglycerides.

17.47 A triglyceride with three saturated fatty acid tails would be a solid at room temperature. The long, straight fatty acid tails would stack with one another because of strong intermolecular and intramolecular attractions.

17.49 Triglyceride with palmitic acid at position 1 and two oleic (cis-9) acyl chains at positions 2 and 3.

17.51 Phosphatidyl diester: positions 1 and 2 are decanoyl ($CH_3CH_2CH_2CH_2CH_2CH_2CH_2CH_2CH_2—C(=O)—O—$) esters of glycerol, position 3 is $CH_2—O—P(=O)(O^-)—O^-$.

17.53 Triglycerides consist of three fatty acids esterified to the three hydroxyl groups of glycerol. In phospholipids there are only two fatty acids esterified to glycerol. A phosphoryl group is esterified (phosphoester linkage) to the third hydroxyl group.

17.55 A sphingolipid is a lipid that is not derived from glycerol, but rather from sphingosine, a long-chain, nitrogen-containing (amino) alcohol. Like phospholipids, sphingolipids are amphipathic.

17.57 A glycosphingolipid or glycolipid is a lipid that is built on a ceramide backbone structure. Ceramide is a fatty acid derivative of sphingosine.

17.59 Sphingomyelins are important structural lipid components of nerve cell membranes. They are found in the myelin sheath that surrounds and insulates cells of the central nervous system.

17.61 Cholesterol is readily soluble in the hydrophobic region of biological membranes. It is involved in regulating the fluidity of the membrane.

17.63 Progesterone is the most important hormone associated with pregnancy. Testosterone is needed for development of male secondary sexual characteristics. Estrone is required for proper development of female secondary sexual characteristics.

17.65 Cortisone is used to treat rheumatoid arthritis, asthma, gastrointestinal disorders, and many skin conditions.

17.67 Myricyl palmitate (beeswax) is made up of the fatty acid palmitic acid and the alcohol myricyl alcohol—$CH_3(CH_2)_{28}CH_2OH$.

17.69 Isoprenoids are a large, diverse collection of lipids that are synthesized from the isoprene unit:

$$CH_2=C(CH_3)-CH=CH_2$$

17.71 Steroids and bile salts, lipid-soluble vitamins, certain plant hormones, and chlorophyll

17.73 Chylomicrons, high-density lipoproteins, low-density lipoproteins, and very low density lipoproteins

17.75 The terms "good" and "bad" cholesterol refer to two classes of lipoprotein complexes. The high density lipoproproteins, or HDL, are considered to be "good" cholesterol because a correlation has been made between elevated levels of HDL and a reduced incidence of atherosclerosis. Low density lipoproteins, or LDL, are considered to be "bad" cholesterol because evidence suggests that high levels of LDL is associated with increased risk of atherosclerosis.

17.77 Atherosclerosis results when cholesterol and other substances coat the arteries causing a narrowing of the passageways. As the passageways become narrower, greater pressure is required to provide adequate blood flow. This results in higher blood pressure (hypertension).

17.79 If the LDL receptor is defective, it cannot function to remove cholesterol-bearing LDL particles from the blood. The excess cholesterol, along with other substances, will accumulate along the walls of the arteries, causing atherosclerosis.

17.81 The basic structure of a biological membrane is a bilayer of phospholipid molecules arranged so that the hydrophobic hydrocarbon tails are packed in the center and the hydrophilic head groups are exposed on the inner and outer surfaces.

17.83 A peripheral membrane protein is bound to only one surface of the membrane, either inside or outside the cell.

17.85 Cholesterol is freely soluble in the hydrophobic layer of a biological membrane. It moderates the fluidity of the membrane by disrupting the stacking of the fatty acid tails of membrane phospholipids.

17.87 L. Frye and M. Edidin carried out studies in which specific membrane proteins on human and mouse cells were labeled with red and green fluorescent dyes, respectively. The human and mouse cells were fused into single-celled hybrids and were observed using a microscope with an ultraviolet light source. The ultraviolet light caused the dyes to fluoresce. Initially the dyes were localized in regions of the membrane representing the original human or mouse cell. Within an hour, the proteins were evenly distributed throughout the membrane of the fused cell.

17.89 If the fatty acyl tails of membrane phospholipids are converted from saturated to unsaturated, the fluidity of the membrane will increase.

17.91 In simple diffusion the molecule moves directly across the membrane, whereas in facilitated diffusion a protein channel through the membrane is required.

17.93 Active transport requires an energy input to transport molecules or ions against the gradient (from an area of lower concentration to an area of higher concentration). Facilitated diffusion is a means of passive transport in which molecules or ions pass from regions of higher concentration to regions of lower concentration through a permease protein. No energy is expended by the cell in facilitated diffusion.

17.95 An antiport transport mechanism is one in which one molecule or ion is transported into the cell while a different molecule or ion is transported out of the cell.

17.97 Each permease or channel protein has a binding site that has a shape and charge distribution that is complementary to the molecule or ion that it can bind and transport across the cell membrane.

17.99 One ATP molecule is hydrolyzed to transport 3 Na^+ out of the cell and 2 K^+ into the cell.

Chapter 18

18.1 **a.** Glycine (gly):

$$\begin{array}{c} COO^- \\ | \\ H_3{}^+N-C-H \\ | \\ H \end{array}$$

b. Proline (pro):

$$\begin{array}{c} COO^- \\ H_2{}^+N-CH \\ | \quad\quad | \\ H_2C\diagdown_{C}\diagup CH_2 \\ H_2 \end{array}$$

c. Threonine (thr):

$$\begin{array}{c} COO^- \\ | \\ H_3{}^+N-C-H \\ | \\ H-C-OH \\ | \\ CH_3 \end{array}$$

d. Aspartate (asp):

$$\begin{array}{c} COO^- \\ | \\ H_3{}^+N-C-H \\ | \\ H-C-H \\ | \\ COO^- \end{array}$$

e. Lysine (lys):

$$\begin{array}{c} COO^- \\ | \\ H_3{}^+N-C-H \\ | \\ H-C-H \\ | \\ H-C-H \\ | \\ H-C-H \\ | \\ H-C-H \\ | \\ N^+H_3 \end{array}$$

18.3 a. Alanyl-phenylalanine:

Structure: $H_3{}^+N-CH(CH_3)-C(=O)-NH-CH(CH_2C_6H_5)-COO^-$

b. Lysyl-alanine:

Structure: $H_3{}^+N-CH(CH_2CH_2CH_2CH_2N^+H_3)-C(=O)-NH-CH(CH_3)-COO^-$

c. Phenylalanyl-tyrosyl-leucine:

Structure: $H_3{}^+N-CH(CH_2C_6H_5)-C(=O)-NH-CH(CH_2-C_6H_4-OH)-C(=O)-NH-CH(CH_2CH(CH_3)_2)-COO^-$

18.5 The primary structure of a protein is the amino acid sequence of the protein chain. Regular, repeating folding of the peptide chain caused by hydrogen bonding between the amide nitrogens and carbonyl oxygens of the peptide bond is the secondary structure of a protein. The two most common types of secondary structure are the α-helix and the β-pleated sheet. Tertiary structure is the further folding of the regions of α-helix and β-pleated sheet into a compact, spherical structure. Formation and maintenance of the tertiary structure results from weak attractions between amino acid R groups. The binding of two or more peptides to produce a functional protein defines the quaternary structure.

18.7 Oxygen is efficiently transferred from hemoglobin to myoglobin in the muscle because myoglobin has a greater affinity for oxygen.

18.9 High temperature disrupts the hydrogen bonds and other weak interactions that maintain protein structure.

18.11 Vegetables vary in amino acid composition. No single vegetable can provide all of the amino acid requirements of the body. By eating a variety of different vegetables, all the amino acid requirements of the human body can be met.

18.13 An enzyme is a protein that serves as a biological catalyst, speeding up biological reactions.

18.15 A transport protein is a protein that transports materials across the cell membrane or throughout the body.

18.17 Enzymes speed up reactions that might take days or weeks to occur on their own. They also catalyze reactions that might require very high temperatures or harsh conditions if carried out in the laboratory. In the body, these reactions occur quickly under physiological conditions.

18.19 Transferrin is a transport protein that carries iron from the liver to the bone marrow, where it is used to produce the heme group for hemoglobin and myoglobin. Hemoglobin transports oxygen in the blood.

18.21 Egg albumin is a nutrient protein that serves as a source of protein for the developing chick. Casein is the nutrient storage protein in milk, providing protein, a source of amino acids, for mammals.

18.23 The general structure of an L-α-amino acid:

$$\begin{array}{c} COO^- \\ | \\ H_3{}^+N-C-H \\ | \\ R \end{array}$$

18.25 A zwitterion is a neutral molecule with equal numbers of positive and negative charges. Under physiological conditions, amino acids are zwitterions.

18.27 A chiral carbon is one that has four different atoms or groups of atoms attached to it.

18.29 Interactions between the R groups of the amino acids in a polypeptide chain are important for the formation and maintenance of the tertiary and quaternary structures of proteins.

18.31 Glycine, Alanine, Valine, Leucine, Isoleucine, Phenylalanine, Proline, Tryptophan, Methionine

(Structures of each amino acid shown with standard zwitterionic form $H_3{}^+N-CH(R)-COO^-$ where R is the appropriate side chain.)

18.33 A peptide bond is an amide bond between two amino acids in a peptide chain.

18.35 Linus Pauling and his colleagues carried out X-ray diffraction studies of protein. Interpretation of the the pattern formed when X-rays were diffracted by a crystal of pure protein led Pauling to conclude that peptide bonds are both planar (flat) and rigid and that the N-C bonds are shorter that expected. In other words, they deduced that the peptide bond has a partially double bond character because it exhibits resonance. There is no free rotation about the amide bond because the carbonyl group of the amide bond has a strong attraction for the amide nitrogen lone pair of electrons. This can best be described using a resonance model:

$$\left[\begin{array}{c} \ddot{O} \\ \| \\ R-C-N-R' \\ | \\ H \end{array} \longleftrightarrow \begin{array}{c} :\ddot{O}:^- \\ | \\ R-C=N^+-R' \\ | \\ H \end{array} \right]$$

The partially double bonded character of the resonance structure restricts free rotation.

18.37 a. His-trp-cys:

$$H_3^+N-\underset{\underset{\underset{\underset{N}{\overset{|}{\underset{H}{N}}}}{\overset{|}{N}}}{\overset{|}{CH_2}}}{\overset{H}{\underset{|}{C}}}-\overset{O}{\underset{\|}{C}}-\underset{H}{\overset{|}{N}}-\underset{\underset{\underset{NH}{\overset{|}{}}}{\overset{|}{CH_2}}}{\overset{H}{\underset{|}{C}}}-\overset{O}{\underset{\|}{C}}-\underset{H}{\overset{|}{N}}-\underset{\underset{SH}{\overset{|}{CH_2}}}{\overset{H}{\underset{|}{C}}}-COO^-$$

b. Gly-leu-ser:

$$H_3^+N-\underset{H}{\overset{H}{\underset{|}{C}}}-\overset{O}{\underset{\|}{C}}-\underset{H}{\overset{|}{N}}-\underset{\underset{\underset{CH_3}{\overset{|}{CH_3}}}{\overset{|}{H-C-CH_3}}}{\overset{H}{\underset{|}{C}}}-\overset{O}{\underset{\|}{C}}-\underset{H}{\overset{|}{N}}-\underset{\underset{H}{\overset{|}{H-C-OH}}}{\overset{H}{\underset{|}{C}}}-COO^-$$

c. Arg-ile-val:

$$H_3^+N-\underset{\underset{\underset{\underset{\underset{NH_2}{\overset{|}{C=N^+H_2}}}{\overset{|}{NH}}}{\overset{|}{CH_2}}}{\overset{|}{CH_2}}}{\overset{H}{\underset{|}{C}}}-\overset{O}{\underset{\|}{C}}-\underset{H}{\overset{|}{N}}-\underset{\underset{\underset{CH_3}{\overset{|}{CH_2}}}{\overset{|}{CHCH_3}}}{\overset{H}{\underset{|}{C}}}-\overset{O}{\underset{\|}{C}}-\underset{H}{\overset{|}{N}}-\underset{\underset{CH_3}{\overset{|}{CHCH_3}}}{\overset{H}{\underset{|}{C}}}-COO^-$$

18.39 The primary structure of a protein is the sequence of amino acids bonded to one another by peptide bonds.

18.41 The primary structure of a protein determines its three dimensional shape because the location of R groups along the protein chain is determined by the primary structure. The interactions among the R groups, based on their location in the chain, will govern how the protein folds. This, in turn, dictates its three-dimensional structure and biological function.

18.43 The genetic information in the DNA dictates the order in which amino acids will be added to the protein chain. The order of the amino acids is the primary structure of the protein.

18.45 The secondary structure of a protein is the folding of the primary structure into an α-helix or β-pleated sheet.

18.47 a. α-Helix
b. β-Pleated sheet

18.49 A fibrous protein is one that is composed of peptides arranged in long sheets or fibers.

18.51 A parallel β-pleated sheet is one in which the hydrogen bonded peptide chains have their amino-termini aligned head-to-head.

18.53 The tertiary structure of a protein is the globular, three-dimensional structure of a protein that results from folding the regions of secondary structure.

18.55

$$H_3^+N-\underset{H}{\overset{H}{\underset{|}{C}}}-COO^-$$
$$|$$
$$CH_2$$
$$|$$
$$S$$
$$|$$
$$S$$
$$|$$
$$CH_2$$
$$|$$
$$H_3^+N-\underset{H}{\overset{|}{\underset{|}{C}}}-COO^-$$

18.57 The tertiary structure is a level of folding of a protein chain that has already undergone secondary folding. The regions of α-helix and β-pleated sheet are folded into a globular structure.

18.59 Quaternary protein structure is the aggregation of two or more folded peptide chains to produce a functional protein.

18.61 A glycoprotein is a protein with covalently attached sugars.

18.63 Hydrogen bonding maintains the secondary structure of a protein and contributes to the stability of the tertiary and quaternary levels of structure.

18.65 The peptide bond exhibits resonance, which results in a partially double bonded character. This causes the rigidity of the peptide bond.

$$\begin{array}{c} -C \\ \diagdown \\ C-N \\ \diagup \quad \diagdown \\ O \quad\quad C- \\ \quad\quad H \end{array} \longleftrightarrow \begin{array}{c} -C \\ \diagdown \\ C=N^+ \\ \diagup \quad \diagdown \\ ^-O \quad\quad C- \\ \quad\quad H \end{array}$$

18.67 The code for the primary structure of a protein is carried in the genetic information (DNA).

18.69 The function of hemoglobin is to carry oxygen from the lungs to oxygen-demanding tissues throughout the body. Hemoglobin is found in red blood cells.

18.71 Hemoglobin is a protein composed of four subunits—two α-globin and two β-globin subunits. Each subunit holds a heme group, which in turn carries an Fe^{2+} ion.

18.73 The function of the heme group in hemoglobin and myoglobin is to bind to molecular oxygen.

18.75 Because carbon monoxide binds tightly to the heme groups of hemoglobin, it is not easily removed or replaced by oxygen. As a result, the effects of oxygen deprivation (suffocation) occur.

18.77 When sickle cell hemoglobin (HbS) is deoxygenated, the amino acid valine fits into a hydrophobic pocket on the surface of another HbS molecule. Many such sickle cell hemoglobin molecules polymerize into long rods that cause the red blood cell to sickle. In normal hemoglobin, glutamic acid is found in the place of the valine. This negatively charged amino acid will not "fit" into the hydrophobic pocket.

18.79 When individuals have one copy of the sickle cell gene and one copy of the normal gene, they are said to carry the *sickle cell trait*. These individuals will not suffer serious side effects, but may pass the trait to their offspring. Individuals with two copies of the sickle cell globin gene exhibit all the symptoms of the disease and are said to have *sickle cell anemia*.

18.81 *Denaturation* is the process by which the organized structure of a protein is disrupted, resulting in a completely disorganized, nonfunctional form of the protein.

18.83 Heat is an effective means of sterilization because it destroys the proteins of microbial life-forms, including fungi, bacteria, and viruses.

18.85 Even relatively small fluctuations in blood pH can be life threatening. It is likely that these small changes would alter the normal charges on the proteins and modify their interactions. These changes can render a protein incapable of carrying out its functions.

18.87 Proteins become polycations at low pH because the additional protons will protonate the carboxylate groups. As these negative charges are neutralized, the charge on the proteins will be contributed only by the protonated amino groups ($-N^+H_3$).

18.89 The low pH of the yogurt denatures the proteins of microbial contaminants, inhibiting their growth.

18.91 An essential amino acid is one that must be provided in the diet because it cannot be synthesized in the body.

18.93 A complete protein is one that contains all of the essential and nonessential amino acids.

18.95 Chymotrypsin catalyzes the hydrolysis of peptide bonds on the carbonyl side of aromatic amino acids.

18.97 In a vegetarian diet, vegetables are the only source of dietary protein. Because individual vegetable sources do not provide all the needed amino acids, vegetables must be mixed to provide all the essential and nonessential amino acids in the amounts required for biosynthesis.

18.99 Synthesis of digestive enzymes must be carefully controlled because the active enzyme would digest, and thus destroy, the cell that produces it.

Chapter 19

19.1 a. Transferase
 b. Transferase
 c. Isomerase
 d. Oxidoreductase
 e. Hydrolase

19.3 a. Pyruvate kinase catalyzes the transfer of a phosphoryl group from phosphoenolpyruvate to adenosine diphosphate.

 b. Alanine transaminase catalyzes the transfer of an amino group from alanine to α-ketoglutarate, producing pyruvate and glutamate.

 c. Triose phosphate isomerase catalyzes the isomerization of the ketone dihydroxyacetone phosphate to the aldehyde glyceraldehyde-3-phosphate.

 d. Pyruvate dehydrogenase catalyzes the oxidation and decarboxylation of pyruvate, producing acetyl coenzyme A and CO_2.

19.5 a. Sucrose
 b. Pyruvate
 c. Succinate

19.7 The induced fit model assumes that the enzyme is flexible. Both the enzyme and the substrate are able to change shape to form the enzyme-substrate complex. The lock-and-key model assumes that the enzyme is inflexible (the lock) and the substrate (the key) fits into a specific rigid site (the active site) on the enzyme to form the enzyme-substrate complex.

19.9 An enzyme might distort a bond, thereby catalyzing bond breakage. An enzyme could bring two reactants into close proximity and in the proper orientation for the reaction to occur. Finally, an enzyme could alter the pH of the microenvironment of the active site, thereby serving as a transient donor or acceptor of H^+.

19.11 Water-soluble vitamins are required by the body for the synthesis of coenzymes that are required for the function of a variety of enzymes.

19.13 A decrease in pH will change the degree of ionization of the R groups within a peptide chain. This disturbs the weak interactions that maintain the structure of an enzyme, which may denature the enzyme. Less drastic alterations in the charge of

R groups in the active site of the enzyme can inhibit enzyme-substrate binding or destroy the catalytic ability of the active site.

19.15 Irreversible inhibitors bind very tightly, sometimes even covalently, to an R group in enzyme active sites. They generally inhibit many different enzymes. The loss of enzyme activity impairs normal cellular metabolism, resulting in death of the cell or the individual.

19.17 A structural analog is a molecule that has a structure and charge distribution very similar to that of the natural substrate of an enzyme. Generally they are able to bind to the enzyme active site. This inhibits enzyme activity because the normal substrate must compete with the structural analog to form an enzyme-substrate complex.

19.19 a.

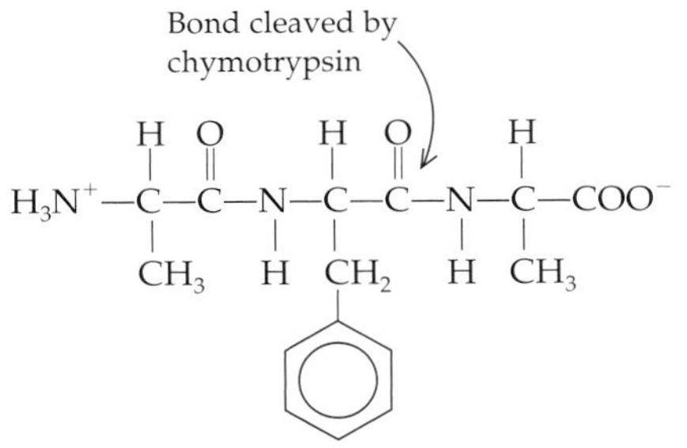

Bond cleaved by chymotrypsin

H_3N^+—C(H)(CH$_3$)—C(O)—N(H)—C(H)(CH$_2$Ph)—C(O)—N(H)—C(H)(CH$_3$)—COO$^-$

ala-phe-ala

b.

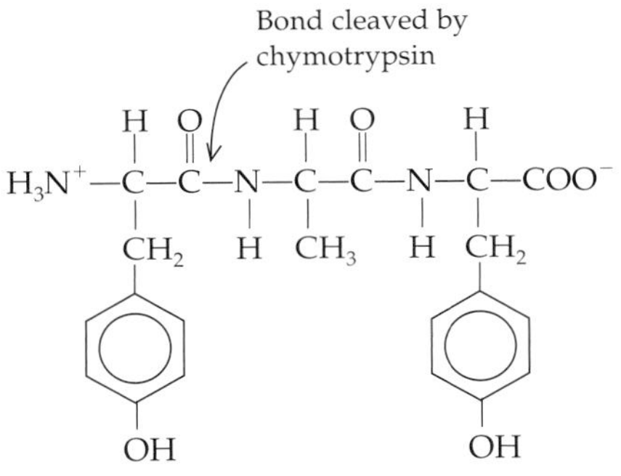

Bond cleaved by chymotrypsin

H_3N^+—C(H)(CH$_2$-C$_6$H$_4$-OH)—C(O)—N(H)—C(H)(CH$_3$)—C(O)—N(H)—C(H)(CH$_2$-C$_6$H$_4$-OH)—COO$^-$

tyr-ala-tyr

19.21 Chymotrypsin   Elastase   Elastase

H_3^+N—C(H)(CH(CH$_3$)$_2$)—C(O)—N(H)—C(H)(CH$_2$Ph)—C(O)—N(H)—C(H)(CH$_3$)—C(O)—N(H)—C(H)(H)—C(O)—N(H)—C(H)(CH$_2$CH(CH$_3$)$_2$)—COO$^-$

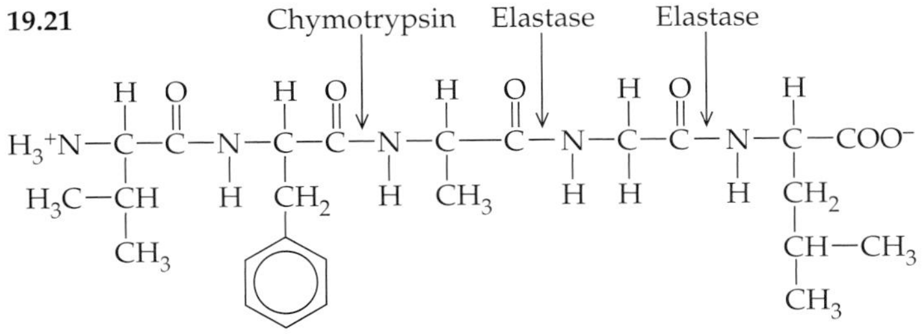

19.23 The common name of an enzyme is often derived from the name of the substrate and/or the type of reaction that it catalyzes.

19.25
1. Urease
2. Peroxidase
3. Lipase
4. Aspartase
5. Glucose-6-phosphatase
6. Sucrase

19.27 a. Citrate decarboxylase catalyzes the cleavage of a carboxyl group from citrate.
b. Adenosine diphosphate phosphorylase catalyzes the addition of a phosphate group to ADP.

c. Oxalate reductase catalyzes the reduction of oxalate.
d. Nitrite oxidase catalyzes the oxidation of nitrite.
e. *cis-trans* Isomerase catalyzes interconversion of *cis* and *trans* isomers.

19.29 A substrate is the reactant in an enzyme-catalyzed reaction that binds to the active site of the enzyme and is converted into product.

19.31 The activation energy of a reaction is the energy required for the reaction to occur.

19.33 The equilibrium constant for a chemical reaction is a reflection of the difference in energy of the reactants and products. Consider the following reaction:

$$aA + bB \rightarrow cC + dD$$

The equilibrium constant for this reaction is:

$$K_{eq} = [D]^d[C]^c/[A]^a[B]^b = [\text{products}]/[\text{reactants}]$$

Because the difference in energy between reactants and products is the same regardless of what path the reaction takes, an enzyme does not alter the equilibrium constant of a reaction.

19.35 The rate of an uncatalyzed chemical reaction typically doubles every time the substrate concentration is doubled.

19.37 The rate-limiting step is that step in an enzyme-catalyzed reaction that is the slowest, and hence limits the speed with which the substrate can be converted into product.

19.39

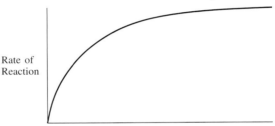

Rate of Reaction vs. Concentration of Substrate

19.41 The enzyme-substrate complex is the molecular aggregate formed when the substrate binds to the active site of an enzyme.

19.43 The catalytic groups of an enzyme active site are those functional groups that are involved in carrying out catalysis.

19.45 Enzyme active sites are pockets in the surface of an enzyme that include R groups involved in binding and R groups involved in catalysis. The shape of the active site is complementary to the shape of the substrate. Thus, the conformation of the active site determines the specificity of the enzyme. Enzyme-substrate binding involves weak, noncovalent interactions.

19.47 The lock-and-key model of enzyme-substrate binding was proposed by Emil Fischer in 1894. He thought that the active site was a rigid region of the enzyme into which the substrate fit perfectly. Thus, the model purports that the substrate simply snaps into place within the active site, like two pieces of a jigsaw puzzle fitting together.

19.49 Enzyme specificity is the ability of an enzyme to bind to only one, or a very few, substrates and thus catalyze only a single reaction.

19.51 Group specificity means that an enzyme catalyzes reactions involving similar molecules having the same functional group.

19.53 Absolute specificity means that an enzyme catalyzes the reaction of only one substrate.

19.55 Hexokinase has group specificity. The advantage is that the cell does not need to encode many enzymes to carry out the phosphorylation of six-carbon sugars. Hexokinase can carry out many of these reactions.

19.57 Methionyl tRNA synthetase has absolute specificity. This is the enzyme that attaches the amino acid methionine to the transfer RNA (tRNA) that will carry the amino acid to the site of protein synthesis. If the wrong amino acid were attached to the tRNA, it could be incorporated into the protein, destroying its correct three-dimensional structure and biological function.

19.59 The first step of an enzyme-catalyzed reaction is the formation of the enzyme-substrate complex. In the second step, the transition state is formed. This is the state in which the substrate assumes a form intermediate between the original substrate and the product. In step 3 the substrate is converted to product and the enzyme-product complex is formed. Step 4 involves the release of the product and regeneration of the enzyme in its original form.

19.61 In a reaction involving bond breaking, the enzyme might distort a bond, producing a transition state in which the bond is stressed. An enzyme could bring two reactants into close proximity and in the proper orientation for the reaction to occur, producing a transition state in which the proximity of the reactants facilitates bond formation. Finally, an enzyme could alter the pH of the microenvironment of the active site, thereby serving as a transient donor or acceptor of H^+.

19.63 A cofactor helps maintain the shape of the active site of an enzyme.

19.65 $NAD^+/NADH$ serves an acceptor/donor of hydride anions in biochemical reactions. $NAD^+/NADH$ serves as a coenzyme for oxidoreductases.

19.67 Changes in pH or temperature affect the activity of enzymes, as can changes in the concentration of substrate and the concentrations of certain ions.

19.69 Each of the following answers assumes that the enzyme was purified from an organism with optimal conditions for life near 37°C, pH 7.
  a. Decreasing the temperature from 37°C to 10°C will cause the rate of an enzyme-catalyzed reaction to decrease because the frequency of collisions between enzyme and substrate will decrease as the rate of molecular movement decreases.
  b. Increasing the pH from 7 to 11 will generally cause a decrease in the rate of an enzyme-catalyzed reaction. In fact, most enzymes would be denatured by a pH of 11 and enzyme activity would cease.
  c. Heating an enzyme from 37°C to 100°C will destroy enzyme activity because the enzyme would be denatured by the extreme heat.

19.71 High temperature denatures bacterial enzymes and structural proteins. Because the life of the cell is dependent on the function of these proteins, the cell dies.

19.73 A lysosome is a membrane-bound vesicle in the cytoplasm of cells that contains approximately fifty types of hydrolytic enzymes.

19.75 Enzymes used for clinical assays in hospitals are typically stored at refrigerator temperatures to ensure that they are not denatured by heat. In this way they retain their activity for long periods.

19.77 a. Cells regulate the level of enzyme activity to conserve energy. It is a waste of cellular energy to produce an enzyme if its substrate is not present or if its product is in excess.
  b. Production of proteolytic digestive enzymes must be carefully controlled because the active enzyme could destroy the cell that produces it. Thus, they are produced in an inactive form in the cell and are only activated at the site where they carry out digestion.

19.79 In positive allosterism, binding of the effector molecule turns the enzyme on. In negative allosterism, binding of the effector molecule turns the enzyme off.

19.81 A proenzyme is the inactive form of an enzyme that is converted to the active form at the site of its activity.

19.83 Blood clotting is a critical protective mechanism in the body, preventing excessive loss of blood following an injury. However, it can be a dangerous mechanism if it is triggered inappropriately. The resulting clot could cause a heart attack or stroke. By having a cascade of proteolytic reactions leading to the final formation of the clot, there are many steps at which the process can be regulated. This ensures that it will only be activated under the appropriate conditions.

19.85 *Competitive enzyme inhibition* occurs when a structural analog of the normal substrate occupies the enzyme active site so that the reaction cannot occur. The structural analog and the normal substrate compete for the active site. Thus, the rate of the reaction will depend on the relative concentrations of the two molecules.

19.87 A structural analog has a shape and charge distribution that are very similar to those of the normal substrate for an enzyme.

19.89 Irreversible inhibitors bind tightly to and block the active site of an enzyme and eliminate catalysis at the site.

19.91 The compound would be a competitive inhibitor of the enzyme.

19.93 The structural similarities among chymotrypsin, trypsin, and elastase suggest that these enzymes evolved from a single ancestral gene that was duplicated. Each copy then evolved independently.

19.95

Bond cleaved by chymotrypsin

$$H_3N^+-\underset{\underset{\underset{\underset{OH}{\bigcirc}}{CH_2}}{CH_2}}{\overset{H}{\underset{|}{C}}}-\overset{O}{\underset{\|}{C}}-\underset{H}{\overset{H}{\underset{|}{N}}}-\underset{\underset{\underset{\underset{N^+H_3}{CH_2}}{CH_2}}{CH_2}}{\overset{H}{\underset{|}{C}}}-\overset{O}{\underset{\|}{C}}-\underset{H}{\overset{H}{\underset{|}{N}}}-\underset{CH_3}{\overset{H}{\underset{|}{C}}}-\overset{O}{\underset{\|}{C}}-\underset{H}{\overset{H}{\underset{|}{N}}}-\underset{\underset{\bigcirc}{CH_2}}{\overset{H}{\underset{|}{C}}}-COO^-$$

tyr-lys-ala-phe

19.97 Elastase will cleave the peptide bonds on the carbonyl side of alanine and glycine. Trypsin will cleave the peptide bonds on the carbonyl side of lysine and arginine. Chymotrypsin will cleave the peptide bonds on the carbonyl side of tryptophan and phenylalanine.

19.99 Creatine phosphokinase (CPK), lactate dehydrogenase (LDH), and aspartate aminotransferase (AST/SGOT)

Chapter 20

20.1 a. Adenosine diphosphate:

[structure of adenosine diphosphate]

b. Deoxyguanosine triphosphate:

[structure of deoxyguanosine triphosphate]

20.3 The RNA polymerase recognizes the promoter site for a gene, separates the strands of DNA, and catalyzes the polymerization of an RNA strand complementary to the DNA strand that carries the genetic code for a protein. It recognizes a termination site at the end of the gene and releases the RNA molecule.

20.5 The genetic code is said to be degenerate because several different triplet codons may serve as code words for a single amino acid.

20.7 The nitrogenous bases of the codons are complementary to those of the anticodons. As a result they are able to hydrogen bond to one another according to the base pairing rules.

20.9 The ribosomal P-site holds the peptidyl tRNA during protein synthesis. The peptidyl tRNA is the tRNA carrying the growing peptide chain. The only exception to this is during initiation of translation when the P-site holds the initiator tRNA.

20.11 The normal mRNA sequence, AUG-CCC-GAC-UUU, would encode the peptide sequence methionine-proline-aspartate-phenylalanine. The mutant mRNA sequence, AUG-CGC-GAC-UUU, would encode the mutant peptide sequence methionine-arginine-aspartate-phenylalanine. This would not be a silent mutation because a hydrophobic amino acid (proline) has been replaced by a positively charged amino acid (arginine).

20.13 A heterocyclic amine is a compound that contains nitrogen in at least one position of the ring skeleton.

20.15 It is the N-9 of the purine that forms the *N*-glycosidic bond with C-1 of the five-carbon sugar. The general structure of the purine ring is shown below:

[purine ring structure with numbered positions]

20.17 The ATP nucleotide is composed of the five-carbon sugar ribose, the purine adenine, and a triphosphate group.

20.19 The two strands of DNA in the double helix are said to be *antiparallel* because they run in opposite directions. One strand progresses in the 5′ → 3′ direction, and the opposite strand progresses in the 3′ → 5′ direction.

20.21 The DNA double helix is 2 nm in width. The nitrogenous bases are stacked at a distance of 0.34 nm from one another. One complete turn of the helix is 3.4 nm, or 10 base pairs.

20.23 Two

20.25 [structure of dinucleotide with cytosine and thymine]

20.27 The prokaryotic chromosome is a circular DNA molecule that is supercoiled, that is, the helix is coiled on itself.

20.29 The term *semiconservative DNA replication* refers to the fact that each parental DNA strand serves as the template for the synthesis of a daughter strand. As a result, each of the daughter DNA molecules is made up of one strand of the original parental DNA and one strand of newly synthesized DNA.

20.31 The two primary functions of DNA polymerase III are to read a template DNA strand and catalyze the polymerization of a new daughter strand, and to proofread the newly synthesized strand and correct any errors by removing the incorrectly inserted nucleotide and adding the proper one.

20.33 3′-TACGCCGATCTTATAAGGT-5′

20.35 The *replication origin* of a DNA molecule is the unique sequence on the DNA molecule where DNA replication begins.

20.37 The enzyme helicase separates the strands of DNA at the origin of DNA replication so that the proteins involved in replication can interact with the nitrogenous base pairs.

20.39 The RNA primer "primes" DNA replication by providing a 3′-OH which can be used by DNA polymerase III for the addition of the next nucleotide in the growing DNA chain.

20.41 DNA → RNA → Protein

20.43 Anticodons are found on transfer RNA molecules.

20.45 3′-AUGGAUCGAGACCAGUAAUUCCGUCAU-5′.

20.47 *RNA splicing* is the process by which the noncoding sequences (introns) of the primary transcript of a eukaryotic mRNA are removed and the protein coding sequences (exons) are spliced together.

20.49 Messenger RNA, transfer RNA, and ribosomal RNA

20.51 Spliceosomes are small ribonucleoprotein complexes that carry out RNA splicing.

20.53 The *poly(A) tail* is a stretch of 100–200 adenosine nucleotides polymerized onto the 3′ end of a mRNA by the enzyme poly(A) polymerase.

20.55 The *cap structure* is made up of the nucleotide 7-methylguanosine attached to the 5′ end of a mRNA by a 5′-5′ triphosphate bridge. Generally the first two nucleotides of the mRNA are also methylated.

20.57 Sixty-four

20.59 The reading frame of a gene is the sequential set of triplet codons that carries the genetic code for the primary structure of a protein.

20.61 Methionine and tryptophan

20.63 The codon 5'-UUU-3' encodes the amino acid phenylalanine. The mutant codon 5'-UUA-3' encodes the amino acid leucine. Both leucine and phenylalanine are hydrophobic amino acids, however, leucine has a smaller R group. It is possible that the smaller R group would disrupt the structure of the protein.

20.65 The ribosomes serve as a platform on which protein synthesis can occur. They also carry the enzymatic activity that forms peptide bonds.

20.67 The sequence of DNA nucleotides in a gene is transcribed to produce a complementary sequence of RNA nucleotides in a messenger RNA (mRNA). In the process of translation the sequence of the mRNA is read sequentially in words of three nucleotides (codons) to produce a protein. Each codon calls for the addition of a particular amino acid to the growing peptide chain. Through these processes, the sequence of nucleotides in a gene determines the sequence of amino acids in the primary structure of a protein.

20.69 In the initiation of translation, initiation factors, methionyl tRNA (the initiator tRNA), the mRNA, and the small and large ribosomal subunits form the initiation complex. During the elongation stage of translation, an aminoacyl tRNA binds to the A-site of the ribosome. Peptidyl transferase catalyzes the formation of a peptide bond and the peptide chain is transferred to the tRNA in the A-site. Translocation shifts the peptidyl tRNA from the A-site into the P-site, leaving the A-site available for the next aminoacyl tRNA. In the termination stage of translation, a termination codon is encountered. A release factor binds to the empty A-site and peptidyl transferase catalyzes the hydrolysis of the bond between the peptidyl tRNA and the completed peptide chain.

20.71 An ester bond

20.73 A point mutation is the substitution of one nucleotide pair for another in a gene.

20.75 Some mutations are silent because the change in the nucleotide sequence does not alter the amino acid sequence of the protein. This can happen because there are many amino acids encoded by multiple codons.

20.77 UV light causes the formation of pyrimidine dimers, the covalent bonding of two adjacent pyrimidine bases. Mutations occur when the UV damage repair system makes an error during the repair process. This causes a change in the nucleotide sequence of the DNA.

20.79 a. A *carcinogen* is a compound that causes cancer. Cancers are caused by mutations in the genes responsible for controlling cell division.
b. Carcinogens cause DNA damage that results in changes in the nucleotide sequence of the gene. Thus, carcinogens are also mutagens.

20.81 A *restriction enzyme* is a bacterial enzyme that "cuts" the sugar–phosphate backbone of DNA molecules at a specific nucleotide sequence.

20.83 A selectable marker is a genetic trait that can be used to detect the presence of a plasmid in a bacterium. Many plasmids have antibiotic resistance genes as selectable markers. Bacteria containing the plasmid will be able to grow in the presence of the antibiotic; those without the plasmid will be killed.

20.85 Human insulin, interferon, human growth hormone, and human blood clotting factor VIII

20.87 1024 copies

20.89 The goals of the Human Genome Project are to identify and map all of the genes of the human genome and to determine the DNA sequences of the complete three billion nucleotide pairs.

20.91 A genome library is a set of clones that represents all of the DNA sequences in the genome of an organism.

20.93 A dideoxynucleotide is one that has hydrogen atoms rather than hydroxyl groups bonded to both the 2' and 3' carbons of the five-carbon sugar.

20.95 Sequences that these DNA sequences have in common are highlighted in bold.
a. 5' **AGCTCCT**GATTTCATACAGTTTCTACT**ACCTACTA** 3'
b. 5' AGACATTCTATCTACCTAGACTATG**TTCAGAA** 3'
c. 5' **TTCAGAA**CTCATTCAGACCTACTACTATACCTTGGG **AGCTCCT** 3'
d. 5' **ACCTACTA**GACTATACTACTACTAAGGGGACTATTC CAGACTT 3'

The 5' end of sequence (a) is identical to the 3' end of sequence (c). The 3' end of sequence (a) is identical to the 5' end of sequence (d). The 3' end of sequence (b) is identical to the 5' end of sequence (c). From 5' to 3', the sequences would form the following map:

Chapter 21

21.1 ATP is called the universal energy currency because it is the major molecule used by all organisms to store energy.

21.3 The first stage of catabolism is the digestion (hydrolysis) of dietary macromolecules in the stomach and intestine.
In the second stage of catabolism, monosaccharides, amino acids, fatty acids, and glycerol are converted by metabolic reactions into molecules that can be completely oxidized.
In the third stage of catabolism, the two-carbon acetyl group of acetyl CoA is completely oxidized by the reactions of the citric acid cycle. The energy of the electrons harvested in these oxidation reactions is used to make ATP.

21.5 Substrate level phosphorylation is one way the cell can make ATP. In this reaction, a high-energy phosphoryl group of a substrate in the reaction is transferred to ADP to produce ATP.

21.7 Glycolysis is a pathway involving ten reactions. In reactions 1–3, energy is invested in the beginning substrate, glucose. This is done by transferring high-energy phosphoryl groups from ATP to the intermediates in the pathway. The product is fructose-1,6-bisphosphate. In the energy-harvesting reactions of glycolysis, fructose-1,6-bisphosphate is split into two three-carbon molecules that begin a series of rearrangement, oxidation-reduction, and substrate-level phosphorylation reactions that produce four ATP, two NADH, and two pyruvate molecules. Because of the investment of two ATP in the early steps of glycolysis, the net yield of ATP is two.

21.9 Both the alcohol and lactate fermentations are anaerobic reactions that use the pyruvate and re-oxidize the NADH produced in glycolysis.

21.11 Gluconeogenesis (synthesis of glucose from noncarbohydrate sources) appears to be the reverse of glycolysis (the first stage of carbohydrate degradation) because the intermediates in the two pathways are the same. However, reactions 1, 3, and 10 of glycolysis are not reversible reactions. Thus, the reverse reactions must be carried out by different enzymes.

21.13 The enzyme glycogen phosphorylase catalyzes the phosphorolysis of a glucose unit at one end of a glycogen molecule. The reaction involves the displacement of the glucose by a phosphate group. The products are glucose-1-phosphate and a glycogen molecule that is one glucose unit shorter.

21.15 Glucokinase traps glucose within the liver cell by phosphorylating it. Because the product, glucose-6-phosphate, is charged, it cannot be exported from the cell.

21.17 Glucagon indirectly stimulates glycogen phosphorylase, the first enzyme of glycogenolysis. This speeds up glycogen degradation. Glucagon also inhibits glycogen synthase, the first enzyme in glycogenesis. This inhibits glycogen synthesis.

21.19 ATP

21.21 Adenosine triphosphate + H_2O ⟶ Adenosine diphosphate + Inorganic phosphate group

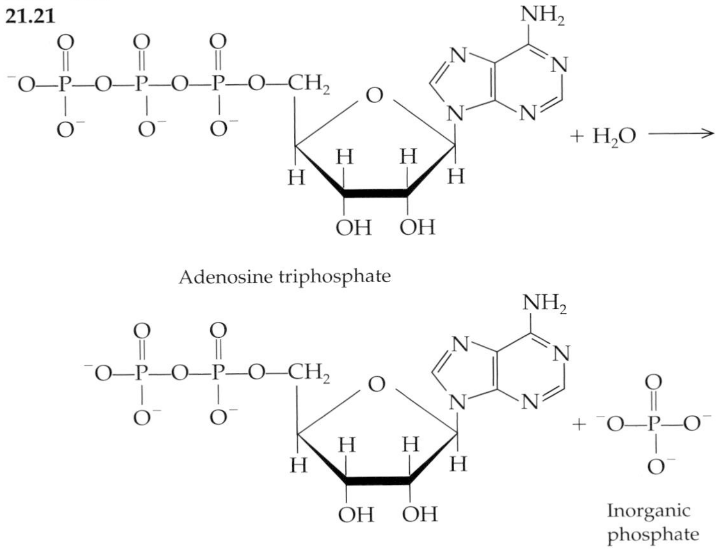

21.23 A coupled reaction is one that can be thought of as a two-step process. In a coupled reaction, two reactions occur simultaneously. Frequently one of the reactions releases the energy that drives the second, energy-requiring, reaction.

21.25 Carbohydrates

21.27 The following equation represents the hydrolysis of maltose:

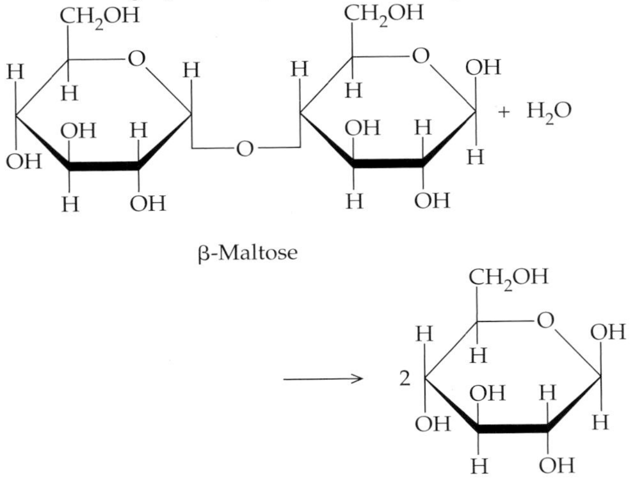

β-Maltose + H_2O ⟶ 2 β-D-Glucose

21.29 The following equation represents the hydrolysis of lactose:

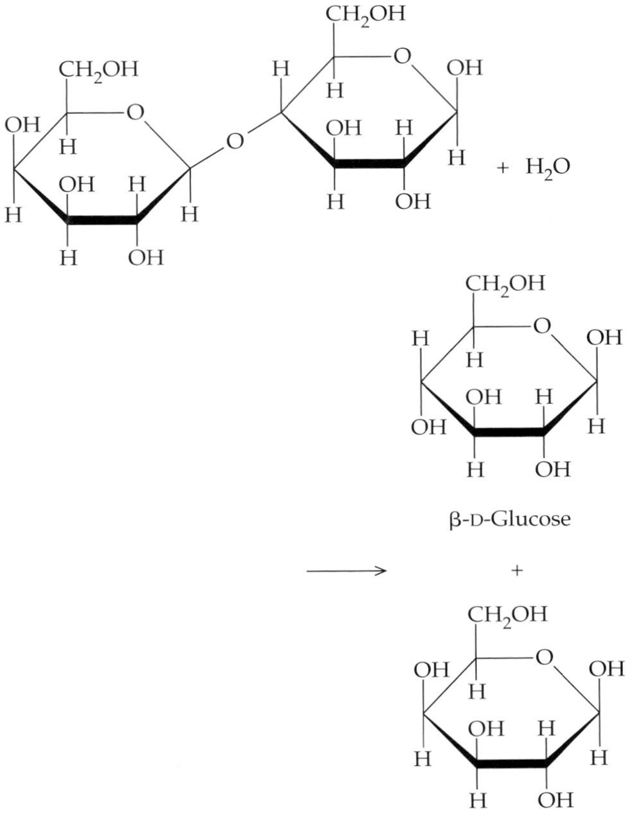

Lactose + H_2O ⟶ β-D-Glucose + β-D-Galactose

21.31 The hydrolysis of a triglyceride containing oleic acid, stearic acid, and linoleic acid is represented in the following equations:

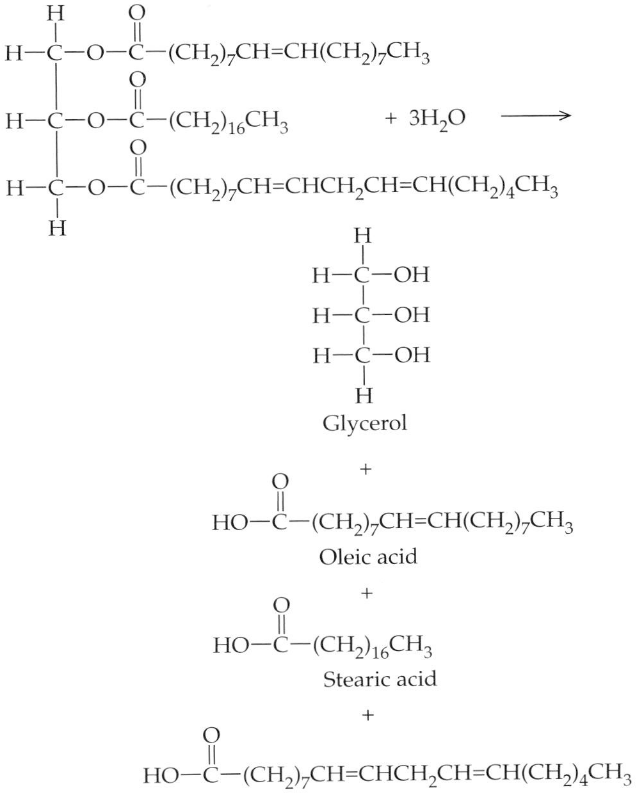

Triglyceride + $3H_2O$ ⟶ Glycerol + Oleic acid + Stearic acid + Linoleic acid

$$HO-\overset{O}{\underset{\|}{C}}-(CH_2)_7CH=CH(CH_2)_7CH_3 \quad \text{Oleic acid}$$

$$HO-\overset{O}{\underset{\|}{C}}-(CH_2)_{16}CH_3 \quad \text{Stearic acid}$$

$$HO-\overset{O}{\underset{\|}{C}}-(CH_2)_7CH=CHCH_2CH=CH(CH_2)_4CH_3 \quad \text{Linoleic acid}$$

21.33 The hydrolysis of the dipeptide alanyl leucine is represented in the following equation:

$$H_3{}^+N-CH(CH_3)-C(=O)-NH-CH(CH_2CH(CH_3)CH_3)-C(=O)-O^- + H_2O \longrightarrow$$

Alanyl leucine

$$H_3{}^+N-CH(CH_3)-C(=O)-O^-$$
Alanine

+

$$H_3{}^+N-CH(CH_2CH(CH_3)CH_3)-C(=O)-O^-$$
Leucine

21.35 Glycolysis is the enzymatic pathway that converts a glucose molecule into two molecules of pyruvate. The pathway generates a net energy yield of two ATP and two NADH. Glycolysis is the first stage of carbohydrate catabolism.

21.37 Glycolysis requires NAD^+ for reaction 6 in which glyceraldehyde-3-phosphate dehydrogenase catalyzes the oxidation of glyceraldehyde-3-phosphate. NAD^+ is reduced.

21.39 Two ATP per glucose

21.41 Although muscle cells have enough ATP stored for only a few seconds of activity, glycolysis speeds up dramatically when there is a demand for more energy. If the cells have a sufficient supply of oxygen, aerobic respiration (the citric acid cycle and oxidative phosphorylation) will contribute large amounts of ATP. If oxygen is limited, the lactate fermentation will speed up. This will use up the pyruvate and re-oxidize the NADH produced by glycolysis and allow continued synthesis of ATP for muscle contraction.

21.43 $C_6H_{12}O_6 + 2ADP + 2P_i + 2NAD^+ \rightarrow$
Glucose
$$2C_3H_3O_3 + 2ATP + 2NADH + 2H_2O$$
Pyruvate

21.45 a. Hexokinase catalyzes the phosphorylation of glucose.
b. Pyruvate kinase catalyzes the transfer of a phosphoryl group from phosphoenolpyruvate to ADP.
c. Phosphoglycerate mutase catalyzes the isomerization reaction that converts 3-phosphoglycerate to 2-phosphoglycerate.
d. Glyceraldehyde-3-phosphate dehydrogenase catalyzes the oxidation and phosphorylation of glyceraldehyde-3-phosphate and the reduction of NAD^+ to NADH.

21.47 Isomerase

21.49 Enediol

21.51 A kinase transfers a phosphoryl group from one molecule to another

21.53 NAD^+ is reduced, accepting a hydride anion.

21.55 Enolase catalyzes a reaction that produces an enol; in this particular reaction, it is phosphoenolpyruvate.

21.57 To optimize efficiency and minimize waste, it is important that energy-harvesting pathways, such as glycolysis, respond to the energy demands of the cell. If energy in the form of ATP is abundant, there is no need for the pathway to continue at a rapid rate. When this is the case, allosteric enzymes that catalyze the reactions of the pathway are inhibited by binding to their negative effectors. Similarly, when there is a great demand for ATP, the pathway speeds up as a result of the action of allosteric enzymes binding to positive effectors.

21.59 ATP and citrate are allosteric inhibitors of phosphofructokinase, whereas AMP and ADP are allosteric activators.

21.61 Citrate, which is the first intermediate in the citric acid cycle, is an allosteric inhibitor of phosphofructokinase. The citric acid cycle is a pathway that results in the complete oxidation of the pyruvate produced by glycolysis. A high concentration of citrate signals that sufficient substrate is entering the citric acid cycle. The inhibition of phosphofructokinase by citrate is an example of *feedback inhibition*: the product, citrate, allosterically inhibits the activity of an enzyme early in the pathway.

21.63

$$CH_2=C(H)-H \xrightarrow[\text{NAD}^+]{\text{NADH}} CH_3CH_2OH$$

Acetaldehyde → Ethanol

21.65 The lactate fermentation

21.67 Yogurt and some cheeses

21.69 Lactate dehydrogenase

21.71 This child must have the enzymes to carry out the alcohol fermentation. When the child exercised hard, there was not enough oxygen in the cells to maintain aerobic respiration. As a result, glycolysis and the alcohol fermentation were responsible for the majority of the ATP production by the child. The accumulation of alcohol (ethanol) in the child caused the symptoms of drunkenness.

21.73 The first stage of the pentose phosphate pathway is an oxidative stage in which glucose-6-phosphate is converted to ribulose-5-phosphate. Two NADPH molecules and one CO_2 molecule are also produced in these reactions. The second stage of the pentose phosphate pathway involves isomerization reactions that convert ribulose-5-phosphate into other five-carbon sugars, ribose-5-phosphate and xylulose-5-phosphate. The third stage of the pathway involves a complex series of rearrangement reactions that results in the production of two fructose-6-phosphate and one glyceraldehyde-3-phosphate molecules from three molecules of pentose phosphate.

21.75 The ribose-5-phosphate is used for the biosynthesis of nucleotides. The erythrose-4-phosphate is used for the biosynthesis of aromatic amino acids.

21.77 Gluconeogenesis is production of glucose from noncarbohydrate starting materials. This pathway can provide glucose when starvation or strenuous exercise leads to a depletion of glucose from the body.

21.79 The liver

21.81 Lactate is first converted to pyruvate.

21.83 Because steps 1, 3, and 10 of glycolysis are irreversible, gluconeogenesis is not simply the reverse of glycolysis. The reverse reactions must be carried out by different enzymes.

21.85 Steps 1, 3, and 10 of glycolysis are irreversible. Step 1 is the transfer of a phosphoryl group from ATP to carbon-6 of glucose and is catalyzed by hexokinase. Step 3 is the transfer of a phosphoryl group from ATP to carbon-1 of fructose-6-

phosphate and is catalyzed by phosphofructokinase. Step 10 is the substrate-level phosphorylation in which a phosphoryl group is transferred from phosphoenolpyruvate to ADP and is catalyzed by pyruvate kinase.

21.87 The liver and pancreas

21.89 *Hypoglycemia* is the condition in which blood glucose levels are too low.

21.91 **a.** Insulin stimulates glycogen synthase, the first enzyme in glycogen synthesis. It also stimulates uptake of glucose from the bloodstream into cells and phosphorylation of glucose by the enzyme glucokinase.
b. This traps glucose within liver cells and increases the storage of glucose in the form of glycogen.
c. These processes decrease blood glucose levels.

21.93 Any defect in the enzymes required to degrade glycogen or export glucose from liver cells will result in a reduced ability of the liver to provide glucose at times when blood glucose levels are low. This will cause hypoglycemia.

Chapter 22

22.1 Mitochondria are the organelles responsible for aerobic respiration.

22.3

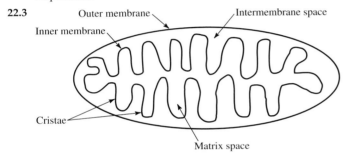

22.5 Pyruvate is converted to acetyl CoA by the pyruvate dehydrogenase complex. This huge enzyme complex requires four coenzymes, each of which is made from a different vitamin. The four coenzymes are thiamine pyrophosphate (made from thiamine), FAD (made from riboflavin), NAD^+ (made from niacin), and coenzyme A (made from the vitamin pantothenic acid). The coenzyme lipoamide is also involved in this reaction.

22.7 *Oxidative phosphorylation* is the process by which the energy of electrons harvested from oxidation of a fuel molecule is used to phosphorylate ADP to produce ATP.

22.9 $NAD^+ + H:^- \rightarrow NADH$

22.11 During transamination reactions, the α-amino group is transferred to the coenzyme pyridoxal phosphate. In the last part of the reaction, the α-amino group is transferred from pyridoxal phosphate to an α-keto acid.

22.13 The purpose of the urea cycle is to convert toxic ammonium ions to urea, which is excreted in the urine of land animals.

22.15 An amphibolic pathway is a metabolic pathway that functions both in anabolism and catabolism. The citric acid cycle is amphibolic because it has a catabolic function—it completely oxidizes the acetyl group carried by acetyl CoA to provide electrons for ATP synthesis. Because citric acid cycle intermediates are precursors for the biosynthesis of many other molecules, it also serves a function in anabolism.

22.17 The mitochondrion is an organelle that serves as the cellular power plant. The reactions of the citric acid cycle, the electron transport system, and ATP synthase function together within the mitochondrion to harvest ATP energy for the cell.

22.19 The intermembrane compartment is the location of the high-energy proton (H^+) reservoir produced by the electron transport system. The energy of this H^+ reservoir is used to make ATP.

22.21 The outer mitochondrial membrane is freely permeable to substances of molar mass less than 10,000 g/mol. The inner mitochondrial membrane is highly impermeable. Embedded within the inner mitochondrial membrane are the electron carriers of the electron transport system, and ATP synthase, the multisubunit enzyme that makes ATP.

22.23 Coenzyme A is a molecule derived from ATP and the vitamin pantothenic acid. It functions in the transfer of acetyl groups in lipid and carbohydrate metabolism.

22.25 Decarboxylation is a chemical reaction in which a carboxyl group is removed from a molecule.

22.27 Under aerobic conditions pyruvate is converted to acetyl CoA.

22.29 The coenzymes NAD^+, FAD, thiamine pyrophosphate, and coenzyme A are required by the pyruvate dehydrogenase complex for the conversion of pyruvate to acetyl CoA. These coenzymes are synthesized from the vitamins niacin, riboflavin, thiamine, and pantothenic acid, respectively. If the vitamins are not available, the coenzymes will not be available and pyruvate cannot be converted to acetyl CoA. Because the complete oxidation of the acetyl group of acetyl CoA produces the vast majority of the ATP for the body, ATP production would be severely inhibited by a deficiency of any of these vitamins.

22.31 An aldol condensation is a reaction in which aldehydes or ketones react to form larger molecules.

22.33 Oxidation in an organic molecule is often recognized as the loss of hydrogen atoms or a gain of oxygen atoms.

22.35 A dehydrogenation reaction is an oxidation reaction in which protons and electrons are removed from a molecule.

22.37 **a.** False
b. False
c. True
d. True

22.39 **a.** The acetyl group of acetyl CoA is transferred to oxaloacetate.
b. The product is citrate.

22.41 Three

22.43 Two ATP per glucose

22.45 The function of acetyl CoA in the citric acid cycle is to bring the two-carbon remnant (acetyl group) of pyruvate from glycolysis and transfer it to oxaloacetate. In this way the acetyl group enters the citric acid cycle for the final stages of oxidation.

22.47 The high-energy phosphoryl group of the GTP is transferred to ADP to produce ATP. This reaction is catalyzed by the enzyme dinucleotide diphosphokinase.

22.49

$$\begin{array}{c} COO^- \\ | \\ C=O \\ | \\ CH_2 \\ | \\ COO^- \end{array} + H_3C-\overset{O}{\overset{\|}{C}}\sim S-CoA + H_2O \longrightarrow$$

Oxaloacetate   Acetyl CoA

$$\begin{array}{c} COO^- \\ | \\ CH_2 \\ | \\ HO-C-COO^- \\ | \\ CH_2 \\ | \\ COO^- \end{array} + HS-CoA + H^+$$

Citrate   Coenzyme A

The importance of this reaction is that it brings the acetyl group, the two-carbon remnants of the glucose molecule, into the citric acid cycle to be completely oxidized. Through these reactions, and subsequent oxidative phosphorylation, the majority of the cellular ATP energy is provided.

22.51 The conversion of citrate to *cis*-aconitate is an example of the dehydration of an alcohol to produce an alkene (double bond). The conversion of *cis*-aconitate to isocitrate is an example of the hydration of an alkene, that is, the addition of water to the double bond, to produce an alcohol (—OH).

22.53 This reaction is an example of the oxidation of a secondary alcohol to a ketone. The two functional groups are the hydroxyl group of the alcohol and the carbonyl group of the ketone.

22.55 It is a kinase because it transfers a phosphoryl group from one molecule to another. Kinases are a specific type of transferase.

22.57 An allosteric enzyme is one that has an effector binding site and an active site. Effector binding can change the shape of the active site, causing it to be active or inactive.

22.59 Negative allosterism is a means of enzyme regulation in which effector binding inactivates the active site of the allosteric enzyme.

22.61 The citric acid cycle is regulated by the following four enzymes or enzyme complexes: pyruvate dehydrogenase complex, citrate sythase, isocitrate dehydrogenase, and the α-ketoglutarate dehydrogenase complex.

22.63 Energy-harvesting pathways, such as the citric acid cycle, must be responsive to the energy needs of the cell. If the energy requirements are high, as during exercise, the reactions must speed up. If energy demands are low and ATP is in excess, the reactions of the pathway slow down.

22.65 ADP

22.67 The electron transport system is series of electron transport proteins embedded in the inner mitochondrial membrane that accept high-energy electrons from NADH and $FADH_2$ and transfer them in stepwise fashion to molecular oxygen (O_2).

22.69 Three ATP

22.71 The oxidation of a variety of fuel molecules, including carbohydrates, the carbon skeletons of amino acids, and fatty acids provides the electrons. The energy of these electrons is used to produce an H^+ reservoir. The energy of this proton reservoir is used for ATP synthesis.

22.73 The electron transport system passes electrons harvested during oxidation of fuel molecules to molecular oxygen. At three sites protons are pumped from the mitochondrial matrix into the intermembrane compartment. Thus, the electron transport system builds the high-energy H^+ reservoir that provides energy for ATP synthesis.

22.75 a. Two ATP per glucose (net yield) are produced in glycolysis, whereas the complete oxidation of glucose in aerobic respiration (glycolysis, the citric acid cycle, and oxidative phosphorylation) results in the production of thirty-six ATP per glucose.
   b. Thus, aerobic respiration harvests nearly 40% of the potential energy of glucose, and anaerobic glycolysis harvests only about 2% of the potential energy of glucose.

22.77 Transaminases transfer amino groups from amino acids to ketoacids.

22.79 The glutamate family of transaminases is very important because the ketoacid corresponding to glutamate is α-ketoglutarate, one of the citric acid cycle intermediates. This provides a link between the citric acid cycle and amino acid metabolism. These transaminases provide amino groups for amino acid synthesis and collect amino groups during catabolism of amino acids.

22.81 a. Pyruvate d. Acetyl CoA
   b. α-Ketoglutarate e. Succinate
   c. Oxaloacetate f. α-Ketoglutarate

22.83

$$\underset{\text{α-Ketoglutarate}}{\begin{array}{c} O \\ \| \\ C-COO^- \\ | \\ CH_2 \\ | \\ CH_2 \\ | \\ COO^- \end{array}} + NADPH + N^+H_4 \longrightarrow \underset{\text{Glutamate}}{\begin{array}{c} N^+H_3 \\ | \\ H-C-COO^- \\ | \\ CH_2 \\ | \\ CH_2 \\ | \\ COO^- \end{array}} + NADP^+ + H_2O$$

(Ammonia)

22.85 Hyperammonemia

22.87 a. The source of one amino group of urea is the ammonium ion and the source of the other is the α-amino group of the amino acid aspartate.
   b. The carbonyl group of urea is derived from CO_2.

22.89 Anabolism is a term used to describe all of the cellular energy-requiring biosynthetic pathways.

22.91 α-Ketoglutarate

22.93 Citric acid cycle intermediates are the starting materials for the biosynthesis of many biological molecules.

22.95 An essential amino acid is one that cannot be synthesized by the body and must be provided in the diet.

22.97

$$\underset{\text{Pyruvate}}{\begin{array}{c} O \\ \| \\ C-COO^- \\ | \\ CH_3 \end{array}} + CO_2 + ATP \longrightarrow \underset{\text{Oxaloacetate}}{\begin{array}{c} O \\ \| \\ C-COO^- \\ | \\ CH_2 \\ | \\ COO^- \end{array}} + ADP + P_i$$

Chapter 23

23.1 Because dietary lipids are hydrophobic, they arrive in the small intestine as large fat globules. The bile salts emulsify these fat globules into tiny fat droplets. This greatly increases the surface area of the lipids, allowing them to be more accessible to pancreatic lipases and thus more easily digested.

23.3 a. Four acetyl CoA, one benzoate, four NADH, and four $FADH_2$
   b. Three acetyl CoA, one phenyl acetate, three NADH, and three $FADH_2$
   c. Three acetyl CoA, one benzoate, three NADH, and three $FADH_2$
   d. Five acetyl CoA, one phenyl acetate, five NADH, and five $FADH_2$

23.5

$$CH_3CH_2CH_2-\overset{O}{\underset{\|}{C}}\!\sim\!S-CoA + FAD \longrightarrow$$

$$CH_3CH=CH-\overset{O}{\underset{\|}{C}}\!\sim\!S-CoA + FADH_2$$

$$\downarrow H_2O$$

$$\longleftarrow CH_3-\underset{H}{\overset{OH}{\underset{|}{C}}}-CH_2-\overset{O}{\underset{\|}{C}}\!\sim\!S-CoA + NAD^+$$

$$CH_3-\overset{O}{\underset{\|}{C}}-CH_2-\overset{O}{\underset{\|}{C}}\!\sim\!S-CoA + NADH$$

$$\downarrow \text{Coenzyme A}$$

$$2\,CH_3-\overset{O}{\underset{\|}{C}}\!\sim\!S-CoA$$

23.7 Starvation, a diet low in carbohydrates, and diabetes mellitus are conditions that lead to the production of ketone bodies.

23.9 (1) Fatty acid biosynthesis occurs in the cytoplasm whereas β-oxidation occurs in the mitochondria.
(2) The acyl group carrier in fatty acid biosynthesis is acyl carrier protein while the acyl group carrier in β-oxidation is coenzyme A.
(3) The seven enzymes of fatty acid biosynthesis are associated as a multienzyme complex called *fatty acid synthase*. The enzymes involved in β-oxidation are not physically associated with one another.
(4) NADPH is the reducing agent used in fatty acid biosynthesis. NADH and FADH$_2$ are produced by β-oxidation.

23.11 The liver regulates blood glucose levels under the control of the hormones insulin and glucagon. When blood glucose levels are too high, insulin stimulates the uptake of glucose by liver cells and the storage of the glucose in glycogen polymers. When blood glucose levels are too low, the hormone glucagon stimulates the breakdown of glycogen and release of glucose into the bloodstream. Glucagon also stimulates the liver to produce glucose for export into the bloodstream by the process of gluconeogenesis.

23.13 Insulin stimulates uptake of glucose and amino acids by cells, glycogen and protein synthesis, and storage of lipids. It inhibits glycogenolysis, gluconeogenesis, breakdown of stored triglycerides, and ketogenesis.

23.15 Bile consists of micelles of lecithin, cholesterol, bile salts, proteins, inorganic ions, and bile pigments that aid in lipid digestion by emulsifying fat droplets.

23.17 A micelle is an aggregation of molecules having nonpolar and polar regions; the nonpolar regions of the molecules aggregate, leaving the polar regions facing the surrounding water.

23.19 A triglyceride is a molecule composed of glycerol esterified to three fatty acids.

23.21 A chylomicron is a plasma lipoprotein that carries triglycerides from the intestine to all body tissues via the bloodstream. That function is reflected in the composition of the chylomicron, which is approximately 85% triglycerides, 9% phospholipids, 3% cholesterol esters, 2% protein, and 1% cholesterol.

23.23 Triglycerides

23.25 The large fat globule that takes up nearly the entire cytoplasm

23.27 Lipases catalyze the hydrolysis of the ester bonds of triglycerides.

23.29 Acetyl CoA is the precursor for fatty acids, several amino acids, cholesterol, and other steroids.

23.31 Chylomicrons are plasma lipoproteins (aggregates of protein and triglycerides) that carry dietary triglycerides from the intestine to all tissues via the bloodstream.

23.33 Bile salts serve as detergents. Fat globules stimulate their release from the gallbladder. The bile salts then emulsify the lipids, increasing their surface area and making them more accessible to digestive enzymes (pancreatic lipases).

23.35 When dietary lipids in the form of fat globules reach the duodenum, they are emulsified by bile salts. The triglycerides in the resulting tiny fat droplets are hydrolyzed into monoglycerides and fatty acids by the action of pancreatic lipases, assisted by colipase. The monoglycerides and fatty acids are absorbed by cells lining the intestine.

23.37 The energy source for the activation of a fatty acid entering β-oxidation is the hydrolysis of ATP into AMP and PP$_i$ (pyrophosphate group), an energy expense of two high-energy phosphoester bonds.

23.39 Carnitine is a carrier molecule that brings fatty acyl groups into the mitochondrial matrix.

23.41 The following equation represents the reaction catalyzed by acyl-CoA dehydrogenase. Notice that the reaction involves the loss of two hydrogen atoms. Thus, this is an oxidation reaction.

$$H_3C-\underset{\underset{H}{|}}{\overset{H}{\underset{|}{C}}}-\underset{H}{\overset{\text{\textcircled{H}}}{\underset{|}{C}}}-\overset{O}{\underset{\|}{C}}\!\sim\!S-CoA$$

$$\overset{FAD}{\underset{FADH_2}{\rightleftarrows}} \Bigg| \; \text{Acyl-CoA dehydrogenase}$$

$$H_3C-\overset{H}{\underset{}{C}}=\underset{H}{\overset{}{C}}-\overset{O}{\underset{\|}{C}}\!\sim\!S-CoA$$

23.43 An alcohol is the production of the hydration of an alkene.

23.45 Six acetyl CoA, one phenyl acetate, six NADH, and six FADH$_2$

23.47 112 ATP

23.49 Two ATP

23.51 The acetyl CoA produced by β-oxidation will enter the citric acid cycle.

23.53 Ketone bodies include the compounds acetone, acetoacetone, and β-hydroxybutyrate, which are produced from fatty acids in the liver via acetyl CoA.

23.55 Ketosis is an abnormal rise in the level of ketone bodies in the blood.

23.57 Matrix of the mitochondrion.

23.59

Acetoacetate: CH$_3$—C(=O)—CH$_2$—C(=O)—O$^-$

β-Hydroxybutyrate: CH$_3$—CH(OH)—CH$_2$—C(=O)—O$^-$

23.61 In those suffering from uncontrolled diabetes, the glucose in the blood cannot get into the cells of the body. The excess glucose is excreted in the urine. Body cells degrade fatty acids because glucose is not available. β-oxidation of fatty acids yields enormous quantities of acetyl CoA, so much acetyl CoA, in fact, that it cannot all enter the citric acid cycle because there is not enough oxaloacetate available. Excess acetyl CoA is used for ketogenesis.

23.63 Ketone bodies are the preferred energy source of the heart.

23.65 Cytoplasm

23.67 Fatty acid synthase

23.69 The phosphopantetheine group allows formation of a high-energy thioester bond with a fatty acid. It is derived from the vitamin pantothenic acid.

23.71 Fatty acid synthase is a huge multienzyme complex consisting of the seven enzymes involved in fatty acid synthesis. It is found in the cell cytoplasm. The enzymes involved in β-oxidation are not physically associated with one another. They are free in the mitochondrial matrix space.

23.73 Glycogenesis is the synthesis of the polymer glycogen from glucose monomers.

23.75 Gluconeogenesis is the synthesis of glucose from non-carbohydrate precursors.

23.77 β-oxidation of fatty acids

23.79 The major metabolic function of the liver is to regulate blood glucose levels.

23.81 Ketone bodies are the major fuel for the heart. Glucose is the major energy source of the brain, and the liver obtains most of its energy from the oxidation of amino acid carbon skeletons.

23.83 Fatty acids are absorbed from the bloodstream by adipocytes. Using glycerol-3-phosphate, produced as a by-product of glycolysis, triglycerides are synthesized. Triglycerides are constantly being hydrolyzed and resynthesized in adipocytes. The rates of hydrolysis and synthesis are determined by lipases that are under hormonal control.

23.85 In general, insulin stimulates anabolic processes, including glycogen synthesis, uptake of amino acids and protein synthesis, and triglyceride synthesis. At the same time, catabolic processes such as glycogenolysis are inhibited.

23.87 A target cell is one that has a receptor for a particular hormone.

23.89 Decreased blood glucose levels trigger the secretion of glucagon into the bloodstream.

23.91 Insulin is produced in the β-cells of the islets of Langerhans in the pancreas.

23.93 Insulin stimulates the uptake of glucose from the blood into cells. It enhances glucose storage by stimulating glycogenesis and inhibiting glycogen degradation and gluconeogenesis.

23.95 Insulin stimulates synthesis and storage of triglycerides.

23.97 Untreated diabetes mellitus is starvation in the midst of plenty because blood glucose levels are very high. However, in the absence of insulin, blood glucose can't be taken up into cells. The excess glucose is excreted into the urine while the cells of the body are starved for energy.

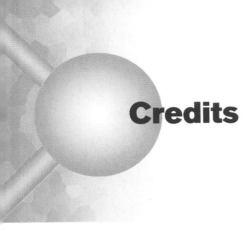

Credits

Design Elements
Microscope: © Vol. 85/PhotoDisc; Stethoscope: © Vol. OS48/PhotoDisc; Eye: © Vol. OS2/PhotoDisc; Pill Bottle: © Vol.168/Corbis; Pills: © Vol. 168/Corbis.

Chapter 1
Opener: © Greg Fiume/NewSport/Corbis; 1.2a: © Geoff Tompkinson/SPL/Photo Researchers, Inc.; 1.2b: © T.J. Florian/Rainbow; 1.2c: © David Parker/Segate Microelectronics Ltd./Photo Researchers, Inc.; 1.2d: APHIS, PPQ, Otis Methods Development Center, Otis, MA/USDA; 1.3a: © McGraw-Hill Higher Education/Louis Rosenstock, photographer; 1.3b: © The McGraw-Hill Companies, Inc./Jeff Topping, photographer; 1.3c: © The McGraw-Hill Companies, Inc./Louis Rosenstock, photographer; 1.4(both): © The McGraw-Hill Companies, Inc./Ken Karp, photographer; 1.6: © Doug Martin/Photo Researchers, Inc; 1.8a-c, 1.10a-d, 1.12: © The McGraw-Hill Companies, Inc./Louis Rosenstock, photographer.

Chapter 2
Opener: © The McGraw-Hill Companies, Inc./Louis Rosenstock, photographer; 2.1: © IBM Corporation–Almaden Research Center; 2.7: © Yoav Levy/Phototake; p. 48: © Earth Satellite Corp./SPL/Photo Researchers, Inc.; p. 49(right): © The McGraw-Hill Companies, Inc./Louis Rosenstock, photographer; p. 49(left): © Dan McCoy/Rainbow; p. 52: © Vol. 2/PhotoDisc.

Chapter 3
Opener: © Chase Swift/Corbis; 3.2: Courtesy of Trent Stephens; 3.14: © Charles D. Winters/Photo Researchers, Inc.

Chapter 4
Opener: © Richard Megna/Fundamental Photographs; 4.1, 4.3: © The McGraw-Hill Companies, Inc./Louis Rosenstock, photographer; p. 144: © Vol. 72/PhotoDisc.

Chapter 5
Opener: Courtesy of Robert Shoemaker; p. 152: © Hulton-Deutsch/Corbis; 5.2a,b: © The McGraw-Hill Companies, Inc./Louis Rosenstock, photographer; 5.5: © Peter Stef Lamberti/Stone/Getty.

Chapter 6
Opener: © Richard Megna/Fundamental Photographs; 6.1: © Kip and Pat Peticolas/Fundamental Photographs; p. 183: © J.W. Mowbray/Photo Researchers, Inc.; 6.2: © The McGraw-Hill Companies, Inc./Louis Rosenstock, photographer; p. 197(left): © RMF/Visuals Unlimited; p. 197(right): Courtesy of Rita Colwell, University of Maryland; 6.5 a-b: © The McGraw-Hill Companies, Inc./Louis Rosenstock, photographer; p. 199: © The McGraw-Hill Companies, Inc./Louis Rosenstock, photographer; p. 201: © AJPhoto/Photo Researchers, Inc.

Chapter 7
Opener: © The McGraw-Hill Companies, Inc./Louis Rosenstock, photographer; 7.9: © The McGraw-Hill Companies, Inc./Ken Karp, photographer; p. 221(both): © The McGraw-Hill Companies, Inc./Louis Rosenstock, photographer; 7.15, 7.16: © The McGraw-Hill Companies, Inc./Ken Karp, photographer.

Chapter 8
Opener: © Richard Megna/Fundamental Photographs; 8.1: © Dr. E.R. Degginger/Color Pic Inc.; 8.3a: © The McGraw-Hill Companies, Inc./Louis Rosenstock, photographer; 8.3b: © Richard Megna/Fundamental Photographs; 8.5a-b: © The McGraw-Hill Companies, Inc./Louis Rosenstock, photographer; 8.7a-b: © The McGraw-Hill Companies, Inc./Stephen Frisch, photographer; p. 257, 8.8, 8.9(left): © The McGraw-Hill Companies, Inc./Louis Rosenstock, photographer; 8.9(right): © Tony Freeman/PhotoEdit; 8.9(middle): © Bonnie Kamin/PhotoEdit; p. 266: © AAA Photo/Phototake; 8.10: © The McGraw-Hill Companies, Inc./Stephen Frisch, photographer.

Chapter 9
Opener: © Kathy McLaughlin/The Image Works; 9.5: © US Dept. of Energy/Photo Researchers, Inc.; 9.6: © Gianni Tortoli/Photo Researchers, Inc.; p. 285: NASA; 9.7: © Blair Seitz/Photo Researchers, Inc.; 9.8b: Bristol-Myers Squibb Medical Imaging; p. 292(bottom): © SIU/Biomed/Custom Medical Stock Photo; p. 292(top): © The McGraw-Hill Companies, Inc./Louis Rosenstock, photographer; 9.9: © US Dept. of Energy/Mark Marten/Photo Researchers, Inc.; 9.10: © The McGraw-Hill Companies, Inc./Louis Rosenstock, photographer; 9.11: © Scott Cazmine/Photo Researchers, Inc.

Chapter 10
Opener: © Corbis Royalty Free; p. 317: © Vol. 31/PhotoDisc; p. 326: © Corbis Royalty Free; p. 331: © Vol. 27/PhotoDisc.

Chapter 11
Opener: © Vol. 44/PhotoDisc; p. 347: © Buddy Mays/Corbis; 11.3(top to bottom): © Larry Lefever/Grant Heilman Photography, Inc., © Michelle Garrett/Corbis, © Walter H. Hodge/Peter Arnold, Inc., © Hal Horwitz/Corbis; 11.5: © The McGraw-Hill Companies, Inc./Ken Karp,photographer; p. 362: © Vol. 67/PhotoDisc; p. 366: © Alan Detrick/Photo Researchers, Inc.

Chapter 12
Opener: © David Young-Wolff/PhotoEdit; 12.3: © Vol. 12/PhotoDisc; p. 388: © Vol. 18/PhotoDisc; 12 4: © Corbis Royalty Free; p. 401: © Vol. 88/PhotoDisc; 12 8: © Corbis Royalty Free; 12 9: © Vol. OS22/PhotoDisc.

Chapter 13
Opener: © Gary Moss/Getty Images; 13.3(almonds): © B. Borrell Cassals, Frank Lane Picture Agency/Corbis; 13.3(cinnamon): © Rita Maas/Getty Images; 13.3(berries): © Charles Krebs/Corbis; 13.3(vanilla): © Eisenhut & Mayer/Getty Images; 13.3(mushroom): © James Noble/Corbis;

13.3(lemon grass): © Corbis Royalty Free; **13.4(all):** © Richard Megna/Fundamental Photographs; **13.5:** © Rob and Ann Simpson/Visuals Unlimited; **p. 426:** © Vol. 18/PhotoDisc; **p. 430:** © Corbis Royalty Free; **p. 433:** © Hanson Carroll/Peter Arnold, Inc.

Chapter 14
Opener: © Christel Rosenfeld/Tony Stone/Corbis; **p. 456:** © Corbis Royalty Free; **p. 457:** © Paul W. Johnson/BPS; **p. 464(pineapple):** © Vol. 19/PhotoDisc; **p. 464(berries):** © Vol. OS49/PhotoDisc; **p. 464(bananas):** © Vol. 30/PhotoDisc; **p. 465(oranges, apples):** © Corbis Royalty Free; **p. 465(apricots):** © Vol. 121/Corbis; **p. 465(strawberries):** © Vol. 83/Corbis; **p. 478:** © Vol. 81/Corbis; **p. 479(left):** © William Weber/Visuals Unlimited; **p. 479(bottom right):** © Norm Thomas/Photo Researchers; **p. 479(top right):** © Vol. 101/Corbis.

Chapter 15
Opener: © Vol. 6/PhotoDisc; **15 3a-b:** © Vol. 94/Corbis; **15.3c:** © Gregory G. Dimijian/Photo Researchers; **p. 500:** © The McGraw-Hill Companies, Inc./Gary He, photographer; **p. 512a:** © Vol. 9/PhotoDisc; **p. 512b:** © Phil Larkin, CSIRO Plant Industry.

Chapter 16
Opener: © The McGraw-Hill Companies, Inc./Louis Rosenstock, photographer; **16.1:** USDA; **16.2:** © Vol. 67/PhotoDisc; **p. 528b:** © Stanley Flegler/Visuals Unlimited; **p. 530:** © Corbis Royalty Free; **p. 546:** © Jean Claude Revy-ISM / Phototake.

Chapter 17
Opener: © Lester Lefkowitz/Corbis; **p. 563:** © Roadsideamerica.com, Kirby, Smith & Wilkins; **p. 574:** © Hans Pfletschinger/Peter Arnold, Inc.; **17.10a:** © James Dennis/PhotoTake; **17.18(all):** © David M. Phillips/Visuals Unlimited.

Chapter 18
Opener: © University of Oxford/Getty Images; **p.599:** © Vol. 18/PhotoDisc; **p. 604:** © Vol. 270/Corbis; **18.14:** © Meckes/Ottawa/Photo Researchers, Inc.

Chapter 19
Opener: © J. Schmidt/National Park Service; **p. 641:** © Phil Degginger/Color-Pic, Inc.

Chapter 20
Opener: © David Young-Wolff/PhotoEdit; **p. 702:** © Dr. Charles S. Helling/US Dept. of Agriculture; **p. 703:** Courtesy of Orchid Cellmark, Germantown, Maryland.

Chapter 21
Opener: © The McGraw-Hill Companies, Inc./Louis Rosenstock, photographer; **p. 728:** © The McGraw-Hill Higher Companies, Inc./Louis Rosenstock, photographer.

Chapter 22
Opener: © PhotoDisc; **22.1a:** © CNRI/Phototake; **p. 749:** © Vol. 10/PhotoDisc.

Chapter 23
Opener: Courtesy of Katherine Denniston; **p. 785:** © Vol. 110/PhotoDisc.

Index

Note: Page numbers followed by B indicate boxed material; those followed by F indicate figures; those followed by T indicate tables.

A

ABO blood types, 546–47B
absolute specificity, of enzyme, 639
acceptable daily intake (ADI), 382B
accuracy, and measurement, 19–20
acetal, 434, 435, 437, 442, 543
acetaldehyde, 267, 430B, 728F
acetamide, 505
acetaminophen, 478B, 506
acetate ion, 99–100
acetic acid
    buffers, 257, 259, 260
    neutralization, 458
    oxidation–reduction reactions, 267
    source and nomenclature of, 453T
    vinegar, 449
    weak acids, 242, 244
acetoacetate, 791
acetoacetyl ACP, 792
acetoacetyl CoA, 789–90
acetone, 791, 794B
acetylcholine, 515–16
acetylcholinesterase, 516, 654B
acetyl CoA
    biosynthesis of biological molecules, 408
    citric acid cycle, 479–80, 718, 750–51, 754F, 756
N-acetylgalactosamine, 546B, 547B
α-D-N-acetylglucosamine, 550B
acetyl group, 480B
acetylsalicylic acid. *See* aspirin
acid(s). *See also* carboxylic acids; citric acid cycle; fatty acids
    Arrhenius theory, 241
    Brønsted–Lowry theory, 241
    buffers, 256–63
    chemical reactions, 252–55
    conjugate, 243–46
    dissociation of water, 246
    pH scale, 246–52
    properties of water, 242
    strength, 242–43

acid anhydrides, 473
acid-base reactions, 131, 458–60
acid chlorides, 470–73
acid hydrolysis, and fatty acids, 561–62
acidosis, 262B, 769B
acid rain, 252, 256–57B
aconitase, 753
acrylonitrile, 365T
actin, 597
actinide series, 55
activated complex, 220
activation energy, 220, 636
active site, and enzyme–substrate complex, 637–38
active transport, 200, 590, 713T, 716. *See also* electron transport system; transport proteins
acyl carrier protein (ACP), 792, 793F
acyl-CoA dehydrogenase, 786
acyl-CoA ligase, 786
acyl group, 476, 480B, 508
acyl group carriers, 792
Adams, Mike, 630B
addition, of significant figures, 20
addition polymers, 364–65, 374
addition reactions, 354, 374, 391
    aldehydes and ketones, 434–36, 441–42
    alkenes, 374, 391
    double bond, 354
    fatty acids, 564
adenine, 670F, 672F
adenosine diphosphate. *See* ADP
adenosine triphosphate. *See* ATP
adipic acid, 455, 456
adipocere, 563B
adipocytes, 781
adipose tissue, 568, 781, 784–85B, 796
ADP, 438, 633B, 634, 714–15, 732
aerobic respiration, 752, 753–56, 757F
African clawed frog, 5B
agarose gel electrophoresis, 696–98
agglutination, 546B
Agricultural Research Service (ARS), 702B, 703B
agriculture, and pH scale, 252. *See also* plants
AIDS (acquired immune deficiency syndrome), 641B, 676–77B

air. *See also* air pollution
    density, 30
    as solution, 178
air bags, automobile, 118B
air pollution, 295B. *See also* acid rain; greenhouse effect
alanine, 598, 601F, 602, 623T, 764
alanine aminotransferase/serum glutamate-pyruvate transaminase (ALT/SGPT), 658B, 660
alanyl-glycine-valine, 602, 603
albumin, 597, 599B, 621
Alcaligenes eutrophus, 456B
alchemists, 266B
alcohol(s). *See also* ethanol
    addition reactions, 374
    boiling points, 493T
    chemical reactions, 391–400, 409
    classification, 389–90
    functional groups, 309
    hydration, 358
    medical uses, 387–89
    nomenclature, 385–87
    oxidation, 427
    physical properties, 384–85
alcohol abuse and alcoholism, 430B
alcohol dehydrogenase, 727, 729
alcohol fermentation, 727–29
alcoholic beverages, 387, 400, 401B, 430B
aldehydes
    chemical reactions, 424–42
    functional groups, 309T
    hydration, 361
    hydrogenation, 392–93
    medical and industrial uses, 424, 425F
    nomenclature, 419–22
    oxidation reactions, 397
    physical properties, 417–19
aldohexose, 535
aldolase, 439, 723
aldol condensation, 438–39, 442, 723B
aldose, 527
aldosterone, 577
aliphatic hydrocarbons, 308
alkadienes, 344
alkali metals, 55
alkaline metals, 55

I-1

alkalosis, 262B
alkanamide, 506T
alkanamine, 495T
alkanes
  alkyl groups, 313–14
  chemical reactions, 327–32
  conformations of, 325–27
  nomenclature, 315–19
  saturated hydrocarbons, 308, 310
  structure and physical properties, 310–13
alkenes
  chemical reactions, 354–65, 374, 391
  functional groups, 309T
  nomenclature, 343–45
  plants as sources of, 352–53
  structure and physical properties, 342–43
  unsaturated hydrocarbons, 308
alkoxy group, 403B
alkylammonium ion, 498
alkylammonium salts, 498, 499
alkyl dihalide, 374
alkyl groups, 313–14, 315T
alkyl halide, 329, 361, 374
alkynes
  chemical reactions, 354–65
  functional groups, 309T
  nomenclature, 343–45
  structure and physical properties, 342–43
  unsaturated hydrocarbons, 308
allosteric enzymes, 649–50, 726, 756
allotropic forms, of elemental carbon, 305
alloys, and solutions, 178
alpha decay, 280
alpha particles, 278, 279
aluminum, 85, 106
aluminum bromide, 86
aluminum nitrate, 129
alveoli, and respiration, 182
Alzheimer's disease, 599B
American Civil War, 604B
American Diabetes Association, 382B
American Medical Association, 388B
Ames, Bruce, 694B
Ames test, 694–95B
amide(s), 309T
  chemical reactions, 506–9, 517
  common functional groups, 309T
  hydrolysis, 509–10, 517
  medically important, 505–6
  nomenclature, 505
  physical properties, 504–5
amide bond, 602
amines
  chemical reactions, 497–502, 517
  common functional groups, 309T
  heterocyclic amines, 502–4
  medically important, 495–97
  nomenclature, 493–95
  physical properties, 489–93
amino acids. See also proteins
  amides and amide bonds, 508B, 510–11
  classes of, 600
  definition of, 407B
  degradation of, 762–66
  genetic code, 686, 687F

  stereoisomers, 598–99
  structure of, 525, 597–98
α-amino acids, 597–98
aminoacyl group, 510–11
aminoacyl tRNA, 510–11, 689–90
aminoacyl tRNA binding site (A-site), 690
γ-aminobutyric acid (GABA), 515
6-aminopenicillanic acid, 507B
amino sugars, 550B
ammonia
  amines, 489
  covalent bonding, 82
  decomposition, 210, 211
  hydrogen bonding, 171
  Lewis structures of covalent
    compounds, 96–97
  molecular structure, 106, 490
  nomenclature, 91
  solubility, 111–12
  synthesis, 223F, 233F
ammonium chloride, 87, 216, 507
ammonium ion, and Lewis structure, 98
ammonium nitrate, 135
ammonium sulfate, 124
ammonium sulfide, 89
amniocentesis, 668B
amorphous ionic compounds, 93
amorphous solid, 171
amphetamines, 495–96
amphibolic pathways, 770–71
amphipathic molecule, 568
amphiprotic property, of water, 242
amphotericin B, 588B, 589B
ampicillin, 507B
amylopectin, 548, 549F
amylose, 548
anabolic reactions, 751
anabolic steroids, 572
anabolism, 713, 770–72
anaerobic metabolism, 762
anaerobic threshold, 727
analgesics, 478B, 496, 504
analytical balance, 24F
analytical chemistry, 3
anaplerotic reaction, 772
Andersen's disease, 740B
anesthetics, 328B, 405, 504
angiogenesis and angiogenesis
    inhibitors, 596B
angiostatin, 596B
angular molecular structure, 107
aniline, 242, 369, 494
animals. See birds; fish; reindeer; skunk
anions, 43, 66
anisole, 369
anode, 45, 268
anomers, 537
Antabuse, 430B
anthracene, 371, 372
antibiotics, 5B, 507B, 588–89B, 699
antibodies
  defense proteins, 597
  drug-carrying liposomes, 586B
  immune response, 616B
α$_1$-antichymotrypsin, 599B

anticodon, 682, 690
antifreeze, 389
antigens, 597, 618B
antihistamines, 489, 514, 524B
antiknock quality, of gasoline, 326B
antiparallel β-pleated sheet, 609, 610F
antiparallel strands, 672
antiport, 587
antipyretics, 478B
antiseptics, 264B, 402. See also disinfectants;
    sterilization
α-antitrypsin, 705B
apoenzyme, 643
aqueous solution, 178
arachidic acid, 559T
arachidonic acid, 559T, 564, 566
archaeology, and radiocarbon dating, 288
arginase, 768
arginine, 601F, 623T, 768, 770
argininosuccinate, 768
argon, and electron configuration, 63
Aristotle, 669
aromatic amines, 494
aromatic compounds, 366, 373
aromatic hydrocarbons, 308, 366–72
Arrhenius theory, of acids and bases, 241
artificial kidney, 201B
artificial radioactivity, 291
artificial sweeteners, 509
aryl group, 383
ascorbic acid, 644T
asparagine, 601F, 623T, 770
aspartame, 508, 509
aspartate, 601F, 623T, 764, 768
aspartate aminotransferase/serum
    glutamate-oxaloacetate
    transaminase (AST/SGOT), 658B,
    659B, 660
aspartate transaminase, 764
aspirin, 315, 478B, 565, 566–67
asthma, and air pollution, 257B
atherosclerosis, 575, 580, 794
atmosphere, and methane, 307B
atmosphere (atm), as unit of
    measurement, 154
atom. See also atomic theory
  definition of, 39
  mole concept, 119–23
  periodic table and size of, 68
  structure of, 39–44, 46–49
atomic mass, 41–42, 55–56
atomic mass unit (amu), 24, 41, 119
atomic number, 39, 55–56, 70F, 277
atomic orbital, 51, 61, 79
atomic theory. See also atom
  Bohr atom, 49–51
  development of, 44–46
  light and atomic structure, 46–49
  modern, 51–52
ATP
  aerobic respiration, 752
  carbohydrate metabolism, 713–15, 718
  citric acid cycle, 480
  energy metabolism, 748–49B
  energy transport, 590

enzymes, 633B, 634, 646, 650
glycolysis, 438, 477–78, 719, 732
oxidative phosphorylation, 757, 760–62
ribonucleotide, 671F
ATP synthase, 747, 760–61
Aufbau principle, 62–64
autoimmune reaction, 618B
autoionization, 246
automobiles, and air bags, 118B
Avogadro, Amadeo, 153
Avogadro's law, 161–62
Avogadro's number, 119–20, 191
axial atoms, 327
AZT (zidovudine), 676–77B

B

Bacillus polymyxa, 588B
background radiation, 293–94
bacteria
    biodegradable plastics, 456B
    chromosomes, 673
    DNA replication, 677–81
    extreme temperatures, 630B, 647–48
    fermentation, 729B
    heat and sterilization, 647
    magnetotactic, 78B
    membrane lipids, 582
    saliva and tooth decay, 528B
    ultraviolet light and DNA damage, 694
balance(s), 24F
balanced ionic equation, 252
balanced nuclear equation, 279–82
balancing, of chemical equations, 132–36
Bangham, Alec, 586B
barbital, 506
barbiturates, 505
barium-131, 290
barium sulfate, 129
barometer, 153–54
base(s)
    amines, 498, 517
    Arrhenius theory, 241
    Brønsted-Lowry theory, 241
    buffers, 256–63
    chemical reactions, 252–55
    conjugate, 243–46
    dissociation of water, 246
    pH scale, 246–52
    properties of water, 242
    strength, 242–43
base pairs, 672
batteries
    human body as, 270B
    voltaic cells, 269
Becquerel, Henri, 276B
beeswax, 577
behavior modification, and weight loss, 785B
Benadryl, 514
Benckiser, Reckitt, 513B
bends, and scuba diving, 183B
Benedict's reagent, 429, 431, 541–43
benzaldehyde, 369, 425F
benzalkonium chloride, 502
Benzedrine, 496

benzenamine, 494
benzene, 308, 372, 374
benzene ring, 368
benzenesulfonic acid, 372, 374
benzodiazepines, 515
benzoic acid, 369, 428, 454, 458
benzoic anhydride, 473
benzopyrene, 371, 372
benzoyl chloride, 471
benzoyl peroxide, 264B
benzyl alcohol, 371
benzyl chloride, 371
beryllium, 58, 59, 63
beryllium hydride, 103–4
Berzelius, Jöns Jakob, 304
beta decay, 280
beta-oxidation pathway, 360
beta particles, 278, 279
big bang theory, 92B
bile, 779
bile salts, 575
bilirubin, 2B
binding energy, of nucleus, 282
binding site, of enzyme, 638
biochemistry, 3. *See also* organic chemistry
biodegradable plastics, 456–57B, 470
bioinformatics, 706
biological systems. *See also* birds; fish;
    human body; life; plants
    DDT and biological magnification, 340B
    disaccharides, 543–47
    effects of radiation on, 293–95
    genetics and information flow in, 681–85
    monosaccharides, 535–43
    most importance elements, 54T
    oxidation-reduction reactions, 267, 400–2
biopol, 457B
biosynthesis, 512–13B, 713T
biosynthetic intermediates, and citric acid
    cycle, 770–72
biotin, 644T
birds, and migration, 78B
birth control, 346B, 576
1,3-bisphosphoglycerate, 724
bleach and bleaching, 264B, 266–67
blimps, 152B
blood. *See also* hemoglobin; red blood cells
    alcohol levels, 401B
    buffer solutions, 256
    control of pH, 262B
    electrolytes, 200
    glucose levels, 543, 738B, 796, 797
    proteins, 599B
    specific gravity, 31
    urea levels, 660
blood clotting, and protaglandins, 565
blood group antigens, 546–47B
blood pressure
    calcium and regulation of, 67B
    sodium ion/potassium ion ratio, 94B
blood sugar, 429
blood transfusions, 546–47B
blood types. *See* blood group antigens
blood urea nitrogen (BUN), 660
B lymphocytes, 618B, 619B

boat conformation, 326–27
body battery, 270B
body fluids, electrolytes in, 200–202
body temperature. *See also* temperature
    brown fat and thermogenesis, 758–59B
    enzymes, 620
    of mammals, 582
Bohr, Niels, 46, 49
Bohr theory, 49–51, 60
boiling point
    alcohols, 384, 493T
    aldehydes and ketones, 417
    alkanes, 311T, 313T, 343T
    alkenes and alkynes, 343T
    amines, 491, 492T, 493T
    carboxylic acids, 449
    concentration of solutions, 192–93
    ethers, 403
    ionic and covalent compounds, 93
    liquids and solids, 112–13
    physical properties, 7, 9
    vapor pressure of liquids, 169–70
    of water, 198
bomb calorimeter, 217
bond angle, 105, 311, 342
bond energies, 100–101
bond order, 101B
bone
    calcium in diet, 67B
    nuclear medicine, 290
Borgia, Lucretia, 652
Borneo, and DDT, 340B
boron, 58, 59
boron-11, 280
boron trifluoride, 105–6
Boyle, Robert, 155, 387
Boyle's law, 155–57, 160, 233
brain
    blood glucose levels, 796
    opium poppy and peptides in, 604–5B
branching enzyme, 739
bread making, and yeast, 729B
breathalyzer test, 401B
breeder reactors, 287
bridging conversion unit, 15
Bright's disease, 31B
British Anti–Lewisite (BAL), 407
bromination, 358, 359F
bromobenzene, 369, 372
o-bromobenzoic acid, 454
2-bromobenzoic acid, 472
2-bromobenzoyl chloride, 472
β-bromocaproic acid, 453
bromochlorofluoromethane, 532
2-bromo-2-chloro-1,1,1-trifluoroethane, 328B
2-bromocyclohexanol, 386
2-bromo-3,3-dimethylpentane, 317
bromoethane, 361
2-bromohexane, 316
2-bromo-3-hexyne, 344
bromomethane, 330
1-bromo-4-methylhexane, 319
2-bromo-4-methylpentanoic acid, 451
2-bromopropane, 316, 361
3-bromopropanoyl chloride, 470

γ-bromovaleraldehyde, 421
Brønsted-Lowry theory, 241
Brookhaven National Laboratory (New York), 291
brown fat, 758–59B
bubonic plague, 340B
buckminsterfullerene, 305
bucky ball, 304
buffer(s), acid-base, 256–63
buffer capacity, 258
buffer solution, 259–61
Bunsen burner, 209
buprenorphine, 512B, 513B
buret, 253
butanal, 421T, 427
butanamine, 492
butane
    boiling point, 384, 403, 417, 449
    major properties of, 306T
    molecular formula, 310, 312F
    staggered conformation, 325F
butane gas, 134–35
butanoic acid, 464B
butanoic anhydride, 475
1-butanol, 467
butanone, 422, 424
butanoyl chloride, 470, 472
1-butene and 2-butene, 348
butylated hydroxytoluene (BHT), 402
butyl group, 314T, 315T
1-butyne and 2-butyne, 343T
butyric acid, 448B, 449, 453T, 464B
butyric acid, butanol, acetone fermentation, 729B
butyryl ACP, 792
butyryl chloride, 472

C

calcium, 67B, 281
calcium carbonate, 127, 128
calcium hydroxide, 87, 124, 140–41
calcium hypochlorite, 264B
calcium oxide, 86
calcium phosphate, 126
calories, 27, 28B, 217, 785B. *See also* kilocalorie
calorimeter, 215
calorimetry, 209
cancer. *See also* carcinogens
    angiogenesis inhibitors, 596B
    calcium in diet, 67B
    mutagens and carcinogens, 693
    Pap smear test, 2B
    radiation therapy, 49B, 289
    sun exposure and skin cancer, 432–33B
Candida albicans, 712B
capillin, 346B, 347B
capric acid, 453T, 559T, 561
caproic acid, 453T
caprylic acid, 453T
cap structure, 684
carbamoyl phosphate, 766
carbinol carbon, 389

carbohydrate(s). *See also* carbohydrate metabolism
    disaccharides, 543–47
    monosaccharides, 527–29, 535–43
    polysaccharides, 548–51
    stereoisomers and stereochemistry, 529–34
    types of, 526–27
carbohydrate metabolism
    ATP, 713–15
    catabolic processes, 715–17
    fermentations, 726–29
    gluconeogenesis, 730–32
    glycogen synthesis and degradation, 733–41
    insulin and glucagon, 797
    pentose phosphate pathway, 730
    regulation, 793–97
carbolic acid, 402
carbon
    aldehydes, 420, 421
    alkyl groups, 313
    atomic mass, 42–43
    chain length and nomenclature, 316T
    chiral carbon, 530
    isotopes, 277, 278F, 280
    organic chemistry, 305–9
    radiocarbon dating, 288
    valence electrons, 58, 59
carbon-14
carbonated beverages, 177F, 181
carbonate ion, and Lewis structure, 98–99
carbon dioxide
    alcohol fermentation, 728B
    Avogadro's law, 162
    blood gases, 168B, 262B
    carbonated beverages, 177F, 181
    covalent compounds, 90
    greenhouse effect and global warming, 166B, 167, 329
    Lewis structure, 95–96
carbonic acid, 242
carbon monoxide, 90, 136B
carbon skeletons, of amino acids, 766
carbon tetrachloride, 110
carboxylases, 635
carboxyl group, 449
carboxylic acid(s). *See also* carboxylic acid derivatives
    aldehydes, 420
    chemical reactions, 457–60
    common functional groups, 309T
    esterification, 460
    foods and food industry, 455–56
    fragrance, 464B
    nomenclature, 420, 451–55
    oxidation and reduction reactions, 400
    physical properties, 449–50
    weak electrolytes, 306B
carboxylic acid derivatives, 461, 470, 478–79B
carboxypeptidase A, 651T
carcinogens, 693, 694–95B. *See also* cancer
cardiotonic steroids, 574B
cardiovascular disease, and nuclear medicine, 290

carnitine, 786
carnitine acyltransferase, 786
Carroll, Lewis, 524B
Carson, Rachel, 340B
carvone, 524B
casein, 597
catabolism, 713, 715–18, 770
catalase, 631
catalyst
    equilibrium composition, 234
    reaction rate, 222–23
catalytic cracking, 326B
catalytic groups, 638
catalytic reforming, 326B
catecholamines, 511
cathode, 45, 268
cathode rays, 45
cations, 43–44, 65, 200
cause-and-effect relationship, and mental illness, 178B
cell membrane
    antibiotics, 588–89B
    lipids, 557, 581–90
    osmotic pressure, 196
cellular metabolism
    acetyl CoA, 751
    insulin and glucagon, 797–98
cellulase, 549
cellulose, 549, 550F, 551
Celsius scale, 25, 157B
Centers for Disease Control, 264B, 430B, 641B
central dogma, of molecular biology, 681
cephalin, 568, 569F
ceramide, 571
cerebrosides, 571
Cerezyme, 661
cetyl palmitate, 577
cetylpyridinium chloride, 502
Chadwick, James, 45
chain elongation, 683, 690
chain reaction, 286
chair conformation, 325, 326F
Challenger (space shuttle), 100–101
champagne, and wine making, 728B
Chargaff, Irwin, 671
charge, of subatomic particles, 39T
Charles, Jacques, 157
Charles's law, 157–59, 160
cheese, and fermentation, 729B
chemical bonding. *See also* covalent bonding; hydrogen bonding; ionic bonding
    definition of, 78
    electronegativity, 83, 84T
    Lewis symbols, 79
    principal types of, 79–83
chemical change, 8, 126
chemical compounds. *See also* aromatic compounds; covalent compounds; inorganic compounds; ionic compounds; organic compounds
    definition of, 9
    naming and writing formulas for, 84–92
chemical control, of microbes, 264B

Index

chemical equations. *See also* nuclear
    equations; rate equations
  balancing, 132–36
  calculations using, 136–45
  chemical change, 126
  chemical reactions, 128–31
  experimental basis of, 127–28
  features of, 127
chemical equilibrium, 227
chemical formulas. *See also* condensed
    formula; molecular formula;
    structural formula
  chemical compounds, 84–92
  formula weight and molar mass, 123–26
chemical properties. *See also* physical
    properties
  matter, 8
  periodic table, 53
chemical reactions, 8. *See also* addition
    reactions; reaction rates; reversible
    reaction; substitution reactions
  acids and bases, 252–55
  alcohols, 391–400
  aldehydes and ketones, 424–42
  alkanes and cycloalkanes, 327–32
  alkenes and alkynes, 354–65, 374
  amides, 506–9, 517
  amines, 497–502, 517
  benzene, 372
  carboxylic acids, 457–60
  citric acid cycle, 753–56
  energy, 209, 215–18
  enzymes, 636–43
  esters, 462–66
  fatty acids, 561–64
  glycolysis, 721, 723–25
  kinetics, 218–25
  β-oxidation, 786–88
  types of, 129–131
  urea cycle, 766–69
  writing, 128–29
chemical warfare, 407
chemistry. *See also* organic chemistry,
    pharmaceutical chemistry
  definition of, 3
  measurement in, 11–16
  models in, 5–6
  scientific method, 3–5
chenodeoxycholate, 575, 779
chiral carbon, 530
chiral molecules, 529
chlorine, 41, 263, 264B
chlorine gas, 264B
chlorobenzene, 372
4-chlorobenzoic acid, 455
4-chlorobenzoyl chloride, 470
2-chloro-2-butene, 344
chloroethane, 328B
chloroform, 328B, 331B
chloromethane, 328B
3-chloro-4-methyl-3-hexene, 344
2-chloropentane, 363
γ-chlorovaleric acid, 453
cholate, 575, 779

cholera, 197B
cholesterol, 575
choline, 502
chondroitin sulfate, 550B, 551B
chorionic villus sampling, 668B
chromic acid, 397
chromosome(s), 673–74, 675F
chromosome walking, 704
chylomicrons, 578–79, 780, 781
chymotrypsin, 622, 651T, 656–57
Cicuta maculata, 346B, 347F
cicutoxin, 346–47B
cigarette smoking. *See also* nicotine
  carcinogens and cancer, 693, 694B
  emphysema, 648B
  nicotine patch, 488B
cimetidine, 373
cinnamaldehyde, 425F
cis-trans isomers, 322–25, 347, 348. *See also*
    geometric isomers
citral, 425F
citrate, 754F, 756, 757F
citrate lyase, 632
citric acid, 455, 456
citric acid cycle
  aerobic respiration, 752, 753–56
  anabolism and biosynthetic
    intermediates, 770–72
  conversion of pyruvate to acetyl CoA, 750
  energy metabolism, 761
  hydration, 359
  oxidation and reduction reactions,
    401–2, 717
  regulation of, 756, 757F
citric synthase, 753
citrulline, 766, 767F, 768
climate, and effect of water in
    environment, 27B
cloning and cloning vectors, 697, 699, 700F
Clostridium perfringens, 464B
coagulation, of proteins, 619
coal, and acid rain, 257B
cocaine
  amines, 499, 500F, 501–2, 503, 504
  DNA fingerprinting of coca plants, 702–3B
  drug abuse, 499, 500F, 501–2, 703B
  medical uses, 504
codeine, 503, 504, 512B
codons, 682, 686, 687F
coefficients
  chemical equation, 132
  equilibrium-constant expressions, 228
coenzyme(s). *See also* acetyl CoA
  citric acid cycle, 480
  functions of, 643–46
  nomenclature, 635
  oxidation and reduction reactions, 400
  thioesters and thioester bond, 407, 479
coenzyme A, 750
cofactors, of enzymes, 643–46
cold packs, 221B
Coleman, Douglas, 778B
colipase, 781
collagen, 551B, 612–13B

colligative properties, 191
colloid(s), 179–80
colloidal suspension, 179–80
colon cancer, 67B
colony blot hybridization, 701F
color(s)
  electromagnetic spectrum, 47
  pH indicators, 253F
colorless solutions, 179
combination reactions, 128
combined gas law, 160–61
combustion, 208B, 265–66, 327–29
comfrey, 574B
common names. *See also* nomenclature
  alcohols, 386
  aldehydes and ketones, 419–22
  nomenclature, 86, 87T, 91
competitive inhibition, 426B, 652–53
complementary strands, 672
complete protein, 623
complex carbohydrates, 526
complex lipids, 577–80
compounds. *See* chemical compounds
compressibility, of liquids, 167
computer(s), and information
    management, 38B
computer-activated tomography, 49B, 296
computer imaging, and diagnostic
    medicine, 296
concentration
  acid and base strength, 242
  properties of solutions, 191–96
  rate of chemical reaction, 222, 232
  of solutions based on mass, 182–87
  of solutions in moles and equivalent,
    187–91
  systems of measurement, 27–28
condensation, 169
condensation polymers, 469–70
condensation reaction, 640, 642F
condensed formula
  alkanes, 310, 311T, 312, 313T, 342
  alkenes and alkynes, 342
conductivity, of metallic solids, 173
conformations, of alkanes and cycloalkanes,
    325–27
conjugate acid, 243–46
conjugate acid-base pair, 243, 245F
conjugate base, 243–46
conservation of mass, law of, 126, 132, 141F
constitutional isomers, 319–21
conversion and conversion factors
  atoms and moles, 122
  chemical equations, 137–41
  mass, 24
  systems of measurement, 13, 15
  temperature, 26
copper
  fireworks, 52B
  as metallic solid, 173
  molar mass, 120F
  oxidation-reaction reactions, 131
  voltaic cells, 267–69
  Wilson's disease, 57B

copper sulfate, 129
copper(II) sulfate pentahydrate, 124
Cori Cycle, 727, 732
Cori's disease, 740B
corrosion, as chemical reaction, 131, 265
cortisone, 576
cosmetics industry, and liposomes, 586B
cost, of energy, 208B
covalent bonding. *See also* polar
    covalent bond
  amino acids, 610
  definition of, 79
  organic compounds, 305, 306
  process and examples of, 81–82
covalent compounds
  definition of, 82
  naming of, 89–92
  properties of, 92–93
  resonance hybrids, 102–3
covalent solids, 172
crack cocaine, 499, 500F, 501
C-reactive protein (CRP), 580
creatine kinase, 748B
creatine phosphate, 748B
creatine phosphokinase (CPK), 658B, 659B
crenation, 196
Crick, Francis, 669, 671, 686
crime, and DNA fingerprinting, 703B
cristae, 747
Crookes, William, 45
crotonyl ACP, 792
crystal lattice, 81
crystalline solids, 93, 171, 172–73
C-terminal amino acid, 602
CT scanner, 49B, 296
curie, as unit of measurement, 297
Curie, Marie & Pierre, 276B, 295B
curiosity, in science and medicine, 5B
curium–245, 281
cyanide, 761B
cyanocobalamin, 644T
cyclic AMP (cAMP), 604–5B
cycloalkanes, 308, 321–32, 347
cyclobutane, 321F
cyclohexane, 321F
cyclohexanecarboxylic acid, 452
cyclohexanol, 391
cyclohexene, 391
cyclooxygenase, 567
cyclopropane, 321F
cysteine, 407, 601F, 610, 623T
cystine, 610
cytosine, 670F, 672F

D

Dalton, John, 44, 209
Dalton's law of partial pressures, 166–67
Dalton's theory, 44
data, 11
DDT, and biological magnification, 340B
debranching enzyme, 733
decanoic acid, 562

decomposition reactions, 128
defense proteins, 597
degenerate code, 686
degradation
  of amino acids, 762–66
  of fatty acids, 782–89, 791–93
  of glycogen, 733–41
dehydration
  alcohols, 394–96, 409
  esters, 463
  ethers, 405, 409
dehydrogenases, 635
dehydrogenation reaction, 786
deletion mutations, 692
α-demascone, 425F
Demerol, 496
denaturation
  of enzymes, 646
  of proteins, 617–22
denatured alcohol, 387
density
  definition of, 28
  of gas, 163
  measurement, 29–30
dental fillings, and electrochemical
    reactions, 266B
dental plaque, 528B
deoxyribonucleotide, 670
deoxyribose, 541, 670F
detergents, 621, 780
deuterium, 92B, 277, 287
dextran, 528B
dextrorotatory, 531
diabetes mellitus
  Benedict's reagent, 542
  blood glucose levels, 738B
  electrolytes, 200
  hemodialysis, 201B
  ketosis, 789
  sucrose in diet, 382B
  urine tests, 31B
diamond, 172, 305
diarrhea, and electrolytes in body fluids, 200.
    See also oral rehydration therapy
diatomic compound, 82B
2,3-dibromobutane, 357
cis-1,2-dibromocyclopentane, 324
2,5-dibromohexane, 316
dibromomethane, 330
1,2-dibromopentane, 358
1,2-dichlorocyclohexane, 323
1,2-dichloroethene, 348
dideoxyadenosine triphosphate (ddA), 705
diet. *See also* estimated safe and adequate
    daily dietary intake; foods and
    food industry
  atherosclerosis, 580
  carbohydrates, 526
  calcium, 67B
  cholesterol, 575
  copper deficiency, 57B
  fiber, 526
  lipids, 556

  proteins, 622–23
  sodium and potassium, 94B
  sucrose, 545
  weight loss, 28B, 784–85B
diethyl ether, 404, 405
differentially permeable membranes, 194F
diffusion
  of gases, 155
  membrane transport, 584–85
digestion and digestive tract
  cellulose, 549, 551
  dietary triglycerides, 779–81
  hydrolysis, 715–17
  proenzymes, 651T
  proteins, 622–23
digitalis, 574B
Digitalis purpurea, 574B
digitoxin, 574B
diglycerides, 567
dihydroxyacetone (DHA), 433B, 439B
dihydroxyacetone phosphate, 439B, 634, 723
diisopropyl fluorophosphate (DIFP), 516
dilution, of concentrated solutions, 188–90
dimensional analysis, 13
dimethylamine, 494
2,3-dimethylbutane, 313T
dimethyl ether, 108, 405
1,1-dimethylethyl, 315T
2,2-dimethyl-3-hexyne, 344
N,N-dimethylmethanamine, 490, 491, 492,
    493, 495T
2,6-dimethyl-3-octene, 344
cis-3,4-dimethyl-3-octene, 349
2,4-dimethylpentanal, 420
2,2-dimethylpropanal, 397
dinitrogen monoxide, 90
dinitrogen tetroxide, 90, 91
dinucleotide diphosphokinase, 755
dipeptide, 602
dipole, 109
dipole-dipole attraction, 417
dipole-dipole interactions, 170
directional covalent bond, 105
disaccharides, 436, 527, 543–47
disease, as chemical system, 128. *See also*
    AIDS; cancer; cardiovascular
    disease; cholera; diabetes mellitus;
    genetic disorders; heart disease;
    medicine
disinfectants, 264B, 402. *See also* antiseptics;
    sterilization
disorder, in chemical systems, 211
dissociation
  polyprotic substances, 255
  solutions of ionic and covalent
    compounds, 93
  strong acids, 242, 244
  water, 246
distillation, 387B
disulfide bond, 407, 408F
disulfiram, 430B
divergent evolution, and enzymes, 656
division, of significant figures, 21–22

DNA (deoxyribonucleic acid)
  cloning, 700F
  damage repair, 695–96
  nucleotides, 669
  purines and pyrimidines, 373, 525
  replication, 674–81
  structure of, 541, 671–74
  ultraviolet light, 694–95
DNA cloning vectors, 697, 699
DNA fingerprinting, 702–3B
DNA helicase, 679F
DNA ligase, 633, 679F
DNA polymerase I, 679F, 680
DNA polymerase III, 678–79, 680, 681F
DNA sequencing, 704–6
cis-7-dodecenyl acetate, 479B
Domagk, Gerhard, 653
dopamine, 500, 501, 511
d orbitals, 62
double bond, 96, 101
double helix, of DNA, 671–72, 673F
double-replacement reaction, 129
drug(s). *See also* antibiotics; pharmaceutical chemistry
  HIV protease inhibitors, 641B
  mechanisms of delivery, 240B, 586B
  pharmaceutical chemistry, 144B
drug abuse
  cocaine, 499, 500F, 501–2
  dopamine levels, 511
  methamphetamine, 496, 500–501B
  opiates, 513B, 605B
dry heat, and sterilization, 647
DuPont Corporation, 364B
Duve, Christian de, 647
dynamic equilibrium, 181, 226–27

E

EcoR1, 696
effective collision, 219
eicosanoids, 564–67
Einstein, Albert, 284–85
elaidic acid, 351
elastase, 622, 651T, 656–57
electrochemical reactions, 266B
electrolysis, 269–70, 271F
electrolytes, 93, 179, 200–202
electrolytic solution, 93
electromagnetic radiation, 47–49
electromagnetic spectrum, 47
electron(s)
  atomic theory, 45
  composition of atom, 39–40
  properties based on structure of, 111–13
electron acceptor, 80
electron affinity, 70
electron arrangement, and periodic table, 56, 58–65
electron carriers, and fatty acid synthesis, 792
electron configuration, 56, 58–65
electron density, and atomic theory, 51

electron donor, 80
electronegativity, 83, 84T
electronic balance, 24F
electronic transitions, and Bohr atom, 50
electron pairs, 95
electron spin, 62
electron transport system, 747, 760
electrostatic force, 80
elements. *See also* periodic table
  biological systems, 54T
  definition of, 9
elimination reaction, 394
elongation factors, 690
Embden-Meyerhof Pathway, 719
emission spectrum, 48, 50F, 51
emphysema, 648B, 705B
emulsification, of lipids, 780
emulsifying agent, 568
emulsion, 468
enantiomers, 524B, 529–30
endosomes, 579
endostatin, 596B
endothermic reactions, 209–10, 214
enediol reaction, 542
energy. *See also* energy metabolism; fossil fuels; petroleum industry
  chemical reactions, 209, 215–18
  cost of, 208B
  definition of, 3
  equilibrium, 226–35
  food calories, 28
  kinetics, 218–25
  lipids, 557, 568
  membrane transport, 590
  systems of measurement, 27
  thermodynamics, 208–14
energy levels, and Bohr atom, 49
energy metabolism
  anabolism and citric acid cycle, 770–72
  degradation of amino acids, 762–66
  exercise, 748–49B
  oxidative metabolism, 757, 760–62
  urea cycle, 766–69
English system, of measurement, 11–12, 13, 14T, 15, 154
enkephalins, 604B, 605B
enolase, 725
enol form, 361, 436, 437, 442
enoyl-CoA hydrase, 787
enthalpy, 211
enthrane, 405
entropy, 211–13
environment. *See also* air pollution; climate, and effect of water in environment; global warming; greenhouse effect
  acid rain, 256B
  biodegradable plastics and waste disposal, 456–57B
  DDT and biological magnification, 340B
  enzymes, 646–48
  frozen methane, 307B
  nuclear waste disposal, 285B
  oil-eating microbes, 317B

petroleum industry and gasoline production, 326B
  plastic recycling, 366–67B
enzyme(s). *See also* allosteric enzymes; coenzymes
  activation reaction of, 636
  body temperature, 620
  cellular functions, 597
  classification, 631–35
  cofactors, 643–46
  definition of, 630
  environmental effects, 646–48
  extreme temperatures, 630B, 647–48
  fatty acid biosynthesis, 792
  inhibition of activity, 652–56
  medical uses, 660–61
  myocardial infarction, 658–59B
  nerve transmission and nerve agents, 654–55B
  nomenclature, 635–36
  oxidation-reduction reactions, 400
  proteolytic enzymes, 656–57
  regulation of activity, 649–52
  specificity, 630–31, 639
  stereospecific, 524B
  substrate concentration and catalyzed reactions, 637
  transition state and product formation, 639–43
enzyme assays, 660
enzyme inhibitors, 652–56
enzyme replacement therapy, 660–61, 722B
enzyme-substrate complex, 637–38, 639
EPA (Environmental Protection Agency), 317B
ephedra, 496–97
ephedrine, 489, 496, 501B
epinephrine, 511, 631
equatorial atoms, 327
equilibrium
  energy, 226–35
  solubility, 181
equilibrium constant, 227–31
equilibrium reactions, 143, 226
equivalence point, 254
equivalent, and concentration of solutions, 191
error, 19–20
erythrulose, 433B
Escherichia coli, 676, 694, 696
essential amino acids, 622, 623, 770B
essential fatty acids, 564
ester(s), 309T
  chemical reactions, 462–66
  common functional groups, 309T
  hydrolysis, 466–68
  nomenclature, 461–62
  physical properties, 461
esterification, 460, 463, 561
estimated safe and adequate daily dietary intake (ESADDI), 57B, 94B
estrone, 576
ethanal, 400, 419, 421T, 424

ethanamide, 505, 506T
ethanamine, 491, 492T, 493T, 495T
ethane
   bond angle, 342
   molecular formula, 310, 312F
   nomenclature, 343
   staggered conformation, 325F
1,2-ethanediol, 386, 389
ethanoate anion, 428
ethanoic acid
   boiling point, 449
   chemical reactions, 457
   hydrolysis, 467, 472
   nomenclature, 451
   oxidation, 400
ethanoic anhydride, 473, 474
ethanoic propanoic anhydride, 474
ethanol
   alcohol fermentation, 728
   boiling point, 491, 493T
   breathalyzer test, 401B
   chemical equations, 140
   functional groups, 309
   hydration, 358
   medical use, 387
   molecular structure, 389
   nomenclature, 386
   oxidation-reduction reactions, 267
   rate equations, 225
ethanoyl chloride, 470, 471
ethene
   bond angle, 342
   dehydration, 394
   nomenclature, 343
   physical properties, 343T
   plants, 352
ethers, 309T, 403–5, 409
ethyl acetate, 461
ethylamine, 494
ethylbenzene, 369
ethyl butanoate, 463, 464B
2-ethyl-1-butanol, 390
ethyl dodecanoate, 562
ethylene, 193, 362
ethyl group, 314T
5-ethylheptanal, 420
3-ethyl-3-hexene, 350
ethylmethylamine, 494
4-ethyl-3-methyloctane, 317
4-ethyloctane, 316
6-ethyl-2-octanone, 422
ethyl pentyl ketone, 423
ethyl propanoate, 466
ethyne, 342, 343
17-ethynylestradiol, 346B
eukaryotes, and genetics, 673–74, 675F, 681, 684F, 685, 700F
evaporation, 169
evolution, and enzymes, 656
exact numbers, 22
excited state, of atom, 49, 50
exercise
   anaerobic threshold, 727
   energy metabolism, 748–49B
   weight loss, 28B, 568, 785B

exercise intolerance, 722B
exons, 684–85
EXO-SURF Neonatal, 556B
exothermic reactions, 209–10
expanded octet, 103
experimental basis, of chemical equation, 127–28
experimental quantities, 23–31
experimentation, and scientific method, 4
exponential notation, 18
extensive property, 9
Exxon Valdez (ship), 317B
eye, and chemistry of vision, 440–41B

F

Fabry's disease, 573B
facilitated diffusion, 585–87
factor-label method, 13
Fahrenheit scale, 25–26, 157B
falcarinol, 346B, 347B
familial emphysema, 648B, 705B
farnesol, 352, 353F
fast twitch muscle fibers, 749B
fat(s)
   brown fat, 758–59B
   digestion of, 716–17
fatty acid(s)
   carboxylic acids, 341F, 449, 455
   chemical reactions, 561–64
   degradation, 782–89, 791–93
   eicosanoids, 564–67
   geometric isomers, 351
   membrane phospholipids, 582
   structure and properties, 558–60
   synthesis, 791–93
fatty acid metabolism, 789–91. *See also* lipid metabolism
fatty acid synthase, 792
feedback inhibition
   of enzymes, 650–51
   of glycolysis, 726
feedforward activation, 726
fermentation, 383, 387, 456, 726–29
fetal alcohol syndrome (FAS), 388B, 401B
fetal hemoglobin, 616
F_0F_1 complex, 760
fibrils, 549
fibrinogen, 599B, 621
fibrous proteins, 608
filling order, for electrons in atoms, 63F, 64
film badges, 296
fireworks, 52B
Fischer, Emil, 524–25, 532, 534, 638
Fischer, Hermann Otto Laurenz, 525
Fischer projection formulas, 532–33
fish
   acid rain, 256B
   migration, 78B
fission, nuclear, 286
flavin adenine dinucleotide (FAD), 645, 646, 755, 761
flavors, and esters, 464–65B
Fleming, Alexander, 4B, 507B
Flotte, Terry, 705B

fluid mosaic structure, of cell membrane, 582–83
fluorine, 58T, 59, 66, 171
3-fluoro-2,4-dimethylexane, 317
folic acid, 644T, 652–53
folklore, and ethylene, 352, 362B
Food and Drug Administration (FDA), 496, 506, 509
foods and food industry. *See also* diet
   calories, 28B
   carboxylic acids, 455–56
   copper, 57T
   esters, 461
   ethylene and fruit ripening, 352, 362B
   fuel value, 217, 218
   hydrogenation, 356
   potassium, 94B
   sodium, 94B
   sugars and sugar substitutes, 382B
f orbitals, 62
forests, and acid rain, 256B
formaldehyde, 421T, 426B. *See also* methanal
formalin, 424
formic acid, 449, 453T
formula unit, 124
formula weight, 124–26
fossil fuels. *See also* oil spills; petroleum industry
   acid rain, 256B, 257B
   combustion, 265–66
   global warming, 329
   nonrenewable resources, 304B
foxglove plant, 574B
fragrances, and esters, 464–65B
Franklin, Rosalind, 671
free energy, 214
free rotation, of carbon–carbon single bond, 325
freeze–fracture, 582
freezing point, and concentration of solutions, 192–93
Friedman, Jeffrey, 778B
fructose, 539–40, 640, 642F
β-fructose, 545
D-fructose, 529
fructose-1,6-biphosphatase, 731
fructose-1,6-bisphosphate, 439, 650, 723
fructose-6-phosphate, 650, 721, 723
fruit flavors, 464B
fucose, 546B, 547B
fuel value, of foods, 217, 218
Fuller, Buckminster, 305
fumarase, 632, 756
fumarate, 754F, 757F, 768
functional group, 308–9
furan, 373
fused rings, 574, 575
fusion, nuclear, 287
fusion reactions, 92B

G

GABA, 515
galactocerebroside, 571
galactose, 540–41, 546B, 547B

galactosemia, 544–45
gallium-67, 291
gallon, units of measurement, 11
gamma production, 280–81
gamma radiation, 48B, 278–79, 280, 289
gamma rays, 278
Gane, R., 362B
gangliosides, 572
gas(es)
    Avogardro's law, 161–62
    blood and respiration, 168B
    Boyle's law, 155–57
    Charles's law, 157–59
    combined gas law, 160–61
    Dalton's law of partial pressures, 166–67
    densities, 163
    ideal gas concept, 153
    ideal gases versus real gases, 167
    ideal gas law, 163–65
    kinetic molecular theory, 154–55
    measurement of, 153–54
    molar volume, 163
    solubility of, 181–82
    states of matter, 7
gas gangrene, 464B, 729B
gasoline, petroleum industry and production of, 326B
gastrointestinal tract, and prostaglandins, 565. *See also* digestion
Gaucher's disease, 573B, 660–61
GDP molecule, 758B, 759B
Geiger, Hans, 45–46
Geiger counter, 296
gene therapy, 668B, 722B
genetic(s). *See also* DNA; genome; molecular genetics; RNA
    DNA repair, 695–96
    DNA replication, 674–81
    genetic code, 685–86, 687F
    information flow in biological systems, 681–85
    mitochondria, 746B
    mutations, 692–93
    polymerase chain reaction, 701–702
    primary amino acid sequences, 606
    protein synthesis, 687–92
    recombinant DNA, 696–701
    structure of DNA and RNA, 671–74
    structure of nucleotide, 669–71
    ultraviolet light damage, 694–95
genetic code, 685–86, 687F
genetic counseling, 668B
genetic disorders
    energy metabolism, 746B
    familial emphysema, 648B, 705B
    glycolysis, 722B
    molecular genetics and detection of, 668B
    obesity, 778B
    pyruvate carboxylase deficiency, 769B
    sickle cell anemia, 448B, 617, 693
    sphingolipid metabolism, 573B
genetic engineering, 699, 701. *See also* recombinant DNA technology
genome, 673, 701, 703–6. *See also* Human Genome Project

genome analysis, 704
genomic library, 704
Genzyme Corporation, 661
geometric isomers, 322, 346–52. *See also* cis-trans isomers
geraniol, 352, 353F, 405, 406F
Glaxo-Wellcome Company, 556B
global warming, 131, 166B, 329. *See also* greenhouse effect
globular proteins, 609–11
α-globulins, β-globulins, and γ-globulins, 599B
glucagon, 794B, 797–98
glucocerebrosidase, 573B, 660, 661
glucocerebroside, 571, 660, 661
glucokinase, 736
gluconeogenesis, 439, 730–32, 793
D-glucosamine, 550B, 551B
glucose
    Benedict's reagent, 429, 542–43
    biological systems, 535–39
    blood levels, 543, 738B, 796, 797
    calculating molarity from mass, 187–88
    condensation reaction, 640, 642F
    conformation of cycloalkanes, 325
    dehydration, 396
    facilitated diffusion, 586–87
    gluconeogenesis, 730–32
    glycolysis, 721
    Haworth projection, 538–39
    osmosis, 194
    urine levels, 542–43
    weight/volume percent, 183
D-glucose, 529, 586
L-glucose, 586
glucose-6-phosphatase, 731
glucose-6-phosphate, 721, 740B
glucose tolerance test, 738B
glucosuria, 542
glucosyl transferase, 528B
α-D-glucuronate, 550B
glutamate, 516–17, 601F, 623T, 764, 770
glutamate dehydrogenase, 764
glutamic acid, 693
glutamine, 601F, 623T
glyceraldehyde, 530, 533
D-glyceraldehyde, 529, 531F, 533, 536, 598
L-glyceraldehyde, 531F, 533, 598
L-glyceraldehyde-3-phosphate, 634, 723
glyceraldehyde-3-phosphate dehydrogenase, 724
glycerides, 567–70
glycerol, 167
glycerol-3-phosphate, 796
glycine, 515, 601F, 602, 612B, 623T
glycogen, 548–49, 733–41
glycogen granules, 733
glycogenolysis, 733–41, 793
glycogen phosphorylase, 652, 733, 735F
glycogen storage diseases, 740B
glycolysis
    aldol condensation, 439
    ATP, 714
    chemical reactions, 721, 723–25
    enzymes, 634

genetic disorders, 722B
    hydroxyl group, 383
    overview of, 719–21
    phosphoenolpyruvate, 438
    phosphoesters, 476–77
    regulation of, 726
glycoproteins, 583, 612
glycosaminoglycans, 550B
glycosidase, 733
glycosides, 543
glycosidic bonds, 436, 543, 545F
glycosphingolipids, 571
glycyl-alanine, 602
gold, 120F, 291
Goldstein, Eugene, 45
Gore-Tex, 364B
gram, 12, 137–41
gravity, 23
Greek language, and nomenclature, 91
greenhouse effect, 131, 166B, 167, 266, 307B. *See also* global warming
ground state, of Bohr atom, 50–51
group(s), and periodic table, 55
Group A elements, 55
Group B elements, 55
Group IA elements, 55, 59
Group IIA elements, 55, 59
Group IIIA elements, 59
Group VIIA elements, 55, 59
Group VIIIA elements, 55
group specificity, of enzyme, 639
guanine, 670F, 672F
guanosine triphosphate (GTP), 732

H

Haber process, 223F, 233F
Halaas, Jeff, 778B
half cells, 268
half-life, of radioactive isotopes, 282–84, 285B, 293–94
halide, 309T, 330
haloalkane, 329
halobenzene, 374
halogen(s), 55
halogenated ethers, 405
halogenation, 329–30, 356–58, 374
halothane, 328B
hangovers, and alcoholic beverages, 400
hard water, 563
Harrison Act (1914), 604B
Haworth projection, 538–39
heartburn, 514–15
heart disease, 94B, 270B, 574B, 658–59B, 660
heavy metals, 622
heavy water, 277
helicase, 678
helium
    combined gas law, 161
    density, 163
    electron configuration, 63
    molar volume, 164
    nuclear fusion, 287
    origin of elements, 92B
    valence electrons, 58T, 59

α-helix, of amino acid, 607–9
heme group, 615, 616F
hemiacetal
  addition reactions, 434, 435, 436, 442
  glucose, 537
  keto-enol tautomers, 437
hemiketal, 435, 436, 441, 540
hemodialysis, 201B
hemoglobin
  blood pH, 262B
  carbon monoxide poisoning, 136B
  human genetic diseases, 448B
  oxygen transport, 597, 616
  sickle cell anemia, 617
hemolysis, 196
Henderson-Hasselbalch equation, 261–63
Henry's law, 181–82
heparin, 551B
2,4-heptadiene, 344
heroin, 503, 504, 604–5B
heterocyclic amines, 488B, 502–4
heterocyclic aromatic compounds, 373
heterogeneous mixture, 10
heteropolymers, 457B, 469
heteropolysaccharides, 550–51B
hexachlorophene, 403
2,4-hexadiene, 344
hexane, 312, 313T
2-hexanone, 427
hexokinase, 634, 721, 726
hexose, 527
hexylresorcinal, 403
high-density lipoproteins (HDL), 578, 579F, 580
high-density polyethylene (HDPE), 366B, 367B
high-energy bond, 407B, 714
high entropy, 211
Hindenburg (airship), 152B
histamine, 489, 514–15
histidine, 601F, 623T
HIV (human immunodeficiency virus), 264B, 641B, 676–77B
HMG-CoA (β-hydroxy-β-methylglutaryl CoA), 790–91
holoenzyme, 643
homogeneous mixture, 10
homopolymer, 469
homostatic mechanisms, 758
hormones. *See also* insulin; steroids
  cellular metabolism, 798
  definition of, 564B
  lipids, 557
hot-air balloon, 159
hot packs, 221B
hot springs, in Yellowstone National Park, 582, 629F, 648, 701
Hughes, John, 604B
human body. *See also* biological systems; bone; brain; digestion and digestive tract; infants; medicine; reproductive system; respiratory tract
  chromosomes, 673
  genome, 701
  lipids, 557
  magnetite in brain of, 78B

Human Genome Project (HGP), 703–6
hyaluronic acid, 551B
hybridization, and recombinant DNA, 697, 698F, 699, 701F
hydrates, 124, 128
hydration, 358–61, 374, 391
hydrocarbon(s), 266, 308, 311
hydrocarbon tails, of membrane phospholipids, 581
hydrochloric acid
  Arrhenius and Brønsted–Lowry theories, 241
  calculation of pH, 248
  concentration of solution, 253–54
  energy involved in calorimeter reactions, 215–16
  hydrohalogenation, 363
  molarity of solutions, 189, 190
  neutralization, 252
  replacement reactions, 129
  as strong acid, 242, 244
hydrogen
  acid-base reactions, 131
  airships and blimps, 152B
  balancing of chemical equations, 134
  chemical equilibrium, 227
  chemical formula, 124
  covalent bonding, 81–82
  electron configuration, 63
  emission spectrum, 50F, 51
  Lewis structure and stability, 100–101
  molar mass, 120
  nuclear fusion, 287
  origin of elements, 92B
  polarity, 109
  valence electrons, 58, 59
hydrogenases, 635
hydrogenation
  alcohols, 392–93
  aldehydes and ketones, 431
  alkenes and alkynes, 354–56, 374
  catalyst, 222
  fatty acids, 564
hydrogen bonding
  alcohols, 384
  aldehydes and ketones, 418F
  amides, 505F
  amines, 491F
  amino acids, 600B, 610
  carboxylic acids, 450F
  liquids, 170–71, 173F
hydrogen chloride, 128, 132, 133
hydrogen fluoride, 82, 110, 171
hydrogen iodide, 231, 234
hydrogen ion gradient, 760
hydrogen peroxide, 132B, 264B
hydrohalogenation, 361–63, 374
hydrolases, 632, 633
hydrolysis. *See also* acid hydrolysis
  amides, 509–10, 517
  digestion, 715–17
  esters, 466–68
  lipids, 780
hydrometer, 31B
hydronium ion, 241, 249

hydrophilic molecules, 385, 468, 600
hydrophobic molecules, 384–85, 468, 600, 601F
hydrophobic pocket, 656
hydroxide ion, 249
L-β-hydroxyacyl-CoA dehydrogenase, 787
hydroxyapatite, 612B
3-hydroxybutanal, 438
β-hydroxybutyrate, 791
β-hydroxybutyric acid, 453, 457B
β-hydroxybutyryl ACP, 792
hydroxyl group, 383
4-hydroxylysine, 613B
4-hydroxy-4-methyl-2-pentanone, 439
4-hydroxyproline, 613B
α-hydroxypropionic acid, 453
β-hydroxyvaleric acid, 457B
hyperammonemia, 769
hypercholesterolemia, 579
hyperglycemia, 740
hypertension, 94B, 575
hypertonic solution, 196
hyperventilation, 262B
hypoglycemia, 740
hypothalamus, 778B
hypothesis, 4
hypotonic solution, 196, 589

I

ibuprofen, 478B, 524B
ice
  characteristics of water, 199B
  endothermic reactions, 214
  molecular solids, 172, 173F
ichthyothereol, 346B, 347B
ideal gas, 153, 163, 167
ideal gas law, 163–65
imidazole, 373, 502
immune system, 618B
immunoglobulin(s), 618–19B, 621
immunoglobulin G (IgG), 619B
incomplete octet, 103
incomplete protein, 623
indicator, of pH, 253
indole, 503
induced fit model, of enzyme activity, 638
industry, and uses of aldehydes and ketones, 424, 425F. *See also* foods and food industry; pharmaceutical chemistry; petroleum industry
inert gases, and octet rule, 65
inexact numbers, 22
infants
  brown fat and thermogenesis, 758B
  essential amino acids, 623T
  premature, 556B
inflammatory response, and prostaglandins, 565, 566
information management, and computers, 38B
infrared lamps, 48B
initiation, of transcription, 682–83, 690
initiation factors, 690, 691F

inner mitochondrial membrane, 747
inorganic chemistry, 3
inorganic compounds, compared to organic compounds, 306–7
insect(s), and pheromones, 479B
insecticides, 328B, 340B
insertion mutations, 692
instantaneous dipole, 170
insulin
    cellular metabolism, 797–98
    diabetes mellitus, 794B
    glucose and glucose transport, 535, 587
    glycogenesis, 735
    thiols and synthesis of, 408F
intensive properties, 8–9
intermembrane space, 747
intermolecular forces, 111
intermolecular hydrogen bonds, 385B
intermolecular reactions, 436
International Union of Pure and Applied Chemistry (IUPAC), 55, 315
intracellular location, of fatty acid biosynthesis, 792
intramolecular forces, 111
intramolecular hemiacetal, 537
intramolecular hemiketal, 540
intramolecular hydrogen bonds, 385B
intramolecular reactions, 436
intravenous (IV) solutions, 589–90
introns, 684–85
iodine, and nuclear medicine, 281, 284, 289, 290T
m-iodobenzoic acid, 454
ion(s)
    composition of atom, 43–44
    concentration of solutions, 190–91
    octet rule and formation, 65–68
    periodic table and size of, 68–69
ionic bonding
    amino acids, 610
    definition of, 79
    inorganic compounds, 306
    process and examples of, 79–81
ionic compounds
    formulas and nomenclature, 84–89
    properties of, 92–93
    solubilities of common, 130T
ionic solids, 172
ionization energy, 69, 70F
ionizing radiation, 279, 289
ion pairs, 80, 81
ion product, for water, 246
iron, 121, 144–45
irreversible enzyme inhibitors, 652
ischemia, 658B
isobutyl methanoate, 465B
isocitrate, 753, 754F, 756, 757F
isocitrate dehydrogenase (ICD), 658B, 753
isoelectric point, 620
isoelectronic ions, 66
isoenzymes, 658–59B
isoleucine, 601F, 623T
D-isomer, 536
isomerases, 633, 634, 721B
isoprene, 352, 364B, 572

isoprenoids, 352, 353F, 572
isopropyl alcohol, 359
isopropyl benzoate, 461
isotonic solutions, 196, 589
isotope
    nuclear structure and stability, 282
    nuclide, 277
    radioactive tracers and nuclear medicine, 289, 290T, 291–93
    structure of atom, 40–43
I.U.P.A.C. Nomenclature System, 315–21. *See also* nomenclature

J

jaundice, 2B
Jeffries, Alec, 702B
joules, 27

K

Kekulé, Friedrich, 368
Kekulé structures, 368
Kelvin scale, 25–26, 157B
keratin, 597
α-keratins, 608, 609F, 668
ketal, 435, 436, 441, 543
α–keto acid, 762, 763
ketoacidosis, 789
keto-enol tautomers, 436–38, 442
ketogenesis, 789–91
α–ketoglutarate
    amino acid degradation, 763, 764
    citric acid cycle, 753, 754F, 755, 756, 757F, 770
ketone(s)
    aldehydes and ketones, 424–42
    common functional groups, 309T
    hydration of alkenes, 361
    hydrogenation, 392–93
    medical and industrial uses, 424, 425F
    nomenclature, 422–24
    physical properties, 417–19
ketone bodies, 789–91, 794–95B
ketoses, 527, 542
ketosis, 789
9-keto-*trans*-2-decenoic acid, 479B
kidneys
    diabetes, 794B
    hemodialysis, 201B
    nuclear medicine, 290T
    prostaglandins, 566
kidney stones, and precipitation reactions, 130
kilocalorie, 27B, 526B
kilopascal, 154
kinase, 631
kinetic(s), and energy, 218–25
kinetic energy (K.E.), 27, 155B
kinetic molecular theory, 154–55, 209
Knoop, Franz, 782
Koshland, Daniel E., 638
Krebs, Sir Hans, 750
Krebs cycle, 750, 753–56

L

lactase, 545
β–lactam ring, 507B
lactate dehydrogenase (LDH)
    heart attack and enzyme assays, 660
    lactate fermentation, 727
    myocardial infarction, 658B, 659B
    oxidoreductases, 631
    reduction of ketones, 434
lactate fermentation, 727
lactic acid, 382B, 456, 528B
lactic acidosis, 769B
lactose, 544–45
lactose intolerance, 543
*lac*Z gene, 699
lagging strand, 679
lakes, and acid rain, 256B
Landsteiner, Karl, 546B
lanthanide series, 55
Latin, and nomenclature, 86
lauric acid, 559T
law(s)
    Avogardo's law, 161–62
    Boyle's law, 155–57, 160, 233
    Charles's law, 157–59, 160
    combined gas law, 160–61
    conservation of mass, 126, 132, 141F
    Dalton's law of partial pressures, 166–67
    Henry's law, 181–82
    ideal gas law, 163–65
    multiple proportions, 136B
    Raoult's law, 192
    scientific method, 5
law enforcement, and breathalyzer test, 401B. *See also* DNA fingerprinting; drug abuse
lead, 120, 186
leading strand, 679
LeBel, Joseph Achille, 532
Leber's hereditary optic neuropathy (LHON), 746B
LeChatelier's principle, 232, 258, 262B, 458
lecithin, 568, 569F
length, units of measurement, 11, 24
leptin, 778B
lethal dose (LD), of radiation, 297
leucine, 601F, 623T
leucine enkephalin, 604B, 605B
leukotrienes, 566
Levene, Phoebus, 669
levorotatory, 531
Lewis, G. N., 79
Lewisite, 407
Lewis structures
    alkanes, 311
    definition of and symbols, 79, 80F
    molecules, 93, 95–97
    octet rule, 103–4
    polarity, 109–11
    polyatomic ions, 97–100
    resonance, 101–3
    stability, multiple bonds, and bond energies, 100–101
    VSEPR theory, 104–9

life. *See also* biological systems
    importance of water, 198B
    origins of, 630B
ligases, 633
light
    atomic structure, 46–49
    fractured solids, 213B
    speed of, 47
limonene, 352, 353F, 524B
linear molecular structure, 105, 107T
line formula, 310–11
linkage specificity, of enzyme, 639
linoleic acid, 559T, 564
linolenic acid, 559T
lipases, 632, 779
lipid(s). *See also* lipid metabolism
    biological functions, 557
    cell membranes, 557, 581–90
    complex lipids, 577–80
    controversies concerning, 556
    diet, 556
    fatty acids, 558–67
    glycerides, 567–70
    nonglyceride lipids, 570–77
lipid bilayers, 581
lipid metabolism, 779–82, 793–97, 798
Lipkin, Dr. Martin, 67
liposomes, 586B
liquid(s)
    boiling and melting points, 112–13
    compared to gases and solids, 153T
    compressibility, 167
    hydrogen bonding, 170–71
    solutions, 179
    states of matter, 7
    surface tension, 168
    van der Waals forces, 170
    vapor pressure, 169–70
    viscosity, 167–68
Lister, Joseph, 264B, 402
liter, 12
lithium, 58T, 59, 63
lithium sulfide, 86
liver
    enzymes and diseases of, 660
    lipid and carbohydrate metabolism, 793
    lipoprotein receptors, 579–80
    metabolism and oxidation-reaction reactions, 267
    nuclear medicine, 290T
lock-and-key model, of enzyme activity, 638
London, Fritz, 170
London forces, 170
lone pair, of electrons, 97
low-density lipoproteins (LDL), 578, 579, 580
low-density polyethylene, 367B
low entropy, 211
lung cancer, 694B
lung diseases, and acid rain, 257B
lyases, 632
lysergic acid diethylamide (LSD), 503, 504
lysine, 601F, 623T
lysosomes, 579, 647

M

McArdle's disease, 740B
McCarron, Dr. David, 67B
McGrayne, Sharon Bertsch, 276B
magainins, 5B
magnesium carbonate, 128
magnetic resonance imaging (MRI), 178B, 292B
magnetism, and migration, 78B
magnetite, 78B
magnetosomes, 78B
magnetotactic bacteria, 78B
malaria, 617
malate
    citric acid cycle, 401–2, 754F, 756, 757F
    glycolysis, 632
    lyases, 632
malate dehydrogenase, 756
maltose, 543–44
mammals, and body temperature, 582. *See also* reindeer; skunk
mannitol, 382B
margarine, 564
marijuana, 511
Markovnikov, Vladimir, 359
Markovnikov's rule, 359, 361, 363
mass. *See also* mass number; molar mass
    concentration of solutions based on, 182–87
    experimental quantities, 23–24
    subatomic particles, 39T
Massachusetts General Hospital (Boston), 178B
mass number (A), 39, 277
mathematical representation, of reaction rate, 224–25
matrix space, 747
matter. *See also* gas; liquid(s); solid(s)
    chemical properties, 8
    classification of, 9–10
    definition of, 3
    physical properties, 7, 153T
measurement
    in chemistry, 11–16
    of gases, 153–54
    pH scale, 246–52
    of radiation, 295–98
mechanical stress, and proteins, 622
medicine. *See also* antibiotics; disease; drugs; genetic disorders; human body; pharmaceutical chemistry
    AIDS virus and nucleotides, 676–77B
    alcohols, 387–89
    aldehydes and ketones, 424
    alkynes, 346–47B
    Ames test for carcinogens, 694–95B
    amides, 505–6
    amines, 495–97
    blood gases and respiration, 168B
    blood pH, 262B
    blood pressure and sodium ion/potassium ion ratio, 94B
    carbon monoxide poisoning, 136B
    chloroform in swimming pools, 331B
    copper deficiency and Wilson's disease, 57B
    drug delivery, 240B, 586B
    enzymes, 654–55B, 658–59B, 660–61
    familial emphysema, 648B, 705B
    formaldehyde and methanol poisoning, 426B
    genetic engineering, 701T
    Gore-Tex, 364B
    hemodialysis, 201B
    HIV protease inhibitors, 641B
    hot and cold packs, 221B
    immunoglobulins, 618–19B
    liposomes, 586B
    monosaccharide derivatives and heteropolysaccharides, 550–51B
    nuclear medicine and radioactivity, 288–93
    oral rehydration therapy, 197B
    oxidizing agents for control of microbes, 264B
    pharmaceutical chemistry, 144B
    polyhalogenated hydrocarbons as anesthetics, 328B
    proteins in blood, 599B
    radiation therapy, 49B, 289
    role of curiosity in, 5B
    role of observation in, 2B
    semisynthetic penicillins, 507B
    sphingolipid metabolism, 573B
    steroids and heart disease, 574B
    urine tests and diagnosis, 31B
melanin, 668
melting point
    alkanes, 311T, 313T, 343T
    alkenes and alkynes, 343T
    fatty acids, 559F, 560
    ionic and covalent compounds, 93
    liquids and solids, 112–13, 171
    physical properties, 7
membranes. *See* cell membrane
membrane transport, 583–90
Mendel, Gregor, 669
Mendeleev, Dmitri, 52–54
Menkes' kinky hair syndrome, 57B
mental illness, 178B
mercury, 25, 30, 154
Meselson, Matthew, 676
messenger RNA (mRNA), 681
metabolic myopathy, 722B
metabolism, and oxidation-reduction reactions, 267. *See also* carbohydrate metabolism; cellular metabolism; energy metabolism; fatty acid metabolism
meta–Cresol, 369
metal(s)
    catalysts, 222
    corrosion, 265
    ionic compounds, 84–85
    periodic table, 55
metal hydroxides, 242

metallic bonds, 172
metallic solids, 172
metalloids, 55
metastable isotope, 280
metastasis, 596B
meta–Xylene, 369
meter, 12
methadone, 513B
methamphetamine, 496, 500–501B
methanal, 397, 419, 424. See also formaldehyde
methanamide, 506T
methanamine, 490, 492T, 493, 495T
methane
    bromination, 330
    carbon monoxide, 136B
    chemical reaction and energy, 209
    chemical reaction with oxygen, 131
    covalent bonding, 82
    crystal structure, 172F
    formula units, 125F
    frozen on ocean floor, 307B
    melting point, 93
    model of, 5–6
    molecular formula, 310
    molecular geometry, 106
    oxidation, 266
methane hydrate, 307B
methanogens, 307B
methanol, 387, 389, 426B, 493T
methedrine, 496
methicillin, 507B
methionine, 601F, 623T
methionine enkephalin, 604B, 605B
Meth lab, 501B
methoxyethane, 403, 417, 449
methyl alcohol, 389
methylamine, 242, 494
methylammonium chloride, 498
N-methylbutanamide, 508
3-methyl-1-butanethiol, 405, 406F
methyl butanoate, 465B
3-methylbutanoic acid, 451
3-methyl-1-butanol, 433
2-methyl-2-butene, 395
3-methyl-1-butene, 395
methylbutyl ethanoate, 465B
methyl butyl ketone, 423
β-methylbutyraldehyde, 421, 423
3-methyl-1,4-cyclohexadiene, 344
3-methylcyclopentene, 345
methyl decanoate, 562
N-methylethanamide, 506T
N-methylethanamine, 493, 495T
methyl ethanoate, 461
1-methylethyl, 315T
methyl group, 108, 314T
6-methyl-2-heptanol, 385
methyl methacrylate, 365T
N-methylmethanamide, 506T
N-methylmethanamine, 490, 492T, 495T
2-methylpentanal, 419, 420
3-methylpentanal, 420
2-methylpentane, 313T

3-methylpentane, 316
2-methyl-2-pentanol, 427
3-methyl-2-pentene, 350
N-methyl-1-phenyl-2-propanamin, 496
N-methylpropanamide, 505, 510
N-methylpropanamine, 494
2-methylpropane, 310
methyl propanoate, 463
2-methyl-2-propanol, 390, 399
2-methylpropene, 343T
1-methylpropyl, 315T
methyl propyl ether, 404
methyl propyl ketone, 423
methylurea, 639
metric system, of measurement, 12–13, 15
Meyer, Lothar, 52–53
micelles, 468, 779, 780
microbes. See also bacteria
    oil spills and oil-eating microbes, 317B
    oxidizing agents for chemical
        control, 264B
microfibril, 608, 609F
microwave radiation, 48B
Miescher, Friedrich, 669
migration, of birds, 78B
milliequivalents, 191
milliequivalents/liter (meq/L), 191
mitochondria, 746B, 747
mixture, definition of, 9–10
models, use of in chemistry, 5–6
molality, 192–93
molarity, 187–88
molar mass, 119–26
molar quantity, and chemical equations, 132
molar volume, of gas, 163
mole(s)
    Avogadro's number, 119–20
    calculating atoms, mass, and, 121–23
    chemical equations, 137–38
    concentration of solutions, 187–91
    conversion of reactants to products,
        138–41
molecular formula. See also nuclear
        equations; structural formula
    alkanes, 310, 311T, 342, 343T
    alkenes and alkynes, 342, 343T
molecular genetics. See also genetic(s);
        genetic engineering; recombinant
        DNA technology
    central dogma of, 681
    detection of genetic disease, 668B
molecular geometry
    Lewis structures, 104–9
    properties based on, 111–13
molecular solids, 172
Molecular Targets Drug Discovery
        Program, 346B
molecule(s). See also molecular formula;
        molecular geometry
    covalent compounds, 89
    Lewis structures, 93, 95–97
    optical activity and structure of, 532
molybdate ion, 292–93

molybdenum–99, 292–93
monatomic ions, 86, 87T
monochromatic light, 530
monoglycerides, 567
monomers, 364, 469, 717
monoprotic acid, 255
monosaccharide(s), 527–29, 535–43
monosaccharide derivatives, 550–51B
morphine, 503, 504, 512B, 604–5B
Morton, Dr. Wiliam, 405
motility, and cellular energy, 713T
movement proteins, 597
Mulder, Johannes, 596
Müller, Paul, 340B
multilayer plastics, 367B
multiple bonds, 100–101
multiple proportions, law of, 136B
multiplication, of significant figures, 21–22
mummies, and soap, 563B
muscle tissue, 796
mutagens, 692, 693
mutations, genetic, 692–93
Mylar, 470
myocardial infarction, 658–59B
myoglobin, 597, 615–17
myoglobinuria, 722B
myosin, 597
myrcene, 352, 353F
myricyl palmitate (beeswax), 577
myristic acid, 559T

N

NAD^+ (nicotinamide adenine dinucleotide
        ion), 401–402, 645–46, 660, 719
NADH (nicotinamide adenine dinucleotide)
    citric acid cycle, 402, 761
    coenzymes, 645–46
    fatty acid synthesis, 792
    glycolysis, 724
    heart attack and enzyme assays, 660
    oxidative phosphorylation, 757
    reduction reactions, 434
NADH dehydrogenase, 746, 760–61
$NADP^+$ (nicotinamide adenine dinucleotide
        phosphate), 645, 646
Na^+-K^+ ATPase, 590
nalbuphine, 512B
naloxone, 512B
naming, of chemical compounds, 84, 86–88.
        See also common names;
        nomenclature
nanometers (nm), 24
naphthalene, 371, 372
naproxen, 478B
NASA (National Aeronautic and Space
        Administration), 307B
National Academy of Science, 94B
National Institutes of Health, 703
National Research Council, 94B
natural radioactivity, 45–46, 277–79, 291–93
negative allosterism, 649
negatively charged amino acids, 600, 601F

neon, 58T, 59, 63
neotame, 509
nerve agents, and enzymes, 654–55B
nerve synapse, 654B
net energy, 220
neurotransmitters, 511, 514–17, 654–55B
neutral glycerides, 567–68
neutralization, of acids and bases, 252–55, 458, 459, 498–99, 517
neutrons, 39–40, 45
niacin, 644T
nicotine, 373, 488B, 503, 504, 515–16. *See also* cigarette smoking
Niemann-Pick disease, 573B
nitrate ion, and resonance hybrids, 102–103
nitration, 374
nitric acid, 242, 256B, 257B
nitric oxide, 103, 257B, 516–17
nitrobenzene, 369, 372, 374
nitrogen
    automobile air bags, 118B
    gas density, 163
    Lewis structure and stability, 100–101
    scuba diving and the bends, 183B
    valence electrons, 58T, 59
nitrogen-16, 280
nitrogen dioxide, 90, 160, 227, 257B
nitrogen monoxide, 91
nitrogen oxides, and acid rain, 256B
Nobel Prize Women in Science (McGrayne), 286B
noble gases, 55
nomenclature. *See also common names; I.U.P.A.C. Nomenclature System*
    acid anhydrides, 475
    acid chlorides, 472
    alcohols, 385–87
    aldehydes, 419–22
    alkanes, 315–19
    alkenes and alkynes, 343–45
    amides, 505
    amino acids, 600T
    aromatic compounds, 368–71
    carboxylic acids, 451–55
    of chemical compounds, 84, 86–88
    cycloalkanes, 322, 324
    enzymes, 635–36
    esters, 461–62
    ethers, 403–4
    ketones, 422–24
    monosaccharides, 527, 529
    stereoisomers, 534
    thiols, 406
2,5-nonadiene, 344
nonane, 316
nondirectional covalent bond, 105
nonelectrolytes, 93, 179
nonessential amino acids, 622–23
nonglyceride lipids, 570–77
nonmetals, 55, 84–85
nonpolar molecules, 109–10
nonshivering thermogenesis, 758B
nonspontaneous reactions, 211

nonsteroidal anti-inflammatory drugs (NSAIDs), 478B, 551B
norepinephrine, 511, 631
norlutin, 576
normal boiling point, 169
19-norprogesterone, 576
novocaine, 496, 499B
N-terminal amino acid, 602
nuclear decay, 281–82
nuclear equations, 279–82. *See also* molecular formula
nuclear fission, 286–87
nuclear fusion, 287
nuclear imaging, 289, 296
nuclear medicine, 288–93
nuclear power, 284–87
nuclear power plant, 286
nuclear reactions, 92B
nuclear reactor, 291
nuclear stability, 282
nuclear symbols, 277
nuclear waste disposal, 285B, 295
nucleoside analogs, 676B
nucleosome, 674
nucleotide, 669–71, 713
nucleus, of atom, 39, 45–46, 282
nuclide, 277, 279
nutrient proteins, 597
nutritional calorie, 217, 218
nystatin, 588B, 589B

O

Oak Ridge National Laboratories (Tennessee), 287F
obesity, 382B, 778B, 784–85B
obligate anaerobes, 729B
observation, in medicine and science, 2B, 4
Occupational Safety and Health Act (OSHA), 279
ocean, and frozen methane, 307B
octadecanoic acid, 564
octane ratings, of gasoline, 326B
3-octanol, 431
2-octanone, 425F
4-octanone, 422
octet rule, 65–68, 79, 95, 103–4
octyl ethanoate, 465B
odd electron molecules, 103
oil-eating microbes (OEMs), 317B
oil spills, 317B
oleic acid, 351, 559T, 564
oligosaccharides, 546B, 547B
opiates, 512–13B
opium poppy, 512–13B, 604–5B
oral contraceptives, 346B, 576
oral rehydration therapy, 196, 197B
order of reaction, 224
organic chemistry
    alkanes, 310–21, 325–32
    carbon, 305–9
    chemical reactions, 327–32
    cycloalkanes, 321–32

organic compounds
    families of, 308–9
    inorganic compounds, 306–7
    nomenclature, 315–19
    origin of, 304B
organic solvents, 621
organophosphates, 654B
ornithine, 766, 767F, 768
ornithine transcarbamoylase, 766
ortho-Cresol, 369
ortho-Xylene, 369
osmolarity, 195, 588
osmosis, 193–94, 587–90
osmotic concentration, 588
osmotic membranes, 193
osmotic pressure, 193–96, 587–88
osteoporosis, 576
outer mitochondrial membrane, 747
oxacillin, 507B
oxaloacetate, 401–2, 634
    citric acid cycle, 401–2, 754F, 756, 757F, 764, 770, 771
    glycolysis, 634
oxidation, of nutrients, 718, 782–83, 786–88
oxidation half-reaction, 263
oxidation reactions
    alcohols, 397, 409, 427
    aldehyde and ketones, 427–28, 440–42
oxidation-reduction reactions, 131, 263–70, 400–402
oxidative deamination, 764–65
oxidative phosphorylation, 719, 752, 757, 760–62
oxidizing agents, 263, 264B, 265
oxidoreductases, 400, 631
oxycodone, 512B
oxygen
    blood gases, 168B
    chemical equations, 134, 137
    chemical reactions, 131
    formulas of ionic compounds, 85
    hemoglobin and transport of, 616
    Lewis structure, 100–101
    valence electrons, 58T, 59
oxygen-16, 280
oxygen gas, 164, 165
oxyhemoglobin, 616
oxymorphone, 512B
ozone, 264B

P

pacemakers, heart, 270B
paired electrons, 62
palmitic acid, 453T, 559T, 788F
palmitoleic acid, 341F, 558, 559T
pancreas, and diabetes, 794B, 795B. *See also* pancreatitis
pancreatic serine proteases, 656
pancreatitis, 660
pantothenic acid, 644T
Papanicolaou, Dr. George, 2B
Papaver somniferum, 512B

Pap smear test, 2B
para–aminobenzoic acid (PABA), 653
para-Cresol, 369
paraffin wax, 577
paragyline, 346B
parallel β-pleated sheet, 609
parasalamide, 346B
para-Xylene, 369
parent compound, 315–16, 385
Parkes, Alexander, 366B
Parkinson's disease, 501B, 511
partial electron transfer, 82
partial hydrogenation, 564
partial pressures, 166
particle accelerators, 291
parts per million (ppm), 185–86
parts per thousand (ppt), 185–86
Pascal, Blaise, 154
pascal, as measurement unit, 154
passive transport, 583–85
Pasteur, Louis, 532
Pauling, Linus, 83
penicillamine, 57B
penicillin, 4B, 315, 507B
Penicillium notatum, 315, 507B
1,4-pentadiene, 344
pentanal, 421T
2-pentanamine, 493
pentane, 316, 343
1,4-pentanedithiol, 406
2-pentanol, 360
2-pentanone, 428
3-pentanone, 422, 431
1-pentene, 343, 350
2-pentene, 355
penthrane, 405
pentose, 527
pentose phosphate pathway, 730
pentyl butanoate, 465B
pentyl group, 314T
1-pentyne, 343
pepsin, 646, 651T
pepsinogen, 622, 651
peptidase, 633
peptide bond, 510, 602–6
peptidyl transferase, 690
peptidyl tRNA binding site (P site), 690
percent yield, 143–45
periodicity, concept of, 53
periodic table
    development of, 52–56
    electron configuration, 56, 58–65
    octet rule, 65–68
    structural relationships, 107–8
    trends in, 68–70
periods, and periodic table, 55
peripheral membrane proteins, 582
permeases, 585
Perrine, Susan, 448B
Persian Gulf War, 317B
petroleum industry, and gasoline production, 326B. *See also* fossil fuels; oil spills

pH
    acid rain, 256B
    blood and control of, 262B
    buffer solution, 260–61
    calculating, 247–51
    color indicators, 253
    definition of, 246–47
    enzymes, 641, 646–47
    importance of, 252
    measuring, 247
    proteins, 620–21
pharmaceutical chemistry, 144B, 524B, 641B. *See also* antibiotics; drugs
pharmacology, 240B
phenacetin, 506
phenanthrene, 371, 372
phenols, 369, 402–3
phenylalanine, 601F, 623T, 770B
2-phenylbutane, 371
3-phenyl-1-butane, 371
phenylephrine, 496
2-phenylethanoic acid, 454
2-phenylethanol, 405, 406F
phenylethanolamine-N-methyltransferase (PNMT), 631
phenyl group, 371
phenylketonuria (PKU), 509
o-phenylphenol, 403
1-phenyl-2-propanamine, 496
3-phenylpropanoic acid, 454
pheromones, 479B
phosphate ion, 191
phosphatidate, 568, 569F
phosphatidylcholine, 568, 569F
phosphatidylethanolamine, 568, 569F
phosphoanhydride, 477
phosphoenolpyruvate, 396, 438, 725
phosphoenolpyruvate carboxykinase, 732
phosphoesters, 476–78
phosphofructokinase, 650, 723, 726
phosphoglucomutase, 734, 736F, 737
phosphoglucose isomerase, 721
2-phosphoglycerate, 633, 725
3-phosphoglycerate, 633, 724, 725
phosphoglycerate kinase, 724
phosphoglycerate kinase deficiency, 722B
phosphoglycerate mutase, 633, 724
phosphoglycerate mutase deficiency, 722B
phosphoglycerides, 568–70
phospholipids, 581
phosphopantetheine group, 792, 793F
phosphorolysis, 733
phosphorus pentafluoride, 104
phosphorus trichloride, 471
phosphoryl, 477B
photosynthesis, 8, 526F
phthalic acid, 454
physical change, 7
physical chemistry, 3
physical equilibrium, 226–27
physical properties. *See also* gas(es); liquid(s); matter; properties; solid(s)
    alcohols, 384–85

aldehydes and ketones, 417–19
alkanes, 310–13
alkenes and alkynes, 342–43
amides, 504–5
amines, 489–93
carboxylic acids, 449–50
comparison of states, 153T
esters, 461
ionic and covalent compounds, 92–93
periodic table, 53
physical state, of reactants, 222
physiological saline, 196
physiology, pH and biochemical reactions, 252
Phytochemistry (journal), 703B
planar molecule, 342
Planck, Max, 49
plane-polarized light, 530–31
plants. *See also* agriculture; biological systems; forests
    alkenes, 352–53
    cellulose in cell wall, 549
    DNA fingerprinting of coca plants, 702–3B
    opium poppy and opiates, 512–13B, 604–5B
    as sources of medications, 574B
plasma lipoproteins, 577–78
plastics. *See also* polymers
    biodegradable, 456–57B, 470
    recycling, 366–367B
platinum, 185
β-pleated sheet, 609
plutonium, 281, 285B, 287, 294
point mutation, 692, 693
poisoning, and carbon monoxide, 136B
polar covalent bonding, 82–83, 110
polar covalent molecule, 110
polarimeter, 531
polarity, and Lewis structures, 109–11
polar neutral amino acids, 600, 601F
polio virus, 426B
pollution. *See* acid rain; air pollution; global warming; greenhouse effect; oil spills; waste disposal
polonium isotope, 295B
polyacrylonitrile, 365T
polyanions, 621
poly(A) tail, 684
polyatomic ions, 86–87, 88T, 97–100
polycations, 621
polyesters, 469–70
polyethylene, 364B, 365
polyethylene naphthalate (PEN), 470
polyethylene terephthalate (PETE), 366B, 367B, 470
polyhalogenated hydrocarbons, 328B
polyhydroxyaldehydes, 527
polyhydroxybutyrate acid (PHB), 456B, 457B
polyhydroxyketones, 527
polylactic acid (PLA), 456B
polymer(s), 469–70. *See also* addition polymers; plastics

polymerase chain reaction (PCR), 701–2
polymethyl methacrylate, 365T
polymyxins, 588–89B
polynuclear aromatic hydrocarbons (PAH), 371–72
polyols, 382B
polypropylene, 365, 366–67B
polyprotic substances, 255
polysaccharides, 527, 548–51
polysomes, 688
polystyrene, 365T, 367B
polytetrafluoroethylene (Teflon), 365T
polyvinyl chloride (PVC), 365T, 366B, 367B
poppy. *See* opium poppy
p orbitals, 61
porphyrin, 373, 503
position emission, 280
positive allosterism, 649
positively charged amino acids, 600, 601F
postsynaptic membrane, 654B
post-transcriptional processing, of RNA, 684–85
potassium, 94B, 200, 281
potassium hydroxide, 242
potassium perchlorate, 52B
potassium permanganate, 397
potassium propanoate, 460
potential energy, 27
pound (lb), 11
precipitate, 181
precipitation reactions, 129–30
precision, in measurement, 20
prefixes
    metric system, 12
    naming of covalent compounds, 90
pregnancy. *See also* infants
    alcohol consumption during, 388B
    hemoglobin and oxygen transport, 616
    premature infants, 556B
preimplantation diagnosis, 668B
premature infants, 556B
premenstrual syndrome (PMS), 67B
pressure. *See also* vapor pressure
    chemical reactions, 233
    gases, 153, 166–67
    solubility, 181
primary alcohol, 389, 427
primary amine, 490
primary carbon, 313
primary structure, of proteins, 606, 613, 614F
primary transcript, 684
primase, 678, 679F
principal energy levels, 60, 61
products
    of chemical equation, 127
    of chemical reaction, 8, 636
proenzymes, 651
progesterone, 575, 576
prokaryotes, 673
proline, 600, 601F, 623T, 770
promoter, and transcription, 682
promotion, and Bohr atom, 49–50, 51
propanal, 419, 421T

propanamide, 505
propanamine, 491, 492T, 493T
1-propanamine, 493
propane
    boiling point, 491
    functional groups, 309
    molecular formula, 310
    nomenclature, 316, 343
propane gas, 134
1,2,3-propanetriol, 389
propanoic acid, 261, 451, 458, 466, 474
propanoic anhydride, 474
propanol, and boiling point, 384, 417, 449, 493T
1-propanol, 393, 403, 417, 449, 467
2-propanol, 359
    classification, 390
    hydration, 359
    hydrogenation, 393
    medical uses, 388–89
    molecular structure, 389
    nomenclature, 386
    oxidation, 398
propanone
    boiling point, 417, 449
    hydrogenation, 393
    industrial applications, 424
    naming of ketones, 422
    oxidation, 398
propene, 343, 394
properties. *See also* chemical properties; matter; physical properties
    definition of, 7
    electronic structure and molecular geometry, 111–13
    intensive and extensive properties, 8–9
    ionic and covalent compounds, 92–93
    periodic table, 53
propionamide, 505
propionibacteria, 729B
propionic acid, 453T
propyl acetate, 462
propyl decanoate, 561
propyl ethanoate, 463
propyl group, 314T
N-propylhexanamide, 505
propyne, 343
prostaglandins, 478B, 565–67
prostate specific antigen (PSA), 599B
prosthetic group, 612
protease inhibitors, 641B
protein(s). *See also* amino acids
    angiogenesis inhibitors, 596B
    blood, 599B
    cellular functions, 597
    collagen, 612–13B
    denaturation, 617–22
    dietary and digestion, 622–23, 716
    DNA replication, 679F
    Fischer's research on, 525
    myoglobin and hemoglobin, 615–17
    overview of, 510–11
    peptide bond, 602–6

    primary structure, 606, 613, 614F
    quaternary structure, 611–12, 614
    secondary structure, 606–9, 613–14
    shape of molecule, 638B
    tertiary structure, 609–11, 614
protein metabolism, 797
protein modification, and enzymes, 651–52
protein synthesis, 510–11, 687–92
proteolytic enzymes, 656–67
protium, 277
protofibril, 608
protons, 39–40, 45
Prozac, 514
pseudoephedrine, 496, 501B
pulmonary disease, and nuclear medicine, 290
pulmonary surfactant, 556B
pure substance, 9
purines
    heterocyclic amines, 502, 503
    heterocyclic aromatic compounds, 373
    structure of DNA, 525, 669, 670F
pyridine, 242, 373, 502, 503
pyridine aldoxime methiodide (PAM), 516, 655B
pyridoxal phosphate, 764
pyridoxine, 644T
pyrimidine, 373, 502, 503, 669, 670F
pyrimidine dimer, 694–95
pyrophosphorylase, 737
pyrrole, 373, 502
pyruvate
    alcohol fermentation, 728F
    citric acid cycle, 750–51, 754F, 756, 757F
    degradation of amino acids, 764
    glycolysis, 447B, 719, 721
pyruvate carboxylase, 732, 771–72
pyruvate carboxylase deficiency, 769B
pyruvate decarboxylase, 727
pyruvate dehydrogenase complex, 750
pyruvate kinase, 725, 726

Q

quantization, of energy, 49
quantum levels, and Bohr atom, 49
quantum mechanical atom, 60–61
quantum mechanics, 61
quantum number, 51
quaternary ammonium salts, 501–2
quaternary carbon, 313
quaternary structure, of proteins, 611–12, 614
quats, 502
quinine, 503, 504

R

Rabi, Isidor, 292B
rad (radiation absorbed dosage), 297
radiation. *See* gamma radiation; radioactivity
radiation therapy, 49B, 289
radioactive decay, 39

radioactive isotopes, 40–41
radioactivity
  balanced nuclear equation, 279–82
  biological effects of radiation, 293–95
  measurement of, 295–98
  medical applications and nuclear medicine, 288–93
  natural, 45–46, 277–79, 291–93
  nuclear power, 284–87
  radiocarbon dating, 288
  radioisotopes, 282–84
radiocarbon dating, 288
radioisotopes, 282–84
radio waves, 48B
radium, 276B, 295B
radon, 295B
Raleigh, Sir Walter, 694B
random error, 19
Raoult's law, 192
rate constant, 224, 225
rate equations, 224, 225
rate-limiting step, and enzyme-substrate complex, 637
reactants
  of chemical equation, 126, 127
  of chemical reaction, 8
reaction rate. *See also* chemical reactions
  factors affecting, 221–23
  mathematical representation, 224–25
  substrate concentration and enzyme-catalyzed reactions, 637
real gases, versus ideal gases, 167
receptor-mediated endocytosis, 579, 580F
recombinant DNA technology, 659B, 661, 696–701. *See also* genetic engineering
recycling, of plastics, 366–67B
red blood cells (RBCs), 262B, 546B, 581, 587
redox reactions. *See* oxidation-reduction reactions
reducing agent, 263, 265
reducing sugars, 541–43
reduction. *See* oxidation-reduction reactions
reduction half-reaction, 263
reduction reactions, of aldehydes and ketones, 431–34, 440–42. *See also* oxidation-reduction reactions
regulation
  of citric acid cycle, 756, 757F
  of enzyme activity, 649–52
  of glycolysis, 726
  of lipid and carbohydrate metabolism, 793–97
regulatory proteins, 597
reindeer, and body temperature, 582
relaxation, and Bohr atom, 49–50, 51
release factors, 691
rem (roentgen equivalent for man), 297
repair endonuclease, 695
replacement reactions, 129
replication, of DNA, 674–81
replication fork, 677
replication origin, 677

representative elements, 55, 58
reproductive system, and prostaglandins, 565. *See also* birth control; pregnancy
resonance, and Lewis structures, 101–3
resonance forms, 102
resonance hybrid, 102–3, 505
respiration. *See also* aerobic respiration; respiratory tract
  blood gases, 168B
  Henry's law, 181–82
respiratory distress syndrome (RDS), 556B
respiratory tract. *See also* lung disease; pulmonary disease; respiration
  acidosis and alkalosis, 262B
  electron transport system, 760
  hemoglobin and oxygen transport, 616
  prostaglandins, 566
restriction enzymes, 696
results, and measurement, 11
retina, of eye, 440–41B
retinol, 352
retroviruses, 676B. *See also* HIV
reverse aldol condensation, 723B
reverse transcriptase, 676B
reversible competitive enzyme inhibitors, 652–53
reversible noncompetitive enzyme inhibitors, 653–54
reversible reaction, 226, 229–30
rhodopsin, 440B
riboflavin, 644T, 646
ribonucleotides, 670, 674
ribose, 541, 670F, 674
ribosomal RNA (rRNA), 682
ribosomes, 687, 688F
ribozymes, 630B
RNA (ribonucleic acid). *See also* mitrochondria; transfer RNA (tRNA)
  classes of, 681–82
  heterocyclic aromatic compounds, 373
  nucleotides, 669
  post-transcriptional processing, 684–85
  structure of, 674
RNA polymerase, 682–83
RNA primer, 678
RNA splicing, 684–85
rock, and equilibrium composition, 234
roentgen, 297
Roman mythology, and soap, 468
rounding off numbers, 22–23
rubbing alcohol, 388
rust, as chemical reaction, 131, 265F
Rutherford, Ernest, 45, 46

S

safety
  acids and bases in laboratory, 246
  radioactivity and radiation, 279, 293–95
saliva, and bacteria, 528B
Salmonella typhimurium, 694B

salt(s), and carboxylic acid, 460. *See also* sodium; sodium chloride
salt bridge, 268
Sanger, Frederick, 704
saponification, 467, 468F, 562–63
Sarin (isopropylmethylfluorophosphate), 654–55B
saturated fatty acids, 559T, 560
saturated hydrocarbons, 308
saturated solution, 181
Saunders, Jim, 702B
schizophrenia, 511
Schröedinger, Erwin, 60–61
science
  role of curiosity in. 5B
  role of observation in, 2B
  scientific method, 3–5, 52
  scientific notation, 18–23
  serendipity in research, 586B
scientific method, 3–5, 52
scientific notation, 18–23
scuba diving, 183B
scurvy, 613B
seasonal affective disorder (SAD), 511
secondary alcohol, 389, 398, 427
secondary amine, 490
secondary carbon, 313
secondary structure, of proteins, 606–9, 613–14
selectable marker, 699
selectively permeable membranes, 194F
selective serotonin reuptake inhibitors (SSRIs), 514
self-tanning lotions, 432–33B
semiconservative replication, 676
semipermeable membranes, 193, 587
semisynthetic penicillins, 507B
Semmelweis, Ignatz, 264B
sensors, carbon monoxide, 136B
sequential synthesis, 144B
serendipity, in scientific research, 586B
serine, 601F, 623T, 651
serine proteases, 657B
serotonin, 511, 514
shared electron pair, 82
shielding, and radiation, 294
shorthand electron configuration, 64–65
Shroud of Turin, 288F
sialic acid, 546B, 547B
sickle cell anemia, 448B, 617, 693
side effects, of drugs, 240B
significant figures, 16–23
silent mutations, 692–93
Silent Spring (Carson 1962), 340B
silicon, and molecular geometry, 106
silicon dioxide, 90
silk fibroin, 609, 610F
silver, 120F, 173, 428, 429F
single bond, 96, 101
single-replacement reaction, 129
single-strand binding protein, 678, 679F
skeletal structure, of compound, 93
skin cancer, 432–33B

skunk, and scent molecules, 405, 406F
slow twitch muscle fibers, 749B
small nuclear ribonucleoproteins (snRNPs), 685
Smithsonian Museum of Natural History, 563B
smoking. *See* cigarette smoking
soaps, 460, 468, 469F, 562–63. *See also* detergents
sodium. *See also* salt(s); sodium chloride; sodium hydroxide
  blood pressure, 94B
  cations in body fluids, 200
  electron configuration, 63
  fireworks, 52B
  formulas of ionic compounds, 85
  ion formation, 66
  molar mass, 119–20
  oxidation and reduction, 263
sodium acetate, 257, 260
sodium azide, 118B
sodium benzoate, 459
sodium bicarbonate, 87
sodium chloride. *See also* salt(s); sodium
  chemical formula, 84, 124
  crystal structure, 172F
  formula units, 125F
  ionic bonding, 79–80, 81F
  major properties, 306T
  melting point, 93
  nomenclature, 86
  osmotic pressure, 195
  vapor pressure and boiling point, 193
  weight/volume percent, 184
sodium hypochlorite, 264B, 266
sodium hydroxide
  neutralization, 252, 253
  polyprotic substances, 255
  reactant quantities, 141–42
  replacement reactions, 129
  strong bases, 242
sodium oxide, 86
sodium propanoate, 261
sodium sulfate, 87, 89, 125
soil, and equilibrium composition, 234
solar energy, 48B
solid(s)
  boiling and melting points, 112–13
  compared to gases and liquids, 153T
  crystalline solids, 171, 172–73
  properties of, 171
  states of matter, 7
  triboluminescence and fracturing of, 213B
solubility. *See also* solutions
  common ionic compounds, 130T
  degree of, 180–81
  electronic structure, 111–12
solutes, 178
solutions. *See also* solubility
  concentration based on mass, 182–87
  concentration-dependent properties, 191–96
  concentration in moles and equivalents, 187–91

electrolytes in body fluids, 200–202
homogeneous mixture, 10
ionic and covalent compounds, 93
properties of, 178–82
water as solvent, 198
solvent, 178, 198
s orbital, 61
sorbitol, 382B
Southern blotting hybridization, 697, 698F
specific gravity, 31
specific heat, 215
specificity, of enzymes, 630–31, 639
spectral lines, and Bohr theory, 51
spectrophotometer, 48B
spectroscopy, 47, 48–49B
speed of light, 47
sphingolipids, 570–72, 573B
sphingomyelin, 570
sphingomyelinase, 573B
sphingosine, 570
Spirea ulmaria, 315
spliceosomes, 685
spontaneous reactions, 211
stability, and Lewis structure, 100–101
staggered conformation, of alkanes, 325
Stahl, Franklin, 676
stalagmites and stalactites, 129–30
standard atmosphere, 154
standard mass, 23
standard solution, 252
standard temperature and pressure (STP), 163
starch, 548
states of matter. *See* gas; liquid(s); matter; solid(s)
Statue of Liberty, 266B
stearic acid, 453T, 559T, 564
stereochemical specificity, of enzyme, 639
stereochemistry, 524B, 529
stereoisomers, 322–23, 524B, 529–34, 598–99
stereospecific enzymes, 524B
sterilization, of medical instruments, 647. *See also* antiseptics; disinfectants
Stern, Otto, 292B
steroids, 572–77
stock system, of nomenclature, 86, 87T
strength, of acids and bases, 242–43
Streptococcus mutans, 528B
Streptococcus pyogenes, 659B
streptokinase, 659B
strong acids, 242–43, 244, 246
strong bases, 242–43, 458, 459
strontium, 52B
structural analogs, of enzymes, 652
structural formula. *See also* molecular formula
  alkanes, 310, 312, 342, 343T
  alkenes and alkynes, 342, 343T
structural isomers, 319–21
structural proteins, 597
strychnine, 503, 504
styrene, 365T
sublevels, and quantum mechanics, 61
Suboxone, 513B

subscripts, in chemical equations, 132B
substituted cycloalkanes, 323
substituted hydrocarbon, 308
substitution reactions, 329, 372
substrate, of enzyme, 635–36, 637. *See also* enzyme-substrate complex
substrate-level phosphorylation, 719
subtraction, of significant figures, 20–21
succinate, 754F, 755, 757F
succinate dehydrogenase, 755
succinylcholine, 516
succinyl CoA, 754F, 755, 756, 757F
succinyl CoA synthase, 755
sucrose, 198, 382B, 545, 640, 642F
suffixes
  covalent compounds, 90
  ionic compounds, 86
sugar. *see also* fructose; glucose; sucrose
  equilibrium and dissolution in water, 226–27
  Fischer's research on, 525
  reducing sugars, 541–43
  tooth decay, 382B, 528B
sugar alcohols, 382B
sulfa drugs, 497, 653
sulfanilamide, 497, 653
sulfatides, 571–72
sulfhydryl group, 383, 405
sulfur, 121, 122
sulfur dioxide, 101–2, 257B
sulfuric acid, 188, 242, 255, 256B, 257B
sulfur oxides, and acid rain, 256B, 257B
sulfur trioxide, 257B
sun exposure, and skin cancer, 432–33B
sun protection factor (SPF), 433B
super hot enzymes, 630B
superoxide radical, 264B
supersaturated solution, 181
surface tension, 168
surfactants, 168
surroundings, and energy stored in chemical system, 209
suspension, 180
sweet tastes, 382B
Sweeting, Linda M., 213B
swimming pools, chloroform in, 331B
symmetrical acid anhydrides, 473
synthesis
  acid anhydrides, 474
  acid chlorides, 471
  esters, 466
synthetic opioids, 512–13B
system(s), definition of, 209, 211B
systematic error, 19
systematic names, 86, 87T
Système International (S.I.), 12

T

Tagamet (cimetidine), 515
target cells, and insulin, 797
Tasmania, and opium poppies, 512B
Tauri's disease, 722B
tautomers, 436

Index

Tay-Sachs disease, 572, 573B
technetium-99m, 280–81, 283F, 290, 292
Teflon, 364B, 365T
temperature. *See also* body temperature; boiling point; freezing point; melting point
    denaturation of proteins, 618–20
    enzymes, 630B, 647–48
    rate of chemical reaction, 222
    solubility, 181
    systems of measurement, 25–27, 157B
    vapor pressure of liquids, 169
    water and moderation of, 199B
terminal electron acceptor, 761
termination, of transcription, 683, 691–92
termination codons, 691
terpenes, 352, 572
tertiary alcohol, 389, 399, 427
tertiary amine, 490, 491
tertiary carbon, 313
tertiary structure, of proteins, 609–11, 614
testosterone, 576
tetrabromomethane, 330
tetrafluoroethene, 364B, 365T
tetrahedral molecular structure, 106, 107T, 311
tetrose, 527
β-thalassemia, 448B
thalidomide, 524B
thallium–201, 290
thebaine, 512B
theoretical yield, 143–45
theory, and scientific method, 4, 52
thermochemical equation, 210
thermocycler, 701–2
thermodynamics, 208–14
thermogenesis, 758–59B
thermogenin, 758B, 759B
thermography, 758B
thermometer, 25–26
Thermus aquaticus, 701
thiamine, 644T
thioester, 407, 478–80
thioester bond, 786
thiolase, 787
thiols, 384, 405–8
thiolysis, 787
thionyl chloride, 471
Third World countries, and oral rehydration therapy, 197B
Thomson, J. J., 45
thorium, 280
Three Mile Island nuclear accident (1979), 294
threonine, 601F, 623T, 651
thromboxanes, 565, 566F
thrombus, 659B
thymine, 670F, 672F
thyroid, and nuclear medicine, 290T
Thys-Jacobs, Dr. Susan, 67B
time
    radiation exposure, 294
    units of measurement, 25
tin, 64
tissue-type plasminogen activator (TPA), 659B

titration, 252, 255F
tobacco. *See* cigarette smoking; nicotine
Tollens' test, 428, 429F
toluene, 369
m-toluic acid, 454
o-toluidine and *p*-toluidine, 494
tooth decay, 382B, 528B
topoisomerase, 678, 679F
Torricelli, Evangelista, 153, 154F
Towson University, 702B
trace minerals, 57B
tracers, and nuclear medicine, 289
transaminases, 631, 762–64
transamination, 762–64, 765F
trans-2-butene, 356–57
trans-2-butene-1-thiol, 405, 406F
transcription, and genetics, 681, 682–83
trans-1,3-demethylcyclohexane, 324
trans-3,4-dichloro-3-heptene, 349
trans-fatty acids, 352
transferases, 631, 634
transferrin, 597
transfer RNA (tRNA), 510–11, 682, 688–90
trans isomer, 348
transition elements, 55, 58
transition state, of enzyme-substrate complex, 640
translation, and genetics, 681, 687–92
translocation, of ribosome, 680
transmembrane proteins, 582
transmethylase, 631
trans-3-methyl-3-pentene, 350
trans-9-octadecenoic acid, 351
transport proteins, 597, 616. *See also* active transport; electron transport system
trans-unsaturated fatty acids, 564B
triacylglycerol lipase, 651–52
triboluminescence, 213B
tribromomethane, 330
3,5,7-tribromooctanoic acid, 451
tricarboxylic acid (TCA) cycle, 750
trichloromethane, 328B
triesters, 468
triglycerides, 567–68, 632, 779–81, 796
trigonal planar molecule, 106, 107T
trigonal pyramidal molecule, 106, 107T
trihalomethanes (THMs), 331B
trimethylamine, 108, 491, 494
3,5,7-trimethyldecane, 316
2,2,3-trimethylpentane, 316
2,2,4-trimethylpentane, 326B
triose phosphate isomerase, 634, 723
tripeptide, 603
triple bond, 101
triple helix, 612
tritium, 92B, 277, 287
true solution, 179
trypsin, 622, 646, 651T, 656–57
tryptophan, 514F, 601F, 623T
tumors, and angiogenesis inhibitors, 596B
Tyndall effect, 180
type I insulin-dependent diabetes mellitus, 542
tyrosine, 512B, 601F, 623T, 651

U

UDP-glucose, 737, 738
ultraviolet lamps, 48B
ultraviolet (UV) light, 2B, 694–95
uncertainty, and measurement, 19–20
unequal electron density, 82
unit, and measurement, 11
U.S. Department of Agriculture (USDA), 702B, 703B
U.S. Department of Energy, 307B, 703
U.S. Department of Health and Human Services, 331B
U.S. Dietary Guidelines, 556
U.S. Geological Survey, 307B
U.S. Surgeon General, 388B
universal energy currency, 713
universal solvent, 198
universe, and big bang theory, 92B
University of Florida, 705B
University of Georgia, 630B
unsaturated fatty acids, 558, 559T, 560, 582
unsaturated hydrocarbons, 308, 341. *See also* alkenes; alkynes; aromatic hydrocarbons; geometric isomers; heterocyclic aromatic compounds
unshared pair, of electrons, 97
uracil, 670F, 674
uranium, 280, 286F, 295B
urea, 304, 660
urea cycle, 766–69
urease, 635, 639
uridine triphosphate (UTP), 737
urine tests, 31

V

vaccines and vaccination, 426B, 618–19B
valence electrons, 58–59, 79, 95
valence shell electron pair repulsion theory. *See* VSEPR theory
valeric acid, 453T
valine, 601F, 623T, 693
van der Waals forces, 170, 610
vanillin, 424, 425F
van't Hoff, Jacobus Hendricus, 532
vaporization, of water, 199B
vapor pressure
    of liquids, 169–70
    of solutions, 192
variable number tandem repeats (VNTRs), 702B
vegetarian diets, 623
very low density lipoproteins (VLDL), 578, 579F, 793
Vibrio cholera, 197B
vinegar, 244
vinyl chloride, 365T
viruses. *See* retroviruses
viscosity, of liquids, 167–68
vision, chemistry of, 440–41B
vital force, 304
vitamin(s)
    coenzymes, 643, 644T
    lipids, 557

vitamin A, 341F, 441B
vitamin B, 503, 644T
vitamin C, 613B, 643
vitamin D, 67B
vitamin K, 341
volatile esters, 464B
voltaic cells, 267–69, 270B, 271F
volume, units of measurement, 11, 24
von Gierke's disease, 740B
VSEPR theory, 104–9, 311

W

waste disposal
    biodegradable garbage bags, 456–57B
    nuclear, 285B, 295
    plastic recycling, 366–67B
water. *See also* acid rain; ice; lakes; ocean; water treatment
    acid-base properties, 242
    boiling point, 198
    chemical equations, 132B
    chemical formula, 124
    climate, 27B
    combination reactions, 128
    covalent bonding, 82, 83
    dissociation, 246
    equilibrium and sugar in, 226–27
    formula weight and molar mass, 125
    hard water, 563
    hydrogen bonding, 171F, 173F
    Lewis structure, 107
    life on Earth, 199B
    nomenclature, 91
    osmosis, 194
    pH scale, 250
    solubility, 111–12
    solvents, 198
    specific gravity, 31B
water treatment, and ozone, 264B. *See also* oral rehydration therapy
Watson, James, 669, 671
waves, and electromagnetic radiation, 47
waxes, 577
weak acids, 242–43, 244, 246, 458
weak bases, 242–43, 498
weight, and mass, 23
weighted average, and atomic mass, 41
weight loss, and diet, 28B, 784–85B
weight/volume percent, 182–84
weight/weight percent, 185
white fat, 758B
Wilkens, Maurice, 671
Wilson, A., 746B
Wilson's disease, 57B
wine and winemaking, 728B
withdrawal syndrome, and morphine, 605B
Withering, William, 574B
Wittgenstein, Eva, 432B
Wöhler, Friedrich, 304, 488
wonder drug, 144B
wood alcohol, 387
World Health Organization, 340B

X

xenon, 281, 290
xeroderma pigmentosum, 696
xerophthalmia, 441B
X-linked genetic disorders, 573B, 722B
X-rays, 48B
xylitol, 382

Y

yard, units of measurement, 11
yeast, and fermentation, 729B
Yellowstone National Park, and hot springs, 582, 629F, 648, 701

Z

Zaitsev, Alexander, 395
Zaitsev's rule, 395
Zasloff, Dr. Michael, 5B
zidovudine (AZT), 676–77B
zinc, 131, 268, 269
zwitterion, 598